UNDERDOWN'S PRACTICAL FIRE PRECAUTIONS

Underdown's Practical Fire Precautions

Third Edition

Ron Hirst

Gower Technical

Published by
Gower Technical
Gower Publishing Company Limited
Gower House
Croft Road
Aldershot
Hants GU11 3HR
England

Gower Publishing Company
Old Post Road
Brookfield
Vermont 05036
USA

First published by Gower Press, Teakfield Limited

First edition 1971
Reprinted 1977

Second edition published by Gower Publishing Company Limited 1979

Reprinted 1982

Third edition 1989

British Library Cataloguing in Publication Data

Hirst, Ron
Underdown's Practical Fire Precautions – 3rd ed.
1. Fire prevention
I. Title
628.9′22

ISBN 0 566 09016 3

Printed and bound in Great Britain at
The Camelot Press Ltd, Southampton

Contents

Contents

Figures

Tables

Preface to the Third Edition

When George Underdown told me that he was about to start work on the third edition of this book, I offered to help. This was because I though that certain aspects of fire engineering, and particularly the ones in which I was interested, were insufficiently covered. I hoped to restore the balance by providing input on, for example, automatic systems other than sprinklers. I knew however that George was enormously painstaking in collecting the information, and meticulous in its presentation: clearly it would be a demanding task.

Sadly George died soon after that, and when the publishers asked whether I would take over the task of preparing the new edition, I was pleased to be able to do so. It was agreed however that work was needed in three areas: the preparation of new material, the removal of some existing parts, and the complete revision of the remainder. It was decided for example that large sections which had been quoted from the FOC Rules for Construction could be removed. Rules are frequently revised and it is unwise for a reader to rely on anything other than the document itself before making any decisions. However, an outline in the book of the contents of such a document can be of value, particularly when it can assist in understanding the original document. For this reason there is still a discussion on the Sprinkler Rules, and also a new section on detector systems, based on the new British Standard. However, the removal of the FOC material and other parts, including for example some of the practical firemanship, has made room for some new topics.

The completely new first part of the book is concerned with combustion technology. At first sight it may seem odd to find several chapters on theory in a book supposedly concerned with practical fire precautions. They are there however for a very practical purpose. No fire engineering problem will ever fit exactly into a set of rules, a standard, or a textbook description. An appeal to basic concepts may be needed before a suitable compromise can be made. A lack of understanding of basic principles may also lead to a failure to recognize a hazard. For example, one practical book which contains some excellent advice on fire surveys describes an explosion as 'a very fast fire'. This is no help at all in assessing the explosion potential on a process plant. On the other hand, some of the more academic books are concerned with mathematical modelling of combustion phenomena, and their applicability to practical problems may be equally remote. It is to be hoped that Part One goes some way towards bridging the gap, in addition to providing an insight into the wealth of information which is available.

As fire engineers we should be aware of the methods which are available for the analysis and quantification of risks. The brief discussion in Part Eight gives an indication of the scope, and the great potential, of these metods although no attempt is made to provide instructions in their use.

People using a book like this will often read a chapter on a particular topic in isolation. It is therefore helpful if each chapter, or each part of the book, is reasonably self-contained. This must necessarily result in some repetition, usually as a brief summary with a reference to the main discussion in another chapter.

A few suggestions for further reading have been made at the end of each part, and within Part Nine. Many of these are standard textbooks and no indication of the date of publication has therefore been given: the latest edition should be studied. This also applies to standards, rules, and codes of practice: the latest edition and its most recent amendments should be obtained.

Ron Hirst

PART ONE

HOW THINGS BURN

1

Flames and glowing combustion

Problems in fire protection can rarely be solved simply by consulting a book and following a list of instructions. The selection of the most effective means of protection must depend on a knowledge of the particular hazard and of the combustion processes which are involved. It is for this reason that Part One of this book contains a brief discussion on combustion technology and on the behaviour of fires and explosions. This information is also essential for a proper understanding of later sections of the book which deal with the design and application of fire protection equipment and procedures. The explanations given here are largely non-mathematical and are therefore relatively simple. They may however lack the precision of a more rigorous approach. For this reason a bibliography is included at the end of this Part in which more detailed discussions of the many topics can be found.

Although a description of ignition processes might appear to be a logical starting point, it was decided to begin instead by outlining the mechanisms by which flaming combustion and other processes can be sustained. This is followed by a discussion on some practical aspects of flame propagation. It seems more reasonable to look then at the methods by which combustion can be initiated and inhibited.

DIFFUSION FLAMES

The familiar candle flame provides a useful starting point for a discussion on flaming combustion. Once it has been lighted, some of the heat radiated from the flame will melt the wax, and the hot liquid can then climb up the wick by capillary action. The continuing flow of heat, by radiation, into the liquid will raise its temperature and cause it vaporize. The vapour will then be raised to higher temperatures as it approaches the reaction zone of the flame. The reaction zone can be seen as a blue sheath around most of the flame, and this is where all the active (heat-producing) combustion processes take place. This zone is only one or two millimetres thick, so that much of the flame is rather like an empty eggshell.

Candle wax is usually a mixture of different waxes: this ensures that it has the right physical and mechanical properties. A main ingredient is paraffin wax, and it will be interesting to see what happens when this wax vaporizes and passes through the flame. The paraffins (or aliphatic hydrocarbons) comprise a large family of chemical compounds which can be distilled from crude oil. The simplest is methane (natural gas) with a molecule which comprises one carbon atom linked to four hydrogen atoms. Other paraffins contain between two and perhaps 100 carbon atoms. Petrol contains a mixture of liquids and typically their molecules will have between five and 12 carbon atoms. Similarly, fuel oil is based on 12 to 18 carbon atoms, and lubricating oils on 15 to 24. Paraffin wax molecules may contain as many as 40 carbon atoms. It is a characteristic of the paraffins that the carbon atoms are all arranged in a long chain, with the hydrogen atoms attached along the length.

When the paraffin wax molecules are heated as they approach the reaction zone, they begin to break up into smaller molecules, which are in turn broken into smaller fragments. If we suppose that the 40 carbon chain has been reduced to one of 10 carbons, (a decane molecule), it can be represented like this:

$$
\begin{array}{ccccccccccccccccccccc}
 & & H & & H & & H & & H & & H & & H & & H & & H & & H & & H & & \\
 & & | & & | & & | & & | & & | & & | & & | & & | & & | & & | & & \\
H & - & C & - & C & - & C & - & C & - & C & - & C & - & C & - & C & - & C & - & C & - & H \\
 & & | & & | & & | & & | & & | & & | & & | & & | & & | & & | & & \\
 & & H & & H & & H & & H & & H & & H & & H & & H & & H & & H & &
\end{array}
$$

This molecule will not be capable of breaking in half to form two completely similar molecules, because of the need for a hydrogen atom at each end of the new chains. It will instead break like this (into a pentane and a butane molecule):

$$
\begin{array}{ccccccccccccc}
 & & H & & H & & H & & H & & H & & \\
 & & | & & | & & | & & | & & | & & \\
H & - & C & - & C & - & C & - & C & - & C & - & H \\
 & & | & & | & & | & & | & & | & & \\
 & & H & & H & & H & & H & & H & &
\end{array}
\quad + \quad C \quad + \quad
\begin{array}{ccccccccccc}
 & & H & & H & & H & & H & & \\
 & & | & & | & & | & & | & & \\
H & - & C & - & C & - & C & - & C & - & H \\
 & & | & & | & & | & & | & & \\
 & & H & & H & & H & & H & &
\end{array}
$$

In the process, one carbon atom has been released, and this happens every time a larger molecule is split into two smaller ones. Six of these spare carbon atoms then link up to form a closed ring, which is a graphitc molecule. These molecules in turn link together into small particles of solid carbon. In the final stage of this disintegration of the wax (which is called pyrolysis – 'loosening by fire') there will be only hydrogen atoms and carbon, but much of the carbon will be in the small solid particles. The hydrogen atoms are very mobile and extremely reactive: they rush ahead into the reaction zone where they combine with oxygen atoms to form water vapour. The carbon particles will be heated nearly to the reaction zone temperatures before they break down into carbon atoms, which can then also react with oxygen, to form carbon dioxide. It is the white-hot carbon particles which radiate the light from the candle flame.

Thus, fuel atoms are streaming into the candle flame reaction zone from one side, and oxygen atoms from the other, the oxygen molecules having also been broken down by pyrolysis as they approach this zone:

$$O_2 \rightarrow O + O$$

The mixing of the reactants in the reaction zone is by diffusion, which at high temperatures is a very efficient process. The rate at which the reactants can burn is deterined by the rate of diffusion. This clearly is a critical process, which is why flames of this kind are called diffusion flames. The diffusion rate cannot be increased in normal combustion except by increasing the flame temperature. Hence, the rate of burning can be increased only by increasing the size of the flame.

The reactions of the hydrogen and the carbon with oxygen in the flame release a great deal of heat: they are exothermic reactions. The other chemical processes – the pyrolysis of the fuel and the breaking of oxygen molecules into atoms – are endothermic, that is they absorb heat. Heat is also needed, as we have seen, to melt and vaporize the wax. Some will also be used to raise the temperature of the nitrogen in the air to the flame temperature, but the nitrogen will contribute nothing to the reactions. If we assume that 20 per cent of the air is oxygen, and the rest is nitrogen, then we can express the burning of paraffin wax like this:

$$C_{40}H_{82} + 60\tfrac{1}{2}O_2 + 242N_2 = 40CO_2 + 41H_2O + 242N_2 + \text{heat}$$

However, this chemical equation tells us nothing about the very complex process involved, nor about the critical heat balance. In any diffusion flame sufficient heat must be released continuously to maintain all the physical and chemical changes that are needed to sustain the combustion processes. In an actual fire, there must also be a surpluse of heat in excess of these requirements if the fire is to spread from one item to another.

The chemical equation above is exactly balanced: there are just enough oxygen molecules to react with all of the fuel, and to produce carbon dioxide and water vapour as the end products. This is the 'ideal' or 'stoichiometric' mixture of fuel and oxygen. It is the mixture which will release the most heat from a given volume and will therefore achieve the highest temperature. The reactions in a diffusion flame are almost invariably stoichiometric. Suppose for example that our candle has burned down to a part of the wax which is more volatile than the rest. For a moment, too much fuel vapour will enter the reaction zone and there will be insufficient oxygen there to react with it. Some unburned fuel vapour will pass out of the zone, but it will have been heated to the flame temperature and will be meeting oxygen at almost the same temperature. These gases will of course react, and the result will simply be an enlargement of the flame. In an enlarged flame the reaction zone will be further away from the wick, which will therefore receive less heat and the evaoration rate will be reduced. The final shape of the flame will clearly be compromise between these conflicting effects. The reaction zone will adjust its area to correspond to the available flow of fuel vapour, so that it can always remain stoichiometric. However, if we look at the overall reaction in a large flame, the total process may not be stoichiometric. The candle flame is just the right size to ensure that everything is burned, but if it were slightly bigger some of the carbon particles would fail to react. The hydrogen atoms will still reach the reaction zone, but some of the carbon would never get near enough to reach the right temperature. The flame would then be open at the top and the unburned carbon would be lost as smoke. Most of the flames which are present at a fire are big turbulent diffusion flames which are producing large volumes of hot smoke.

At one time candle flames would become large and smoky as they burned down because the wicks were too long. The length of the wick was then reduced with snuffing scissors. A wick was then developed which bent over in the flame. Whenever this wick grows long enough to enter the reaction zone, its end is burned away. A candle is indeed a neat, efficient and beautiful mechanism for the production of light. It is an interesting example of a diffusion flame, of which the main characteristics are:

- The reactants are gaseous.
- The reactants are mixed by diffusion.
- Diffusion is rate-controlling.
- The reaction is usually stoichiometric.

- The flames are luminous due to the presence of carbon particles.
- Large flames are open at the top and unburned carbon is released as smoke.
- The flame must always be located close to the source of fuel.
- Most fires comprise large turbulent diffusion flames.

Diffusion flame temperatures

The light from the candle flame comes from white hot particles of carbon. If you heat a piece of steel it begins to glow with a dull red colour just above 500°C and is white hot at about 1300°C. The light from the candle is slightly yellow and does not have quite the brilliance of the hot metal; also the radiation from small particles behaves differently from that of a large mass. However, the carbon particles must be at a temperature of well over 1000°C, and the maximum flame temperature must be higher than that. This is because the particles are gaining heat from the flame by conduction and radiation, but at the same time they are losing heat (and light) by radiation. The flow of heat away from the particles by radiation increases as their temperature increases (the loss is proportional to the fourth power of the absolute temperature). So the particle temperature will adjust itself until there is an exact balance between the heat gained from the flame and the heat lost by radiation. Clearly this temperature must be lower than the flame temperature; otherwise there could not be a heat flow from the flame to the particles. This suggests a flame temperature of perhaps 1500°C, which indeed can be measured. If a fine thermocouple is used to explore the temperature distribution in a candle flame, readings similar to those in Figure 1.1 can be obtained. However, it must be remembered that, like the carbon particles, the thermocouple will be at a temperature at which the heat lost by radiation (and by conduction along the wires) exactly balances the heat gained from the flame. We are therefore measuring the thermocouple temperature and not the flame temperature, and a correction of up to 200°C should be added to the higher values. In some parts of the flame it is difficult to obtain a meaningful reading because of soot deposits on the probe, but this is not usually a problem in the lower parts of the flame.

These flame temperatures may be surprisingly high, particularly when they are compared with some published values which can be as low as 1000°C or even 500°C. Flames cannot exist at temperatures as low as this, except in very unusual laboratory situations where combustion proceeds in a number of separate steps, each of which produces a 'cool flame'. The 1000°C measurement could have been obtained with a large thermocouple which was coated with soot, but it is difficult to understand how measurements as low as 500°C could be obtained. Flame temperature can be

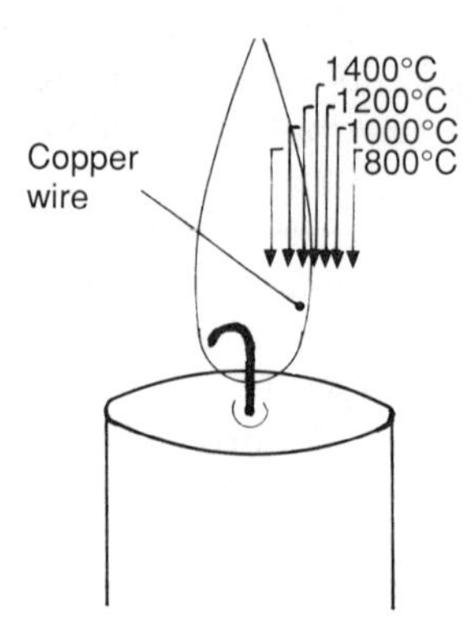

Figure 1.1 Candle flame

measured more accurately by complicated optical methods, which confirm the existence of temperatures higher than those measured by a thermocouple probe. However, there is still some debate about the exact interpretation of the results. Temperature is a measure of the total kinetic energy of the molecules, atoms and ions in the flame. The energy can be present as the random movement, the vibration and the rotation of these species and also as different states within their structure. At flame temperatures it is difficult to relate one form of energy with the total energy because things happen so fast that nothing is ever quite in equilibrium. Again, the chemical formula shows correctly what you start with and what you have at the end. It tells you nothing about the highly complex processes in between.

However, we can be certain that most of the flames encountered at a fire will have temperatures at least as high as those in Figure 1.1, and some will be considerably hotter. The existence of temperatures over 1000°C in the candle flame can be readily demonstrated. If a thin copper wire is removed from an electrical flex and is placed in the position shown in the figure, the end will be melted into a spherical blob. The melting point of copper is 1083°C and for the reasons given above the flame temperature must be higher than this. The wire is brought in through the flame, as in the figure, to reduce the heat losses by conduction.

Diffusion flames from gases

A diffusion flame can burn quite readily supplied by gas from say a crack in a pipeline. The size of the flame will be determined by the flow of gas. The reactants, gas and air, will have to be heated to the flame temperature as they enter the reaction zone, but no additional heat will be needed to ensure a continuous flow of fuel. Thc positioning of the flame relative to the fuel source is therefore not critical and this may account for a reduced tendency of flames in this situation to produce smoke.

Diffusion flames from liquids

Before a liquid can burn it must be vaporized. If this is done on a wick, as with the candle or in paraffin lamps, then the flame is usually small, stable and closed (not smoking). A small cup of liquid fuel will also burn steadily with either an open or a closed flame, depending on the fuel area. The flames over a larger area of fuel will become wrinkled and will be pulsating, and on pools with diameters greater than about 1 m the flames will be fully turbulent. Once the flame is fully established, the rate at which the fuel is consumed will be constant for a particular pool size but the actual rate per unit area will vary with the size of the pool. For example, on a burner only 5 mm in diameter petrol is consumed at 15 mm/min. A pool 10 cm in diameter burns at only 0.5 mm/min, but when a turbulent flame is established on any pool greater than 1 m in diameter the rate is steady at about 4 or 5 mm/min. Other hydrocarbons, like kerosene and even butane, have similar burning rates. Once steady burning has been established, the surface of the fuel is at a temperature just below its boiling point. The temperature drops quite quickly beneath the surface and the heated layer is only a few centimetres deep (however, this does not apply to certain fuels, particularly crude oils, which produce a 'heat wave' – see below under Boilover). If attempts are made to burn a very thin layer of fuel, the heat losses may be too great for a flame to be sustained. This is why a slick of crude oil from which the more volatile fractions have been lost cannot be ignited even though it is still combustible.

Liquids will also burn as running fires: typically when a leak has occurred at a high level on a process plant. Again diffusion flames are involved, but they may be particularly difficult to extinguish.

The burning of droplets of fuel is of great interest to designers of furnaces and jet engines. It becomes of practical importance to fire engineers when spray fires and fuel mists are involved (although not necessarily as diffusion flames; this is discussed later).

Not all diffusion flames are luminous. The best known exceptions are those of the alcohols methanol and ethanol. These both have pale blue flames which, because little heat is lost by radiation, have appreciably higher temperatures than those in hydrocarbon flames. However, heat transfer to the liquid surface by radiation is very low and for this reason the flames are very close to the surface – so close that much of the heat needed for evaporation is probably transferred by conduction. The low luminosity of a methanol flame together with the comparatively high latent heat of vaporization account for the low burning rate of a pool fire: about one-fifth the rate of a hydrocarbon fire. Despite this, methanol fires are unusually difficult to extinguish.

Diffusion flames from solids

Some solid materials, like paraffin wax, will melt in a fire and then burn as a liquid. Most solid fuels do not melt and the volatiles needed to feed a diffusion flame are released by pyrolysis. Wood for example breaks down into a wide range of products which include simple things like methanol, carbon monoxide and hydrogen, together with very complex tars. The cellulose in the wood decomposes at between 200°C and 350°C, but the lignin, the hard part of the wood, comprising about 10 per cent of the total weight, requires 280° to 500°C. The surface temperature of burning wood and most other solid materials is in excess of 350°C, so there will be considerable heat loss by radiation. It is for this reason that a heap of logs, in which mutually radiating surfaces are close together, will burn more efficiently than a single log.

As the volatiles are released from burning wood some of the carbon, particularly from the lignin, is left behind as a solid porous mass (charcoal). As this char forms on the surface, heat transfer to the unburned wood will be reduced, but the volatiles will continue to escape through cracks in the surface which penetrate as fissures as the depth of char increases. The cracks gradually widen to form a pattern which is called 'crocodiling'. When all the volatiles have been burned about 15 to 25 per cent of the original wood will remain as char. When the flames have gone, the hot carbon surface will be in contact with the oxygen in the air, with which it can react. The slow oxidation process which results is called glowing combustion or smouldering. However, this process will not continue for long on an exposed surface. This is because as the mass of char cools down a temperature is reached at which heat released by the slow combustion process cannot keep pace with heat losses by radiation, conduction and convection. Smouldering can continue only if additional heat can be supplied to the burning surface. A charcoal fire will continue to burn because of the mutually radiating surfaces within the heap, but combustion will cease if the pieces of charcoal are separated.

The burning rate of wood is directly dependent on the flow of heat to the burning surface. At one time it was thought that the duration of a fire could be determined by measuring the depth of char on exposed wooden beams. A burning rate of 0.025 in/min (0.6 mm/min) was assumed and this was based on the results of standard fire tests. In fact in a fully developed room fire the burning rate could be as high as 4 or 5 mm/min.

GLOWING, SMOULDERING AND DEEP-SEATED FIRES

Glowing combustion occurs as the direct surface oxidation of carbon char at temperatures of about 500° to 750°C. The overall reaction can be represented by:

$$C + O_2 = CO_2 + \text{heat}$$

However, if the glowing surface is watched carefully it is usually possible to see the pale blue flame of carbon monoxide, indicating a two-stage process:

$$\begin{aligned} 2C + O_2 &= 2CO + \text{heat} \\ 2CO + O_2 &= 2CO_2 + \text{heat} \end{aligned}$$

The combustion of wood char in this way finally results in a light grey ash which contains mainly the inorganic constituents of the wood. As much as 33 per cent of the heat released in the total combustion of wood comes from the char.

Smouldering can also occur in unburned materials. For this to happen, the fuel must be porous, like a heap of sawdust or a foamed plastic material. If we consider a layer of combustible dust to which an ignition source has been applied at one end, then a smouldering zone will spread gradually throughout the layer. This is illustrated in Figure 1.2. The temperature within this zone will be 600°C to 750°C and the combustion will be similar to that of charcoal: the direct oxidation of a hot carbon surface. Heat will be transferred to the unburned fuel ahead of the combustion zone and as soon as the temperature there reaches 200° to 300°C, most organic materials will begin to decompose. This pyrolysis will continue until a char is formed, which will then begin to oxidize and the smouldering process will continue. The volatiles which are released will be very similar to those produced during flaming combustion, but in this situation they will not be burning. It is not usually possible to ignite these volatiles because the critical flow needed to sustain a flame has not been achieved. The smouldering process can continue, unlike that on an exposed surface, because heat losses from the reaction zone are low: it is insulated by the residual ash on one side, and the unburned fuel on the other. There is however a critical depth for a thin layer of fuel below which the heat losses to the surroundings are too great for the process to continue. For example, the minimum depth for the

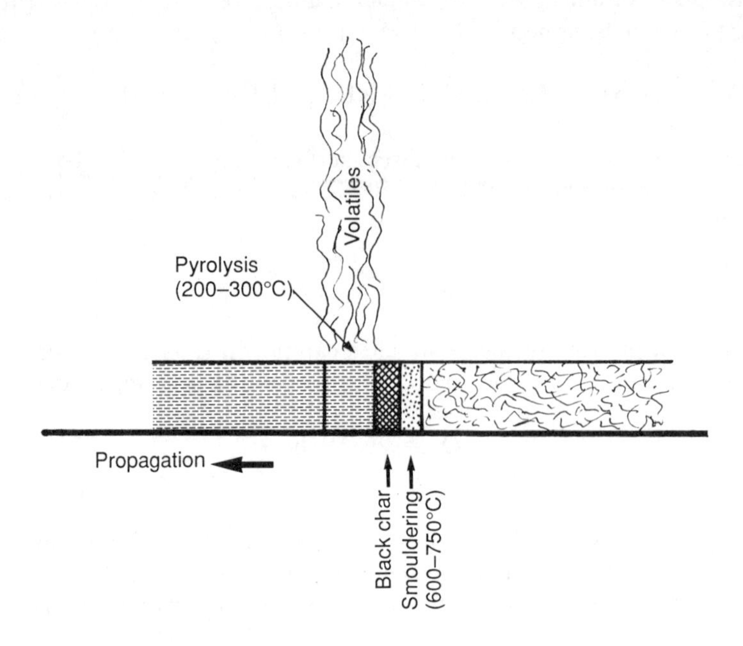

Figure 1.2 Smouldering combustion

sustained smouldering in still air of deal sawdust with a mean particle size of 1 mm is about 30 mm.

Smouldering of this kind can happen only in fuels which produce a rigid char. If a lot of tar is produced, or if the unburned material shrinks away when it is heated, then the process cannot continue. Smouldering can progress unnoticed within a mass of fuel for a considerable time. It was found experimentally for example that it took three months for smouldering to spread from the bottom to the top of a box filled to a depth of 8 feet (2.4 m) with sawdust. This very slow rate of burning within a mass of particulate or fibrous fuel results from the need for air to diffuse inwards to the reaction zone against the flow of combustion products coming out. The excellent thermal insulation ensures that the char remains sufficiently hot to react despite the slow reaction rate. In the sawdust tests it was found that the time taken for combustion to travel from the bottom to the top of a heap was proportional to the square of the depth (thus, for a depth of 0.85 m the time was only 10 days). The presence of the smouldering did not become obvious until it was close to the surface. A damp patch appeared, followed by a musty smell and then by the appearance of charring.

Smouldering will sometimes pass over to flaming combustion, particularly if a slight air flow is present. The increased supply of oxygen to the char surface will increase the flow of volatiles so that thc critical valuc needed to sustain a flame is exceeded. Ignition can then result if the volatiles are blown back over the heated surface, or may occur spontaneously.

A large number of materials of vegetable origin are capable of smouldering. These include paper, fabrics, fibreboard, sawdust, cork dust and latex rubber foam. A number of synthetic foamed materials will also smoulder including polyurethane, polyisocyanurate, and phenolformaldehyde.

It is possible for a liquid fuel to smoulder if it is deposited as a thin layer on the surface of a fibrous material which forms an insulating mass. A well known example in the chemical industry is known as a lagging fire. The lagging is a layer of an inorganic fibre which is used as a thermal insulator on vessels and pipelines containing hot liquids. Typically, a combustible liquid of high molecular weight (perhaps a heavy fuel oil) escapes from a joint in a pipeline and soaks into the lagging. If the temperature is above about 100°C, slow oxidation of the oil will occur and there will be a local increase in temperature. The process continues for several days or even weeks until smouldering begins. Flaming combustion can then by established and this can be a hazard in a plant which is handling flammable liquids.

A fire which begins as flames on the outside of a heap of fibrous or particulate material may soon penetrate into the heap as smouldering. If the flames are then extinguished, the smouldering may continue and it may not be possible to stop it without breaking up the heap to expose the hot surfaces. This is known as a deep-seated fire. Such fires can of course begin within the heap, either spontaneously or as a result of an ignition source. The very slow sawdust fires were deep-seated and would have been difficult to extinguish completely. Deep-seated fires are again discussed in Chapter 25.

PREMIXED FLAMES

The burning of a gas in a diffusion flame has already been discussed. The reaction zone is established where the fuel and the air, approaching from opposite sides, are mixed by diffusion. However, suppose that the gas is leaking freely from a hole in a pipeline but in the absence of an ignition source. It will still be able to mix with the air by diffusion (but more slowly than at flame temperatures) to form mixtures which are combustible. Suppose that this has happened inside an enclosure and that the entire volume has become filled with such a mixture. If now a sufficiently energetic ignition source appears, say at the centre of the volume, then a flame will propagate away from the source and will continue to burn until all of the combustible mixture has been consumed. In this particular situation the flame would be spherical until it touched a wall.

For obvious reasons flames like this are called premixed flames. These flames are nearly always a bright blue colour with none of the luminosity of a diffusion flame. This is because the intimate premixing of the reactants leads to short overall reaction times and precludes the formation of carbon particles. The polymerization of the carbon in a diffusion flame is a relatively slow process and takes place in the absence of oxygen. In a premixed flame the carbon atoms will encounter an oxygen atom at least as frequently as another carbon atom and will therefore have little chance of forming graphite rings. The lack of luminosity means that little heat will be lost by radiation and this, together with higher reaction rates, accounts for higher maximum flame temperatures in premixed flames than in a diffusion flame of the same fuel.

Burning velocity

The burning velocity – the velocity at which the flame propagates into the unburned mixture – is a characteristic of the particular mixture. The maximum values for most fuels are usually around 0.5 m/s, which may seem surprisingly slow. Some typical burning velocities are shown in column G of Table 1.1. If the decomposition of a fuel is exothermic, then this heat is added to the heat of combustion and both the flame temperature and the burning velocity will be high. Acetylene is a good example of such a fuel: the table shows the burning velocity as 1.58 m/s. The rate-controlling step in a premixed flame is the transfer of heat, and also of active species, into the unburned mixture. The relatively high temperature of an acetylene flame enhances the heat (and particle) transfer and therefore ensures a higher burning velocity. (Burning velocity should not be confused with flame speed. If the gas mixture through which a flame is propagating is itself in motion, then the flame speed is the sum of the burning velocity and the velocity of the gas. That is, the flame speed is the velocity of the flame relative to a fixed point in space whereas the burning velocity is relative to a point in the unburned gas.)

Limits of flammability

Diffusion flames are nearly always stoichiometric, but premixed flames can burn in mixtures which are not chemically balanced. However, there is a limited range of mixture ratios over which this is possible. The minimum percentage of a fuel in air which is just capable of supporing a premixed flame is called the lower limit of flammability. The maximum fuel concentration which is capable of burning is at the upper limit. For most

Table 1.1
Combustion properties of some gaseous fuels

A	B	C	D	E	F	G	H	I	J	K	L
		Lower limit %	Upper limit %	Stoichiometric ratio %	C/E	Maximum burning velocity m/s	% fuel at maximum velocity %	Adiabatic flame temperature °C	Expansion ratio	Minimum ignition energy mJ	Quenching distance mm
Methane	CH_4	5.0	15.0	9.5	0.53	0.45	10.0	1875	7.4	0.29	2.0
Ethane	C_2H_6	3.0	12.5	5.6	0.54	0.53	6.3	1895	7.5	0.24	1.8
Propane	C_3H_8	2.2	9.5	4.0	0.55	0.52	4.5	1925	7.6	0.25	1.8
Butane	C_4H_{10}	1.9	8.5	3.1	0.61	0.50	3.5	1895	7.5	0.25	1.8
Hexane	C_6H_{14}	1.2	7.5	2.2	0.55	0.52	2.5	1948	7.7	0.25	1.8
Benzene	C_6H_6	1.4	7.1	2.7	0.52	0.62	3.3	2014	7.9	0.22	1.8
Acetylene	C_2H_2	2.6	80.0	7.7	0.34	1.58	9.3	2325	9.0	0.02	
Hydrogen	H_2	4.0	75.0	30.0	0.13	3.50	54.0	2045	8.0	0.02	0.5

fuels these limits are quite close together, but fuels like acetylene which release heat in addition to the heat of combustion have wider limits. Some examples of limits of flammability are shown in columns C and D of Table 1.1

The existence of the limits can be explained in terms of heat balance. At the lower limit there is an excess of air which is heated to the flame temperature without then taking part in the reactions. This excess air inflicts a thermal load on the flame which reduces its temperature to a critical value at which heat is lost faster than it is generated. The combustion processes can then no longer continue. For many fuels this critical lower limit temperature is around 1400°C.

Column E of Table 1.1 shows the percentage of fuel which is present in the stoichiometric fuel/air mixture. Column F shows the result of dividing this value into the lower limit value. For many fuels, including several in the table, the lower limit concentration is found in this way to be around 0.55 of the stoichiometric concentration. Since the amount of heat generated in a flame must be proportional to the concentration of fuel, it follows that the heat released by a lower limit flame must be about 0.55 of that released by a stoichiometric flame. This in turn indicates that the thermal load needed to reduce the flame temperature to the critical value is 0.45 of the maximum available heat. In other words, if we can extract 45 per cent of the heat from a stoichiometric flame, then the flame will be extinguished. The upper limit of flammability depends similarly on the presence of a thermal load imposed by an excess of fuel. However, the decomposition of most fuels is endothermic (unlike acetylene) so some heat will be used in this way as the fuel approaches the reaction zone. Also there may be carbon formation at this limit, resulting in heat losses by radiation. The critical conditions at the upper limit are therefore less well defined, but the critical temperature will not be less than 1400°C.

So, the flame temperature is at a maximum when the fuel/air mixture is stoichiometric, and falls to a critical value of around 1400°C at the limits. The burning velocity follows a similar pattern: it does not fall to zero at either limit and increases towards the stoichiometric ratio. However, the maximum value is always slightly on the fuel-rich side of stoichiometric, (see column H of Table 1.1). This is because the reaction zone of a rich mixture will contain an excess of hydrogen atoms. These very active species will diffuse rapidly into the unburned gas and enhance the rate of reaction (the rate-determining step is the transfer of both heat and active species). The great mobility and reactivity of the hydrogen atoms also accounts for the wide limits of flammability of this gas: 4 to 75 per cent.

Effect of temperature

The limits of flammability are widened if the temperature is increased. This can readily be explained by the existence of a critical temperature at the limits. At the lower limit, for example, if heat has already been supplied to the mixture then less fuel will have to be burned to achieve the critical temperature. If less fuel is needed, this simply means a lower concentration at the limit. Similarly, less air will need to be burned at the upper limit, so the concentration of fuel at this limit can be increased. (The values shown in Table 1.1 were obtained at normal ambient temperatures.)

Effect of pressure and oxygen concentration

A reduction of pressure has very little effect on the limits down to about 0.1 bar, where combustion may cease. However, this pressure should not be regarded as limiting. Under carefully controlled conditions flames can be established at lower pressures than this. At pressures above atmospheric the lower limit widens only slightly but there can be a considerable effect on the upper limit. For example the limits for methane at atmospheric pressure are 5 and 15 per cent (Table 1.1) but at 200 bar they widen to 4 and 60 per cent.

An increase in oxygen concentration (above the 21 per cent normally present in air) has very little effect on the lower limit, simply because oxygen is just as effective a thermal load as is nitrogen. There is however a marked effect on the upper limit, where any additional oxygen increases the amount of fuel which can be burned. Again taking methane as an example, in pure oxygen the lower limit remains at 5 per cent but the upper limit is widened to 60 per cent. The similarlity with the effect of increased pressure is not coincidental. When the pressure of air in a vessel is increased, then the partial pressure of oxygen must necessarily be increased proportionally. The effect on combustion processes is the same as that of oxygen enrichment by the addition of oxygen. The hazardous situation which results from an increase in pressure is sometimes overlooked. The addition of gases other than oxygen to fuel/air mixtures will usually narrow the limits of flammability and this is discussed later under Inhibition.

For many fuels increased pressure has very little effect on burning velocity. In fuels like methane, with a normal burning velocity of 0.45 m/s or less, an increase in pressure will slightly reduce the velocity. Between 0.45 and 1.0 m/s there is no change, and for gases with normal burning velocities above 1.0 m/s there is a slight increase at higher pressures. However it should be noted that even though the burning velocity of a stoichiometric methane/air mixture may drop to 0.44 m/s at 5 bar, the mass rate of burning (in g/s of fuel), and therefore the rate of heat production, will increase nearly five times. There will therefore be an increase in flame temperature and indeed an increase is necessary to transfer sufficient heat into the unburned gases to maintain even the reduced burning rate (being now five times the density of a mixture at 1 bar, these gases will need five times as much heat to achieve the flame temperature).

Explosions and expansion ratios

If a premixed flame is propagating through a flammable mixture in a closed vessel, there will be an increase of pressure within the vessel. This is due mainly to the heat which is released by the flame, but it may also happen that the volume of the combustion produces (at atmospheric pressure) is greater than the volume of the original mixture. This will result in an additional increase in pressure. Starting with a stoichiometric mixture at atmospheric pressure in a completely sealed vessel, the final pressure for many fuels will be around 8 bar. This maximum pressure will exist for only a short time because the hot gases will be cooled by the walls of the vessel.

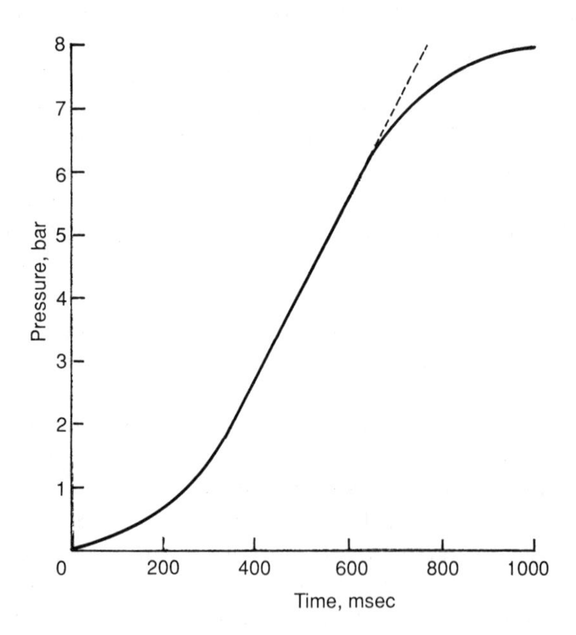

Figure 1.3 Pressure/time curve from the combustion of a stoichiometric methane/air mixture in a 2.5 m^3 cubical vessel.

A plot showing the change of pressure in the vessel with time will look like Figure 1.3. This shows the pressure recorded following the ignition of a stoichiometric methane/air mixture in a roughly cubical vessel with a volume of 2.5 m^3. It will be seen that the pressure rises slowly initially but that the rate then increases rapidly. This effect can be explained if we consider the growth of a spherical flame at the centre of thc vessel. We know that the burning velocity is about 0.45 m/s, so this is the rate at which the radius of the sphere is increasing. The hot gases which have been produced by the flame are inside the sphere and their volume can be calculated from $^4/_3\, {}^{P}r^3$. So if we assume that the radius is increasing linearly with time (because the burning velocity is constant) then the volume of the burned gases must increase with the cube of the time. The pressure will also increase at the same rate, so the shape of the first part of the curve in Figure 1.3 follows a cube law. This is not exactly true because the burning velocity is not constant. It is affected very little by the increase in pressure, but it is increased by the expansion of the hot gases inside the spherical flame. In Figure 1.3 the pressure rises less rapidly after about 0.6 s. This is because at this time the spherical flame has reached the walls of the cubical container (Figure 1.4). The total area of the flame is then reduced as it burns into the corners of the container. At this time, 0.6 s, the flame has travelled about 0.7 m from the centre of the cube. The average flame speed is therefore in excess of the burning velocity of methane (0.45 m/s) and this illustrates the effect of the expansion of hot gases within the spherical flame.

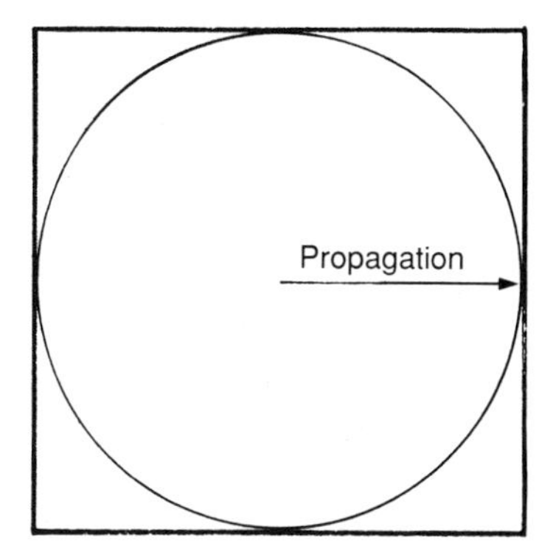

Figure 1.4 Propagation of a premixed flame in a cubical container.

If the flammable mixture were not confined in a vessel but were free to expand, then instead of the pressure increasing eight times, the volume would increase by the same amount. This potential of a stoichiometric mixture to produce an increase in pressure or volume is expressed as the expansion ratio. Values for a number of fuels are shown in column J of Table 1.1. It is usually assumed that for most stoichiometric hydrocarbon/air mixtures the expansion ratio will be 8, and for hydrocarbon/oxygen mixtures it will be 16. It may be surprising that hydrogen and acetylene, which have higher flame temperatures than the other fuels listed, have only slightly higher expansion ratios. This is because with these fuels there happens to be a reduction in the volume (at ambient conditions) of the burned gases compared with that of the original mixture. This can be seen from the chemical equations, because for each of the molecules of gas shown there we can assume that one volume (one litre for example) of gas was involved. (Avagadro's hypothesis says that equal volumes of gases, at the same temperature and pressure, contain equal numbers of molecules.) So, for hydrogen:

$$\begin{array}{ll} 2H_2 + O_2 & = 2H_2O \\ \text{3 volumes} & \text{2 volumes} \end{array}$$

three volumes of mixture are burned to produce two volumes of water vapour. In the same way it can be shown that seven volumes of an acetylene/air mixture will produce six volumes of end products. If however we look at hexane (a typical hydrocarbon with properties similar to those of petrol) we find:

$$\begin{array}{ll} C_6H_{14} + 9\tfrac{1}{2}O_2 & = 6CO_2 + 7H_2O \\ 10\tfrac{1}{2}\text{ volumes} & \text{13 volumes} \end{array}$$

Expansion ratios with values around 8 result from the combustion of stoichiometric mixtures. The ratios falls towards zero as the limits are approached, but the ignition of any mixture within the limits of flammability in a closed vessel will result in a pressure rise. If the pressure is sufficiently high then the vessel will rupture, and we are of course describing an explosion. In strict combustion terms the premixed flame is the explosion (also called a deflagration) and what happens to the vessel is a pressure burst. It is normal practice in industry to construct a vessel so that it can withstand about four or five times its working pressure. A stoichiometric explosion would certainly burst such a vessel.

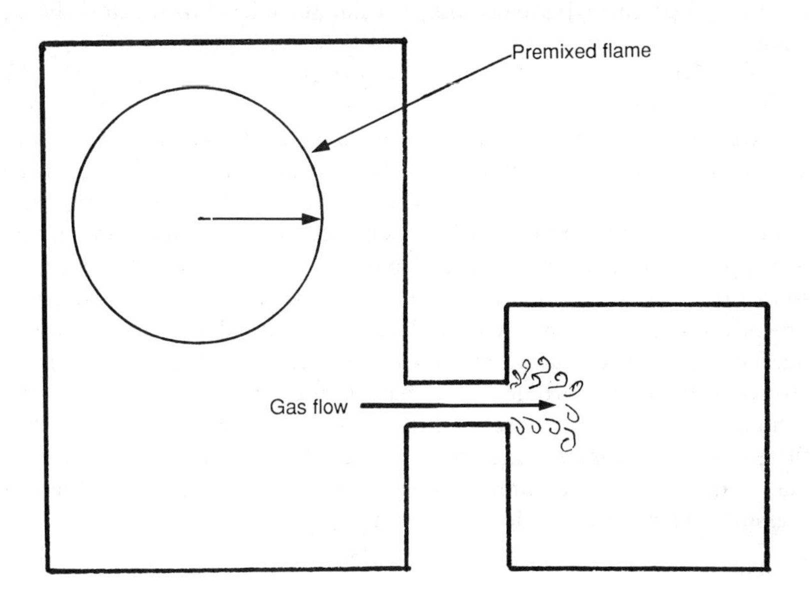

Figure 1.5 Turbulence and pressure piling.

Turbulence and pressure piling

If turbulence is present when a flammable mixture is ignited then the rate at which the mixture is burned will be greater than in still conditions. This is because the reaction zone will be distorted and will be mixed into the unburned gas. The effect will be an increase in the surface area of the flame and an increase in the rate at which heat and active species are transferred. In a closed volume the final pressure will be the same as in quiescent conditions but the rate of rise of pressure will be increased. This can result in higher local pressures; for example a pipeline or a piece of equipment in a room could be damaged by a momentary increase in the pressure acting on the surface nearest to the flame front. In a slower explosion there would be more time for the pressure on either side to equalize. Turbulence can be created by the passage of a premixed flame through a cluttered area of plant. It can also be caused by the situation illustrated in Figure 1.5. Here there are two vessels both filled with a flammable mixture. Ignition has occurred in the one on the left, and a flame is travelling through the mixture. There is therefore an increase of pressure in the vessel and some of the unburned gas will be pushed through the opening, causing turbulence as it enters the other volume.

The situation illustrated in the figure can result in another serious effect. Suppose that the mixture filling the two volumes is stoichiometric and has an expansion ratio of 8. If the two are completely sealed and initially at atmospheric pressure, then by the time the flame reaches the opening the pressure in the second vessel will have increased. The amount by which it has increased at that moment will depend on the size of the opening, but let us suppose that it has gone up to 3 bar. We are now applying an ignition source to a mixture at 3 bar and with an expansion ratio of 8, so the final pressure could be 24 bar. This will not be achieved because some of the burned gases can go back through the opening, thus relieving the pressure. However, the initial turbulence will have reduced the time taken to burn the contents of the second vessel and these losses will therefore be reduced. If there is a series of interconnected vessels, as there might well be in a process plant, then even a near limit explosion can result in very high final local pressures. This effect is called pressure piling.

Detonations

If a flammable mixture has been ignited at the closed end of a pipeline, as in Figure 1.6, then a premixed flame will travel down the tube. The expanding hot gases behind the flame front will set in motion the unburned gas, and the flame speed will be the sum of its fundamental burning velocity and the gas

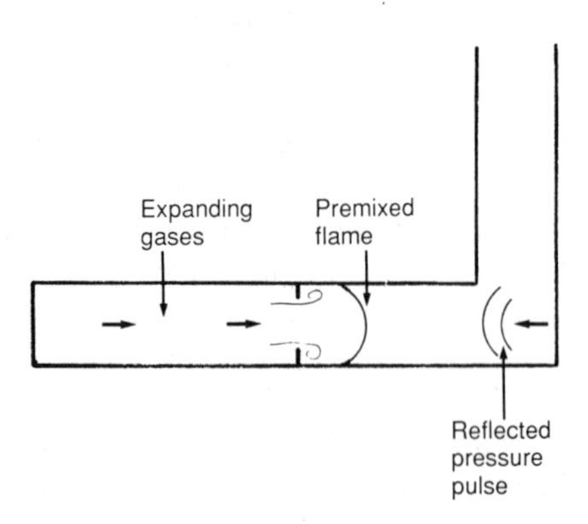

Figure 1.6 Transition to detonation.

velocity. In ideal conditions it is possible for the flame to progress at a constant speed throughout the length of the pipeline. However, if there are some obstructions, as in the figure, then turbulence can cause a sudden change in the flame speed. This will produce a momentary change in pressure, and a pressure pulse will travel down the pipe at the velocity of sound. If it meets a bend in the tube, or another obstruction, it will be reflected back and will pass through the flame front. The increase in pressure, and the corresponding increase in temperature, will kick the flame into another fluctuation in speed. If the burning mixture is sufficiently energetic, these effects can build up until a shock wave is generated – a pressure pulse in which there is an extremely rapid increase in pressure (as in a sonic boom). The reaction zone then travels with the shock wave. Very high reaction rates are possible in the conditions created by the shock of high temperature and pressure, and conversely a great deal of energy is available to sustain the shock wave. Extremely high velocities are possible, corresponding theoretically to the velocity of sound in the heated gases. Measurements have been made at between 1000 and 3500 m/s (2200–7800 mph). The velocity of sound in normal ambient conditions is about 330 m/s, so these values are clearly supersonic. The propagation of a reaction zone in association with a shock wave at supersonic velocity is called a detonation.

Detonations can happen only when there is sufficient energy available from the combustion processes, and the limits of detonability are therefore usually well inside the limits of flammability. For example, the flammability limits for hydrogen/air mixtures are 4 to 75 per cent and the detonation limits are 18 to 59 per cent. The corresponding vaues for acetylene/air are 2.6 to 80 per cent and 4.2 to 50 per cent. However, at slightly elevated pressures pure acetylene will itself detonate (it decomposes exothermically into carbon and hydrogen). At 5 bar it has a detonation velocity of just over 1000 m/s, rising to 1600 m/s at 30 bar. The detonation velocity for stoichiometric hydrogen/air is 2200 m/s, and this value falls to 900 m/s at the limits. Stoichiometric hydrogen/oxygen detonates at 3500 m/s. The maximum detonation flame temperatures are in excess of the maximum explosion flame temperatures: methane/air 2470°C, hydrogen/air 2680°C, acetylene/air 2830°C, hydrogen/oxygen 3430°C, acetylene/oxygen 3930°C.

As might be expected, the expansion ratio of a stoichiometric detonation is greater than that for an explosion in the same mixture, and is usually taken to be about 20. However, the pressure effects are directional, the greatest damage being inflicted in the direction of travel of the shock wave. For this reason a detonation which has failed to split open a straight run of pipe will punch a hole at a sharp bend.

Some books provide a list of gases which are capable of producing detonable mixtures. This can be misleading since it is becoming apparent that almost any flammable mixture can be detonated if a sufficiently energetic source is used (like a high explosive). Indeed, the observed detonation limits may also be directly related to the ignition energy. However, if we are concerned with the transition from deflagration to detonation as described in the beginning of this discussion, then only a limited number of fuel/air mixtures have been known to do this. They include in addition to hydrogen, acetylene and methane: acetone, benzene, butane, chloroform, cyclohexane, n-decane, ether, ethylene, methanol, naphthalene. Some have very narrow detonation limits: for both butane and ether the values are 2.8 to 4.5 per cent. The distance travelled by an explosion before it passes over to a detonation is called the run-up distance. This can vary from a few metres to a few hundred metres, depending on the mixture and on the diameter of the pipe. The distance is less in pipes of small diameter and is therefore sometimes given as so many pipe diameters. For methane, butane and n-decane (and other paraffin hydrocarbons) the run-up is likely to be 60 diameters. For more energetic fuels like hydrogen, acetylene and ethylene it will be substantially less. There may be no run-up at all if a detonation is initiated by a high explosive, but where a run-up does exist it provides a possible method of averting a detonation: it is possible to stop the deflagration before it has completed the distance (ways of doing this are discussed under Explosion venting and Explosion suppression).

If a detonation is travelling down a pipe which is connected to a vessel of large diameter, the detonation will almost certainly revert to a deflagration as it enters the vessel. So-called spherical detonations (spreading into a volume like a deflagration) can happen only in very reactive mixtures, or possibly in the unusual conditions of an OFCE (discussed below). If the deflagration enters another pipeline downstream of the vessel, then after the run-up distance it may again change to a detonation.

Characteristics of premixed flames

We can summarize the main characteristics of premixed flames like this:

- The reactants are gaseous.
- The reactants have already been mixed before ignition occurs.
- Heat transfer and the diffusion of active species are rate-controlling.
- Reaction can occur in any mixture which is within the lower and upper limits of flammability. Typically the limits are respectively from 0.5 to 2.5 of the stoichiometric ratio.
- The flames are usually non-luminous and are coloured blue.

- The flames propagate freely throughout the premixed volume at a velocity which is a characteristic of the particular mixture.
- The occurrence of premixed combustion in a closed vessel results in an increase of pressure which may rupture the vessel.
- Under normal conditions a premixed flame is an explosion.
- Under exceptional conditions a premixed flame will propagate as a detonation at supersonic velocity.

Premixed flames from gases

Premixed conditions can be created very readily by a leak of a flammable gas into an enclosure. If the discharge velocity is high then mixing will be facilitated by turbulence. A slow discharge of a high-density gas may result in a layer of gas on the floor, or conversely a low-density gas will collect at the ceiling. In either situation premixing will proceed by diffusion and a large explosive volume may eventually be formed. In the open air a plume of high-density gas may spread downwind over the ground, but diffusion processes will form a zone around the plume in which there is a transition from pure gas to pure air. Within this zone will be a layer which contains gas/air mixtures within the limits of flammability (see Figure 1.7). If an ignition source occurs within this layer, a premixed flame will spread throughout the layer. The gas remaining in the plume will then burn as a diffusion flame which will move back towards the source. The figure

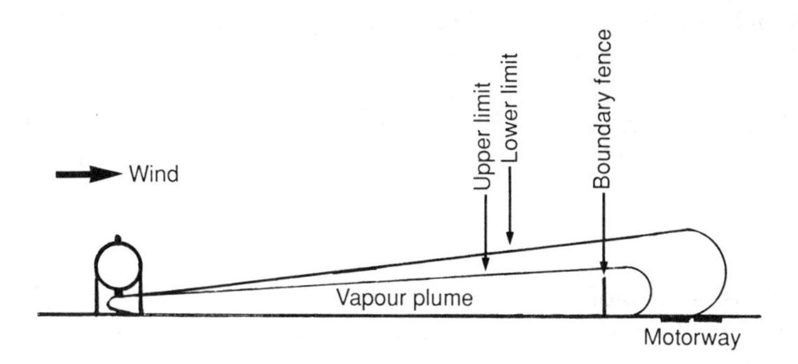

Figure 1.7 Plume of vapour from leaking propane (*Feyzin, 1966*).

illustrates the beginning of a disastrous fire at the Rhone-Alps Refinery in Feyzin, France (1966). Propane was leaking beneath a sphere and the vapour plume had passed through the boundary fence. It was probably ignited on a motorway 160 m from the sphere. It is possible to calculate the maximum downwind ignition distance, that is the greatest distance at which a lower limit mixture can be formed, from a probable leak and in the worst climatic conditions. A safe distance for a boundary fence can then be determined. A fire like this in which there is a rapid initial spread by premixing, but without any pressure effects, is called a flash fire. Another example of such a fire occurred at Los Alfaques in Spain (1978). An overfilled propane tanker, with no relief valve, ruptured when the contents expanded through solar heating. The vapour cloud had spread about 200 m, engulfing a camping site, before it found an ignition source.

The flames which are burned in domestic cookers and in many industrial furnaces are premixed. Diffusion flames are occasionally used industrially to ensure the maximum heat transfer by radiation, but a similar effect can be obtained if the ceramic lining of a furnace is raised to a high temperature by a number of premixed flames. In the simplest design of premixed burner the combustion air is entrained by a venturi inductor and then mixed by turbulence in a tube. The flame is then established at the end of the tube. The Bunsen burner shown in Figure 1.8 was an early example of this method. The air is entrained through the hole (usually variable) at the bottom of the tube and the premixed flame is the bright blue cone on the top. A burner like this is incapable of inducing more than about half the air which is needed for complete combustion, and the remainder has to be obtained from the air surrounding the flame. There are in fact two flames: the blue premixed flame with a larger and less luminous diffusion flame surrounding it. The premixed flame remains stationary because the velocity of the gases in the tube is greater than the burning velocity of the flame. If more air is entrained and the burning velocity is increased, then the area of the flame will be reduced. It is possible to calculate an approximate burning velocity if the gas velocity is known and the area of the flame is measured:

$$AV = FS_u$$

where A is the area of the tube, V is the gas velocity, F is the area of the flame and S_u is the burning velocity (the subscript u in S_u simply indicates that we are talking about the speed of the flame relative to the unburned gases). If the velocity of the gas mixture is reduced until it is exactly equal to S_u, the flame will be a flat disc. At gas velocities below this the flame will flash back into the tube. The opposite effect is called blow-off, when the gas velocity is

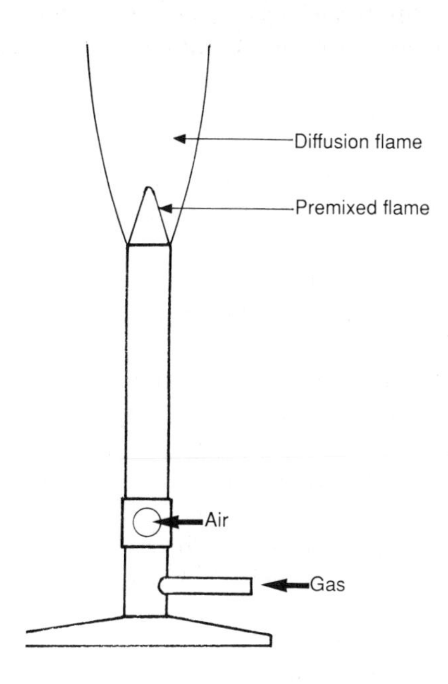

Figure 1.8 Bunsen burner.

sufficiently high to lift the flame off the burner. A premixed flame established in this way on a burner is sometimes (correctly) called a stationary explosion.

Premixed flames from liquids and flash point

If a volatile liquid like petrol is spilled in a room, vapour/air mixtures within the limits of flammability will be formed very quickly. If the room is completey sealed then the vapour pressure of the petrol is sufficiently high that the concentration of vapour in air will eventually be about 30 per cent. The limits of flammability are 1.4 to 7.6 per cent so this concentration is well above the upper limit and is non-flammable. This is why there are very few explosions involving petrol storage taks. They can happen only when the over-rich mixture has been diluted with air, typically after a tank has been emptied; or when the tank is at a temperature of about –20°C. The vapour/air mixture in a kerosene storage tak is non-flammable, but in this case it is because the concentration is below the lower limit. The vapour pressure of kerosene at 20°C is about 1 m bar, giving a concentration of 0.1 per cent (it is convenient here, and subsequently, to assume that atmospheric pressure is 1 bar). This is well below the lower limit of 0.7 per cent. To achieve a lower limit concentration the vapour pressure of kerosene must be increased to 7 m bar by raising its temperature to about 54°C. This temperature is the flash point of kerosene.

Flash points are measured by heating a sample of the fuel in a small cup, which may be open or closed depending on the particlar apparatus. As the temperature of the fuel is slowly increased, a small flame is applied near to the surface at regular intervals until a flame flashes through the small premixed layer which has formed (hence the name flash point). If the temperature is raised a few degrees above the flash point, the flow of volatiles from the surface will exceed the critical value needed to sustain a diffusion flame and continuous burning can be established. This is the fire point, which can be measured only in the open-cup apparatus. It is not always realized that when a fuel is heated to above its flash point it can be just as hazardous as petrol. Heavy fuel oil is a black viscous material which is difficult to ignite. It is heated to around 100°C before being burned to reduce its viscosity and increase its volatility. This temperature is above the flash point for many grades of fuel oil, so the pre-heating has significantly increased the fire hazard. Thus, any combustible liquid can create an explosive atmosphere in an enclosure if it is at a sufficiently high temperature. This must be remembered when any container of liquid fuel (like a storage tank) is heated in a fire.

Although a reduction in pressure has little effect on the lower limit of flammability, it does strongly affect the flash point. It was noted above that kerosene has a vapour pressure at 20°C of 1 m bar, insufficient to achieve the lower limit concentration of 0.7 per cent. However, if the pressure is reduced to 143 m bar then kerosene vapour, with a partial pressure of 1 m bar, will comprise 0.7 per cent of the atmosphere, which will now be explosive. In other words, at a pressure of 143 m bar, the flash point of kerosene is 20°C. This is important because a pressure of 143 m bar exists at an altitude of about 14 000 m. The vapour space of an aircraft fuel tank containing kerosene could enter the lower limit of flammability at this altitude, and could be stoichiometric at about 18 000 m (71.5 m bar). A stoichiometric explosion, with an expansion ratio of 8, could cause a pressure of 0.57 bar in the tank. (These values will of course be different at temperatures above and below 20°C, and for fuels with different vapour pressures.)

A spill of volatile fuel on the ground will release a plume of partially premixed vapour downwind similar to those produced in the two propane incidents described above. In both cases the propane, which has a boiling point of –42°C, was being stored as a liquid under pressure in the tanks.

Liquids with higher boiling points than this but at temperatures above their flash points will generally produce plumes with a shorter downwind ignition distance than that for propane under similar conditions.

Liquids are also capable of supporting a premixed flame if they form a mist of fine droplets. If the droplets have a diameter of less than about 0.01 mm then they will be vaporized completely by the heat radiated from an advancing flame front. The combustion may then be truly premixed even though the mixture may not be uniform in composition. Droplets larger than this may burn surrounded by an individual diffusion flame, but the overall effect will be similar to premixing because a flame front will propagate through the mixture. The fine suspension (of droplets less than 0.01 mm) has limits of flammability which can be calculated from the normal vapour phase limits. For example, taking the vapour density of petrol as 3.15 kg/m^3 and the lower limit as 1.4 per cent we can calculate a lower limit droplet concentration of 44 g/m^3, which corresponds closely to the experimental value. The burning velocity will also be similar, but slightly lower because of the heat needed to vaporize the droplets. The observed limiting concentrations for larger drops are lower than 44 g/m^3 and clearly cannot be related to the vapour phase lower limit. The probable explanation is that since each drop has an individual diffusion flame, the limit is established by the maximum separation between the drops over which combustion can be passed from one drop to the next. As might be expected, the burning velocity is lower for large drops and the flame front is much thicker.

It is important to note that the liquid forming a mist of droplets does not have to be above its flash point for the mist to propagate a premixed flame. A kerosene mist will explode as readily as one containing petrol. Explosions sometimes happen in the crankcases of large engines when lubricating oil has been vaporized on a hot bearing. The oil condenses out as a mist inside the crankcase and may then ignite if the bearing is sufficiently hot. If an aircraft crashes and the fuel tanks are ruptured, the fuel may be released as a cloud of fine droplets. Even if the fuel is kerosene, this spray can be ignited by a friction spark or by ingestion into an engine. Premixed flames will then flash back through the cloud and may possibly engulf the aircraft. Much research has been applied to the development of fuel additives which prevent spray formation and hence reduce the probability of a crash fire. These additives produce negative thixotrophy (or rheopexy) in the fuel, causing it to thicken when it is sheared (the opposite effect to the stirring of a jelly paint, which causes it to liquefy).

Premixed flames from solids

If a solid material is capable of burning by the release of volatiles, it is capable of producing an explosive atmosphere when it is heated in the absence of an ignition source (fuels like coke and charcoal which have lost their volatiles clearly cannot do this). If a fire is burning in an enclosed volume, there may be insufficient air to complete the combustion. If additional air is then admitted, perhaps by the opening of a door, mixtures of fuel vapour and air can form which are within the limits of flammability and which are then ignited by the fire. The result is an explosion, usually a mild one because it is in a near-limit mixture, but still capable of injuring the person who opened the door. Firemen are well aware of this hazard, called a flashback, and take appropriate precautions when they have to enter a burning building. A flashback can also happen when a furnace has been banked back and the door is opened.

There is always the possibility of an explosion if solid fuels are heated in an enclosure by a fire, or by cutting and welding operations. A tragic accident happened at Dudgeon's Wharf in London (1969) when a storage tank was being demolished. The tank had contained myrcene (similar to turpentine) which had left behind a resinous deposit. This was ignited when a cutting flame was used on the roof, and was extinguished by the Fire Brigade. Unfortunately sufficient vapour continued to be released from the hot material to form an explosive mixture in the tank. Firemen were standing on the roof when the mixture was ignited, and they were killed.

Dust explosions are possible whenever a finely divided combustible solid is suspended in air at a sufficient concentration. Larger particles may burn in a diffusion flame like drops of liquid fuel, but if the particle size is less than about 0.005 mm the combustion will be very similar to a premixed gas/air flame. Despite the need to pyrolyse most solid fuels, burning velocities can be high. This is because the high absorptivities of solids (compared to those of liquids) increase the efficiency of heat transfer to the particles by radiation. Some turbulence must be present to cause the initial dispersion of the dust. It frequently happens that a relatively small explosion in process plant initiates a much larger explosion in the building. This happens because the pressure pulse from the rupture of the plant disturbs layers of dust which have collected on horizontal surfaces throughout the building. An extreme example of this effect can happen in a coal mine. If deposits of coal dust have not been adequately treated to prevent their dispersion, a methane/air explosion at the coal face can disturb the nearby deposits and a coal dust explosion can propagate throughout the mine.

Standard laboratory tests are used to find out whether or not a particular dust is explosible. Measurements are made of the minimum concentration which can be ignited, and the maximum pressure and the maximum rate of pressure rise in the most explosive concentration. The rate of rise is

Table 1.2
Explosion properties of dusts

Dust	Minimum explosible concentration g/m^3	Maximum explosion pressure bar	Maximum rate of rise of pressure bar/s
Aluminium, flake	45	8.8	1380
Barley, milled	50	9.6	732
Carbon, activated	100	6.4	118
Charcoal	140	6.9	125
Coal, 25% volatiles	120	4.3	28
Coal, 43% volatiles	50	6.4	138
Cocoa powder	65	5.0	93
Flour	40	7.1	141
Magnesium	30	8.0	1035
Malt	50	6.6	304
Paper	30	7.4	449
Polystyrene	20	6.9	483
Sawdust	–	6.7	138
Soap	20	5.4	194
Sugar	15	7.6	345
Yeast	50	8.5	242

important in designing adequate explosion venting. No attempt is made to measure an upper limit because it is difficult to obtain any meaningful results. Tale 1.2 shows examples of these measurements for some typical dusts. It will be seen that some of the pressures are comparable to those from stoichiometric gas/air explosions, and also that metal dust explosions develop high maximum pressures and rates of pressure rise. Typical dust handling plant will fail at 0.15 bar and the walls of a building will collapse when subjected to about half this pressure. Substantial damage can therefore be caused by the ignition of quite small amounts of dust. The oxides of most metals have melting (or sublimation) points higher than the flame temperature so they appear as solid particles within the flame and are intensely luminous. This effect is exploited in fireworks and in photographic flash bulbs. It is also partly responsible for the high burning velocity (reflected in the rate of rise of pressure), and it increases the probability that fires will be started by a metal dust explostion. Transition from deflagration to detonation has not been observed in dust explosions although high flame speeds have been recorded in turbulent conditions. It is possible however that detonations have occured in coal mines following the initiation of a coal dust suspension by a high explosive.

Premixing is already present in certain specialized solids, notably the propellants and high explosives. Sometimes these are mixtures of fuels and oxidants, but frequently the oxygen is present within the molecule of the substance. Propellants are materials which deflagrate with the release of hot gaseous products. These gases may be used to drive a shell out of a gun, or to propel a rocket. Cordites, which are widely used for these purposes, comprise mainly nitrocellulose and nitroglycerine (NC and NG). Both of these substances are formed by the nitration of the organic materials by the action of nitric acid. Like many explosives, their molecules contain nitrate groups ($-ONO_2$) which are the source of oxygen within the molecule. The combustion of cordite progresses in three separate steps. The first takes place just beneath the burning surface with the exothermic rupture of the nitrate groups and the break-up of the molecules. In the 'fizz zone', on the surface, some of the molecular fragements are oxidized, but some of the oxygen still remains in –NO (nitro) groups. There is a delay before the NO reacts because a higher temperature is necessary, and there is therefore a 'dark zone' between the cordite surface and the final premixed flame. Because of the delay in the dark zone there is actually time for some carbon formation in the flame zone, and this ensures good heat transfer back to the unburned fuel to initiate the pyrolysis processes. Burning takes place in layers parallel to the surface in much the same way that a lump of coal is burned, but at a faster rate: perhaps 2 cm/s instead of 2 cm/hr. The burning rate increases with pressure, mainly because the dark zone becomes thinner and the flame zone moves in closer to the surface. The mass burning rate – the rate at which gas is released – depends also on the propellant surface area, so the pressure within a rocket motor depends on the area of the nozzle through which the gas is ejected and the burning area of the cordite. Special propellant shapes are needed if the pressure is to remain constant during the burning time.

Thus, cordites and other propellants are gas generators and burn by deflagration. Most of them cannot be detonated. Detonation is possible in the high explosives and the process is the same as in gaseous detonation: a premixed flame travelling with a shock wave. The velocity of sound, and therefore the detonation velocity, is much greater in solid materials than in gases. Typical values for a number of high explosives are these: TNT 6950 m/s; PETN 8300 m/s; RDX 8500 m/s; HMX 9124 m/s; TETRYL 7500 m/s; NC 7010 m/s; NG 2500–7000 m/s (both NC and NG are detonable; but a colloidal suspension of NG in NC, in cordites, is not). The actual detonation velocity is of importance because it affects the brisance of an explosive: its ability to shatter solid materials.

The first four explosives listed above are insensitive and detonate only when they are initiated by a shock wave from another detonating explosive. The special explosives used to provide the shock wave are called initiators or primary explosives. They detonate when they are heated (perhaps electrically) or subjected to mechanical shock (as in a percussion cap). Examples are mercury fulminate, lead azide and lead styphnate. TETRYL, unlike most other high explosives, is sensitive to friction and percussion and is sometimes used to prime the less sensitive materials. An electric detonator for example might contain a fusehead made by coating a mixture of potassium chlorate and charcoal (called a flashing composition) around a thin bridge wire. When this composition is ignited by the fusing of the wire, the flame ignites a priming charge, which could be lead azide. This detonates, and the detonation is passed to a small base charge of high explosive like PETN. It is the detonation emerging from this base charge which has sufficient brisance to initiate the main high explosive charge. Explosive materials have been classified here as: propellants which deflagrate; high explosives which detonate but are insensitive; and primary explosives which detonate and are sensitive to heat and percussion. The classification of a particular material may depend on its physical condition, and NG for example can fall into all three groups. All clearly present serious problems in a fire. The propellants burn very fiercely and have their own oxygen supply; the high explosives may burn relatively slowly but may eventually detonate; the primary explosives, which fortunately are usually stored in small quantities, will always burn to detonation. The fire problems associated with explosives and other unstable materials are discussed later.

Some high explosives can be premixed on site as required, and do not therefore need to be stored. ANFO, a mixture of ammonium nitrate and a fuel oil similar to diesel fuel, is often used for quarrying. The stoichiometric mixture containing about 5 per cent of fuel can be prepared quite readily, even by hand, and in quarrying it is simply poured down the boreholes. These holes may be typically 10 cm in diameter, as much as 18 m deep, and located 3.5 m apart and 3.5 m back from the face of the quarry. The amount of explosive needed can be calculated on the assumption that 1 kg will break about 9 tonnes of rock (about 70 kg per borehole in this example). Another 'safe' mining explosive is made by mixing liquid oxygen with charcoal. This mixture has the advantage that it will burn to detonation and does not need to be initiated with a detonator. Also, if there is a misfire it is only necessary to wait for the oxygen to disperse and the charcoal can then be removed safely.

2

Special kinds of combustion

BLEVE

In the Feyzin fire which was mentioned above, a plume of propane vapour travelled downwind from a storage sphere through the boundary fence and was ignited 160 m away on a motorway. Liquid propane continued to escape and fed a fierce fire which burned on the ground beneath the sphere. Liquid propane also spread away from this area to threaten other neighbouring storage vessels. Despite its very low boiling point the propane would exist as a liquid until it had received sufficient heat from its surroundings to enable it to vaporize. The other vessels comprised a further three 1200 m^3 capacity propane spheres and four of 2000 m^3 containing butane. The fire brigade's efforts were directed at preventing their involvement in the fire, but eventually four of these and the original propane sphere were lost. The pressure relief valve on the first vessel had opened early in the fire because the temperature, and therefore the vapour pressure of the liquid propane, had increased. The firemen thought that because the vessel was being vented, it would be safe. Sadly, it was not. Large flames completely engulfed the sphere but initially heat transfer to the shell was reduced by a layer of thermal insulation on the outside. As the fire progressed and the layer disintegrated, the heat flow to the shell increased. The lower part of the sphere was cooled on the inside by contact with the liquid propane, and the shell temperature in this region would not be much above the boiling point of the propane (which would itself be dependent on the pressure within the vessel). The upper part of the shell was not cooled in this way and its

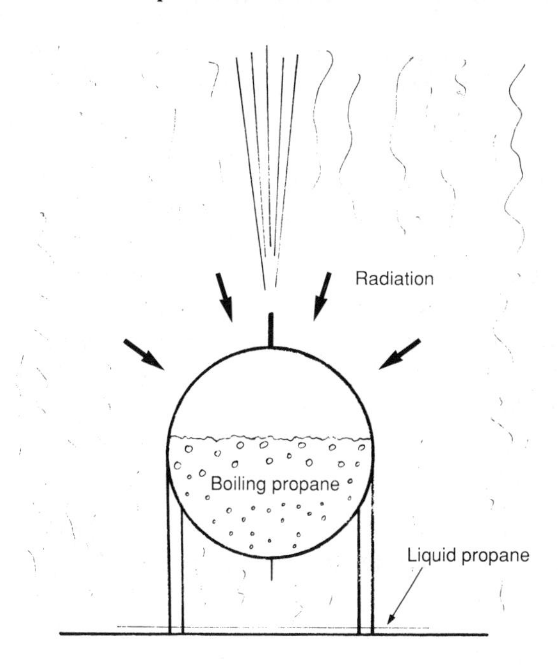

Figure 2.1 The conditions which caused a BLEVE (*Feyzin, 1966*).

temperature steadily increased (see Figure 2.1). Steel completely loses its load-bearing strength at about 500°C, and when the temperature approached this value (after the fire had been burning for about 90 minutes), the metal failed. There was still a substantial pressure within the vessel despite the open relief valve, and it ruptured violently. The sudden release of pressure would leave the liquid propane at a temperature well in excess of the ambient boiling point and the bulk of it would flash into vapour. There was thuse a violent expansion into the fire of a cloud of hot vapour and mist. It was the resulting fireball which killed 18 of the firemen. Fragments of metal from the ruptured sphere cut the legs of the next sphere, which fell over. Liquid propane then escaped from its relief valve, adding to the conflagration.

The release of a fireball from the ruptured vessel was a BLEVE: a boiling liquid expanding vapour explosion. This is likely to happen whenever a sealed or partially vented vessel containing a flammable liquid is heated in a fire. It can usually be averted by cooling the vessel with water. When a rupture does occur much of the hot liquid will flash into vapour, the heat of vaporization being obtained by cooling the liquid to its normal boiling point. Any remaining liquid is dispersed as a spray and may obtain sufficient heat to vaporize from the surrounding air. Air will be mixed violently into the expanding vapour, producing a partially premixed cloud. The characteristics of a BLEVE are therefore a large and raidly expanding fireball (flame speeds as high as 50 m/s have been observed), a blast wave, and (usually) missiles. If the liquid involved is not flammable, then damage can still be caused by the blast wave and the missiles. Many of the reported occurrences of BLEVEs have involved road or rail tankers. Some have happened as little as 10 minutes after the start of the fire but a delay of 30 to 60 minutes is more usual. Relatively small BLEVEs can happen if cylinders of liquid butane or propane are involved in a fire. Aerosols can also cause a small BREVE, particularly when they contain a flammable preparation like a hair lacquer and butane has been used as the propellant.

OFCE and OFCF

The disastrous fire at the Nypro plant at Flixborough, UK (1974) was caused by the failure of a temporary pipeline. Cyclohexane (flash point –33°C, boiling point 80°C) was being processed at 155°C and 9 bar. The pipeline failure released about 30 tonnes of cyclohexane at around 1 t/s. It has been estimated that half of the released material would flash off immediately as vapour and that the remainder would form a mist. Air would be entrained into the high velocity jet of vapour and mist, and some of the droplets would vaporize. There was no immediate ignition and a premixed cloud formed, above the plant, which could have been 200 m in diameter and up to 100 m high. About 45 seconds after the discharge began the cloud was ignited. Since the cloud was completely unconfined one might expect that a premixed flame would travel through it, at a speed dependent on the amount of turbulence, and that the effect at ground level would be a simple pressure pulse and some high levels of radiation. What happened had the appearance of a detonation. The blast caused extensive damage to the cyclohexane plant, its surrounding plants, and to storage tanks. Large volumes of flammable liquids were released and a devastating 40 000 m^2 fire ensued which burned for several days. Twenty-eight people were killed, 18 of them in the wrecked control room. Most of the houses in two villages about 1 km away were damaged, and some buildings up to 3 km away were also affected. The TNT equivalent of the aerial explosion was estimated at between 15 and 45 tonnes. It is unlikely that a true detonation developed in the cloud, but the estimated ground level overpressures of between 0.7 and 1 bar indicate flame speeds in excess of 100 m/s.

The explosion which started the Flixborough disaster was an open flammable cloud explosion (OFCE), which is also called an unconfined vapour cloud explosion (UVCE). If the cloud had burned as a flash fire without any damaging pressure effects it would have been an OFCF, the final F indicated a fire. There had been well over 100 large OFCEs before Flixborough, perhaps the most notable one being in an oil refinery at Pernis, Netherlands (1968). A cloud of about 140 t of hot hydrocarbon vapour shot upwards when the roof of a slops tank was torn off. When it ignited, the resulting blast wave caused extensive damage to the surrounding plant and buildings.

There are many gaps in the available data on OFCEs, but the conditions under which they are likely to happen can be predicted. The released fuel may be a gas, a volatile liquid or a refrigerated liquid, but a large vapour cloud is most likely to be formed by a flashing superheated liquid (as happened at Flixborough). The fuel must be ejected with sufficient violence to form a premixed cloud, and there must be a delay before the cloud is ignited (otherwise the result will be a BLEVE, which would probably be much less destructive). It has been estimated that the same amount of damage would have resulted at Flixborough if the delay had been only 10 seconds. This may well approximate to the minimum delay for any OFCE, and the maximum reported delay (in one explosion) was 15 minutes. The most likely interval is one or two minutes, during which time the cloud will not have drifted more than a few hundred metres. It has been suggested that the minimum size of cloud needed to ensure than an OFCE can occur is about 10 to 15 tonnes. However one explosion has been reported at Beek, Netherlands (1975) involving only 5.5 tonnes, and some experiments with large balloons filled with flammable mixtures produced violent explosions with less than one tonne. It would clearly be unwise to attempt to define a critical quantity, but there seems to be little doubt that the probability that an OFCE will develop decreases with the size of the cloud.

From this discussion on OFCEs it is clear that the particular event which follows the release and ignition of a quantity of fuel will depend on its reactivity and physical properties, on the environmental conditions, and on the way in which it is dispersed and ignited. If we consider the release of a few tonnes of a flammable liquid, it could burn as a pool fire for several minutes, as a flash fire for ten seconds to one minute, as a BLEVE for one or two seconds, as an OFCE for less than one second, and as a detonation for a few milliseconds. The rate of release of energy over this range of combustion modes could vary from perhaps 10^2 to 10^5MW.

BOILOVER and SLOPOVER

When a flammable liquid is burning, sufficient heat must be radiated from the flame to the fuel surface to maintain an adequate flow of volatiles into the reaction zone. The surface temperature is usually the boiling point of the fuel. The temperature falls rapidly beneath the surface and the original ambient temperature may be present only a few centimetres down. This is true particularly of relatively pure fuels which are also transparent. Conditions are very different if the fuel comprises a mixture of fractions with widely differing boiling points, and if the fuel is opaque. Examples of such fuels are crude oil and heavy fuel oils.

When these fuels are burning a fractionating process occurs at the surface. The lighter products, which have the lowest boiling points and greatest volatility, vaporize first and are consumed in the flames. The remaining heavier fractions, with higher boiling points, have a greater density than the bulk of the liquid, and they begin to sink. Some of the heat which this material acquired at the surface (by radiation from the flames) is transferred to the surrounding liquid. The result is that the heavier fractions increase in density and the surrounding liquid becomes less dense, thus continuing the process of separation. As the fire continues and more of the lighter fractions

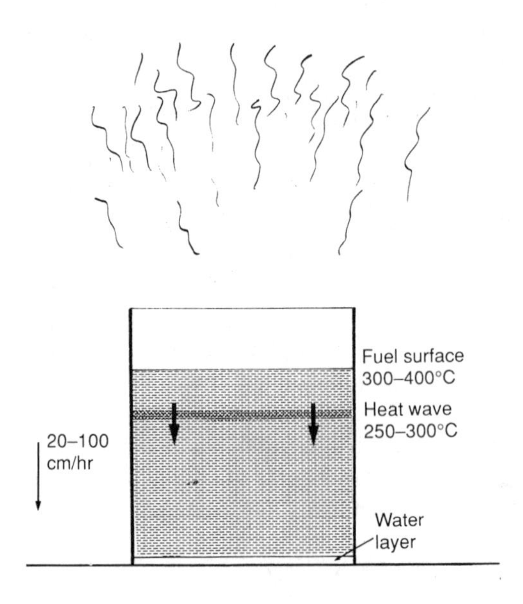

Figure 2.2 Heat wave in a tank of heavy fuel oil.

are lost, the surface temperature will increase, remaining always at the boiling point of the mixture. The heavy fractions will now have a higher density and a higher temperature than the ones already sinking into the tank, which they will eventually overtake. This hot layer, known as a heat wave, progresses downwards at between 20 and 100 cm/hr, depending on the type of fuel (Figure 2.2). This velocity is usually considerably in excess of the burning rate of the fuel which may be as low as 10 cm/hr. As the fractionation continues, the surface temperature can be as high as 370°C with a corresponding heat wave temperature of 300°C.

The real problem is the presence of a layer of water in the bottom of the tank. This is inevitable with stored petroleum fuels and can never be completely drained away. When the first contact is made between the heat wave and the water there will be local boiling, which stirs up the contents of the tank. Rapid heat transfer from the hot oil then causes the bulk of the water to flash into steam. This results in the violent ejection of the entire contents as a highly expanded flammable foam at a temperature close to its flash point. A very large area around the tank and well beyond the bund wall will then be engulfed in the blazing foam.

The only way to prevent a boilover is to put out the fire. If this cannot be done, warning of an impending boilover can be obtained by determining the position of the heat wave. This can sometimes be done by the careful application of a small jet of water to the side of the tank, moving from the bottom upwards. The position of the heat wave will be shown by the sudden appearance of steam. A thermal imager would give a clear indication of its position, as would a vertical stripe of temperature-sensitive paint on the side of the tank. As soon as it is suspected that a boilover is inevitable, everyone should be withdrawn from the area. If a large tank on a sloping site is involved then half a mile would be an appropriate distance.

By comparison a slopover is a small event, but again associated with a heat wave. It usually happens in the early stages of a fire when the heat wave is not far from the surface. Foam which is applied to the fire, or cooling water, can flash into steam if it penetrates to the heat wave. A relatively small volume of burning fuel foam will slop over the edge of the tank and burn in the bund. If this spill comprises the bulk of the hot fuel, then the fire in the tank will be easier to extinguish. In fact a deliberate slopover caused by jets of water is a possible firefighting method.

Boilovers and slopovers do not happen in fires involving petroleum fractions with narrow boiling ranges like petrol and diesel fuel. As we have seen, the heated layer beneath the surface of these liquids is only a few centimetres thick and it does not progress downwards to form a heat wave.

FLAMES IN AIRSTREAMS

If a wind is blowing over a pool fire the flames will be bent over towards the liquid surface and the heat transfer will be increased. The flame can cope with the increased flow of volatiles because of turbulent mixing and because its length can increase downwind of the pool. The result is an increase in the burning rate and also in the probability that fire will spread to other materials. However, if the wind velocity increases above a critical value the flames can be swept off the fuel surface and the fire may be extinguished. In practice such an effect is unlikely because of the presence of obstructions to which flames can be attached. Very high velocities would then be needed to dislodge these flames.

An idealized situation in which a flame is anchored to an obstruction is shown in Figure 2.3. This illustrates a laboratory test in which a laminar (non-turbulent) air flow is passing over a small pool of flammable liquid located behind a vertical metal strip. Much of the flame is smooth and lies roughly parallel to the fuel surface. There is a gap of about 2 mm between the metal strip and the front of the flame. The flame is then coloured blue for the first 2 or 3 cm and the remainder is brightly luminous. This is another example of a diffusion flame co-existing with a premixed flame. The premixing is explained if we follow the path of the single streamline which has been drawn in the figure. After passing over the metal strip, it forms a small turbulent eddy behind it, in which the air mixes with the vapour which

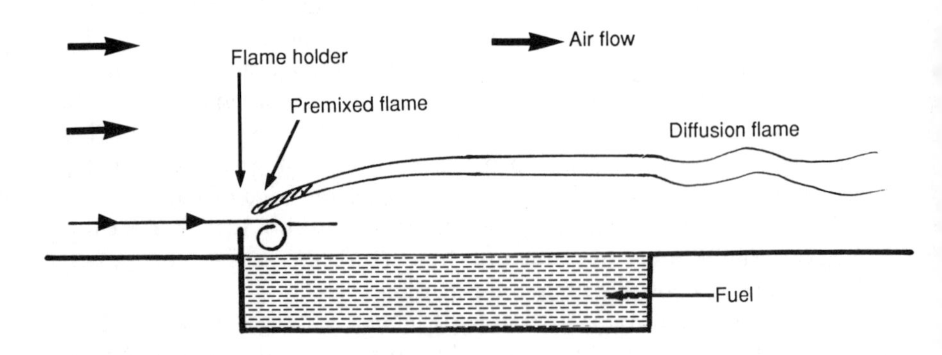

Figure 2.3 Liquid surface diffusion flame in an air flow.

has been released from the surface of the fuel. Eddies like this are formed whenever a layer of air passes over another layer which is either stationary or is moving at a lower velocity: they behave rather like roller bearings. The premixed nose of the flame is located at a point where its burning velocity exceeds the local velocity of the premixed gases.

In the particular test illustrated in the figure the fuel being burned was kerosene. With no obstruction at the front of the pool the blow-off velocity was about 5 m/s, but with an obstruction only 3 mm high the critical velocity increased to about 20 m/s. The optimum obstruction height in this particular apparatus was 15 mm, for which the blow-off velocity was in excess of 100 m/s. At heights greater than this the velocity was less, presumably because the flame was too far from the fuel surface for adequate heat transfer.

The simple strip of metal in these tests was acting as a flame holder, or 'baffle'. In aerodynamic terms an object in an airstream which has a recirculation pattern in its immediate wake is a bluff body. Flame stabilization by bluff bodies is used in the after burners (reheat systems) of aircraft gas turbines, and in ram-jet engines. These flame holders are often hollow cones, and an indication of the streamlines around one is shown in Figure 2.4. They are located downstream of the position where the fuel is injected, usually as a fine spray of liquid, and are therefore supplied with a partially premixed suspension. Vaporization, additional mixing, and flame holding occur in the recirculation zone, and a premixed flame is established within or near to the mouth of the cone. The rest of the flame, which may partially comprise a highly turbulent diffusion flame, fills the rest of the combustion chamber downstream of the baffle. In a gas turbine engine, flame holding usually depends on a recirculation zone which is created by injecting a fraction of the fuel/air mixture into the main stream. This injected stream is either normal to the main flow, or directly opposed to it.

Since flame holding of this kind depends on the establishment of a premixed flame in a flowing environment, the burning velocity must have an important effect on the stability of the flame. Engine testing has shown that the most stable conditions exist at a mixture ratio slightly richer than stoichiometric, which indeed corresponds to the requirements for maximum burning velocity.

When automatic extinguishing systems are designed for industrial situations where high airflows are involved it may be necessary to take into account the presence of premixed flames, which require a higher concentration of extinguishant than do diffusion flames of the same fuel (this is discussed later).

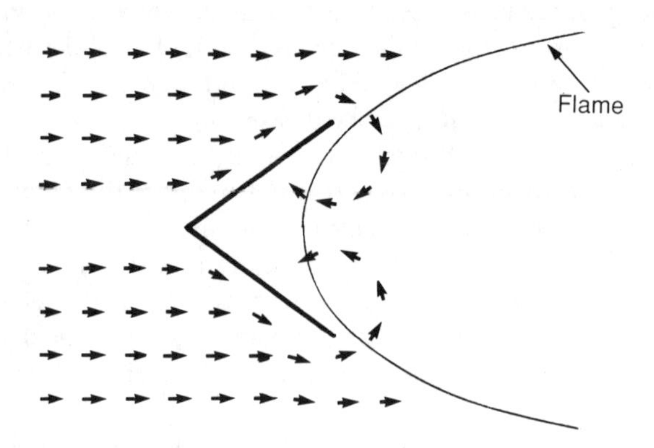

Figure 2.4 Combustion behind a flame holder in a high velocity air flow.

EXPLOSIONS AND DETONATIONS

These are mentioned again because of their practical significance. It may be only of academic interest to know which of the two phenomena is responsible for a particular event because the results may be much the same. If an explosion happens inside a building containing plant or machinery, the flame is likely to be accelerated as it progresses through the clutter to a velocity close to that of a detonation. The main differences in the effects of the two processes are that damage and injury are likely to be inflicted at a greater distance from a detonation (by both pressure effects and missiles); and that fires are more likely to be started by an explosion because it is slower event. There are however marked differences between the results of a deflagration and the detonation of a high explosive.

Explosions are possible whenever a flammable gas or a liquid at a temperature above its flash point is being stored or processed within a building. Indeed it is not possible to burn these materials without starting with an explosion (however small), since ignition can be established only in a fuel/air mixture which is within the limits of flammability. The greater the time between the loss of containment and the moment of ignition, the larger will be the premixed volume. The combustion of this volume will cause damage to the plant or the building if there is a sufficient increase in pressure. The pressure generated will depend on the ratio of the burned volume to the building volume, and on the stoichiometry of the mixture. The rate of rise of pressure will affect local pressure differences, so turbulence may become an important factor, as also may the effects of pressure piling.

The release of flammables in the open air is likely to cause only flash fires and pool fires except in the very unusual conditions which lead to an OFCE; or if heavy vapour spreads across the ground and enters an enclosed volume.

The vapour space within a fixed roof storage tank may contain an explosive mixture if the fuel is above its flash point. Fires in these tanks usually start after a vapour/air explosion has blown off the roof. As we have seen, petrol is sufficiently volatile to produce a mixture in the vapour space at ambient temperatures which is above the upper limit of flammability. However a movement of petrol out of the tank will cause air to be drawn in through the vent, while the heavy vapour layer remains substantially intact and moves down with the liquid. Flammable mixtures will then be formed by the diffusion of vapour into the air and explosions may be possible for several hours. Eventually, the whole volume again becomes too rich to burn. A petrol tank is particularly dangerous for some time after it has been completely emptied. The remaining vapour will gradually be diluted as air diffuses through a free flow vent, or is breathed through a pressure/vacuum vent during natural temperature cycling. It is possible eventually for the whole tank to be filled for a time with a stoichiometric mixture. An ignition then would result in a particlarly violent explosion.

Many of the liquids contained in fixed roof tanks are at temperatures below their flash points and are therefore relatively safe. However, explosive conditions can still be created if product is added at a high temperature (perhaps due to a process malfuncton), or is overheated by a faulty immersion heater, or if a small volume becomes overheated. This latter event can happen for example if an area of wetted wall is heated from the outside by exposure to a fire, or to cutting and welding operations.

Methanol is an unusually hazardous material in a fixed roof tank. Its flash point, corresponding to a lower limit concentration of 6.7 per cent, is 11°C. The unusually high upper limit of 36 per cent will be reached when the liquid temperature is 38°C. Thus in the UK there can be an explosive atmosphere in a methanol storage tank for about half the year.

The conditions under which a premixed flame can propagate through a fuel mist have already been discussed. It is possible for a similar flame to be sustained in an air-blown fuel foam. As with the mist explosions, it is necessary to achieve a critical concentration of fuel in air. It is likely that as the radiation from the flame reaches a layer of foam, some will vaporize immediately and the rest will collapse into a suspension of droplets.

Explosions have also occurred in which the available fuel has been a thin layer of oil on the inside of a compressed air pipeline. This oil is released from a (usually faulty) reciprocating compressor, and the ignition may have resulted from high temperatures at the compressor outlet, possibly in excess of 250°C. Again sufficient oil would have to be present to achieve the lower limit concentration at the working pressure. It seems likely that the oil film would be converted into a mist ahead of the advancing explosion.

3

Fire spread

The spread of flames in combustible gaseous mixtures has already been discussed in the section on premixed flames. We are concerned now with the spread of flames over liquid and solid fuels, and with the spread of fire within a building.

SPREAD OF FLAMES ON LIQUIDS

If an ignition source is applied to a pool of liquid which is at a temperature just above its fire point, then a premixed flame will spread throughout the combustible mixture which has been formed close to the surface. Sustained burning will then be established over the whole pool, as a diffusion flame. If the liquid in the pool had been at a higher temperature, the rate of spread of the premixed flame could have been higher. This flame speed reaches a maximum above a pool at a temperature at which the vapour pressure corresponds to a stoichiometric vapour/air mixture. The flame speed could then be up to five times the fundamental burning velocity, indicating that the expanding combustion products are pushing the unburned gases over the surface.

If the liquid in a pool is at a temperature below its flash point, it will be difficult to ignite. Ignition can be achieved only by the prolonged application of a flame to the surface. Once a diffusion flame has been established in this way, its rate of spread across the surface will be comparatively slow. It might be expected that the spread would be dependent on the rate of heat transfer to the surrounding fuel by radiation from the flame. Clearly this must be an important factor, but there is an additional interesting effect. When the liquid beneath the flame is heated (eventually to the boiling point), its surface tension is reduced. This causes a layer of hot fuel to flow out away from the flame over the cooler surface layers. The surface temperature in a region perhaps 10 cm wide from the edge of the flame is then between the flash point and the fire point. Blue premixed flames can be seen pulsating over this zone for a time, until the temperature increases to the fire point, and the diffusion flame can then advance into the zone. (This is illustrated in Figure 3.1.) The rate of spread can be around 50 mm/s, and is to some extent dependent on the factors, discussed below, which affect flame spread over solid fuels.

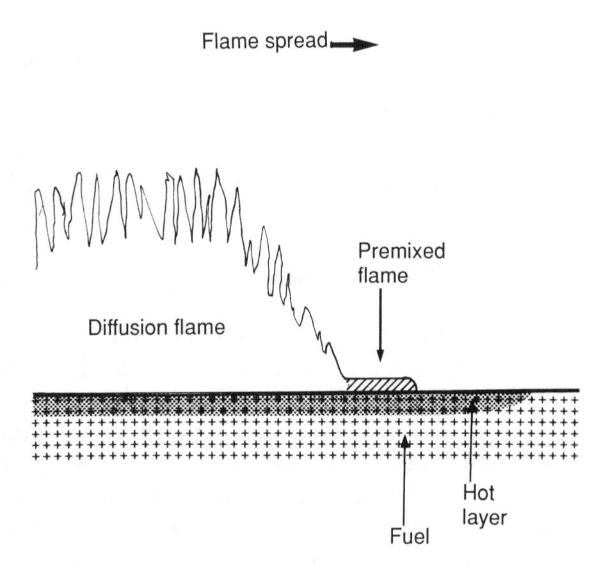

Figure 3.1 Spread of flame on a liquid surface.

SPREAD OF FLAMES ON SOLIDS

If a diffusion flame is to be sustained above a solid fuel, then there must be a sufficient flow of heat to the fuel to achieve the critical flow of combustible volatiles. In other words, the surface temperature must be raised to the fire point. If a flame is to spread over the surface of the fuel, then the unburned material adjacent to the flame must also be raised to this temperature. The flame is both the heat source and the ignition source. The rate of spread over the surface is dependent on a number of factors outlined below.

Physical and chemical properties

The surface layers of the fuel adjacent to the flame will receive heat by radiation. The time taken for the surface to reach the fire point temperature (and to start burning) must depend on the absorptivity, the thermal conductivity, the specific heat, and the density of the material. To these physical properties must be added the heat requirements for the pyrolysis processes. The combined effects of low thermal conductivity and low density can increase both the ease of ignition and the rate of flame spread: this accounts for the hazardous properties of some foamed plastics used in furniture.

Thickness

Thin sections of fuel can be heated by radiation more rapidly than thick ones. This is because of reduced losses by conduction, and because of the high surface to volume ratio. As the thickness is increased, the rate of spread is reduced until a critical size is reached beyond which the rate remains constant.

Orientation

The most rapid flame spread over a solid surface is obtained when the surface is vertical and the flame is travelling upwards. This is because of the enhanced heat transfer resulting from the close proximity of the flame to the surface. The flame becomes very long because it is receiving air from one side only and also because the pyrolysis zone, which is supplying volatiles to the flame, increases in length. The burning rate can be 50 times that of a flame burning vertically downwards over the same fuel, and about three times the rate of horizontal spread.

The rate of vertical spread increases exponentially with time, so that the time required for the flame height to double remains constant. Suppose for example we have a cardboard box 0.5 m high and we light the bottom edge of one vertical face. We note that the time taken for the top of the flame to move from half way up the face to the top edge is say 20 seconds. This is the doubling time, and we can use it to calculate that the flame would reach the top of a stack of boxes 16 m high in another 100 seconds.

Geometry

The width of a vertical fuel surface has no effect on the rate of downward spread of flame, but there is an increase in the upward spread with increased width. This can be related to an increase in the height of the flame, which results from the increased area of burning.

Flames will spread more rapidly along an edge or at a corner because here the fuel is being heated from both sides.

If the fuel is discontinuous, the rate of spread of fire throughout the array of fuel will depend on the separation between the pieces, and on the bulk density of the fuel. The most rapid spread will occur when: separation is small enough to allow adequate heat transfer by radiation; the thickness of individual pieces is small; and the bulk density is relatively low. Forest fires spread much more rapidly in the brush on the ground, or in the crowns of the trees, than by the involvement of the tree trunks and branches.

Temperature

If the unburned fuel has been heated, flame spread will be enhanced because less time will be needed to heat the surface layers to the fire point. The heating of unburned fuel by radiation in a room fire eventually causes the dramatic spread of a flashover (discussed below). The importance of the effect of radiation on flame spread is recognized in the various standard tests which are used to classify materials according to their fire behaviour.

Air flow and oxygen concentration

A flow of air over a surface in the direction of burning causes an increase in the rate of spread up to the blow-off velocity. This is because the flame is bent over towards the surface, and combustion becomes more efficient: both effects increase the heat transfer to the unburned fuel. Similarly, an increase in oxygen concentration increases the temperature of the flame and at the same time reduces its thickness, again enhancing heat transfer.

SPREAD OF FIRE IN ROOMS

In the early stages of a fire in a building, when only a small amount of fuel is involved, the rate of spread of fire is independent of the size of the enclosure. A fire may start for example in an upholstered chair and a flame will spread to involve most of its surface. As radiation from the flames increases, other items of furniture which are sufficiently close will become involved. If the fire is burning freely (in an adequate supply of air) about one-third of the heat generated by the flames is radiated to the surroundings and will cause this initial fire spread. The remainder of the heat is removed by convection in a buoyant plume which rises to the ceiling. This smoke layer increases in concentration, depth and temperature as the fire progresses, and radiates an increasing amount of heat back into the room. This enhances burning rates,

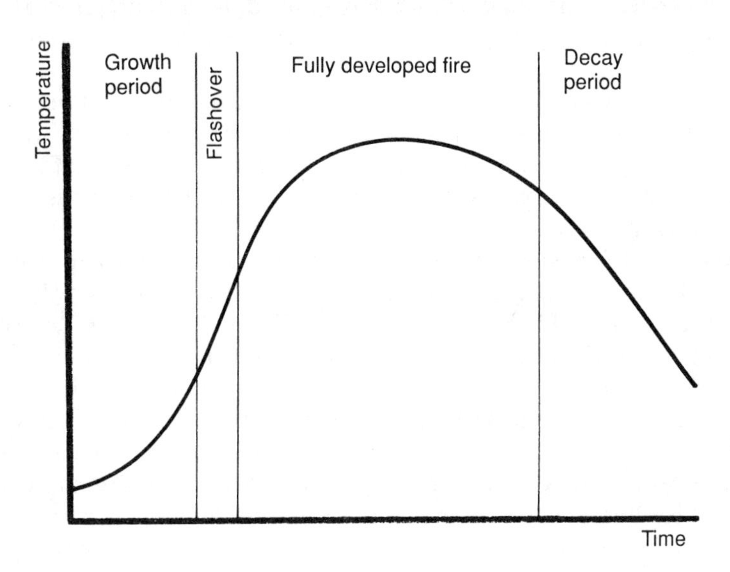

Figure 3.2 Variation of average temperature with time in a room fire.

and promotes the spread of flame. Heat reaching uninvolved fuel increases its temperature until volatiles are released, and these can often be seen as white smoke. As soon as a critical flow of volatiles has been achieved from this fuel there will be a very rapid spread of fire to involve the whole room. This is a flashover. Figure 3.2 shows the variation of the average temperature with time in a room in which a flashover has occurred during a fire. The progress of the fire has been divided into three stages:

- the pre-flashover or growth stage in which the average temperature is low and the fire is still burning quite close to its origin;
- the post-flashover or fully-developed fire in which all the fuel is involved and the compartment appears to be filled with flames;
- the decay period during which both the burning rate and the temperature are falling.

It should be noted that in a room fire involving modern foamed plastic upholstery, the time to flashover may be only a few minutes. The radiation reaching the floor at that time is likely to be 20 kW/m^2, with a ceiling temperature of 600°C.

Flashover is not an event, like ignition, but it is a transition, frequently a very rapid one, from the growth stage to a fully developed fire. It also marks the transition from a potentially survivable fire situation to one from which there is no escape.

Not all fires will progress to flashover. The item first involved may be sufficiently isolated that fire cannot spread to other items. Even when the potential for spread is present, it may not occur if there is not sufficient air. If there is inadequate ventilation the fire may continue to burn very slowly, or may self-extinguish.

If the initial fire is so large that the unrestricted flame height would be greater than the ceiling height, flames as well as hot gases will spread beneath the ceiling. The temperature there will rapidly increase, and so will the radiation down to the room. The result will be reduction in the time to flashover. The height of the fuel may be important in determining whether or not the flames will reach the ceiling. The mode of ignition will also have an effect. A large source in the centre of a mass of fuel will cause a rapid initial growth and a reduction in the flashover time. The presence of combustible lining materials in a room will also reduce this time, but the effect is not significant unless the lining is involved in the early stages. The physical properties of the ceiling and walls will also affect the amount of radiation into the room. However the most significant factors affecting the time to flashover concern the properties of the fuel and the available air, and these are discussed below.

Fuel

The nature, bulk density and distribution of the fuel all significantly affect both the spread and severity of a fire. For example wood shavings will ignite more readily and burn more fiercely than the solid block from which they are cut. Similarly, paper and cardboard in the form of books will behave very differently from the same materials when they are used to package goods in cardboard cartons. The burning rates of these materials, estimated from their behaviour in large fires, are 7 g/s/m^2 of fire for books and 48 g/s/m^2 for cardboard cartons. Items of furniture burn at about the same rate as books, at between 25 and 60 g/s, although one modern chair made from a block of polyurethane achieved a maximum rate in a test of 151 g/s. A chair burning at 25 g/s would release heat at about 370 kW, but this latter one developed about 2500 kW. Items burning at the same rate as cardboard cartons, at 100 to 250 g/s, include stacked polyurethane foam, plastic goods and drums of flammable liquids. The maximum rate of heat release from them could be betwee 1500 and 4000 kW. Intermediate between these lower and higher ranges are fuels like sawn timber, and items packed in solidly made wooden crates.

The height to which the fuel is stacked affects not only the total fire load but also the rate of burning. This is of course because flames will spread much more rapidly vertically than horizontally. The separation between stacks of fuel greatly affets the horizontal spread of fire. However these factors have no effect on the rate of burning once flashover has occurred.

Ventilation

A fully ventilated fire is fuel-controlled, in the sense that the rate of burning is determined by the rate at which flammable vapours are released from the fuel. Such fires have minimum flashover times and achieve maximum burning rates and temperatures (up to 800° to 1000°C in the hot gases). If insufficient air is available to achieve these maxima, the fire is air-controlled: the burning rate is dependent on the rate at which air is flowing into the room. It has been found that the burning rate is strongly related to both the shape and area of the ventilating opening, which is usually a window or a door. The height of the opening has an important effect. The results of a number of large-scale tests, in which ventilation was through a single opening, showed that the relationship between the burning rate and the size of the opening could be given approximately by:

$$R = 5.5A\sqrt{H}$$

where R = burning rate in kg/min, A = area in m^2, and H = height in m.

An air-controlled fire, although of limited intensity, may cause additional problems. The heated fuel continues to release flammable vapours, only part of which can react with the available air. The remaining hot flammable

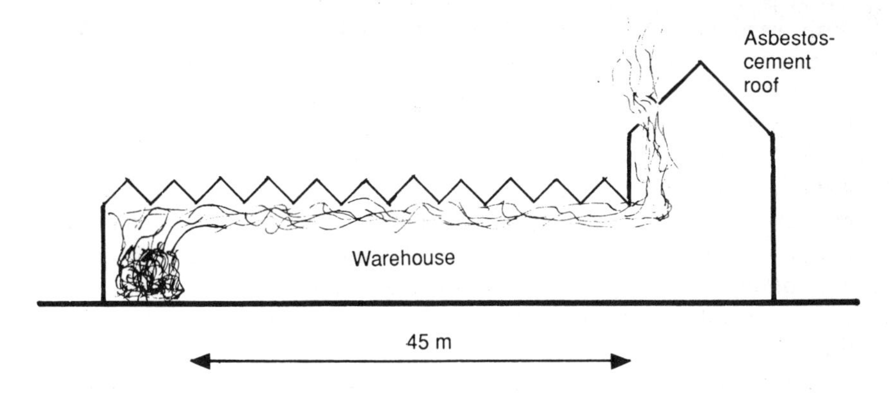

Figure 3.3 Spread of hot gases in a warehouse fire.

gases may then spread a considerable distance before they can find an additional supply of air. They may do this for example by melting through a plastic roof light. The hot gases emerging into the air may still be sufficiently hot to ignite. The flames, remote from the original fire, may cause additional fire spread. A situation like this is illustrated in Figure 3.3. In this particular fire, the hot gases broke through a corrugated asbestos-cement roof which was 45 m from the fire.

The sudden ventilation of an air-controlled fire can cause a dangerous flashback, and this was discussed in Chapter 1 under Premixed flames. Unfortunately this effect is sometimes called a flashover, which may be confusing.

SPREAD OF FIRE IN BUILDINGS

Many domestic fires in the UK are confined to the room of origin: the combined results of built-in fire resistance and short fire brigade attendance times. However, when a fire in a room has become fully developed, there is clearly the probability that it will spread to other rooms and that eventually the whole building may be involved. The spread can result simply from the release of hot gases through an opening in a wall, so that conditions leading to a flashover are established in the next room. Spread can also happen in the same way that growth occurred in the pre-flashover phase: by the sequential involvement of objects in the room. Flames will also spread over combustible floor coverings, and particularly over combustible surfaces on walls and ceilings. The rapid fire spread which contributed to the deaths of 48 people in the Stardust Disco fire in Dublin, Ireland (1981), was due partly to the presence of carpet tiles on the internal walls. These had been glued over nearly all of the wall surface in the ballroom and the foyer. The fire started in the back row of some polyurethane foam seats which were in a tiered area. Flames soon involved the tiles on the wall behind the seats, and then spread in two ways. First, molten burning droplets ignited the tiles below. Then the combined flames from the titles (which released about 400 kW/m^2) and the upholstery spread below the low ceiling, which was only 2.4 m high at the back of the area. The radiation from the flames and hot gases caused a flashover involving the remaining seats and tables. This was followed by a rapid spread to the main ballroom. Some of the people who were still dancing, or were watching the fire, were unable to escape in time.

The presence of voids can result in very rapid penetration of fire throughout a building. It is very convenient for example to run the cables, pipes and trunking supplying a multi-storey office block through a vertical service duct. Unless the openings into the duct at each floor are properly sealed, it will be possible for flames to enter the duct, and then to rise to a considerable height because of the stack (or chimney) effect. Lateral spread from the duct will then be possible at higher levels. Fire can also spread laterally through voids below the floor or above the ceiling (again used for services, particularly for example in computer suites). Extremely rapid spread can also occur over combustible surfaces behind wall panels, which is what happened in the early stages of the Summerlands fire in the Isle of Man (1973). The fire originated in a disused kiosk outside the main building and then spread to the Galbestos cladding. Once this had been penetrated, flames spread unnoticed for about 15 minutes in the gap between the cladding and the decorative inner panelling. When the smoke and flames did eventually emerge into the main building, fire spread was so rapid that there was insufficient time for all of the nearly 3000 people to escape. Fifty of them died in the fire. However, as in most disasters of this kind, there were other contributory factors, including the extensive use of transparent plastic panels which melted and dripped, the presence of a large quantity of wood, and number of locked exit doors.

Again in the fire at the Bradford City Football Club (1985), the spread was so rapid in the timber stand that 56 spectators perished. Flames spread initially over combustible rubbish in a void beneath the wooden floor. Only five minutes after its discovery in this void the entire stand was involved. Flames also spread rapidly over the woodwork and the roof felting, and intense radiation resulted from flames and hot gases beneath the roof. Again unfortunately gates on escape routes were locked.

Fire spread in voids, and over decorative finishes, again contributed to the deaths of 164 people in the Beverley Hills Supper Club in Southgate, USA (1977). Another contributory factor in this disaster was the overcrowding: the room in which most people died had three times as many occupants as it could safely accommodate. Combustible decorations and overcrowding were also significant factors in the worst of these multiple-death building fires: the Coconut Grove Night Club in Boston, USA (1942) in which 492 people died.

Very large voids may be present as an architectural feature in a building. Some of the older multi-storey departmental stores have a large well in the centre of the building, usually extending from the ground floor to a glazed structure on the roof. It was not uncommon at one time to see long lengths of material suspended in the well. Modern atrium buildings also contain large wells, but precautions are taken to prevent the spread of fire and the accumulation of smoke.

Flames released on the outside of a building through the failure of a window can cause the vertical spread of fire. This will be faclitated by the failure of windows immediately above as a result of heat transfer from the flames. Radiation from these flames, or radiation passing through the windows of a burning room, may cause the fire to spread to a neighbouring building. This is particularly likely if the separation is small and if there are readily ignited materials near to the windows of the unaffected building. Fire spread to other buildings can result from flying brands: burning fragments which are carried by the wind. These are particularly likely to cause roof fires. Fire spread by this means is a serious problem in countries where wooden shingles are used for roofing.

Fires can spread within a building, and from one building to another if a large quantity of a material with a low melting point, like butter, is involved. Fires will also spread rapidly if the building was damaged by an initial explosion, or if there has been structural failure as a result of the fire.

4

Smoke

SMOKE PRODUCTION

Most of the materials involved in a fire are capable of burning completely to form gaseous products, mainly carbon dioxide and water vapour. However this is rarely achieved in a diffusion flame unless it is very small. Most of the flames which are present in a fire are open at the top and are emitting smoke. The smoke is a mixture of hot gases, small droplets of liquid (perhaps 0.001 mm in diameter), and solid carbon particles. The gases comprise nitrogen; the oxides of carbon and hydrogen (already mentioned) and those of sulphur, nitrogen, and other elements; and a wide range of pyrolysis products. The liquids present are also complex pyrolysis products but of high molecular weight. As the smoke leaves the flames and begins to cool, these will condense out as extremely small droplets which will then coalesce into a mist. They may eventually be deposited on to a solid surface as an oily liquid. The carbon particles, as we have seen earlier, are the result of incomplete combustion in the flames. Smouldering combustion will produce a predominance of pyrolysis products, and indeed the smoke may be white or grey because of the presence of a mist. Diffusion flames burning in insufficient air will generate large quantities of carbon and perhaps also some liquid products. It has already been noted that these constituents of smoke are combustible, and are capable of causing hazardous conditions. The smallest amounts of both solid and liquid products will be emitted by fully aerated flames in a fuel-controlled fire.

A very small number of fuels can burn with non-luminous diffusion flames which do not produce smoke. These include formaldehyde, formic acid, metaldehyde, methyl alcohol, carbon monoxide and hydrogen (but only when they are pure). If the molecules of a fuel contain oxygen, then the fully aerated flame will give less smoke than the same flame from a chemically similar fuel which has no oxygen. Thus, ethyl alcohol (CH_3CH_2OH) produces less smoke than ethane (CH_3CH_3), and acetone (CH_3COCH_3) produces less than propane ($CH_3CH_2CH_3$). In the same way wood, which comprises many oxygenated molecules, and polymethylmethacrylate ('Perspex': ($CH_2{:}C(CH_3)CO_2CH_3$) both burn with less smoke than hydrocarbon polymers which contain no oxygen in their molecules, like polyethylene ('Polythene') and polystyrene.

The structure of the molecule also affects the amount of smoke which is produced by a fuel. For example, the hydrocarbons (which comprise only hydrogen and carbon atoms) can be listed in order of increasing tendency to smoke production like this:

- n-alkanes, the straight chain paraffin series which include many of the petroleum fractions;
- iso-alkanes, the branched isomers of the paraffin series;
- alkenes, which contain a double carbon–carbon bond ($C = C$);
- alkynes, which contain a triple bond ($C \equiv C$);
- aromatics, based on a ring of six carbon atoms (the benzene ring);
- polynuclear aromatics, which have more than one benzene ring.

The main pyrolysis product of polystyrene is the monomer from which it is built: styrene. This is an aromatic molecule, which explains why burning polystyrene produces substantially more smoke than any of the other materials mentioned in the previous paragraph.

HAZARDS OF SMOKE

The majority of fire deaths result directly or indirectly from the presence of smoke. Statistics originating in the UK and the USA show that over 80 per cent of all the recorded fatalities were caused by smoke inhalation. These deaths could have resulted from a deficiency of oxygen, or injury caused by the high temperature, or the presence of toxic and corrosive gases, or the high concentration of solid matter. Most were due to a combination of these causes. Indirectly, the loss of visibility in smoke may delay or prevent escape, and people are then killed by the fire or by prolonged exposure to the smoke. As we have seen, smoke can also increase the hazard to people by spreading the fire locally, and by producing flames remote from the original source. Its presence can also greatly affect the efficiency of firefighting and rescue operations within a building.

At onc timc thc toxicity of smoke was attributed mainly to the presence of carbon monoxide (CO), which results from the incomplete combustion of carbon. It still causes or contributes to many fire deaths, and is a normal finding in the victims' blood analysis. However, many more toxic materials are released in fires involving particularly the modern plastics. Even one of the earliest plastics, celluloid, was capable not only of very fierce combustion but also of the production, in closely confined conditions, of toxic gases. Like cordite, celluloid is a nitrocellulose, but containing fewer nitrate ($-ONO_2$) groups. It burns in the same way as cordite, in three separate steps (see Chapter 1), and both materials are capable of the first-stage exothermic decomposition without a flame (a kind of premixed smouldering). If celluloid is closely packed and is heated in a fire it can decompose in this way with the production of large volumes of nitrogen dioxide (NO_2) and other oxides of nitrogen. NO_2 is a strong pulmonary irritant which can cause immediate death as well as delayed injury (usually involving septic pneumonia). It was the cause of 123 fatalities when nitrocellulose X-ray films were exposed to a fire in the Cleveland Clinic, USA (1929). Oxides of nitrogen are also produced in small quantities by the combustion of fabrics. The short-term (10 min) lethal concentration (STLC) has been estimated as near to but not less than 200 ppm.

Hydrogen cyanide (HCN) is produced by the combustion of polyurethane, nylon and other plastics, and also of wool, silk and paper. It is a rapidly fatal asphyxiant with an estimated STLC of 350 ppm. However, HCN is flammable and is likely to be present only in air-controlled fires (as might occur for example in an aircraft cabin).

Acrolein is a very potent irritant with an estimated STLC of 30 to 100 ppm. It is produced by the pyrolysis of some plastics and some cellulose materials.

The halogenated gases are all strong irritants. Hydrogen chloride (HCl), which is released in the pyrolysis of PVC and other chlorinated materials, is the most common, but HF and HBr are also found. All form strong acids when they dissolve in the moisture on the surface of mucous membranes. The carbonyl halides may also be formed in fire conditons: phosgene ($COCl_2$) and fluorophosgene (COF_2). All these materials have estimated STLC values between 100 and 400 ppm.

Sulphur dioxide (SO_2) is a strong irritant which, like the hydrogen halides, is intolerable to breathe at concentrations well below the STLC (of about 500 ppm). It can be produced by the combustion of any sulphur compound.

The isocyanates are potent irritants with a STLC as low as 100 ppm. They are the major irritants in the smoke produced by burning urethane isocyanate polymers.

Ammonia is present in generally low concentrations in the smoke from wool, silk, nylon, melamine and other materials. It is an irritant to the nose and eyes, but quite high concentrations, over 1000 ppm, are needed to cause hazardous conditions.

Carbon monoxide, like hydrogen cyanide, reduces the oxygen concentration in the blood. Its effect will be more marked if the victim is already breathing air which is deficient in oxygen. However, comparatively high concentrations are needed: over 4000 ppm for a STLC. Carbon dioxide is also toxic, but as much as 12 per cent is needed to cause death within a few minutes. However, the presence of only 2 per cent will cause the rate of breathing to double, which will increase the damage caused by other poisons. The solid particles in smoke, mainly soot, may cause blockage of small bronchioles in the lungs, but they can also enhance the effects of irritants which have been adsorbed to the surface. The STLC of substances like the hydrogen halides may therefore be reduced in the presence of soot.

The toxicity of the gases released by a particular material in a fire will depend on the way in which it has been pyrolysed: it could be heated in the absence of air, or it could be smouldering, or burning. In some tests with small animals, measurements were made of the minimum weights of various substances which would create LC_{50} conditions in the apparatus (LC_{50} is the lethal concentration which will kill 50 per cent of the animals). The fuels were either pyrolysed or burned, and the relative toxicities were quite different in the two series of tests. Taking the results of just four of the 14 fuels tested, the number of grams needed to produce the LC_{50} conditions from pyrolysed materials were these: red oak 9.0; cotton 10.0; polystyrene 31.0; and wool 60.0. When the same materials were burned the results were quite different, and wool became the most hazardous (of all the fuels tested) instead of being least hazardous: wool 0.4; cotton 2.7; red oak 3.6; and polystyrene 6.0. The dangerous properties of wool in a fire may be surprising, particularly when we look at the values for two of the other materials used in the burning tests: PVC 1.4 and polyurethane foam 1.7. (The source of this information is given in the Bibliography at the end of this Part.)

SMOKE MOVEMENT

Smoke, like any other fluid, will move whenever there is a difference in pressure between two points within its bulk. The most obvious force acting on the smoke is the buoyancy created by the heat from the fire, but movement within a building can result from a number of other effects. The atmosphere within the building may already be in motion because of buoyancy resulting from the difference between the inside and outside temperatures, or because of pressure differences created by wind, or because of air-handling systems in the building.

The stack effect can influence the movement of smoke within a tall building. This is illustrated in Figure 4.1. It must be remembered that the buildings are never completely sealed and there can be a substantial air flow through the gaps around windows and doors, even when they are closed, through other openings. If the temperature inside the building in the figure is higher than that outside, air will flow in through the lower stories and into the central core (which could be a stair well, with lift shafts, or an atrium). It will then rise up the core and be discharged to the outside through the upper stories. The neutral plane shown on the drawing is simply the level at which the pressure inside the building is equal to that outside. Air enters the building below this plane and leaves from floors above it. If there is a fire on one of the lower floors, then hot smoke entering the core will enhance the stack effect and will be carried to the top of the building. Here it will flow outwards, and one or more of the upper floor may become smokelogged.

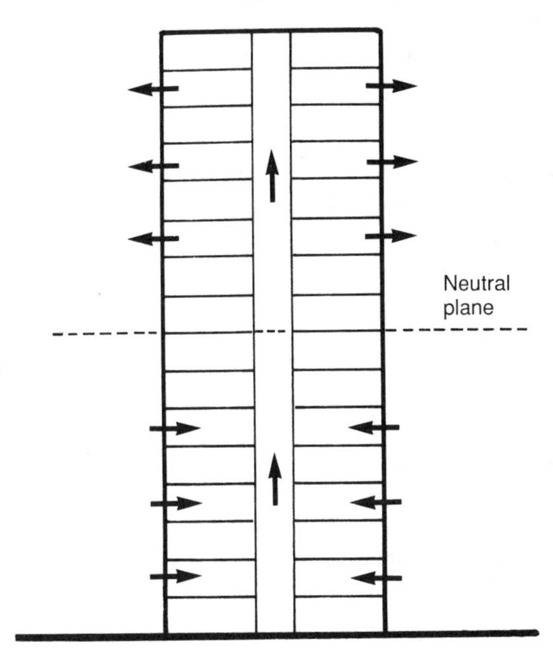

Figure 4.1 Stack effect in a tall building.

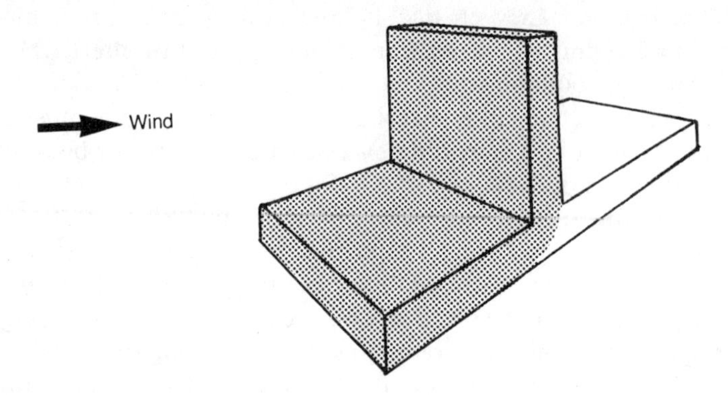

Figure.4.2 Positive pressure on the outside of a building caused by wind.

Figure 4.2 shows a typical modern building comprising a single-storey area, typically used as a shopping mall, with a high-rise office block in the centre. When the wind is blowing from left to right, the whole of the shaded area will be at a slight positive pressure compared with that inside the building. The actual pressure will not be uniform over the whole surface and will be highest near to the tower. The unshaded surfaces, and particularly those downwind of the tower, will be at a negative pressure relative to that inside. In these conditions it is possible that smoke generated by a fire in the left-hand part of the shopping mall will not be vented locally through openings in the roof but will be carried internally to the other end of the mall.

Smoke can also be moved around a building by a heating and ventilating system, or will move within the ducting when the system has been shut down. In some industries it is necessary to maintain certain areas at a positive pressure and others at a negative pressure (to ensure for example that gases cannot travel from the latter to the former). Clearly such systems will greatly affect smoke movement while they are operating. The possibility of designing systems so that they can extract smoke generated by a fire is discussed later.

Smoke will normally spread horizontally in a building in a layer beneath the ceiling, and will rise into any part of the ceiling which is higher than the rest. If for example smoke is flowing from a single-storey corridor into a two-storey entrance hall, the hall will act as an inverted reservoir, which will fill before the smoke can continue into another single-storey area.

5

Ignition

The ignition of a mixture of a flammable gas and air which is within the limits of flammability can take place in two ways. In pilot ignition there is a direct transfer of heat and possibly active species from an ignition source to the mixture. In spontaneous ignition the bulk of mixture is raised to a temperature at which a flame can be established independently of a source. The ignition of liquid and solid fuels can follow the same routes, except that they may need to be heated first to release flammable volatiles. However it is necessary to heat only the surface of these fuels, either by a pilot flame or more usually by radiation, to a critical temperature. The process by which smouldering combustion can occur within the bulk of a particulate solid fuel is dealt with separately.

Thus, pilot ignition from a number of different sources is possible near to either a liquid at a temperature above its flash point, or a discharge of flammable gas. For solid fuels, or for liquids at temperatures below the flash point, a pilot ignition source must initially supply sufficient heat to the fuel to establish a critical flow of volatiles, which it must then ignite. It is also possible for the volatiles to ignite spontaneously if a mixture with air of the right composition is heated to a sufficiently high temperature.

PILOT IGNITION

Column K of Table 1.1 shows the minimum ignition energy for a number of gaseous fuels. These surprisingly low values were obtained in mixtures which were slightly on the rich side of stoichiometric. They were obtained by passing an electric spark through mixtures contained in a glass vessel, and noting whether or not a flame travelled away from the electrodes. The spark resulted from the discharge of a capacitor and the energy was calculated from

$$E = \tfrac{1}{2}CV^2$$

where C is the capacity and V is the voltage. Capacative sparks cause ignition at lower energies than inductive sparks. These latter result from the collapse of a magnetic field in an inductance (an electric motor for example) where

$$E = \tfrac{1}{2}LI^2$$

and L is the inductance, I is the current. Capacative sparks are much shorter in duration and are therefore at a higher temperature. They also cause the ionization of the gases through which they pass, and will therefore generate some of the chain carriers which are responsible for the flame reactions.

When some of the early measurements of minimum ignition energies were made it was noticed that if the electrodes were too close together higher energies were needed. This was because of heat losses from the small initial flame to the electrodes. Flanges made from glass discs were then attached to the ends of the electrodes so that the flame had to travel out between the flanges before the bulk of the flammable mixture could be ignited. The minimum gap between the flanges through which a flame was able to travel is the quenching distance, and observed values for the various fuels are given in column L of Table 1.1. It will be seen that hydrogen has an unusually low value of 0.5 mm. For most hydrocarbons the quenching distance is between 1.8 and 3 mm.

The minimum ignition energy for hydrogen is also unusually low at 0.02 mJ. It is convenient to remember that the value for most hydrocarbons is about 10 times this, at 0.2 mJ, and that for most hydrocarbon/oxygen mixtures it is a tenth of the hydrogen/air value at about 0.002 mJ. All the values quoted here and in the table are for capacative sparks. Since it is necessary for the spark electrodes to be separated by at least the quenching distance of the particular mixture, it is also necesary for the voltage to be sufficiently high to jump this gap. For a gap of 3 or 4 mm the voltage must be at least 5 kV, and it is for this reason that the spark produced by breaking a low-voltage electric circuit may not be capable of igniting a flammable mixture even when the energy released is more than the minimum requirement. However, if a large current is available from the circuit, it may be possible for an arc to form (in which the current is flowing in metal vapour). If this happens, as the contacts are separated, an ignition may occur.

Thus, pilot ignition of gases can result from direct contact with a flame, or from the release of a critical amount of energy into a minimum volume of the flammable mixture. It seems likely that the energy requirements are minimal in a spark discharge and that this is because ionized particles are generated as well as heat. Other forms of pilot ignition, requiring greater amounts of energy, are discussed later.

A flashover is the result of the pilot ignition of solid (sometimes liquid) fuels when the surface has been heated by radiation to the fire point. In some tests in which wood was exposed to radiation with a pilot flame close to the surface, it was found that ignition occurred when the heat flux was about 12 kW/m^2. This would raise the surface temperature to between 300° and 400°C. If the pilot flame was actually in contact with the surface, the minimum heat flow for ignition was only 4 kW/m^2. These values may be compared with the maximum radiation intensity from the sun in the UK: about 0.6 kW/m^2. This must be increased about 20 times to achieve pilot ignition conditions, which can be readily done with a lens or even a glass bottle. A radiation flux of about 10 kW/m^2 on the skin can be quite painful after a few seconds, so it is reasonable to assume that if the radiation is painful it is also hazardous (particlarly since some materials, like fibreboard, ignite more readily than wood).

SPONTANEOUS IGNITION

Gases

Suppose we have a room which has been filled with a flammable mixture at ambient temperature. It might be thought that no reactions could be taking place between the molecules of the fuel and oxygen, but this is not strictly true. All the molecules will be in motion, changing their direction whenever they hit a wall or collide with each other. Their average velocity will depend on the temperature (and indeed the true definition of temperature is related to this velocity). The vast majority of the molecules will be moving at velocities close to the average, and collisions between them will not be sufficiently energetic to cause them to react. However, there is a velocity distribution within the population of molecules and a very small number of them will be travelling at a sufficiently high velocity for reaction to be possible. If a fast oxygen molecule encounters a fast hydrogen molecule it is possible for two OH radicles to be formed, with the release of energy. These radicals are chain carriers in the combustion reactions. If the whole system was adiabatic (that is, there were no heat losses at all to the surroundings) the energy released by this and other reactions would gradually raise the temperature of the gases in the room. The average velocity of the molecules would increase proportionally, and also the number of fast molecules. This in turn would result in an increase in the rate of reaction, which would approximately double for every 10°C increase in temperature. Eventually a temperature would be reached at which a premixed flame propagated throughout the volume, consuming all the remaining reactant molecules. Fortunately for us such an event is not possible because the conditions could never be truly adiabatic, and heat would always be lost from the system faster than it could be generated by the reactions. However, if there were a substantial increase in the temperature of the gases, conditions could be created in which heat was gained from the reactions faster than it could be dissipated to the surroundings. There would then be a rapid transition to flaming combustion. The temperature at which this happens is the auto-ignition temperature, AIT. This is sometimes referred to as the autogenous ignition temperature, or the spontaneous ignition temperature, SIT, although the latter is sometimes used only in connection with solid fuels. In the adiabatic conditions postulated above, the AIT was the ambient temperature, and could in theory have been any temperature above absolute zero. In practical conditions, the AIT can be established quite closely: it is the temperature at which ignition occurs (after a brief delay) and below which there is no apparent reaction. In the standard method of measuring AIT, the gas

Table 5.1
Minimum auto-ignition temperatures (°C)

Carbon monoxide	609
Methane	601
Benzene	560
Ethylene	490
Acetone	465
Propane	450
n-Butane	405
Hydrogen	400
Methanol	385
n-Heptane	280
Kerosene	210

mixture is admitted to a small spherical glass vessel at a known temperature. In a series of tests, the temperature is increased until an ignition occurs. Typical values are shown in Table 5.1. As might be expected from the discussion above, the values obtained depend very much on the technique and the apparatus, particularly the size and shape of the vessel, the nature of the surface, (which could be catalytic), and the fuel/air ratio (the stoichiometric mixture does not usually provide the lowest value). For ignition to occur it is necessary for a certain minimum volume of the mixture, located at a distance from the wall which is greater than the quenching distance, to be heated to the critical temperature. The surface area of the heated vessel will clearly have an effect on the results, and the smaller the vessel the higher will be the observed AIT. This can be seen from Table 5.2 where AIT values are shown for some of the fuels in Table 5.1, but in vessels of five different sizes.

The marked effect of vessel volume should be noted, and also that none of the values is in complete agreement with those obtained by the standard test method (Table 5.1). The conclusion to be drawn here is that it would be unwise to assume that a flammable mixture can be handled safely at a temperature just below its published AIT value. For example, the AIT of low octane petrol is given as 280°C, but if some is spilled in a crash, and a vapour/air mixture enters a hot exhaust pipe at about 200°C, then it is possible for ignition to occur after a delay of perhaps five minutes. During this time some of the petrol may have been pyrolysed, particularly in contact with soot deposits, and the resultant mixture may have a lower AIT than the original petrol. In general however the higher the molecular weight of a hydrocarbon, the lower will be its AIT. This can be seen by comparing the values for methane, propane, butane and heptane in Table 5.1. The values for materials like heavy fuel oil are even lower, and auto-ignition on

Table 5.2
Minimum auto-ignition temperatures in spherical vessels of different sizes (°C)

Fuel	Volume of vessel, ml				
	8	35	200	1000	1200
Benzene	668	619	579	559	
Methanol	498	473	441	425	386
n-Heptane	255	248		233	
Kerosene	283	248	233	227	210

oil-soaked lagging has been known to happen, after a delay of perhaps several days, on pipelines at little over 100°C. However, as was mentioned earlier (Chapter 1), this results initially in smouldering rather than flaming combustion.

Liquids and solids

In the example given above, the initiation of combustion in oil-soaked lagging, ignition probably occured in the vapour/air mixture which was present within the voids of the fibrous structure. The normal auto-ignition of liquids certainly happens in this way, a hot surface providing heat both the vaporize the liquid and to raise the temperature of the vapour/air mixture to the AIT. Measurements of AIT can be made by allowing a few drops of liquid to fall into a small glass flask in an oven. It is sometimes found that the minimum AIT is recorded when sufficient liquid has been added to form a mixture in the vessel which is about ten times the stoichiometric concentration. However, it is reasonable to assume that the small volume of mixture in which ignition started was within the limits of flammability.

The effect of enclosure on AIT can be readily demonstrated if a liquid fuel is dripped on to a metal hotplace to which a thermocouple has been attached. The lowest temperature at which an ignition occurs will probably be 100°C or more above the AIT measured in the standard apparatus. If a ring of metal is placed on the hotplace, giving a shallow enclosure, ignition will be obtained at a lower temperature. The temperature will decrease further as the height of the ring is increased. Finally, if a lid is placed on top of the ring and the liquid is dripped in through a small hole, an AIT will be recorded which is close to the published value (it may well be lower if the metal surface is slightly catalytic).

Any reference to the AIT of a solid fuel must apply to the surface temperature; the temperature within the bulk is immaterial. The temperature must be higher than that required for pilot ignition because the volatiles, having been diluted with air to a flammable mixture, must still be hot enought to ignite. Quite low temperatures were recorded in an experiment in which the fuel was heated in a tube through which there was a slow current of air. Some of the results are shown in Table 5.3. These values are lower than the temperatures which are recorded in static conditions because the method ensured good mixing of the volatiles with air which had been preheated, and this was followed by a stay time of several seconds in the hot zone. It is more usual for auto-ignition to result from the heating of the fuel surface by radition. The intensity of the radiation may then be of greater significance than the surface temperature. In the same series of experiments

Table 5.3
Minimum auto-ignition temperatures for solid fuels in an airstream

Fuel	AIT (°C)
Coal	125–130
Hay	172
Newspaper	184
Sawdust	192–220
Jute	193
Cotton	228
Rayon	234
Magnesium	507

which provided the values for pilot ignition quoted above, the minimum heat flux for auto-ignition of wood was found to be 28 kW/m^2 with a corresponding surface temperature of 600°C. It is interesting to note that when the surface of the wood was heated by convection, that is by a rising plume of hot air, the critical surface temperature was reduced to 490°C. This confirms the effect just discussed in connection with Table 5.3, although this temperature is much greater than that shown in the table for sawdust. In the latter measurement, as we have seen, the low value resulted from the combined effects of preheated air, good mixing, confinement, and stay-time. If the stay-time could have been prolonged to several minutes, by having a very long heated tube, there is no doubt that even lower values would have been recorded. Again the important point is that published AIT values cannot be regarded as an absolute limit below which a material is safe. However the values shown in Table 5.3 cannot be far from the minimum values which could occur in practical conditions.

As a very rough guide to the auto-ignition behaviour of solid and liquid fuels we can select these temperatures:

- Below 100°C (the temperature of a pipeline containing saturated steam at a pressure below 1 bar). No danger with ordinary combustible materials, although cellulose may decompose at 82°C.
- Between 100° and 150°C (steam pressure 1 to 4.8 bar). Some ordinary combustible materials will decompose and may ignite after a long delay. High molecular weight hydrocarbons and other combustible liquids may ignite in oil-soaked lagging, but again after a long delay.

- 150° to 200°C (steam at 4.8 to 17 bar). Ignition of many ordinary combustible materials, both liquid and solid, after a relatively short delay.
- Over 200°C (steam at more than 17 bar). Immediate ignition of most common combustible liquids and solids.

It should be noted that these comments apply only to fuels heated in favourable conditions, and also that there are many pure combustible materials, some of which are included in the tables, which are safe at temperatures well in excess of 200°C.

IGNITION SOURCES

We are concerned here with methods by which flammable gaseous mixtures may be ignited. The causes of fires are dealt with later.

Electric sparks

Ignition by electric sparks has been discussed in some detail. Sparks can occur particularly when a live circuit is broken. Incendive sparks frequently result from the breaking of an inductive circuit. Sparks can also result from the release of a static charge, and this is an important source of ignition in many processes.

Static charges

Static charges are formed whenever two dissimilar surfaces are in contact. Electrons located in the outer orbits of some of the molecules, which are not strongly attached, can drift away from one surface and into the other substance. This substance will then have a negative charge and there will be a corresponding positive charge on the other surface. If the two surfaces are separated and both are good conductors, the electrons will flow back freely before separation is complete, and no charges will remain. However, if one or both of the materials are poor conductors there will be little or no current, and both surfaces will retain their charge. Voltages between 1.0 V and 1.0 kV can result from a simple separation, but in flowing conditions, when separation is continuous, very high voltages can build up on the surfaces involved. Hazardous static charges can result from the separation of solid, liquid and gaseous surfaces in any combination other than gas/gas. This exception is because gases are completely miscible with each other, so that charges would be rapidly dissipated, and indeed this applies also to charges generated by two miscible liquids. Thus charges can be produced by the flow of solids (as particles), liquids and gases, in pipelines; by liquid sprays;

by the mixing of immiscible liquids; by the settling of drops of one liquid through another; by movement in fluidized beds; by the movement of belt drives, conveyor belts, and sheets of paper and plastic; and by the human body. A liquid of low conductivity flowing through a pipeline can generate a flow of electrons at between about 0.001 and 1.0 μA, and a powder being discharged from a grinding mill can produce currents up to 0.1 μA. These very low current flows are characteristic of static generation, but when the electrons can accumulate on a surface from which they cannot escape, very high voltage can result. If a flow of liquid or powder is charging at 1.0 μA and the material is being filled into a container which is insulated from earth, it is possible for the potential of the container to increase at the rate of 1.0 kV/s. After 10 seconds the charge could therefore be at 10 kV, and it could rise beyond this. Eventually the flow into the surface will be balanced by the leakage current to earth (which will increase as the voltage increases), or the charge will be dissipated to earth in a spark. The high voltage will ensure that the spark length is well above the quenching distance for a flammable mixture, and there is frequently sufficient energy available for ignition. For example, a man can readily acquire a charge of 10 kV, sometimes simply by walking about in low humidity conditions. The capacity of the surface of a man's body is about 200 pF, and we can use $E = \frac{1}{2}CV^2$ to show that a discharge at 10 kV would release 10 mJ: well above the minimum ignition energies for gases (quoted in Table 1.1), and for many dusts (as low as 0.3 mJ).

Corona discharge

If an object which is acquiring a static charge is isolated, the charge will not be able to escape either by a spark or a leak to earth. There may then be a corona discharge, which can be seen in the dark as a faint blue glow, possibly accompanied by a hissing sound. This can continue for some time, unlike a spark which lasts for only a few milliseconds. The air is partially ionized by the discharge but the rate of release of energy is low, as is the probability of ignition. In the days of sailing ships the corona discharge which was sometimes seen at the tops of the masts was called St Elmo's Fire.

Mechanical sparks

Sparks can be caused by the impact of one hard material against another: a hammer blow against a rock, a metal object being dropped on to concrete, an aircraft in a wheels-up landing. The results in Table 5.2 indicate that the smaller the area of a heated surface, the higher will be the temperature needed for ignition. A small particle must therefore be at a very high temperature initially if it is to ignite a flammable mixture during the short time for which it is in contact. Its temperature is in any case limited to the melting point of the material. Some particles, those produced by grinding a metal for example, may achieve high temperatures by reacting with oxygen in the air. However, they are then surrounded by a layer of gas which is deficient in oxygen. For all these reasons, the mechanical sparks which occur on industrial sites are rarely responsible for the ignition of flammable mixtures, with the exception of those gases with very low ignition energies: acetylene, carbon disulphide, ethylene, hydrogen. However, the mechanical sparks caused by a crashing aircraft may well be incendive because a great deal of kinetic energy is converted to heat, and the sparks comprise relatively large fragments of metal (and possibly stone).

Hot surfaces

From the discussion on auto-ignition it will be clear that a surface may need to be at a temperature well in excess of the AIT before it will cause an ignition in practice. If the flammable mixture can penetrate into a heated enclosure, ignition may occur (probably after a long delay) at temperatures lower than the measured AIT. A hot wire, because of its small dimension, will need to be at a considerably higher temperature than a heated enclosure. Ignition may be difficult if a hot surface is catalytic because flameless oxidation reactions can take place at the surface, which will then be surrounded by a layer of inert gases. Thus it is difficult to predict the conditions which can occur in practice and which will lead to the ignition of a flammable mixture. Certainly the published AIT vaues should be used only as a guide.

Compression

In a diesel engine, fuel is injected into the cylinder towards the end of a compression stroke. Ignition occurs because the air in the cylinder has been heated by the compression. In the same way a flammable mixture may be ignited if a rapid compression raises its temperature to the particular AIT value which occurs at the maximum pressure. (Rapid compression is necessary to reduce the heat losses).

Pyrophoric materials

Most metals will oxidize readily (and of course exothermically). If the surface area of the metal is large compared to its mass, as happens when the metal is a very fine powder, then high reaction rates are possible. If the powder is dispersed as a cloud in the air, the very rapid heating can result in a spectacular flash. This is a potential ignition source for other materials. Metals and other substances in this condition are said to be pyrophoric (the name means 'fire-carrying'). The most frequently found pyrophoric substance in industry is iron sulphide. This is formed by a reaction between hydrogen sulphide, which is present in crude oil and some petroleum products, and rusty steel:

$$2H_2S + Fe_2O_2 \rightarrow 2FeS + 2H_2O + \tfrac{1}{2}O_2$$

When plant is dismantled, perhaps for annual maintenance, and deposits of iron sulphide are disturbed, they may soon become red hot. Other materials which can ignite spotaneously in air include phosphorus, aluminium alkyls and phosphine. The word pyrophoric is also applied by firemen to smouldering fuel which is left behind after a fire. However, this material is reactive only when it is hot.

Chemical reactions

Combustion reactions release sufficient heat to cause incandescence. Some other reactions are equally energetic and can cause the ignition of combustible materials. The thermite reaction has been the cause of both fires and explosions in industry:

$$Fe_2O_3 + 2Al = Al_2O_3 + Fe$$

The maximum temperature achieved by this reaction between iron oxide and aluminium powder is 2500°C, sufficient to melt the iron which is one of the reaction products. The thermite mixture, containing some additional iron powder, was once used to weld tramlines in the road, but a pyrotechnic device was needed to start the reaction. However, both the ingredients of thermite are present when rusty metal has been covered with aluminium paint, and the heat released by an impact is sufficient to start the reaction. This results in a shower of highly incendive sparks.

Other materials will react violently when initiated by impact or friction. Mixtures of organic fuels with potassium chlorate will behave in this way, as indeed will matches. Some chemical combinations are sufficiently reactive to ignite on contact, and some of these have found use as self-igniting (hypergolic) rocket propellants. Such bipropellant systems include hydrogen peroxide (about 80 per cent)/hydrazine and nitric acid/aniline.

Radiation

Flammable mixtures are relatively transparent to radiation and it is usually assumed that it is not possible to increase their temperature to the AIT in this way. However, when the ignition of a solid fuel has been caused by radition, as in flashover conditions, there is no doubt that the volatiles will have received some heat by the passage of the radiation through them. Also, it is possible to ignite flammable mixtures by the intense radiation of a laser.

Spontaneous heating

The hardening of paint, which happens after all the volatiles have dried out, is due to oxidation. There are oils (mainly vegetable) and resins in the paint which readily accept oxygen atoms and then polymerize into a hard mass. The oxidation is exothermic, but the heat is readily lost to the air and the painted surface, and the temperature rise is negligible. However, the situation is very different if rags which have been used to wipe the brushes and to clean up spilled paint are left in a heap. There will now be a thin and very extensive coating of paint on the cloth fibres, and this will react with the air trapped within the heap. The heat which is released in the middle of the heap will be dissipated only slowly because the rags provide good thermal insulation. There can then be a substantial rise in temperature, and with it an increase in the rate of reaction (although the diffusion of air into the heap may well be the rate-controlling process). If the conditions are right, with both an adequate supply of reactants and good insulation, the temperature will rise until the materials at the centre of the heap begin to smoulder. In this particular situation transition to flaming combustion is very probable, and indeed many fires are caused by the spontaneous heating and ignition of many different fuels.

The process by which the paint rags heated up and then smouldered was the same as that already described for oil-soaked lagging. There is an important difference however: the paint rags were initially at room temperature. The hydrocarbon oils of the paraffin series are not sufficiently reactive to behave like this. Chemically they are said to be saturated, because they do not form addition compounds. Oils of animal and vegetable origin are unsaturated, and several of them are hazardous because they are capable of spontaneous heating. These include linseed oil (which is frequently present in paint) and corn, cottonseed, fish, lard, olive, pine, soybean, tung and whale oils. These oils become hazardous when they have spread over the surface of a fibrous or particulate material, so that a large surface area is exposed to the air. There is a minimum size of heap of the oil-soaked material which will heat up to the ignition temperature. If the heap is too small, heat will be lost to the surroundings faster than it is generated by the oxidation reactions. As the size of the heap is increased above this minimum, the ignition delay time may increase and the rate of burning may be reduced, both being controlled by the rate of diffusion of air. As we have seen earlier, for smouldering combustion to be established it is necessary for a solid char to form.

A number of solid fuels are capable of spontaneous combustion when they are stored in bulk. Again the conditions for ignition are: a sufficiently reactive surface, a heap which is sufficiently porous to allow air to permeate while still providing good thermal insulation, and the ability to form a rigid char. Coal can behave in this way because a freshly exposed surface reacts quite readily with oxygen at ambient temperatures. The conditions under which a stack of coal will begin to smoulder depend on the type of coal (that is, its affinity with oxygen), the size of the lumps, the time since they were broken up, the moisture content, the size of the heap, and the ambient temperature. In general, lumps larger than about 25–75 mm give little trouble, and stacks of less than 200 tonnes, and less than 2.5 m high, have not been known to ignite.

Hay is perhaps the best-known of the materials which are capable of spontaneous heating, although large stacks are now rarely seen. The heating progresses in two stages, the first of which is unusual in this context in that micro-organisms are involved. Optimum conditions for this activity occur when there is a high water content in the stacked hay, between 60 and 90 per cent by weight (however, ignition is less likely with a high water content because of heat losses by conduction). Most biological processes are exothermic, so the activities of the organisms soon raise the temperature at the centre of the stack. The optimum temperature for their growth is around 40°C, and most of them will die at temperatures above 65°C. However, once the temperature has reached 30°C, thermophilic bacteria (the word means 'heat-loving') will begin to take over. Their optimum temperature is 60°C and they can survive up to about 75°C. Some of the degraded materials left behind by the micro-organisms are readily oxidized, and the heat released from these reactions raises the temperature within the stack to the 172°C quoted in Table 5.3, or whatever temperature is needed to initiate smouldering.

Other solid fuels which are capable of spontaneous combustion when they are stored in bulk include charcoal, fish meal, grain, manure, rags, paper, sawdust and wool. A comprehensive listing, and a guide to the conditions under which heating is probable, are given in some of the books mentioned in the Bibliography.

6

Extinction

MECHANISMS OF EXTINCTION

The extinction of premixed flames

Premixed flames provide a convenient point at which to begin a discussion on extinguishing mechanisms because of their characteristic constraint within the limits of flammability. In the earlier description of these flames we saw that the existence of the limits can be explained mainly in terms of heat losses. These must be sufficient to reduce the flame temperature to a critical value (around 1400°C). It was argued that at the lower limit a critical thermal load is applied by the excess air which is heated to the flame temperature without then taking part in the reactions. Similarly, excess fuel imposes a thermal load at the upper limit. It was also shown that it was necessary to remove only 45 per cent of the heat released by a stoichiometric reaction to reduce its temperature to the lower limit value.

A thermal load can be imposed equally well by the use of an unreactive gas like carbon dioxide or nitrogen. A flammable mixture can be inerted (made incapable of ignition) by the admixture of a sufficient amount of one of these gases. The concentration which is needed to do this varies with the composition of the mixture. Figure 6.1 shows the concentration of both carbon dioxide and nitrogen needed to inert n-hexane/air mixtures over the whole of the flammable range. The data from which the curves like this one are plotted can be obtained in the same apparatus which is used to measure limits of flammability. The decision on whether or not a particular mixture

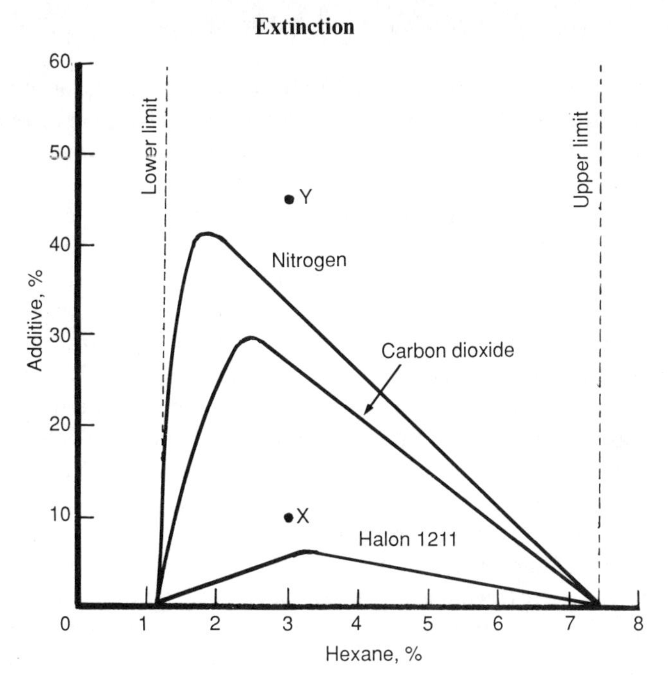

Figure 6.1 Flammability envelopes, showing peak concentrations for carbon dioxide, nitrogen, and BCF, in hexane/air mixtures.

is flammable can be made either by passing a spark through the mixture in a glass vessel and observing the propagation of a premixed flame, or by sparking a mixture in a closed vessel and observing a pressure rise. In the figure mixtures within the envelope formed by the curves and the base line are flammable. For example, the point *X* represents a mixture of 10 per cent of either of the inerts and 3 per cent of fuel (the remaining 87 per cent being of course air). This mixture can be ignited, but the mixture at point *Y*, containing 45 per cent of inert gas, cannot. In the figure, the minimum concentrations of the two additives needed to inert any fuel/air mixture are 29 per cent of carbon dioxide and 41 per cent of nitrogen. These values are sometimes called the inhibitory factor, but more usually the peak concentration (since they occur at the top of the flammability envelope). The concentrations of nitrogen and fuel at the peak can be used to calculate the limiting oxygen concentration for the particular fuel. The two concentrations are respectively 41 per cent and about 2 per cent, so the remaining 57 per cent must be air. Approximately one-fifth of the air is oxygen, so the limiting oxygen concentration for n-hexane is about 11 per cent. A knowledge of this value would be useful if for example a fixed-roof storage tank of n-hexane was being inerted with nitrogen, and the atmosphere in the tank was being continuously monitored for oxygen concentration. The detector could be arranged to alarm at some agreed fraction of 11 per cent.

A stationary flame on a burner can be extinguished if a critical concentration of inert gas is added to the air feeding the flame, but it is not normally practicable to prevent the propagation of premixed flames with an inert gas, simply because of the time needed for injection and mixing. Methods by which an explosion can bc stopped are discussed later.

If the inhibitory effect of carbon dioxide and nitrogen results from their effectiveness as thermal loads, the two peak concentrations should be closely related to their respective thermal capacities. This is indeed so, and it is also true for other inert gases like helium and water vapour. However, this relationship cannot explain the superior effectiveness of Halon 1211 (BCF, $CBrClF_2$), which has a peak concentration against n-hexane/air of only 5.5 per cent (the lower curve on Figure 6.1). Like other Halons it can interfere directly with the combustion reactions, in addition to a less important function as a thermal load. A brief excursion into combustion chemistry may provide a worthwhile explanation of this important effect.

The simplest of all combustion reactions, the burning of hydrogen to produce water vapour, can be expressed like this:

$$2H_2 + O_2 = 2H_2O$$

This equation correctly shows that you start with hydrogen and oxygen and finish up with water, but it would be wrong to conclude that the reaction is dependent on a series of collisions between two hydrogen molecules and one oxygen molecule. Collisions like this must indeed happen, but this 'three-body encounter' is statistically a rare event, and cannot account for the very rapid reactions in a hydrogen/oxygen explosion. All flame reactions depend on the activities of ions: very mobile free radicals which exist for short times but in high concentrations at flame temperatures. The only ions which can be derived from the three molecules in the equation are hydrogen and oxygen atoms, H and O, and hydroxyl radicals, OH. These are the active species on which the combustion reactions depend. They must be generated at a sufficiently high concentration and temperature by the ignition source. This is why an electric spark is such an efficient source: it is capable of ionizing the gases in addition to raising their temperature. These free radicals can react with the hydrogen and oxygen molecules like this:

$$H + O_2 \rightarrow OH + O$$
$$O + H_2 \rightarrow OH + H$$
$$OH + H_2 \rightarrow H_2O + H$$

Other reactions are possible but in these three steps we can see the consumption of the molecules of the two reactants, and the generation of the end product, water. This is a chain reaction since each step depends on a radical produced by the previous one. It is a branching chain reaction since two chain carriers are produced by the reaction of a single carrier in each of the first two steps.

When Halon 1211, $CBrClF_2$, molecules are released into a flame, they too will be ionized. The carbon atom will simply add to the fuel and this may account for the slight displacement of the peak of the flammability envelope (Figure 6.1) to the lean side of stoichiometric. The shift to the lean side is even more marked with methyl bromide, CH_3Br, which can clearly contribute a higher proportion of fuel atoms. It is the halogens (from either of these agents) which interfere with the combustion reactions, and particularly the bromide atoms, Br. An obvious inhibiting reaction is the removal of H atoms: this will have a profound effect on the combustion processes since hydrogen atoms are responsible for the production of the other two chain carriers:

$$H + Br \rightarrow HBr$$

However, calculations based on experimental evidence suggest that a second reaction may have an equally important chain-stopping effect:

$$HBr + OH \rightarrow H_2O + Br$$

The interesting point about the two reactions is that they show the removal of two chain carriers and the regeneration of species responsible for this, the Br atom. Since the bromine is now free to continue the process, it can rightly be regarded as a negative catalyst. In hydrocarbon combustion it does not matter which of two reactions is dominant because of the existence of this equilibrium:

$$OH + CO \rightleftharpoons H + CO_2$$

An excess of either OH or H will simply shift the reaction in the direction which favours its depletion.

Chemical inhibition of flame reactions can also be accomplished by fine powders. These introduce a chemically active surface, which is large in total area, into the reaction zone of a flame. Chain-terminating reactions can take place at the surface, which in these equations is indicated by M:

$$H + M \rightarrow HM$$
$$H + HM \rightarrow H_2 + M \quad \text{or}$$
$$OH + HM \rightarrow H_2O + M$$

The important thing here is that chain carriers are removed, and that the heat generated by the last two reactions is absorbed by the solid. When sodium and potassium compounds are used, these reactions are also possible if fine particles are vaporized:

$$Na + OH \rightarrow NaOH$$
$$K + OH \rightarrow KOH$$

There is no doubt that flammability envelopes could be plotted to demonstrate the peak concentrations of powders. This is not normally attempted because of the difficulty in obtaining a completely uniform distribution of power in the apparatus.

The physical and chemical methods of extinction described in this section are equally effective against diffusion flames. Their behaviour against premixed flames has been treated separately because of their measurable effect over the range of flammable mixtures, and in order to discuss the significance of peak concentrations. These can be measured with reasonable accuracy and are clearly a means of comparing the effectiveness of different extinguishing agents. The values can also be used in calculating design concentrations for automatic systems (discussed in Part Six).

An extinguishing method which is unique to premixed flames has already been mentioned: the effect of quenching. The failure of a premixed flame to propagate through a gap which is less than a characteristic minimum dimension can be accounted for by heat losses and the loss of active species to the walls of the gap. A knowledge of the quenching distance is of practical importance in the design of flame arrestors and flameproof electrical equipment.

The extinction of diffusion flames

If a diffusion flame is burning on a liquid or a solid fuel, it is dependent on receiving a flow of volatiles in excess of a critical vaue. If the temperature of the fuel surface is reduced to below the fire point, the flow of volatiles will be less than the critical value and combustion can cease. The fuel surface can usually be cooled by the application of water and this is the method by which the vast majority of fires are extinguished. The method cannot be used against liquid fuels with a fire point which is lower than the temperature of the water, and it is not normally practicable to use water against oil fires if the fire point is less than about 60°C. The normal method of tackling this kind of fire is with a high-velocity spray operating at about 4 bar, with application rates of between 0.16 and 1.0 $l/m^2/s$. Many fixed installations operate at about 0.4 $l/m^2/s$. Fires in water-miscible fuels, like the alcohols, can be extinguished by diluting the fuel with water until its vapour pressure is reduced to the critical fire point value.

Surface cooling by water is a surprisingly efficient way of extinguishing a fire, or of preventing its spread. For example, the heat transferred to a wooden surface at the minimum radiation intensity for spontaneous ignition, 28 kW/m^2, can be completely removed by the application of water at 0.001 $l/m^2/s$. The rate needed to reduce the surface temperature of burning wood to the fire point is only slightly higher, at 0.0016 $l/m^2/s$, and this is about 2 per cent of the standard application rate for normal sprinkler risks (0.082 $l/m^2/s$). However, with sprinklers and indeed all other practical firefighting methods there is the need to ensure that water well in excess of the theoretical minimum rate is able to penetrate to the burning surface. If, for example, a sprinkler is located directly above a fire, drops of water have to penetrate the rising plume of hot gases to wet the burning surface.

The spread of fire to tanks containing flammable liquids can also be prevented by cooling the walls and roof of the tanks so that the contents remain at a temperature below the flash point. If the fuel is already above its flash point, at ambient temperature, then cooling will keep the temperature in the vapour space below the AIT value. Typical application rates are between 0.1 and 0.2 $l/m^2/s$, which again for practical reasons are well above the theoretical rates.

It might be thought that a fine spray of water could be used to apply a thermal load, in order to extinguish a diffusion flame. Unfortunately, this is not a practicable firefighting method. Droplets of a sufficient size to penetrate a flame, projected from normal firefighting equipment, are incapable of abstracting enough heat for extinction, although the flame volume and the burning rate may be considerably reduced. A spray used in this way can also reduce the temperature of the plume of hot gases rising from the fire, and this can result in a significant reduction in the rate of spread of fire.

The chemical reactions in a diffusion flame are very similar to those in a premixed flame (not exactly the same because of carbon formation), and they are equally susceptible to inhibition by chemical extinguishants: indeed the critical concentration of an agent in the air feeding a diffusion flame is less than the peak concentration of the same agent for a premixed flame of the same fuel. This is mainly because flame temperatures and reaction rates

Table 6.1
Flame-extinguishing concentrations at ambient temperature (see also Table 6.3)

	Halon 1211 BCF	Halon 1301 BTM	Halon 1202 DDM	Halon 2402 DTE	Halon 1011 CB	Halon 1001 MB	Carbon Dioxide CO_2	Nitrogen N_2
n-Heptane	3.8	3.5	2.4	2.1	4.5	5.1	20.5	30.2
Acetone	3.8	3.5	2.2	2.3	4.4	4.6	18.3	29.2
Benzene	2.9	2.9	3.2	1.9	4.0	4.4	20.1	29.7
Diethyl ether	4.4	3.9	2.5	2.2	4.9	5.2	22.2	35.0
Ethanol	4.5	3.9	2.6	2.4	5.1	5.3	21.8	33.3
n-Hexane	3.7	3.3	2.3	2.1			19.5	31.0
Methanol	8.2	7.3	4.2	4.2		7.8	26.2	39.0
n-Pentane	3.7	3.3	2.3	2.1	4.2	5.0	19.0	31.5

are both lower in a diffusion flame (and this happens because diffusion is rate-controlling). Table 6.1 shows the flame extinguishing concentrations of a number of halogenated agents, and carbon dioxide and nitrogen, against eight liquid fuels. They are directly comparable with peak concentration values (like those in Figure 6.1) because, with very few exceptions, diffusion flames are stoichiometric. The values were obtained by adding increasing amounts of agent vapour to the air flowing up a vertical glass chimney, which contained a small cup of fuel (Figure 6.3). Laboratory measurements like this may appear to have little relevance in practical firefighting, but the values obtained were in excess of the concentrations needed to extinguish large pool fires of the same fuels. This is because the small laboratory flames were burning in ideal conditions: they were closed (no release of smoke) and were stabilized aerodynamically by the airflow around the edge of the cup. Despite this, the values in the table do not necessarily represent the maximum possible concentrations for such flames. It was found for example that if the fuel in the cup was heated, so that the flow of volatiles was increased, the critical concentration for many fuels was also increased, but for some there was actually a decrease in critical concentration. These effects can be attributed to changes in the relative velocities of the fuel vapour and air, which in turn affected the premixing and flame holding at the edge of the cup. When the apparatus in Figure 2.3 was used to measure critical extinguishing concentrations for flames burning in air-streams, values were obtained which were very close to those for premixed flames. This is hardly surprising since the part of the flame immediately downstream of the obstruction was indeed premixed. The same explanation can be applied to the cup burner results, although here there was only a very small premixed region and considerably lower velocities were involved. It would clearly be safer to use peak concentration values in designing automatic systems, but there are some situations in which the flame extinguishing concentration is acceptable: this is discussed in Part Six.

Small flames, like those on a match or a candle, can be blown out. Larger flames are increasingly difficult to extinguish in this way, but it is the main mechanism by which large oil well fires are extinguished by the detonation of a high explosive charge. What happens in blow-out is that the reaction zone is distorted and stretched, so that it becomes thinner. The stay-time of the reactants in the zone is therefore reduced, and the reactions are incomplete. Less heat is released by the reactions, and the flame temperature drops below the critical value. A contributing effect is the shifting of the reaction zone into areas where there is either too much or too little fuel, so that the reaction is no longer stoichiometric.

The concept of the triangle of fire is used by firemen to illustrate the methods by which they will normally tackle a fire. The three sides of the triangle are heat, fuel and air. When all three are present, a fire is possible. If any side is removed, the triangle collapses and the fire goes out. Heat is removed by cooling, fuel by starving and air by smothering. Cooling is used here not in the sense of a thermal load, but as a means of reducing the temperature of a burning surface to the fire point, or of preventing surrounding fuel from reaching this temperature. The very potent effect of chemical extinguishants is ignored, and one might claim that this should represent a fourth side to the triangle. However, chemical agents are rarely used against full-scale fires. The other two methods in the triangle, starving and smothering, are really the same process: the separation of the reactants. Starving is usually achieved by the physical removal of the fuel, perhaps with a bulldozer. Smothering usually implies the spreading of a layer of foam over the surface of a flammable liquid. This stops, or greatly reduces, the flow of volatiles from the surface, and also acts as a barrier to prevent the penetration of radiation to the surface.

The extinction of burning metals

Nearly all metals are capable of burning, but the less reactive ones need to be finely divided. Even steel, in the form of fine wool, can be ignited. Some metals will react sufficiently violently with water to achieve their AIT, but with the remainder the heat generated by this reaction, and by oxidation, is dissipated as rapidly as it is released, and auto-ignition does not occur. The metals which can be readily ignited are usually called the combustible metals, and these include calcium, hafnium, lithium, magnesium, plutonium, potassium, sodium, thorium, titanium, uranium, zinc and zirconium. Flame temperatures are high and the hot metal surfaces are very reactive. They will react exothermically with all the common extinguishing agents: with water to form the oxide, releasing hydrogen which will burn; with carbon dioxide to form the oxide and carbonate; with nitrogen to form the nitride; and violently with halogenated agents to form the halides. If a metal is burning in an enclosed space it is possible to displace the air with argon or helium: these very inert gases do not react even with hot metals. In the open it may sometimes be possible to cool a burning metal with a massive application of water, but this is clearly a hazardous procedure, particularly if molten metal is present. The only reliable method is smothering: the application of a thick layer of an inert powder, or one which either fluxes or sinters, to form a coating which restricts the penetration of air to the metal surface.

EXTINGUISHING AGENTS

Having outlined the methods by which flames may be extinguished we can conveniently look at the agents which are employed to do this. The physical, chemical, and where appropriate, toxicological properties of each are discussed, together with an indication of the ways in which they can be used. The various pieces of equipment used to apply the agents are discussed later in the sections dealing with first aid appliances, automatic systems, and fire brigade equipment. A final section of this chapter deals with practical methods of comparing the effectiveness of various agents, particularly when they are discharged from hand extinguishers.

Water

Water is the most frequently used of all the extinguishing agents. Fortunately it is also the most abundant and most readily available substance on our planet. Its physical properties are ideally suited to its main function of removing heat from surfaces which are involved, or are about to be involved, in a fire. Its specific heat is unusually high at 4.18 kJ/kg/°C. This should be compared with, for example, the values for fuel oil at 2.3, wood at about 3.0, and steel at 0.46 kJ/kg/°C. The heat of vaporization is also very high at 2260 kJ/kg (the corresponding value for petrol is 293 kJ/kg). Thus, to heat a litre of water to its boiling point from 20°C, and then to vaporize it completely will require (4.18 x 80) + 2260 = 2594 kJ. In addition this volume of water will produce about 1700 litres of steam, although in most situations this will contribute very little to the firefighting. Water has quite a high density which ensures that a jet has high momentum, and therefore a long trajectory and good penetration. It has a high surface tension at ordinary temperatures (72 dynes/cm) which helps in preventing the break-up of a jet, and also enables uniform and stable sprays and mists to be formed. There is very little change in viscosity over the normal temperature range, so that the delivery from a pump, and the pressure losses in hoses, are predictable. It also has good chemical stability and does not decompose at flame temperatures (other than those of some burning metals). Perhaps its only serious disadvantage is that it goes solid at 0°C.

The calculation above, which shows that about 2600 kJ are needed to turn a litre of water into steam, can be used to demonstrate that a single jet at a fire, delivering about 450 l/min, could absorb all the heat from a 19 MW fire. As we have already seen, in normal firefighting water is used to cool the surface of the burning fuel, and this flow of water would be capable of extinguishing a substantially larger fire when used in this way. However, it is sometimes necessary to assume that a fire will not be readily extinguished and that water will be used for some time simply to absorb the heat of combustion. This might happen in a large refinery fire, where it may take several hours to extinguish the flames, and meanwhile it is necessary to prevent the collapse of exposed structural steelwork. Water application rates should then be based on the assumption that the water will be heated in contact with the hot metal to a temperature well below the boiling point: if water is allowed to boil, the cooling process will be hampered by the layer of steam at the surface of the metal. A typical application rate for cooling is 300 $g/m^2/s$.

We saw earlier that the minimum application rate of water needed to extinguish burning wood had been found by calculation and in laboratory experiments to be about 1.6 $g/m^2/s$. As might be expected, application rates considerably greater than this are needed in an actual fire, partly because it is not possible to apply all of the water to all of the burning surface. Indeed it is very likely that access to at least part of the surface will be obstructed. More water will also be needed if there are mutually radiating surfaces: these will be at a higher temperature because the normal heat losses are reduced. A fire will also be particularly intractable if it has become deep-seated.

Some experiments were made at the UK Fire Research Station (FRS) in which simulated room fires were extinguished by hose reel equipment. The room had a volume of $50m^3$ and contained 360 kg of mock furniture. Fires were tackled when they were fully developed. It was found that they could be controlled by only 32 litres of water and extinguished by about 76 litres. (It is difficult to determine the exact moment when a fire is completely extinguished. A 'control point' can be arbitrarily established as the time at which the radiation has been reduced to 10 per cent of the maximum value. This point is easily detectable and corresponds to a situation in which extinction is certain if agent application is continued at the same rate). The minimum application rates needed to achieve control in these tests were between 30 and 100 g/m^2/s, that is at least about 20 times the theoretical value. The spread of the results was due partly to differences in the time taken for fires to become fully developed. If some of the simulated furniture had collapsed during the growth period, additional water was needed to deal with the deep-seated fire. Estimates of water application rates needed to control very large fires (up to about 10 000 m^2) in both the UK and USA range from 300 g to 6 kg/m^2/s, and exceptionally up to 60 kg/m^2/s. This reflects the difficulty in achieving maximum effectiveness in practice and also the need to use water for cooling structures which are exposed to radiation from large fires.

The tests at FRS were conducted partly to assess the effect of pressure on the efficiency of both jets and sprays. It was found that when all the rsults were examined statistically, there was no measurable effect, and also that there was no difference between the results obtained by jets and sprays. This means that in these particular tests, which simulated a full-scale room fire, it did not matter how the water got there so long as it was applied to the fire. Many firemen will dispute this because they believe that high pressure increases the effectiveness of water application. High pressure will certainly increase the penetration of a jet into a large fire, and in this sense the fireman's observation is correct. The results of the tests also showed that when sprays were used to control the fires the droplet size was not important. The spray tests also showed that steam production, which could be considerable, had no measurable effect on the volume of water needed for control.

Although water is very nearly the ideal firefighting agent for most common fires, there are some applications for which an improvement in its effectiveness can be made by the use of suitable additives:

- *Wetting agents.* These increase the ability of water to penetrate into, for example, a hay stack or a thatched roof. Some improvement may also be found when 'wet water' is used against heather fires, but in general these materials provide only a slight advantage and are rarely used.
- *Fire retardants.* The addition of chemicals like sodium ammonium phosphate to water does improve its firefighting effectiveness. Portable appliances containing solutions of this kind are sometimes described

as 'loaded stream' extinguishers. However, the improvement is not very great, and the use of retardants is usually restricted to forest fires, where water is at a premium.

- *Freezing point depressants.* Calcium chloride is the most widely used depressant, but a corrosion inhibitor may also be needed. The addition of about 140 g to a litre of water will reduce the freezing point to –10°C. Ethylene and propylene glycols can also be used but they do reduce the specific heat of the water, and at high concentrations the mixture is flammable.
- *Thickeners.* Bentonite clay, sodium alginates and carboxymethyl cellulose have all been used to increase the viscosity of firefighting water. Friction losses in hoses are then increased but the reach of a jet is extended. The important effect is that a thick layer of water which drains only slowly can be applied to exposed surfaces, for example in forest fires. As little as 0.2 per cent of additive can double the effectiveness of water in these fires.
- *Ablatives.* If a sufficient amount of a thickening agent is added to water it will form a gel on an exposed surface. The layer will be very thick and there will then be no drainage. Protection against ignition can therefore be provided for a considerable time. However these gels are difficult to apply and are less cost-effective than thickened water.
- *Friction-reducing agents.* The addition of as little as 0.003 per cent of polyethylene oxide polymer (PEO) to water can enable a reduction of about 70 per cent to be made in the friction factor used in hydraulic calculations. Less energy will be needed to pump this 'slippery water' down small diameter hoses, and both the height and reach of the jet will be increased. Greater benefit can be derived from the use of PEO by US fire brigades than by UK brigades because the latter tend to use larger-diameter hose to feed smaller nozzles.
- *Foam concentrates.* Foam is an extinguishing agent in its own right and is dealt with separately.

Naturally occurring water, and particularly sea water, is sufficiently conductive to present a hazard when it is used against fires involving live electric circuits. If a jet used in normal firefighting operations encounters a domestic supply circuit it is unlikely that injury will result, but clearly it is important to isolate the mains supply whenever possible. A spray will not conduct electricity and may be used safely in this case, but special precautions are needed if high-voltage equipment is involved. Another hazard in the practical use of water results from the jet reaction of a nozzle. The magnitude of the reaction can be calculated approximately from:

$$R = 1.57Pd^2$$

where R = reaction in kg force, P = pressure in bar, and d = nozzle diameter in mm. Some of the backward thrust of a branchpipe will be taken by the hose and the ground on which it is resting, but much will have to be absorbed by the man holding it. Calculation shows that the reaction from a 15 mm nozzle operating at its optimum pressure of 3.5 bar is 12.3 kgf, and that from a 25 mm nozzle at its optimum pressure of 7 bar is 68.7 kgf. The first could be held quite safely by a man, but the second could not. If a man does lose control of a branchpipe it will whip violently and can cause serious injury.

Foam

Foam is generated by first dissolving a foam concentrate in water and then aerating the solution so that it is expanded into a mass of bubbles. It is the purpose of the concentrate to enhance bubble formation and to maintain the stability of the foam. The made foam can be projected from suitable equipment on to a fire, where it can spread over a surface (usually that of a burning liquid). Foam produced in this way is sometimes called air foam or mechanical foam to distinguish it from foam which is generated by a chemical reaction. Typical reactants for the production of chemical foam are solutions of aluminium sulphate (which is acidic), and sodium bicarbonate with foam stabilizers. The solutions are mixed near to the point of application and the bubbles are formed by the carbon dioxide which is released by the reaction. Chemical foam systems are very rarely seen now and the distinction is really unnecessary. However not all foams are produced by aeration. Some high-expansion foams contain nitrogen, and experiments have been made with foams containing Halon vapour.

Three of the physical properties of foam are used in specifications, as a means of quality control, and as an indication of the behaviour of the foam in use. These properties depend on the amount of concentrate which is added to the water, on the pressure at which the solution is pumped, and on the particular foam-making appliance which is used. There are however standard test procedures.

- *Expansion.* This is simply the ratio of the volume of the foam to the volume of the solution from which it was made. Low expansion (LX) foams have expansions between about 5 and 15, but more usually between 8 and 10. Medium expansion (MX) foam is usually around 50, and high expansion (HX, or HiEx) may have an expansion up to 1000.

- *Drainage.* This is expressed either as the volume of water which drains out of a known volume of the foam in a certain time, or as the time taken for a certain percentage of the water to drain from the foam.
- *Shear stress.* The stress exerted by a rotating vane in shearing a volume of foam is used as a measure of the mechanical strength of the foam.

The two latter measurements give an indication of the stability of the foam, which is itself related to the work which was done in forming the bubble structure. There are two important performance measurements which can be made by means of test fires:

- *Critical application rate.* This is the minimum rate of application of foam to the surface of a particular fuel which is just capable of controlling the fire. It is usually expressed as $1/m^2/min$.
- *Burn-back resistance.* If a layer of foam on the surface of a flammable liquid is exposed to the radiation from flames burning on an uncovered area of the liquid, it will eventually break down and permit the flame to spread back. The resistance to burn-back is usually expressed as the time in minutes before its occurrence in a particular fire test.

Many different kinds of foam concentrate are available. The main types are detailed below together with an indication of their characteristics. Most fire brigade equipment is designed to use concentrate at about 6 per cent, but many manufacturers supply concentrates which are intended to be used at 3 per cent and sometimes as low as 1 per cent.

Protein foam

This is the most widely used concentrate. It is a solution of hydrolysed protein, usually derived from hoof and horn meal, together with a number of additives which improve its storage and performance properties. It is effective against many flammable liquid fires, and the critical application rate against petrol is about $1.0\ l/m^2/min$ (this is the flow rate of the solution, not the made foam). It is ineffective against water-miscible fuels, like alcohol, which rapidly break down the bubble walls. It may also be incompatible with some firefighting powders. It has the advantage of reasonable burn-back resistance and it is cheap, but is less effective than some of the more expensive concentrates.

Alcohol-resistant foam

This is usually a modified protein foam. It may also be called all-purpose, or multi-purpose, foam. Some types are specially formulated so that a barrier is formed between the base of the foam and the fuel surface. It is necessary to apply them very soon after mixing, so the solution cannot be pumped through long lengths of hose. The application rate is usually higher than that for ordinary protein foam against hydrocarbon fires, even when the alcohol-resistant foam is used against these fuels. The water-miscible fuels against which these formulations are suitable include acetone, acrylonitrile, alcohols (including ethyl, methyl and isopropyl alcohols), amines, anhydrides, ethyl and butyl acetates, lacquer thinners, and methyl ethyl ketone (MEK).

Synthetic foam

Foam concentrates derived from protein have a limited storage life (which may however be several years). Completely synthetic concentrates have the advantage of a much longer life, and consistent quality. They can be used to make LX, MX and HX foam, usually at concentrations of between 1.5 and 3 per cent. The foam usually gives rapid knock-down of flames but has relatively poor burn-back resistance.

Acqueous film forming foam (AFFF)

This is also known as light water because of its ability to produce a thin film of solution which can float on the surface of a burning fuel. This effect can occur only when the surface tension of the solution plus the interfacial tension between the solution and the fuel are together less than the surface tension of the fuel. (Molecules in the bulk of a liquid are mutually attracted, the attraction for an individual molecule being equal in all directions. For molecules at the surface there is no attraction on the exposed side, and the unbalanced pull from the liquid side causes the surface to shrink to its minimum area. The force which causes this shrinkage is the surface tension). The surface tension of water has already been quoted as 72 dynes/cm. That of ordinary petroleum fuels is around 24 dynes/cm, and the interfacial tension is between about 1.0 and 6.0 dynes/cm. Thus, it is necessary for the AFFF solution to have a surface tension no more than 35 to 40 per cent that of water if it is to form a film. This reduction is achieved by the available concentrates, but it may not necessarily be adequate to promote film formation on all fuels. The AFFF concentrates are mixtures of synthetic surfactants containing fluorine molecules. Some are alcohol-resistant. The film-forming ability ensures that solution draining away from foam will spread out to cover any area not protected by the foam. The film is self-healing: any break made by stirring or splashing will be rapidly recovered. It is not necessary to generate foam to produce the film. In conditions where projected foam is likely to be carried away by wind or by the up-draught from a fire, a jet of AFFF solution may be able to penetrate. The film can then run down from a wetted surface. Such a situation could for example occur on an offshore platform if a helicopter crash fire has to be tackled in a high wind. When AFFF foam is applied to a burning fuel, knock-down is rapid but the burn-back resistance is not as good as that of a protein foam. Because of the considerable reduction in surface tension produced by AFFF concentrates, the solution can be used effectively against fires involving materials like hay and thatch. The solution is then being used as 'wet water', which has already been discussed.

Fluoroprotein foam

This is a protein foam concentrate to which has been added perfluoro surfactants similar to those used in AFFF concentrates. The foam has some of the advantages of both protein and AFFF foams: good burn-back resistance and rapid knock-down. It is not capable of producing a film on fuels like petrol, but there is some indication that it may do so on materials like heavy fuel oil. The foam can also be used effectively for base injection, in which made foam is pumped (usually through a product line) into the base of a burning storage tank. The foam rises to the surface of the fuel where it can extinguish the fire. Protein foam cannot be used in this way because it picks up enough fuel to become flammable by the time it reaches the surface. Fluoroprotein concentrates are more expensive than ordinary protein concentrates, but the critical application rate is lower: about $0.7\ l/m^2/min$.

LX foams can be applied gently to the surface of a burning liquid, for example from a fixed foam-pourer on a storage tank. This is the most efficient way to use foam. It can also be projected from a suitable branchpipe or cannon. Solution application rates as high as 5000 l/min can be achieved with the latter, with a throw of over 50 m. Ideally the projected foam should strike a vertical surface at the back of the fire from which it can run down on to the fuel, or it may be possible to apply it to the ground just in front of a burning pool. In either case, some foam will inevitably be wasted. There will also be losses in the method already mentioned: base injection. Whichever method is used, the objective is the same: to cover the whole of surface of the fuel with a layer of foam, or with an aqueous film from AFFF, so that the flow of flammable vapour into the flame is stopped. A layer of foam has the advantage over a liquid film of greatly reducing the amount of radiation reaching the fuel. The mechanism of extinction by foam, as defined by the triangle of fire, is that of smothering.

MX and HX foams are also used for surface sealing, but they are difficult to project and certainly cannot be used for base injection. In the absence of a wind it possible to project MX for up to 10 m, but HX is very light and is usually conveyed to a fire in large-diameter ducting (like for example large lay-flat polythene tubing). MX can be made in a special branchpipe, but HX is formed by blowing air through a mesh over which a surfactant solution is flowing.

The great advantage of HX foam is its ability to fill completely a complex volume. Thus if the volume cannot be entered and the exact location of the fire is not known, foam pumped from the outside should be able to penetrate to the fire. Indeed this kind of foam was developed in the UK for firefighting in coal mines, where it is not usually possible to approach sufficiently close to the fire to make a manual attack. There are, however, some limitations to the use of HX. There is a limit to the distance through which the foam will flow: eventually the continued addition at the inlet will simply keep pace with the mechanical breakdown of the compressed foam. Also, if flames have reached the ceiling before a room is filled with foam, the radiation may break down the upper layers of foam faster than they can be replaced. There may be problems in removing the foam after a fire, and it is certainly hazardous to enter a foam-filled room. The eventual inhalation of a dangerous amount of water from the foam cannot be avoided, and even if breathing apparatus is worn, a man can become disorientated because of the complete loss of vision and hearing.

Carbon dioxide, nitrogen, etc.

The flame-extinguishing concentrations for carbon dioxide and nitrogen against n-heptane are 20.5 per cent and 30.2 per cent. This superior effectiveness of carbon dioxide is sometimes attributed to slight chemical inhibition of the flame reactions. If the thermal capacities of the two gases are compared it will be found that they are respectively 54.3 and 32.7 J/mol/°C at 1000°C. If we take these values into account, the performance of the two agents on a molar basis is very similar, indicating as we might expect that both are acting as a thermal load in the flame. This does not alter the fact that carbon dioxide is superior on a volume basis. Also its physical properties make it more attractive as a practical extinguishing agent.

Carbon dioxide can be liquefied quite readily simply by compression. At 15°C a pressure of about 52 bar is needed and the liquid can then be stored in a steel cylinder at this pressure. If the temperature is increased the vapour pressure will also increase: allowance must be made for this, and for the expansion of the liquid, in the design of the cylinder. Carbon dioxide can be stored as a liquid up to its critical temperature of 31°C (where its vapour pressure is about 75 bar). Above this temperature it can exist only as vapour. These values mean that the quantity of carbon dioxide needed to flood a typical enclosure can be conveniently stored and transported in pressure vessels which can be manhandled. The construction of a large pressure vessel capable of holding the liquefied gas at ambient temperatures would be very expensive. For this reason a large volume of carbon dioxide (for example that needed to flood an extensive computer installation) should be refrigerated, so that is vapour pressure is reduced. At −17°C for example the vapour pressure is reduced to about 20 bar. The cost of a large container designed for these conditions, together with the necessary refrigeration equipment, is still considerably less than that of a similar container designed for ambient conditions.

Carbon dioxide has the unusual physical property of a triple point which is at a pressure higher than atmospheric: about 5 bar. At lower pressures than this it cannot exist as a liquid, but only as solid or gas. Solid carbon dioxide is available commercially and is used as a simple refrigerant. It sublimes (releases vapour without melting into a liquid) at −78°C. This triple point pressure must be remembered when carbon dioxide systems are designed. Liquid flow down a pipeline causes a drop in pressure and if this falls below 5 bar, the pipeline, or its nozzles, may become blocked with solid carbon dioxide. The pressure behind the nozzles must always be higher than this so that liquid flow is maintained. As the liquid emerges from the nozzle some of it will immediately flash off as vapour. The latent heat needed for this will reduce the temperature of the remaining liquid to −78°C, so that it freezes. About 50 per cent of the discharge becomes solid (usually called snow) and the nozzles must be carefully designed to ensure that the snow can remove sufficient heat from the air in order to sublime.

Nitrogen has a critical temperature of −147°C and cannot therefore be liquefied at ambient temperatures. This makes it unattractive as a total flooding agent because very high pressures are needed if an adequate quantity is to be stored in a reasonable volume. Also, since there can be no liquid flow to the nozzles, discharge times may be unacceptably long unless large-diameter (and therefore expensive) pipelines are used. Nitrogen is conveniently used for blanketing: the provision of an inert atmosphere in plant or storage vessels. Nitrogen may be readily available on site, and may

Table 6.2
Physiological effects of nitrogen and carbon dioxide

Concentration of added nitrogen (%)	Resulting concentration of oxygen (%)	Effect
20	17	Impaired coordination
40	13	Headache. Dizziness. Fatigue
50	10	Paralysis. Unconsciousness
70	6	Death in a few minutes
Concentration of added carbon dioxide (%)		
2	20.6	Breathing rate x 2
4	20.2	Headache. Breathing rate x 4
8	19.3	Dizziness. Stupor
12	18.3	Death in a few minutes, even with added oxygen

be less soluble than carbon dioxide in the process liquids. However, nitrogen is not completely inert and will react with burning metals. Plants in which combustible metals are processed may be protected with argon, helium, krypton, neon or xenon (all of which can be stored only as gas).

Table 6.2 shows the physiological effects resulting from the inhalation of various concentrations of both nitrogen and carbon dioxide. As shown in the table, 40 per cent of nitrogen in the air reduces the oxygen concentration from the normal 21 per cent to 13 per cent. The same effect, that is, the same reduction in the partial pressure of oxygen, happens at an altitude of 4 km (12 600 ft). The physiological effects at this altitude are the same – the onset of headaches, dizziness and fatigue – unless the ascent is very gradual. The more dangerous symptoms of paralysis and unconsciousness occur after an exposure to a 50 per cent nitrogen concentration. It will be noted from the lower half of the table that headaches can be caused by exposure to only 4 per cent of carbon dioxide, and that 8 per cent will cause stupor. It is evident therefore that carbon dioxide is toxic and is not, like nitrogen and the other inert gases, simply an asphyxiant. This is also shown by the fact that 12 per cent of carbon dioxide is rapidly fatal even if oxygen has been added to the mixture to restore the original concentration of 21 per cent. The toxic effects of carbon dioxide are enhanced because exposure to low concentrations stimulates breathing and hence increases thc intake of the gas. Since the minimum design concentration for a carbon dioxide total flooding system is more than twice this lethal concentration, strict precautions must be taken against the discharge of a system into an occupied area.

The Halons

The word Halon is an abbreviation of halogenated hydrocarbon. It is usually followed by a number which indicates the molecular structure of the particular agent. The numbering system was developed after the Second World War by the US Army Corps of Engineers as a more convenient alternative to the use of chemical formulae. The first digit gives the number of carbon atoms in the molecule. For all the currently available Halons this is either 1 or 2, indicating the derivatives of either methane or ethane. The second digit shows the number of fluorine atoms, the third the number of chlorine atoms, and the fourth the number of bromine atoms. Hydrogen atoms are not shown, and a final 0 is omitted. Thus BCF, $CBrClF_2$, is Halon 1211; CB, CH_2BrCl, is Halon 1011 (the two hydrogen atoms are not indicated); and tetrafluoromethane, CF_4, is Halon 14, not 1400.

It has already been mentioned in the first part of this chapter that the bromine atoms play the most active part in the combustion-inhibiting reactions. Fluorine and chlorine atoms do contribute, but only to a minor extent. It is possible to predict the efficiency of a Halon as an extinguishant by awarding marks for each of the halogen atoms and then adding up the score. The values used are 1 for a fluorine atom, 2 for a chlorine, and 10 for a bromine atom. Thus for the three Halons mentioned above the scores are respectively 14, 12 and 4. The cup burner apparatus (already mentioned earlier, and discussed later in this chapter) provides accurate and reproducible measurements of flame-extinguishing concentrations, and these can be used to show a close correlation between a Halon's score and its efficiency as an extinguishant. Table 6.3 shows the scores for 12 Halons (only those in colums A to F are normally used as extinguishants) and also the flame-extinguishing concentrations against n-heptane (from Table 6.1). These are plotted together (on a log-log scale) in Figure 6.2, and it will be seen that there is reasonable agreement. Halon 14 is less efficient than is indicated by its score and this is probably due to its high chemical stability, even at flame temperatures.

The first of the Halons to be recognized as an extinguishant was CTC, Halon 104 (column G, Table 6.3). This happened during the last century but

Table 6.3
Properties of commercially available Halons

Halon number	(A) 1211	(B) 1301	(C) 1202	(D) 2402	(E) 1011	(F) 1001	(G) 104	(H) 113	(I) 14	(J) 122	(K) 251	(L) 131
'Arcton' number	12B1	13B1						11		12	115	13
Name	BCF	BTM	DDM	DTE	CB	MB	CTC					
Formula	$CBrClF_2$	$CBrF_3$	CBr_2F_2	$C_2Br_2F_4$	CH_2BrCl	CH_3Br	CCl_4	CCl_3F	CF_4	CCl_2F_2	C_2ClF_5	$CClF_3$
Boiling point (°C)	–4.0	–57.6	24.4	47.5	67.8	4.5	76.6	23.8	–128	–29.8	–38.0	-81.4
Concentration* (%)	3.8	3.5	2.4	2.1	4.5	5.1	6.7	7.0	16.7	8.3	7.4	7.7
Score	14	13	22	24	12	10	8	7	4	6	7	5

* Flame-extinguishing concentration against n-heptane at ambient temperature.

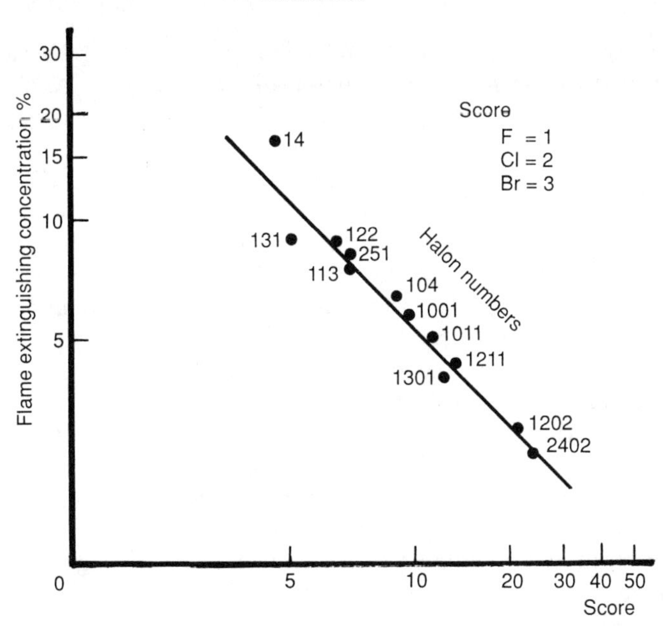

Figure 6.2 Flame extinguishing concentration plotted against score for 12 Halons.

but the agent was not readily available until the advent of cheap chlorine from the electrolytic process for the production of caustic soda from brine. The first of a very large number of CTC hand pumps was sold at around the beginning of this century.

The main advantage of the chemical extinguishants over those which can simply cool or dilute is their efficiency: a lot of firefighting potential can be contained in a small space. They can be used to protect against fire hazards where space and weight must be limited. An obvious application is in the automatic systems which protect aircraft engines against in-flight fires. CTC was successfully demonstrated in such a system in 1926. Petrol was poured over the engine of an SE5 aircraft while it was flying over the Royal Aircraft Establishment at Farnborough. The petrol was ignited but the fire was rapidly extinguished by a CTC discharge. Further experimental work led to the choice of MB (column F of Table 6.3) as the extinguishant for aircraft use because of its greater efficiency. MB became the standard agent for military aircraft from 1938 until after the Second World War.

The decision to change from CTC to MB is interesting because their Halon scores are quite close: 8 and 10. It might therefore be expected that MB would provide only a 25 per cent improvement in performance. The actual improvement was considerably greater, and this can be explained by the difference between the boiling points of the two agents. As shown in the table, CTC boils at 76.7°C and MB at 4.5°C. In order to interfere chemically the Halons must enter the reaction zone of a flame as a vapour (or indeed as a pyrolysed vapour), but all except the smallest droplets of a CTC spray can pass right through a flame and be wasted (apart possibly from providing a small amount of cooling). On the other hand, the higher volatility of MB ensures that in normal use the whole of a discharge is vaporized as it approaches the flame. The extinguishing concentrations and the Halon scores quoted in the table apply only to the completely vaporized agents and must therefore represent the maximum potential of the agent, which it may not be possible to achieve in practice.

Although MB achieves a reasonable performance as a chemical extinguishant, it has the disadvantage of being extremely toxic. It is used for killing vermin in ships' holds. The Germans considered it too toxic for use on aircraft and by the end of the Second World War they had developed systems based on CB (column E in the table). This has a good potential as an extinguishant with a score of 12, but it has the disadvantage like CTC of a high boiling point: 67.8°C. The Germans overcame this problem by mixing it with 35 per cent of liquid carbon dioxide, to make a sufficiently volatile agent which they called Dachlaurin. CB is certainly much less toxic than MB, and slightly less than CTC, but the vapours of both CB and CTC can be hazardous.

So by about 1945 three reasonably efficient Halon extinguishants were in use, but all had their disadvantages. MB had good chemical and physical properties, but was unacceptably toxic. CB and CTC were considerably less toxic but had a poor performance because of their high boiling points. As it happens, by that time the provision of an entirely new range of low toxicity agents had become possible. The chemical industry had developed methods of controlling the violent reactions between fluorine and the hydrocarbons. The end products of these very exothermic reactions are extremely stable: the energy released in their formation must be put back before they can be decomposed. These fluorinated hydrocarbons found applications as aerosol propellants, refrigerants, solvents and anaesthetics. Most of them were derivatives of methane and ethane in which one or more of the hydrogen atoms had been replaced by fluorine atoms, and the remainder by chlorine atoms. The materials in columns H, J, K and L of the table are typical. They are all used as refrigerants, and two or more can be blended to provide a range of aerosol propellants. They are sold under the 'Arcton' trade mark with a numbering system which is simpler than the Halon one, and with similar numbers under the 'Freon' and 'Iceon' trade marks. All of them are chemically inert, very stable, non-corrosive, and of very low toxicity. The point of interest to us is that the presence of one or more bromine atoms in one of these molecules should result in a chemical extinguishant possessing the same desirable properties.

As they became available, a wide range of potential extinguishants was evaluated, some of which included iodine atoms in their molecules. On the Halon scoring system an iodine atom earns 16 marks, so these agents had a very good performance. Unfortunately they were all toxic, and also expensive (iodine is a rare element), and none was considered suitable for development as a practical extinguishing agent. A number of bromine compounds were also eliminated because of unsuitable chemical or physical properties. The four which were developed and are now commercially available are shown in columns A, B, C and D of the table. It should be noted that only the first two, Halons 1211 and 1301, have low boiling points. The other two, Halons 1202 and 2402, both achieve high scores because of the presence of two bromine atoms in their molecules. Their high boiling points make it very difficult to realise their potential in practice but they do find application in explosion suppression systems (described in Chapter 26). They may also have an advantage over the more volatile agents in local application systems which are operated in high air flows (outdoor systems, or those protecting exposed engines on vehicles).

Halon 1211 was originally developed in the UK as a replacement for MB on military aircraft. It is now employed almost exclusively on both UK military and civil aircraft. In the USA Halon 1301 was chosen for this use, although its higher vapour pressure resulted in slightly heavier systems. In both countries the use of the Halons was gradually extended to the protection of other vehicles, and to a wide range of industrial equipment. It was found that the physical properties of Halon 1211 were idealy suited to its use in hand extinguishers and wheeled appliances. When it is discharged at pressures around 6 bar through a suitably designed nozzle, a throw of 10 m or more can be achieved. The mixture of vapour and spray approaching a fire is then readily vaporized. The vapour pressure of Halon 1301 is rather too high for it to be used effectively in hand appliances: the throw, like that of carbon dioxide, is only a few metres. For the same reason, Halon 1211 is better suited to use in local application systems. Both agents can be used very effectively for total flooding but in this situation Halon 1301 has the advantage of lower toxicity. Its higher volatility may also ensure a better distribution, but the violence of the discharge can be a minor disadvantage.

The toxicity of the extinguishing agents was originally assessed by tests in which small animals were exposed to different concentrations of their vapours. Tests like this can be used to establish the LC_0 value: this is the highest concentration which, after an agreed exposure time, will not have caused the deaths of any of the test animals. The LC_{50} concentration causes 50 per cent fatalities, and LC_{100} will kill all the animals (LC is an abbreviation for lethal concentration). Table 6.4 shows the approximate 15 minute LC_0 values for rats which were obtained for six of the agents, and for carbon dioxide.

It may be surprising that the LC_0 value for carbon dioxide is lower than that for Halon 1301, but as we have already seen, carbon dioxide is not simply an asphyxiant but is toxic in its own right. The high LC_0 values for Halons 1301 and 1211, which are some five to ten times the concentrations used for total flooding, suggested that these agents could be safely used for this application in the presence of people. Indeed, some of the early Halon 1211 systems were demonstrated to customers by discharging the agent in their presence, and with no untoward effects. However during the preparation of the NFPA standards for automatic systems using these agents (NFPA 12A and 12B) the manufacturers were asked to conduct tests in which human beings were exposed to low agent concentrations. These tests showed that symptoms could be produced by quite brief exposures to concentrations much lower than the animal LC_0 values, and it is quite likely that similar effects were experienced by the rats. The earliest symptoms recorded by the people in the tests were slight dizziness, lack of coordination and reduced mental acuity. With Halon 1211, prolonged exposure to concentrations lower than 4 per cent produced minimal, if any, effects. The symptoms became apparent after an exposure of one minute to concentrations between 4 and 5 per cent. Nothing was experienced during the first 30 seconds and the effects of a one minute exposure were not incapacitating.

Table 6.4
Approximate LC_0, 15 min, values for rats

Agent	Halon number	Approximate LC_0, 15 min
BTM	1301	80
BCF	1211	30
DTE	2402	9
DDM	1202	5.5
CB	1011	2.8
CTC	104	1.3
CO_2		65

At concentrations between 5 and 10 per cent there was a risk of unconsciousness, and death was possible after a prolonged exposure. The effects of a brief exposure were similar to those which would be produced by an anaesthetic. They persisted for only a brief time following exposure, and recovery was rapid and complete. Similar tests using Halon 1301 produced no symptoms from five-minute exposures to concentrations below 7 per cent. The symptoms detailed above became apparent after several minutes at between 7 and 10 per cent. Concentrations of 10 per cent were incapacitating after exposures greater than one minute. There was a risk of unconsciousness and possibly death at concentrations between 10 and 15 per cent. Again no effect was felt for the first 30 seconds even at the higher concentrations, and again recovery was rapid and complete after an exposure.

Although the symptoms produced by short exposures to low concentrations of the agents were not incapacitating, it was wisely decided that the maximum acceptable intake of an agent would be that which produced no symptoms at all. These exposures are one minute at 4 per cent for Halon 1211, and one minute at 10 per cent for Halon 1301. The minimum design concentraton for both agents permitted in the NFPA standards is 5 per cent, which is just above the safe limit for Halon 1211 and below the limit for Halon 1301. For this reason, the NFPA standards do not permit the use of Halon 1211 total flooding in normally occupied areas. Halon 1301 can be used in these areas but not in concentrations greater than 10 per cent.

There is an additional hazard from the inhalation of Halon vapour: its ability to sensitize the heart to the effects of adrenalin. Adrenalin is likely to be released into the bloodstream in the stressful conditions of a fire and its enhanced effect on the heart, due to the presence of Halon vapour, may cause irregularities (cardiac arrhythmias, including ventricular fibrillation). These effects may be fatal. Cardiac sensitization is not unique to the Halons. It can be caused by any lipid soluble vapour, for example that of petrol. Human testing to establish the exact conditions under which these effects occur has not been possible because of the danger. It may be that some people – the elderly, for example, or those with heart disease – may be particularly sensitive. For this reason it may be inadvisable to discharge Halon 1301 into an area to which the general public has access. Certainly the British Standard on Halon 1301 total flooding systems provides stricter controls over its use than does the NFPA standard, and these are in close agreement with the guidance provided by the Health and Safety Executive. The precuations needed for the safe use of both agents are discussed in more detail in Parts Six and Seven (in chapters dealing with both automatic systems and hand extinguishers). However it should be emphasized that only in very exceptional circumstances will any advantage be gained by discharging any automatic system into an occupied area. This too is discussed in more detail later.

It will be remembered that in order for a Halon to act as a chemical inhibitor it is necessary that it decomposes as it enters the reaction zone of a flame. The halogen atoms, which are the active species, are thereby released. They can exist in atomic form for only a brief period and must be released from the flame, either during or after its extinction, as part of a new molecule. The formation of molecules of fluorine, bromine and chlorine (F_2, Br_2 and Cl_2) might be expected, but very few are formed. The halogen halides, HF, HBr and HCl, comprise a very high proportion of the breakdown products. The formation of carbonyl halides is theoretically possible, COF_2, $COBr_2$, $COCl_2$ (the latter is phosgene), but again these are very rarely found, and then only in the presence of fuel-rich combustion. Cleary the breakdown products are considerably more toxic than the undecomposed Halon, but fortunately in all normal uses of the Halons they are produced in very low concentrations (perhaps a few parts per million). They are extremely pungent, so there is a built-in warning of their presence, and a hazardous concentration is intolerable to breathe. Breakdown products do not therefore constitute a toxic hazard in normal practice. Surprisingly they very rarely cause corrosion problems either. Solutions in water of the three hydrogen halides produce hydrofluoric, hydrobromic and hydrochloric acids, and clearly a mixture of these is likely to attack a metallic surface. Fortunately the hydrogen halides are not very reactive as the warm dry gases which are present after a fire and they are usually dispersed harmlessly by normal ventilation.

Powders

Firefighting powders inhibit the combustion processes by presenting a chemically active surface within the reaction zone of a flame. To a lesser extent they may also quench the flame by acting as an omni-directional flame trap. Powders must also be responsible for a small amount of cooling, and some formulations also affect the ability of a solid fuel surface to release flammable volatiles.

The earliest powder to be used for firefighting was sodium bicarbonate. It was originally thought that it decomposed in the fire to release carbon dioxide, which extinguished the flames. In fact very little decomposition occurs, and even when it does it is likely that the released sodium atoms are the only effective inhibitors. Sodium bicarbonate powders are still widely used, usually with the addition of small amounts of materials like silicones and stearates. These make the powder free-flowing and also reduce the tendency to cake. More effective (although more expensive) powders are made from potassium bicarbonate. When this material was first introduced by the US Navy it was called Purple K, because of the flame colouration (due to the presence of potassium atoms). Purple K is about twice as effective as sodium bicarbonate. Another potassium salt, the chloride, has a similar performance to Purple K and is called Super K. It is not widely used, possibly because it is more corrosive than the bicarbonate when it is damp. All these powders will produce very rapid flame knock-down and are excellent agents against flammable liquids and gases. They are just as effective against the flames of solid fuels, but do not inhibit smouldering, and the fuel is therefore capable of reverting to flaming combustion after the discharge. Smouldering can be inhibited by materials like ammonium phosphate, which melt to form a sticky layer on the surface. Powders based on these materials are also effective against flames and are marketed as 'General purpose' or 'ABC' powders. Some of these materials do appear to be capable of chemical inhibition of smouldering combustion. A possible mechanism is the linking together of carbon atoms into graphite rings, so that volatile carbon compounds cannot be formed. At the same time H and OH radicals which are stripped off the cellulose and lignin molecules are combined catalytically at the surface of the fuel to form water.

The chain-terminating reactions which inhibit flaming combustion take place at the surface of the powder. It follows therefore that the greater the specific surface of the powder, the more efficient it will be. The specific surface can be increased simply by reducing the size of the particles. Suppose for example that a 1 mm cube is cut up into eight smaller cubes each measuring 0.5 mm; this will have increased the total surface area from 6 mm^2 to 12 mm^2. Thus the efficiency of a firefighting powder can be approximately doubled simply by halving the particle size (only approximately because the particles may not be cubical). Unfortunately the degree of comminution is limited by ballistic considerations: the powder must be projected for some distance from a nozzle, and must then penetrate the updraught around the flame. In practice this limits the minimum particle size to around 40 microns. However a powder has been developed which is capable of increasing its specific surface as it enters the flame. 'Monnex' is a reaction product of potassium carbonate and urea. It therefore has the superior performance of a potassium compound, but in addition it has an unstable crystal structure derived from the urea. This ensures that when a particle is heated as it approaches the flame, it shatters (or decrepitates) into smaller particles. Particles which were originally around 40 microns may be reduced to sub-micron size. 'Monnex' is about five times as effective as a sodium bicarbonate powder, but it should be noted that a reduction in size from 40 to less than one micron should increase the effectiveness at least 40 times. Clearly the effect of decrepitation has not been fully exploited and it may be that only the smallest particles of 'Monnex' are able to shatter as they enter a flame (Table 6.5 summarizes typical data on powders).

All powers suffer from the disadvantage that they become compacted if they are left in an extinguisher for a long time, particularly in the presence of vibration. They may not then discharge satisfactorily. Extinguishers should therefore be checked regularly to see that the powder is still free-flowing. The ideal arrangement is to withdraw all extinguishers from use at regular intervals and to have them discharged against practice fires as a part of a training programme. It is then possible to check the maximum safe exposure time on a particular plant. It will be found that powder in a stored pressure extinguisher has a longer life than in one that is cartridge-operated (extinguisher design is discussed later). The reason for this is that the gas in a stored pressure extinguisher will have penetrated into the voids between the powder particles, so that there is the same pressure within the whole volume of the extinguisher. When the control valve is opened there will be a local drop in pressure as the contents begin to flow out of the nozzle. This will create a pressure difference across the bed of powder, with the highest pressure in the voids. This will tend to separate the particles, breaking any adhesion that has occured with time. When a cartridge-operated extinguisher

Table 6.5
Typical physical and chemical properties of powders

Type	Standard	Purple K	Super K	'Monnex'*	All-purpose (ABC)
Main ingredient	Sodium bicarbonate	Potassium bicarbonate	Potassium chloride	Potassium carbamic	Monammonium phosphate
Density g/ml	2.15	2.17	1.98	1.98	1.80
Specific surface typical cm^2/g	3000	4200	3200	5500	4250
'Median' particle size micron	40	34	41	30	40
Approx. Critical application rate for n-heptane g/m^2s	40	18	22	7.4	40

* Trade mark of ICI

is actuated, there will be a sudden increase of pressure at the gas discharge points. As the gas flows into the voids in the powder bed, the pressure differential will be in the opposite direction, tending to compact the powder, resulting in a possible increase in adhesion.

Agents for metal fires

If it is possible to approach a metal fire, it can usually be extinguished by heating a dry non-combustible powder over it. Sand is sometimes recommended, but it can seldom be found completely dry. Also, some metals can react with the oxygen in the silicon dioxide which is present in sand. Grahite can be used, despite being combustible, and both salt and soda ash are frequently used because they are available in industry in tonnage quantities (sometimes in tankers, from which they can be discharged through large-diameter hoses). A number of powders are available commercially and these are effective either because they sinter into a crust or melt to form a flux. TEC powder (ternary eutectic chloride) fluxes at a low temperature (the eutectic). The three chlorides in the mixture are those of potassium, sodium and barium, and the latter is extremely toxic. TMB (trimethoxy boroxine) is a liquid. When it is applied to the hot metal it decomposes into methanol, which burns, and boric oxide which fluxes on the metal. TMB is also toxic, but many powder formulations are available commercially which are safe to use.

It should be noted that in the US the term 'dry powder' is frequently used for metal fire agents and 'dry chemical' for powders which can be used against normal fuels.

Inert gases

This is listed separately from carbon dioxide and nitrogen (which are indeed inert gases) because the name refers to a mixture of gases produced by an inert gas generator. These generators are found particularly on ships. They are designed to burn the same fuel oil as the ship's propulsion system, but more efficiently, so that the oxygen concentration in the exhaust gases is well below the limiting oxygen concentratons for most combustion processes. At the UK Fire Research Station in the 1950s some extensive tests were conducted with a generator based on a gas turbine aircraft engine. In normal operation the exhaust gases contained 17 per cent oxygen, but this was reduced to 14 per cent in an after-burner, and then to 7 per cent by the injection of a water spray which was vaporized into steam. The appliance was capable of generating 1500 m^3/min of inert gas and was used to extinguish a number of large fires in buildings. The results were very encouraging but the project was eventually abandoned. The high cost of the equipment was one obvious disadvantage. The fire brigade had also found that it was difficult to work in a building filled with gas at 100°C.

There is a possible application for this technique offshore. On production platforms electrical power is generated by alternators driven by gas turbines, and these are usually the industrial versions of large aircraft engines. In an emergency the stand-by gas turbine could be used to produce large volumes of inert gas which could be injected into a burning module.

EXTINGUISHANT PERFORMANCE

Flame-extinguishing concentrations

The cup burner apparatus used in obtaining measurements of flame-extinguishing concentrations has already been mentioned. It is shown in Figure 6.3. Before a measurement can be made it is necessary to check that the air flow is adequate for the flame, but not at a sufficiently high velocity to affect its stability. Increasing amounts of extinguishing vapour or gas are

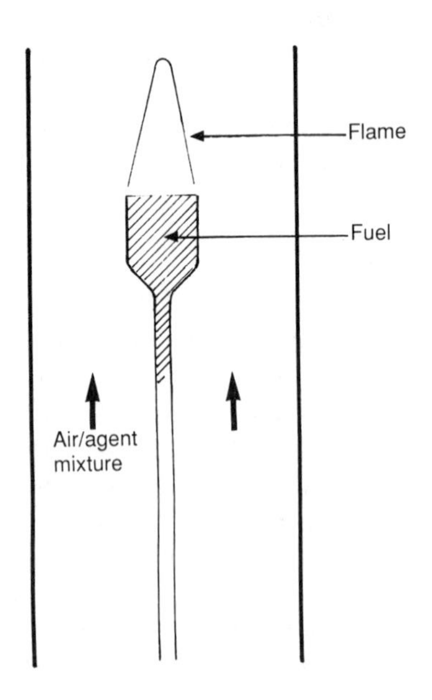

Figure 6.3 Cup burner

then metered into the known air flow until the flame is extinguished. The critical concentration can then be calculated. In practice it has been found necessary to provide a means of heating the cup so that a fuel which has not previously been tested can be checked to see whether a higher agent concentration is needed at elevated temperatures. The values obtained can be used in calculating design concentrations for total flooding systems.

Inerting concentrations

Figure 6.4 shows two kinds of apparatus which have been used to measure inerting concentrations. The one on the left was developed by the US Bureau of Mines. It comprises a vertical glass tube about 5 cm in diameter and 1.5 m long, with two spark electrodes inserted near to the bottom. In use, the end of the tube is closed and then the whole apparatus is evacuated. Fuel vapour, agent and air are then admitted to form a mixture at atmospheric pressure, and the gases are circulated for some time by a pump to ensure thorough mixing. The closure is then removed from the bottom of the tube and a spark is passed between the electrodes. If a flame is observed to travel

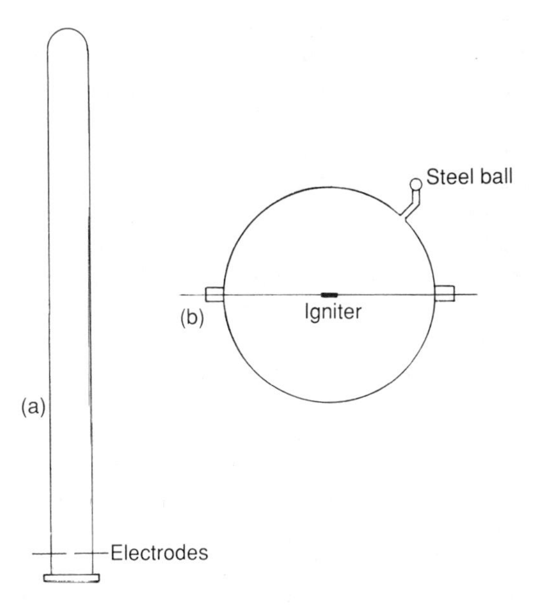

Figure 6.4 Apparatus for measuring inerting concentrations: (a) Bureau of Mines flame tube; (b) 6-litre vessel.

the length of the tube, the mixture is reported as being flammable. If the flame stops half-way up the tube, the mixture is outside the limits of the flammability envelope (figure 6.1). It may seem contradictory that a mixture through which a flame has propagated for a short distance can be non-flammable, but it was the intention of this test to determine the ability of a particular mixture to propagate a flame, not the ability of an ignition source to initiate some combustion. The partial propagation of the flame resulted from an excess of energy in the spark above the minimum energy needed for ignition. However, a large ignition source may occur in practice, and it may therefore be necesary to consider the consequences of whatever combustion it is able to support in a near-limit mixture. For this reason, the measurements of peak concentration obtained by this method may not be suitable for the calculation of total flooding design concentrations. The method also has the disadvantage that it is difficult to obtain reproducible results. This is due mainly to the need to remove the bottom closure before ignition. Many fuel/air/agent mixtures have a density higher than that of air, and will therefore begin to drop out of the tube, and to be replaced by air. The mixture present at the electrodes when the spark is passed may not then be the same as the original mixture.

When an explosion happens in a process vessel or within a building, damage is caused by the increase in pressure. It is reasonable therefore to relate the critical concentration of an agent to an acceptable pressure rise. This increase in pressure could result from a partial flame propagation initiated by an energetic source. In other words, the flammability envelope is so drawn that it encloses all the mixtures in which a pressure rise above an agreed value has occured following the discharge of an ignition source of known value. The apparatus on the right of Figure 6.4 is used for these measurements. It is a spherical vessel formed from two stainless steel hemispheres, and has a capacity of about six litres. A pressure transducer can be used to detect the pressure rise, but a very simple indicator is shown in the figure, which comprises a steel ball resting on a small hole. A number of tests were conducted in which the ball became dislodged from its seating at 22 m bar, and in which the ignition source was a surface discharge on a carbon rod which released about 12 J (several thousand times the minimum ignition energy at the conventional limit). Peak concentrations were recorded for a number of fuels which were well in excess of those found with the Bureau of Mines apparatus. For example, the peak concentration for Halon 1211 against n-hexane/air was 5.8 per cent in the flame tube and 8.01 per cent in the spherical vessel. These values may also be compared with the flame-extinguishing concentration shown in Table 6.1, which is 3.7 per cent. There is unfortunately no constant relationship for the two measurements over a range of fuels, but it is usual to find that the inerting concentration (by the spherical vessel method) is at least twice the flame-extinguishing concentration. The values obtained for inerting concentration by this method are also close to the critical concentrations for flames burning in airstreams mentioned in Chapter 2 (obtained in the apparatus shown in Figure 2.3).

Measurements of flame-extinguishing and inerting concentrations are important in providing design data for total flooding systems. They also provide an accurate comparison of the effectiveness of the various agents against a particular fuel. However the discussion above on the Halons showed that these tests measure the maximum potential of a completely vaporized extinguishant. In practice where the agent is being applied to a fire manually, its physical properties are equally important. This will now be discussed.

Fire trials

A practical fire test is often the only convincing way of comparing the effectiveness of extinguishing agents, or of different extinguisher designs. For example, if a works fire brigade is offered a new powder extinguishant, a sample will be filled into two or three extinguishers and discharged against a standard fire (probably the fire which is used for training the works employees). A comparison may be made by measuring the time taken to extinguish the fire and checking that against the performance of their usual powder, or they may rely on a subjective assessment. It is unlikely however that an accurate measurement of the relative effectiveness of the two agents will be obtained. To do this it is necessary to make a number of tests with each of the agents, and to plot the results.

Suppose we arrange a standard fire, perhaps a large tray of flammable liquid, and attempt to extinguish a series of fires by powder applied at different rates. This can be achieved by maintaining the extinguisher at a constant pressure and by having a number of nozzles of different sizes, which can be screwed into the end of the discharge hose. The actual rate can be determined by noting the loss of weight for a given discharge time. If the time taken to extinguish the fire is plotted against the powder application rate, a curve like the one in Figure 6.5 will be obtained. The line encloses the shaded area in which extinctions were obtained. There is a critical application rate below which the fire cannot be extinguished (the 'optimum' and 'preferred' rates are exlained later). There is also a minimum time to extinguish, which results from the need to sweep the powder from side to side, progressing from front to back, over the surface of the fuel. With an experienced firefighter this time can be quite reproducible.

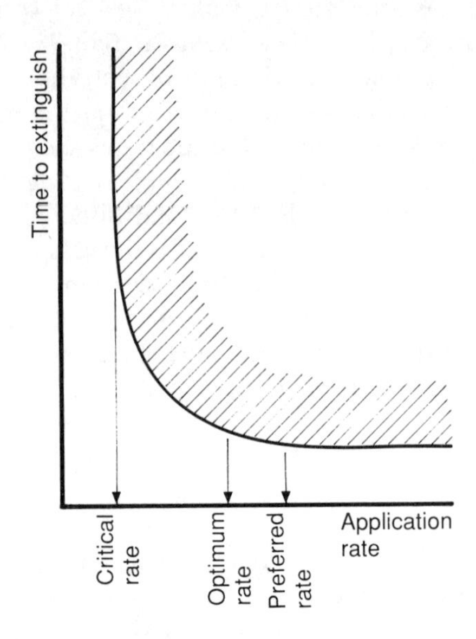

Figure 6.5 Time to extinguish a fire plotted against agent application rate.

The data used in plotting the curve in Figure 6.5 can be presented in another way. From a knowledge of the times needed to extinguish the fire at different application rates we can calculate the quantity of powder that was used. We can then plot a quantity/rate curve, like the ones in Figure 6.6 (where *A* corresponds to Figure 6.5). It is now possible to determine not only the critical application rate but also (as for curve *A*) an optimum rate at which the minimum quantity of powder is used. If curves *A* and *B* represent the results obtained for different powders, then they show that powder *A* is about twice as effective as powder *B*. If now we return to the works fire brigade which is checking the powder, we can see why they may fail to recognize the magnitude of this advantage. A good fireman will always overkill a fire: he will apply an extinguishing agent at a sufficient rate and in a sufficient quantity to make quite certain that the fire is put out. He will probably therefore be applying the powders at the rate marked *X* on the figure, and here he may see only a 20 per cent advantage of *A* over *B*. Indeed he may not be able to detect any difference at all.

The only way to make a reliable comparison between two agents is to conduct a sufficient number of tests against a standard fire to enable the plotting of at least the optimum parts of the Q/R curves. The curve for the chosen agent can then sometimes be used to select a 'preferred' application

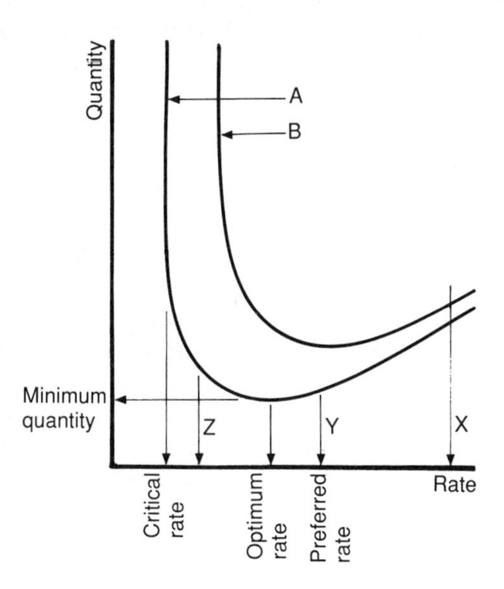

Figure 6.6 Quantity/rate curve.

rate. This could for example by *Y* for powder *A*, which is perhaps three or four times the critical rate. The optimum rate would of course be more economical in the use of powder, but it is rather too close to the critical rate to be used safely for practical firefighting. Also, the preferred rate will give a shorter extinguishing time (see Figure 6.5).

Q/R plots can usefully be drawn for any firefighting agent. In the UK Home Office's *Manual of Firemanship* for example there is a Q/R curve for fluoroprotein foam against a medium size test fire. This shows a critical rate of about 1.5 $l/m^2/min$, and a preferred rate is indicated at between 4 and 6 $l/m^2/min$.

A set of Q/R curves should provide a prospective user with all the information he will need in comparing the performance of a new extinguishing agent with that of the existing agents, Despite this, the manufacturer of the new product usually finds it necessary to provide a practical demonstration. He will usually choose to do this with a number of different fires, each of which is burned twice. In each test the new product and its nearest competitor are discharged at the same rate, and the fire is extinguished with the first but not the second. To do this the manufacturer must select a rate corresponding to *Z* on Figure 6.6, which is close to his optimum rate but below his competitor's critical rate. Unfortunately it is well away from the preferred rate for his product, and quite close to its critical rate. Problems can then arise because of the inevitable variation in the test conditions, due to changes in wind velocity and direction, extinguisher behaviour, and firefighting technique. We can plot the probability that the fire will be extinguished against the application rate. This has been done for the two powders in Figure 6.7, where the probability increases from 0 = it never happens to 1 = it always happens. In this particular plot thc chosen application rate will give a probability of 0.95 that powder *A* will succeed, but with a probability of 0.05 for *B*. If the manufacturer conducts a large number of demonstrations he can expect that on average powder *A* will put out 95 per cent of the fires, but powder *B* will succeed against 5 per cent. Unfortunately in one in 400 tests, *B* will put out a fire and *A* will not. He will regard these failures as a manifestation of Murphy's Law, which indeed it is (if it is possible for something to go wrong, sooner or later it will go wrong). More importantly, this situation demonstrates the significance both of a Q/R plot and of the preferred rate which is derived from it.

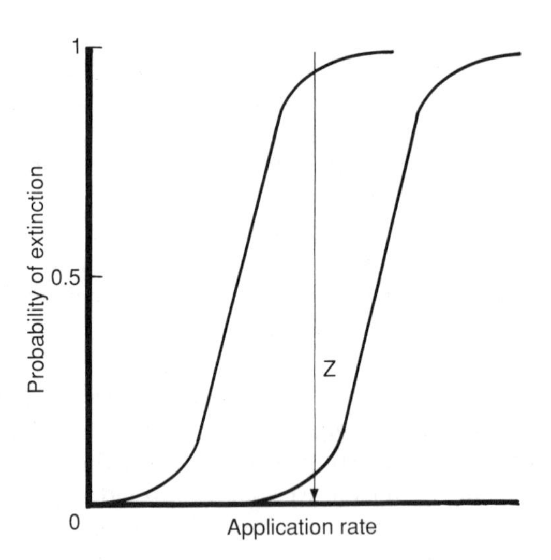

Figure 6.7 Probability of extinguishing a fire plotted against agent application rate.

FURTHER READING, Part One

Chapters 1 to 6 contain a brief and simplified outline of combustion technology. A certain amount of precision has inevitably been lost, and some of the books listed below should be studied for more rigorous explanations of the various phenomena. They usualy employ a mathematical approach, which has been largely avoided in this book. The first two are generally accepted as providing excellent coverage of the basic concepts. The one by Drysdale is concerned particularly with combustion processes in fires.

Lewis, B. and Von Elbe, G., *Combustion, flames and explosions of gases.* Academic Press Inc., New York.

Gaydon, A. G. and Wolfhard, H. G., *Flames, their structure, radiation and temperature.* Chapman and Hall, London.

Drysdale, D. D., *An introduction to fire dynamics.* John Wiley & Sons, Chichester.

Wharry, D. M. and Hirst, R., *Fire technology: chemistry and combustion.* IFE, Leicester.

Spalding, B., *Combustion and mass transfer.* Pergamon Press, Oxford.

Harris, R. J., *Gas explosions in buildings and heating plants.* (British Gas) Spon, London.

Palmer, K. N., *Dust explosions and fires.* Chapman and Hall, London.

Field, P., *Handbook of powder technology: dust explosions.* Elsevier, Amsterdam.

Fordham, S., *High explosives and propellants.* Pergamon Press, Oxford.

Davis, T. L., *The chemistry of powder and explosives.* John Wiley & Sons, London.

Tuve, R. L., *Principles of fire protection chemistry.* NFPA, Boston.

PART TWO

SAFETY OF LIFE

7

Escape from fire

Fire-damaged buildings and plants can be replaced but people cannot. This is why much the legislation associated with fire is concerned with protecting people against its effects. The main concern is that people should be able to escape safely from a building in which a fire has started. If this is to be achieved:

- fire detection must be rapid and reliable;
- warning of the fire must be given immediately;
- the warning must be understood;
- the people must know how to escape;
- the escape route must lead to the open air;
- the escape route must be unaffected by the fire.

This latter requirement cannot be absolute: if the fire becomes large, all the escape routes will eventually be involved. What is important is that the time taken for the fire to penetrate an escape route must be greatly in excess of the total time required for detection, alarm and escape. Indeed from this point of view the main purpose of passive fire protection is simply to buy time so that safe escape is possible. This applies particularly when a large number of people are involved, or where escape is not simply a matter of walking away (in a hospital for example). The total time needed for escape can be controlled by limiting the number of people in a building, as is done in theatres and similar places.

Legislation and standards covering escape routes are based on practical observations which have been made of the rate at which people can pass along corridors, up and down stairs, and through openings. However, the success of an evacuation must depend also on the way in which people respond to a fire warning, to written and spoken instructions, and possibly to the presence of smoke and fire. Very little research has been conducted in the past on human behaviour in fires, and although much information is now becoming available, its application is disappointingly slow. Indeed the only allowance made by the legislators for the probable behaviour of the people they are trying to protect is to assume that they will panic. One important finding of the recent research is that very little evidence exists anywhere to support this commonly held assumption.

The popularly held concept of panic is that in the presence of smoke or fire, or simply in response to a shout of 'fire', people will act illogically and irrationally in their attempts to escape. Their behaviour will become violent, non-human and impossible to control. This belief is illustrated by phrases like 'Under such conditions people do not behave like thinking human beings' and 'panic overcomes man's capacity for rational thinking'. It is thought that panic is particularly likely to affect a crowd. This and the supposed animalistic behaviour are described as 'A pack of animals obsessed by a frenzied desire to escape they know not what'. This quotation is from a journal article which has the title 'Adequate exits – antidote for panic', suggesting that panic is a kind of contagious disease.

Multiple deaths will sometimes occur because it is not possible for all the occupants of a stricken building to escape. The press will almost invariably attribute the deaths to 'panic', and this can be grossly misleading. One hundred and sixty-four people died in a fire at the Beverly Hills Supper Club, Kentucky, USA, in May 1977. The headlines in three UK papers were 'A Killer called Panic' (*Daily Express*), 'Panic and 300 stampede to death' (*Daily Mail*), 'Panic Kills 300' (*The Sun*) (note also the exaggerated numbers). When the NFPA investigated the fire they interviewed 630 people and received completed questionnaires from a further 1117. One of their conclusions was 'Evacuation . . . was calm and orderly without the disorderly behaviour often termed "panic" until thick black smoke entered the areas where occupants were exiting. Panic is not considered a major contributing factor to the large loss of life'. They added that 'such behaviour probably did occur when people knew that they could not escape' but there is no conclusive evidence to support this statement. What is quite clear is that no panic at all existed while between 2400 and 2800 people were escaping from the building, although many of them had seen flames or smoke. People die in situations like this because the alarm is delayed, or the fire spreads too quickly, or there are too many people, or the exits are blocked, or the building is badly designed, or the staff has not been trained, or some combination of these and other causes. They do not die because they panic. At the Beverly Hills Supper Club there were about 1350 people in the room where the fatalities occurred although the maximum safe occupancy was 536. These people were not told about the fire until 20 minutes after it had started, and only a few minutes before smoke entered the room.

'Panic' makes a good headline, and the story describing a tragedy due to 'panic' may be convincing because of the frequent repetitions of this fantasy by the media. This has the unfortunate effect of deflecting our attention from the truth. Panic, and indeed the victims themselves, are being used as scapegoats, forced to accept the blame for the tragedy. Four hundred and eighty-eight people died in the Coconut Grove Night Club fire (USA, 1942), due mainly to overcrowding, poor exits, and very flammable decorations. One report, which appeared after the fire had been investigated, claimed that the owners and some local officials had been incorrectly blamed, because panic was largely responsible for the deaths. Again this was supposition, unsupported by any evidence.

The behaviour of people leaving a burning building is often orderly and considerate, sometimes heroic. If there is an urgent need for rapid escape, because of advancing smoke or flames, people will begin to push and may even trample over others who have fallen. They may all try to escape from the same exit, although others are available. To an outside observer this may all be evidence of panic. However there is nothing irrational about the anti-social behaviour: this is simply an attempt to find the quickest way out. There is nothing illogical about trying to get out by the same route that you used coming in, particularly if you are not familiar with other exits and are not convinced that they lead to safety. The need to escape from imminent death by burning will drive people to quite desperate actions, like jumping out of a window, but this is not a blind animalistic response. It is often the only rational thing to do in the circumstances and with the knowledge and information then available to the victim.

A fire engineer should not need to plan fire precautions on the assumption that people will panic. He should not hesitate to arrange for messages over a public address system which include the word 'fire'. He should certainly not include in a list of actions to be taken in case of fire the instruction 'Do not panic'.

SMOKE: THE KILLER

The composition and toxicology of smoke are discussed in Chapter 4 where it is shown that the great majority of deaths in fires are caused by smoke.

Death results either directly from the inhalation of toxic gases (and from lack of oxygen) or indirectly because escape is prevented by loss of vision The reason for providing secure and smoke-free escape routes is obvious.

Death is preceded by a period of unconsciousness, and before this there is a time during which the victim loses coordination and mental acuity. Burns do not normally occur before unconsciousness, unless the person has had to run through flames to escape.

The toxic gases likely to be present in the smoke from a fire include carbon monoxide, carbon dioxide, hydrogen chloride, hydrogen cyanide, acrolein, sulphur dioxide, and ammonia. All these, and the solid and liquid particles in the smoke, can be filtered out by a simple absorbent respirator. Even a wet handkerchief will remove some of these products but it provides no protection against carbon monoxide. Neither the handkerchief nor the respirator can do anything to mitigate the effects of oxygen deficiency (shown in the upper half of Table 6.2). Only self-contained breathing apparatus can provide adequate protection against the toxic effects of smoke, and will normally be worn by firemen before they enter a smoke-logged building. The apparatus comprises a leak-proof face mask which is provided with a visor and sometimes a microphone. Air is delivered as required by the wearer through a special demand valve, from a cylinder of compressed air. Earlier equipment supplied oxygen instead of air but this is rarely used now.

Although self-contained breathing apparatus is the only safeguard against poisonous gases and lack of oxygen during a fire, it must be emphasized that there are very considerable dangers from the use of such apparatus. Wearers must be carefully trained under professional instructors with the full facilities of specially designed buildings. Not only must the training be very stringent but regular drill with the apparatus must be undertaken. The use of self-contained breathing apparatus in fire must be strictly controlled from a nearby base which has full information of the number of men using the apparatus, the time they have worn it, the state of the charge of the cylinders and other data which are kept constantly under review during action. Men wearing breathing apparatus should never enter a smoke-logged building alone. By the use of breathing apparatus firemen can penetrate further into a burning building than without such aid and consequently can run into danger without realizing what is happening. There is a psychological reaction to being in complete darkness in smoke and with all sorts of unusual noises. Before being allowed to wear breathing apparatus at a fire the wearer should have had considerable experience of firefighting.

EVACUATION SIGNALS

This section contains a general discussion of fire warning systems. Their use is covered by the BS Code of Practice for Fire Detection and Alarm Systems (BS 5839: Part 1, 1988). This is discussed in detail in Chapter 18.

The first consideration for the safety of life from fire is that once a dangerous situation is known to have arisen, it must be possible to communicate that information to other occupants of the premises. This is to warn them that they are in danger and that arrangements, which have been made previously to cope with such a situation, are to be put into effect. In this connection the expressions 'fire bell' and 'fire alarm' are not very definite and it is better to refer to 'sounders', and for that part of the installation which is concerned with initiating the alarm as the 'fire warning system'.

Whatever means are provided, the person initiating the alarm by operating the system must not be placed in jeopardy by delaying his own escape; neither must there be any doubt as to how the alarm is to be operated. To avoid confusion there should be only one method of giving warning in case of fire. The evacuation signal must be distinctive and it must not be possible for it to be confused with any other sounder or signal. It must be familiar to anyone who could be in danger.

Fire triangles and hand-operated gongs

A triangle or bar of steel when beaten with a metallic bar will emit a loud and very distinctive sound. It is possible that these instruments might be used as fire warning signals on caravan sites, etc., but they should not be used indoors because someone has to stand and beat the bar on the triangle. The sound is of very limited range and further extension of the range would mean the operation of a second signal.

The hand-operated gong is a development of the triangle, in which a handle is turned causing a hammer to beat on a heavy gong. While this might again be suitable for outdoor use, it is still unsatisfactory indoors because it means that someone must stay by the machine to turn the handle. This person may then be in jeopardy or may not wait long enough for the signal to be heard and appreciated by everyone concerned. Modern developments of self-contained electrical alarm systems are probably almost as cheap and far more effective.

Internal telephones

Internal telephones are often suggested for use for initiating an alarm of fire and devices similar to the 999 dialling system are incorporated which will enable the local operator to give the call priority. The system is not satisfactory for a number of reasons. An excited person might forget the code, the telephone operator might be busy or might not notice the signal, or the message might be misunderstood, especially if the person giving the alarm was excited. In any case whoever initiated the call has to wait until the message has been received, understood and acknowledged. However, an internal telephone is often a convenient method of confirming the presence of a fire following an automatic detection. It may also be acceptable as a means of raising the alarm in for example a large chemical works. Control room staff can be trained to pass the message to the works fire brigade, and this has the advantage that the brigade can be advised of the size and nature of the fire. It will then be normal practice for the works brigade to pass the message by telephone to the local authority brigade.

Public address systems

These can be of particular value in places like shopping malls where people are unlikely to understand the meaning of signals by bells or sirens. These messages may only be partly heard at first but the warning can be reinforced by staff, who will already have responded to this or other alarms. Once people have realized the situation they will listen carefully and are likely to respond in an ordery manner to a calm and reassuring message, which should be repeated frequently. Public address systems are also useful in hotels where, although people are much more likely to respond to an alarm, it is very helpful to provide them with information. (A very common response of people to a fire alarm is to try to find out if there really is a fire.) In large shops the system may be used first to give a preliminary warning to staff that an evacuation may be necessary. A coded message is used, for example, 'Mr Redhead to the office please'. If the alarm is sounded after that the staff will be in no doubt that they must immediately follow their instructions in guiding customers out of the building. Again they can be assisted in assuring a safe evacuation by appropriate messages through the system.

SIMPLE ELECTRICAL WARNING SYSTEMS

The simplest warning system comprises a number of call points, some sounders, a power supply, and a suitable circuit. People are very good fire detectors and will often be aware of a fire long before it is sensed by a detector. Call points are therefore an essential means by which the human detector can feed information into a warning system. If the system connects both call points and detectors to an automatic control panel, then because of the frequency of false alarms from the detectors it is usual to arrange that procedures for shut-down and automatic extinguishing are not initiated until two detectors have responded. On the other hand, the actuation of a single call point is accepted as a true warning. Thus, in whatever type of circuit call points are used, their integrity and reliability is of considerable importance. (Detector systems are discussed in Chapter 18.)

Manual fire alarm points

Manual fire alarm call points are usually of the break-glass pattern and the following factors should be borne in mind when looking at a piece of well designed apparatus.

1 No live parts should be exposed in such a manner that they can be touched by a person giving an alarm.
2 The operating button should be spring-loaded and held in the 'off' position by a glass cover in such a manner that the breaking of the glass releases the button and initiates the alarm. In exceptional circumstances manual call points might be required for key operation.
3 It must not be necessary for the person giving the alarm to have to stand and operate the device. That is, once started, it must give a constant signal until another action is initiated to stop the signal. Generally this is by replacement of the glass. But if the cell point has been key-operated it should require a special tool (not the key) to stop the signal.
4 The cover could with advantage be opaque: it is less likely then that people will think that it is necessary to stand and press the button. With a simple circuit this could result in suppressing the call, but a control panel could contain a latching mechanism which would hold the initial alarm.
5 The instructions for use, for example 'Fire alarm – break glass', should be clearly indicated by a concise instruction on the case. The marking should be in such a manner that there is no interference with the breaking of the glass. For instance, a piece of paper stuck on the glass would not be acceptable. It is an advantage if the glass is scored to give a clean break, although if it is too deeply scored the glass might be so weakened as to result in cracking and the possibility of false alarm signals being given.
6 There must be means for testing the action of the alarm. This varies with different manufacturers.
7 The alarm point should be coloured red and it is advisable to number each point for easy reference. The number can be marked on the box

after it has been installed. With numbered boxes it is possible to keep in store an 'out of order' plate for each call point. In the event of a particular call point being out of order, the glass can be replaced by the metal plate marked with the nearest alternative methods of initiating an alarm. If any such call point is known to be out of order, a suitable reminder system should be instituted to draw the attention of the executive reponsible for fire precautions to the fact. This could consist of a large notice to stand on his desk, a notice to hang on the back of his office door or below the clock, or any system that will constantly remind him that a serious and dangerous condition exists.

8 A well designed manual call point will be adaptable for either open or closed circuit working.

9 Facilities may be available within the call point to enable a telephone to be plugged in. This can be very useful when large numbers of alarm points have to be tested. But, if provided, the operation of any call point must override the telephone circuit and be able to transmit an alarm without interference.

10 Manual call point boxes are available in several forms: a surface fitting type suitable for factory or general use, a flush fitting type with a high class finish for offices and similar situations, a weather-proof type for outdoor use, a key-operated type for special situations and flame-proof types for use in flammable and explosive atmospheres.

Call points similar to those shown in Figure 7.1 are quite satisfactory for most situations both indoors and out. There are however some cases where

Figure 7.1 The KAC manual call point.
(KAC Alarm Company)

a more robust design may be necessary. For example the actuation of a call point on an offshore platform can result in an expensive shut-down of the whole platform. This could be caused by mechanical damage, perhaps on the drilling floor where a lot of heavy equipment is frequently moved about. In situations like this it may be necessary to consider a call point which requires two positive actions, like opening a door and then turning a switch.

Methods of testing

The method of testing the Gent manual call point is by unscrewing the front plate. If the call point is connected into a closed circuit system or an open circuit monitored system the alarm signal will be given immediately. If it is connected into an open circuit system the alarm will not be transmitted until the test button is pressed. The test button is not apparent until after the front plate has been opened.

It should be noted, however, that if the glass is broken an alarm is transmitted immediately, because the test button is held in the 'on' position by the closed front plate.

The method of testing the Simplex manual call point is by the insertion of a special test probe into a small aperture at the front of the box. When the probe is pressed the whole switch is pushed away from the glass cover allowing the operating button to move forward. The operating button controls a change-over switch which reverses the contacts. Because of this reversal of the contacts an alarm is transmitted immediately the switch moves, whether the switch is in an open circuit system or a closed circuit system.

Recently a new type of manual call point has been introduced and is known as the KAC type (see Figure 7.1). In this type, instead of the operating button pressing on the centre of the glass cover, the pressure is exerted on the edge. This means that the pressure is taken on the strongest plane of the glass which is therefore less likely to break from excessive pressure, and false alarm signals are eliminated from this cause. The glass can also be thinner and is then easier to break from a direct blow on the surface. Because the operating button is resting on the edge of the glass it is sufficient if the glass breaks inwards. Thus a thin film of plastic material can be attached to the inside of the glass which will prevent fragments of glass from falling outwards, so removing any danger of cuts to the operator from splinters of glass. It only requires firm pressure of the tumb on the glass to break it and cause the point to operate (see Figure 7.1a).

The operating button controls a simple change-over switch so the call box can be used in either an open or closed circuit system.

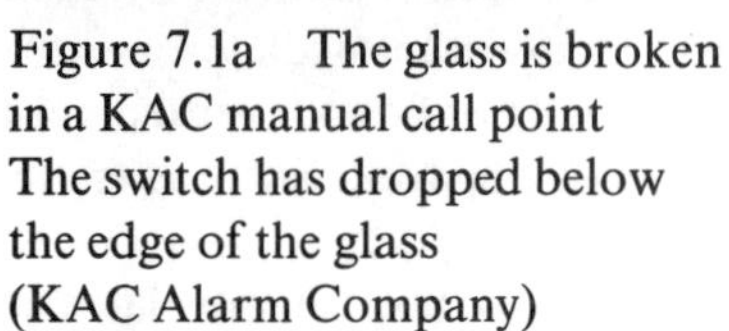

Figure 7.1a The glass is broken in a KAC manual call point The switch has dropped below the edge of the glass (KAC Alarm Company)

Figure 7.1b The test probe in position in a KAC manual call point When pressed the glass moves and allows the switch to operate (KAC Alarm Company)

Testing is carried out by inserting a special probe into the side of the box (see Figure 7.1b). The insertion of the probe causes the glass to tilt allowing the operating button to move down and causing the circuit to be changed to the alarm condition.

Positioning of call points

Manual call points must be provided in all parts of the building. They should be fixed at a height of 1.5 m from floor level, in an easily accessible, well illuminated and very conspicuous position. If there is any possibility of damage because they project from the wall, then a flush type box should be used. It should be part of the daily inspection of the premises to see that no call points are obstructed in any way.

When siting call points they should be fixed on exit routes leading to safety. There should be a point on every landing at the top of the stairs leading downwards and there should be one near every final exit. (A final exit is one that opens out to safety and is usually at street level.) There should also be one near the telephone switchboard within reach of the telephonist without moving from her position at the switchboard. After having set out these positions on a plan, additional call points must be included so that no person can travel more than 30 m forwards towards a final exit without passing a call point. It must never be necessary for the operator to go away from the route to safety so as to operate an alarm signal.

The method of raising an alarm and the method of operating all call points in a building should be the same.

Call points are again discussed in Chapter 18 in the section dealing with the BS code of practice on fire detection and alarm systems.

SOUNDERS

Sounders is the term used for any method of giving warning in case of fire even though no sound is produced.

When comparing the various types of machines that produce noise, the loudness is determined in decibel units abbreviated to dB. However, when comparing decibel readings there can be considerable variation unless the conditions under which the readings were taken are known.

For example, readings taken in the open vary from those taken indoors. Table 7.1 shows some typical noises and their rating.

It must not be possible for the fire warning given by sounders to be confused with any other form of warning or for any other purpose. This applies especially to the use of bells. If it is intended to use bells as sounders a careful investigation must be carried out to ensure that bells are not used, nor likely to be used, for any other purpose. Today it is better to avoid the use of bells for giving warning of fire conditions.

Whichever type of fire sounder is selected, all fire warning sounders should be the same throughout the building, unless it is necessary to use special types for exceptionally noisy places or where special conditions prevail such as in the intensive care unit or the operating theatre of a hospital. The essential feature is that the sound must be loud enough to attract attention and to waken a sleeper if necessary. Everyone in the building must be able to hear it and they must be able to recognize the sound and identify its meaning. On the other hand, sounders that are too loud or too strident can have an adverse effect. It is generally recognized that a sound of 75 decibels is required at the bed-head to awaken a normal sleeper. The human ear varies immensely with different individuals so that it is very difficult to accept any judgement of what is sufficient sound for fire warning purposes. A simple sound meter such as is shown in Figure 7.2 can be used as a guide to the volume of sound at a particular location. If the sounders are then turned on the noise level should rise by between five and ten decibels. It is important to realize that this can only be a general guide as different people hear at different sound levels.

Although some guidance can be obtained from decibel figures provided by manufacturers, the only test of a satisfactory production of sound must be made under working conditions. It is for this reason that although fire

Table 7.1
Some typical noises and their decibel ratings

Noise	Decibels	Relative energy	Sound pressure pascal	Typical examples
	120	1000000000000	20	
Deafening	110	100000000000		Jet aircraft at 150 m Inside boiler-making factory 'Pop' music group Motor horn at 5 m
	100	10000000000	2	
Very loud	90	1000000000		Inside tube train Busy street Workshop Small car at 7.5 m
	80	100000000	0.2	
Loud	70	10000000		Noisy office Inside small car Large shop Radio set – full volume
	60	1000000	0.02	
Moderate	50	100000		Normal conversation at 1 m Urban house Quiet office Rural house
	40	10000	0.002	
Faint	30	1000		Public library Quiet conversation Rustle of paper Whisper
	20	100	0.0002	
Very faint	10	10		Quiet church Still night in the country Sound-proof room Threshold of hearing
	0	1	0.00002	

Figure 7.2 A simple sound meter.
(Dawe Instruments)

authorities will indicate where they require a call point to be situated they will not specify the type or position of sounders.

When there is insufficient volume of sound to overcome the background noise it is better to increase the number of sounders rather than to fit louder units. In the present day when ear muffs are worn to enable people to work comfortably in noisy situations, it might be possible to fit hearing aids in the muffs. In the case of men working within steel-lined cabinets such as for shot or sand blasting, small loud speakers have been fitted in the helmets. In one factory the employees concerned appreciated the reception of radio when working under the claustrophic conditions.

Sounders should always be of a type specially made for fire warning purposes and in particular should not be fitted with devices to prevent interference with radio and television. The reason for this is that the noise from entertainment machines might drown the sound of a warning, whereas if the sounder does not have a suppression device, the very interference could give a warning. However, not all sounders will interfere with entertainment machines and in these case or where there is any doubt that the warning might not be heard against a background noise, a relay should be inserted into the power lead to the loud speakers so that they will be shut off in the event of a fire warning being given.

In some cases it might be possible to adopt a similar expedient and put a relay into the power circuit of noisy machinery so that on the giving of a fire signal the power would be cut off. In any case if evacuation is ordered the machinery should be shut down.

In a factory where the occupants are well trained and disciplined in evacuation procedure, it is sometimes necessary to give advance warning that an evacuation signal may be expected in order that plant may be shut down progressively. The first warning should be given by an intermittent signal and the continuous signal should then be used for an evacuation signal. Although this is the recommended procedure, many people are of the opinion that the signals should be reversed as it has been found that a broken signal especially if the breaks gradually increase in speed have a psychological effect of making people hurry.

If any advance warning systems are introduced it is most important to ensure that a 'stand down' signal is also available in case the full evacuation signal is not required.

Sounders can be considered under five headings:

- Bells
- Sirens
- Electronic devices
- Reed vibrators
- Coloured lights

Bells

Bells can be obtained that will operate from 6 to 50 VDC and from various AC voltages up to 240 volts. Similarly there are variations in the size of the gong from 100–250 mm or more.

The standard distance for comparisons of noise inside buildings is usually made at 3 m from the source of the noise. At this distance

- a 100mm bell will produce about 80 dB
- a 150mm bell will produce about 85 dB
- a 200mm bell will produce about 90 dB
- a 250mm bell will produce about 90–95 dB

When fixing bells the hammer should always point downwards.

Sirens

The sound produced by a siren results from the pulsation of air as it passes through revolving vanes. The pitch of the sound will depend on the speed at which the vanes are rotated. The volume depends on the size of the siren, which must be matched to the power consumption of the motor by which it is driven. Although sirens are available from 6 VDC up to 24 VAC the low voltage models have a very high current consumption. As with many electric motors, sirens require a very high current when starting. By allowing the current supply to rise and fall the speed of the motor will be affected giving rise to the now familiar wailing noise. Unfortunately this unmistakable sound has been used for many other purposes and has now lost much of its unique value as a fire warning. Some sirens especially in the 240 VAC category can reach as high as 110 dB at 3 m.

Electronic devices

Electronic devices consist of solid state circuitry so designed that a distinctive sound is produced. The sound is very penetrating and especially suitable for use in places where there is a high ambient noise level.

By an arrangement of the internal circuit a second frequency can be selected which will vary the pitch. Another alternative arrangement can be made to produce a pulsed tone which varies between the two frequencies. This is a very useful function where it is required to given an 'alert' signal before the full evacuation signal.

The decibel reading at 3 m with a suitable 24 VDC model can exceed 100 dB but the power consumed is very high.

Reed vibrators

Reed vibrators are similar in design to many motor-car horns and produce a distinctive raucous sound. It is particlarly easy to adjust the volume of the sound, a facility that is not so easy to apply to other sounders.

When used with a 240 VAC mains supply the dB reading can be as high as 100 dB at 3 m.

Coloured lights

The use of coloured lights as fire warning signals should be restricted to special situations. In a noisy working area a flashing red or amber light, particularly a strobe light, is a useful back-up to the normal fire alarm. Warning lights may be used on their own in hospitals, especially in operating theatres and intensive care units. On some off shore platforms 'status lights' are displayed throughout the occupied areas. Four lights may be used to convey information of this kind: green = normal operation; amber = fire alert and partial shut-down; red = fire alarm and full shutdown; blue = platform evacuation. The same information as that indicated by the amber, red and blue lights will also be given by sounders and by the public address system, but the lights provide a back-up in noisy areas where ear protection is worn, or when the circuitry of the other systems has been damaged. The green light is reassuring to people who may for example be working in the open in a high wind.

Lighting by master switch

An excellent method of waking people is by switching on the room light and in the event of an emergency in an hotel or boarding house it has the additional advantage of assisting occupants to vacate their rooms.

Figure 7.3 shows the circuit wiring to enable a series of lights to be switched on by one master switch. When the master switch is operated all lights on the circuit that are not already switched on will be switched on and no light on the circuit can be switched off until the master switch has been opened. Two-way switches are required for each individual light but where two-way switching is already in use, intermediate switches can still be connected to the master switch (see Figure 7.3).

The master switch can be in the form of a contactor operated by a relay connected to the evacuation warning signals.

To enable deaf people to become aware of evacuation signals during the day the use of strobe light is recommended. During the night use can be

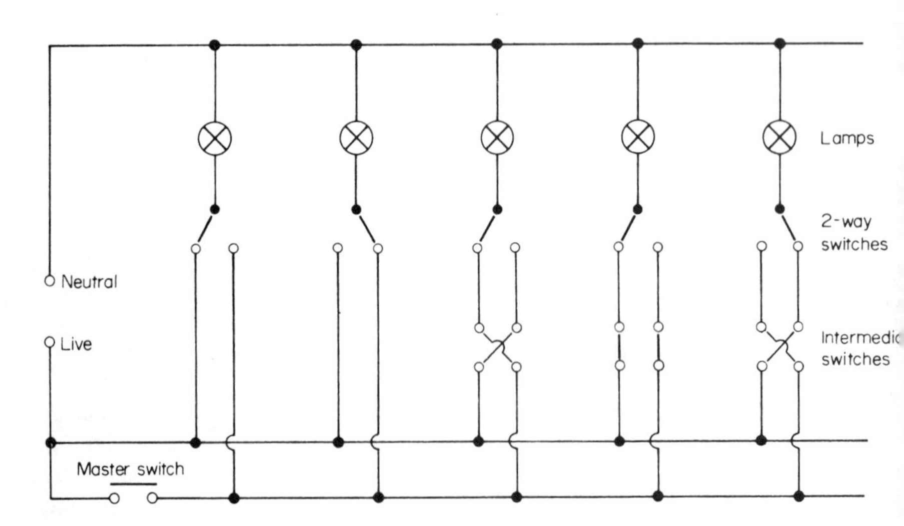

Figure 7.3 Lighting circuits controlled by a master switch.

made of a special vibrator placed underneath the mattress of the bed. The vibrator is rated at 24 VDC and can be interconnected with a 24 volt alarm system. To increase the efficiency of the vibrator it can be pulsed by a circuit interrupter. At the same time it is advisable also to install lights bearing the word '*Fire*' on them.

POWER SUPPLIES

The weakest part of a fire warning system is probably the power supply. It is laid down in British Standard 5839 that the power supply for a fire alarm system shall be exclusive to that system, there being a similar rule for intruder alarm systems. However, there is a growing opinion that it should be permissible for these two systems to have a common power supply. This is especially when dealing with domestic alarm systems. In a small private house where the only warning required from either system is to alert the occupants to danger from either source and where the number of false alarms is immaterial, it does not seem reasonable to go to the expense of duplicating sounders, power supplies and to some extent wiring.

The power supply for fire alarm purposes can consist of the following:

1. Mains with a secondary battery kept in a fully charged condition from the mains supply.
2. Mains supply with a primary battery as stand-by.
3. Mains only.
4. Primary battery only.

Mains with a secondary battery

An alarm system which contains both automatic as well as manual call points should be powered with a secondary stand-by battery as well as an electric mains supply, although an exception can be made in the case of domestic fire alarms. The system should operate at 24 VDC from a rectified mains source. The battery itself should also be 24 volts.

The battery must be capable of supplying the entire load of the system in its quiescent state for a period of 72 hours after the disconnection of the mains supply and if it is used to operate sounders, it must then be able to maintain these in operation for 30 minutes. Warning must be given of a failure of the mains supply and a failure of the charging equipment.

The power supply must be such that the system can still function normally with one of these events:

- the mains disconnected or

- the battery disconnected or
- with the battery connected in a discharged condition.

The charging equipment must be capable of recharging a completely discharged battery within eight hours, at the same time maintaining the total output required including that of sounders if any are connected. Battery chargers must be well designed to ensure that the battery is kept fully charged but not overcharged. Overcharging will quickly ruin the best of batteries.

It is permissible to separate the sounders from the main alarm system bringing them into operation by means of an ancillary device from the main alarm system.

Mains supply with a primary battery as a stand-by

If the power supply is required for a manual alarm system in a small building, it may consist of a step-down transformer with a dry battery used as a standby. Change-over to battery must be automatic on failure of the mains supply, and a faulty warning must also be given when the mains fails. If the system is a closed circuit or an open circuit monitored system, the battery must be capable of maintaining the standing load for 72 hours, at the end of which time it should have sufficient power to operate the sounders for half an hour. A permanent warning that the battery has been used should be given in case the system is left unattended for more than 72 hours in which case the system might have operated and the battery be fully discharged.

Mains only

In the case of very small premises in which only one or two call points are required, the system might be operated only from the mains. It is unlikely that this would be approved where legal requirements specify that an alarm system is to be installed.

Primary battery only

If no mains supply is possible, a small manual alarm system might be run from a primary battery provided that no standing current was required and that another similar battery could be held in reserve.

It is essential that if primary batteries are used as a power supply or as a stand-by for other systems, the batteries should be changed every two years even if they have not been used in any way. This is because primary batteries deteriorate in stock and cannot be relied upon after two years. The batteries on test may ring the sounders immediately and for a few seconds but they might not have sufficient power to maintain the sounder long enough for everyone in the building to hear and appreciate the signal.

In all cases where a supply of electricity is to be taken from the mains, the supply must be in accordance with the wiring rules of the Institution of Electrical Engineers. The actual connection should be taken directly from the main bus-bar by means of a switch-fuse (definitely not from a switch socket). The switch-fuse should preferably be behind a special break-glass cover painted read and marked 'FIRE ALARM – DO NOT SWITCH OFF'.

CIRCUITRY

In older systems and in some small modern systems where only a very few calls points and sounders are required, it is common to connect up the varous parts of the system by 'open' circuit wiring. Occasionally a heat detector may also be a included but when heat and certainly when smoke detectors are required a safer form of wiring is used. Some details of these circuits are described in Chapter 18.

'Open' circuit wiring means that there is no current flowing in the circuit until a call point is operated as compared with a 'closed' circuit in which current is continually flowing until a call point is operated. There are many overlapping variations of each type of circuitry.

In designing any circuit it must be remembered that every additional extra element of test or control adds a further chance of malfunction or breakdown.

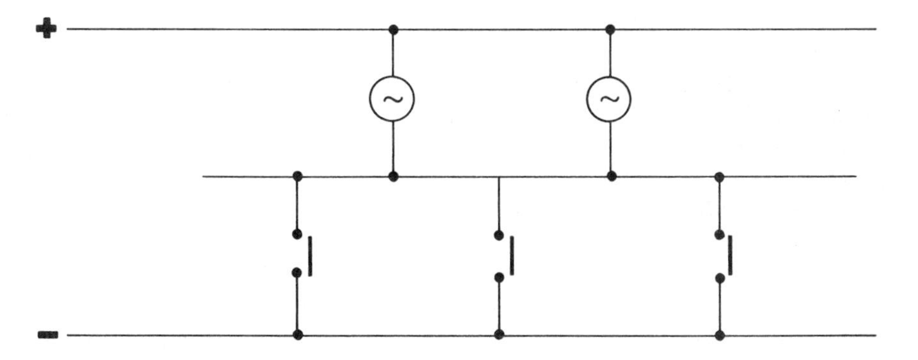

Figure 7.4 A simple open circuit fire warning system.

The open circuit system shown in Figure 7.4 has the following advantages:

- It is very simple.
- No current is consumed other than when sounders are operating and it is therefore cheap to run.

However, it has the following disadvantages:

- Some faults may render all or part of the system out of order without any warning. Such faults might not be discovered until a complete test of the whole system is undertaken – usually on an annual basis. Such a situation could have very serious consequences.
- Some faults would create false alarms. While one such fault might not have any serious consequences, repetition of false alarms tends to make people distrust and ignore the warning.

The silencing button

To avoid the continuous operation of the sounders because of a fault or after action on the alarm has been taken, and until the call point has been restored to normal, it is permissible to use a silencing button. The silencing button or switch, as its name implies silences the sounders but transfers the warning to a fault signal, usually a loud sounding buzzer. If there is more than one circuit in the alarm system there should be a silencing button for each circuit, so that as few call points as possible will be out of order.

As soon as any call point is known to be out of order, an 'out-of-order' notice should be placed over it. The notice should give the details of alternative means of raising the alarm.

Following any false alarm an immediate test of all call points should be made so as to discover whether any other points are also out of order.

It is preferable that the silencing switch should be under the observation of a responsible official and preferably be behind a break-glass protection box. The operation of the silencing switch when no sounders are operating should not put the system out of order. This is necessary because operation of the silencing switch while a fault is still present should reactivate the sounders. This facility can sometimes be used as a fire warning while the call points are out of order.

When the circuit has been restored to normal working, the silencing switch facility should change the supervisory buzzer to a different sound. A small bleep signal is particularly useful as it is a totally different sound and draws attention to the need for resetting the system and for the removal of the out-of-order plates.

CONDUCTORS AND CABLES

The current-carrying capacity of a cable depends on the cross-sectional area of the metal conductor. The smaller this area, the higher will be the resistance per metre of the cable. The higher the resistance, the greater will be the voltage drop along a length of cable for a given current flow, and the higher will be the heat generated in the cable. The voltage drop can be quantified by using Ohm's Law: $E = IR$, where E is the voltage, I is the current in amps, and R is the resistance in ohms. The heat generated, in watts, is given by $W = EI$, and Ohm's Law can be used to convert this to $W = I^2R$.

Thus, if a very small cable is used to carry a large current, it will get hot. It is possible to ignite the insulation, and many fires are caused in this way by overloaded electrical circuits. It is unlikely that an alarm circuit would be so badly under-designed for this to happen, but it is quite possible for the sounders' efficiency to be impaired by the use of unsuitable cables. This would happen if the total resistance of the circuit was so high that insufficient current was flowing. The sounders would then not operate, or would do so at reduced volume. Information on the current needed to operate the sounders, and on the resistance per metre of cables, can be obtained from the manufacturers. Calculations can then be made to see what voltage loss would result from the use of a particular cable, and clearly this should be kept as low as is practicable.

The insulation of the conductors is important not only to prevent short circuiting between conductors or from conductors to earth but it must be capable of withstanding heat from a fire which has occurred and for which warning is required.

A modern and very suitable cable is made from silicone rubber insulated conductors laid up with one or more non-insulated tinned copper conductors laid in the interstices and in contact with a surrounding aluminium foil strip (see Figure 7.5). A hard grade PVC compound is applied over the aluminium strip formed into a tube to give the cable good mechanical properties. Although this cable is intended to operate at 70°C it is suitable for continuous operation at a temperature of 150°C provided that the sheath can be terminated at a point where the temperature does not exceed 70°C.

The cable has withstood flame resisting tests at 750°C for three hours and also a test which required it to resist a temperature of 1000°C for a period of 20 minutes. (It still functioned after one hour at this temperature.)

Its value as a fire-resisting cable is well established but in addition it has a unique design which shows a very significant saving in installation costs. The savings mainly come about because it does not need wiremen to be

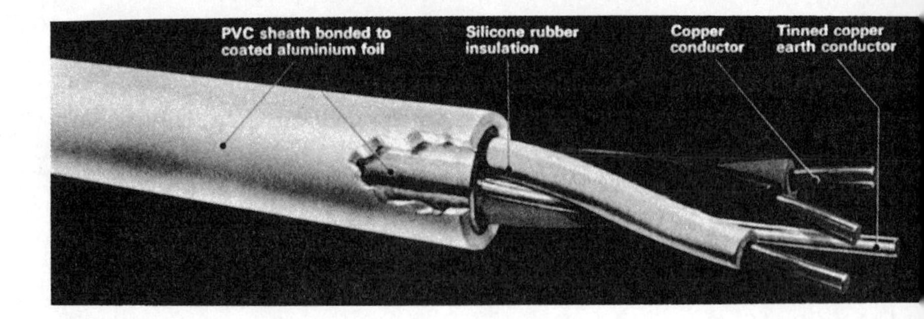

Figure 7.5 Fire-resistant cable.
(Pirelli General Cable Works)

specially trained and it does not require the use of special tools, seals or sealing compound. It has another interesting feature in that when the silicone rubber is stripped it exposes a bright copper conductor which gives a low resistance termination, which is particularly useful with the low voltage used in some fire alarm systems.

The cable can be obtained having 2, 3, 4, 7, 12 and 19 conductors, each with an earth continuity conductor. The cables may be surface-laid or embedded in plaster.

Until recently and for many years MIMS (mineral insulated metal sheathed) and earlier known as MICS (mineral insulated copper sheathed) cable was used in fire warning systems. It does, however, require special tools and special training for wiremen using the system.

Installations

The cables must be laid in accordance with the wiring rules of the Institution of Electrical Engineers and where cables pass a point at which mechanical damage could occur the cables are to be protected by a suitable metal strip or by running them through ducting. It is better, however, if such places can be avoided altogether.

Cables should not be run overhead unless it is impossible to find an alternative route. Consideration of the shortest route should not be allowed to override this requirement.

Fire warning cables that are run underground should be in exclusive ducts and well protected from mechanical damage. Additional precautions should be taken against the entry of moisture or dampness into the cable duct.

DOMESTIC OR SELF-CONTAINED FIRE ALARM SYSTEMS

Self-contained fire warning systems can be very useful where the fire risk is not high and particularly for home use and for other small premises. In the case of premises requiring registration or a fire certificate, it is essential to consult the fire authority before making any final decision as to the suitability of the type of fire warning system it is proposed to use.

There is a number of self-contained systems available, many of which have variations in construction in an endeavour to improve the inherent weaknesses of a particular type. But as previously mentioned these variations must inevitably complicate the working and provide further sources of trouble.

The manual break-glass self-contained system

The battery-operated version of this system is so simple that it is almost a basic model. It has the great advantage that it contains only a minimum number of parts and provided that the battery is regularly changed there is practically nothing that should go wrong. A test key enables a regular weekly test to be made and the switch can be used to silence the bell after evacuation is complete.

The mains version is equally simple and has the advantage that the sounder can be much more powerful and the possibility of battery failure is eliminated. Against these advantages must be set the remote possibility of a mains failure at the crucial moment when the alarm is required.

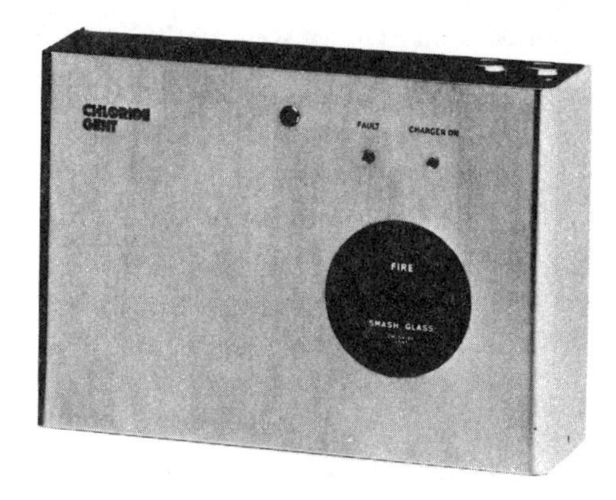

Figure 7.6 A typical self-contained fire alarm system with mains supply and secondary battery.
(Chloride Gent)

Another variation of the system has a secondary battery and small charging unit. This overcomes any problems of a failture of the electric mains or of battery failure. It is slightly more expensive to buy and to install (see Figure 7.6).

All three self-contained units can be interlinked with similar units of their own kind and heat detectors can be connected to them. When interlinking self-contained alarm systems, the connecting wires should be of a heat-resisting type. Regular testing is very important and in the battery version the batteries need regular changing.

One of the weaknesses of these systems is that the sounder cannot be placed in the most favourable position, that is high on the wall, because if this was done it would not be possible to operate the call point. The objection is overcome if extension bells are fitted.

Self-contained optical smoke detectors

Self-contained units consisting of optical smoke detectors are available and suitable for use as fire warning systems in private houses. Two models are available:

Battery operated optical smoke detector units

These units consist of the detector which acts as a call point, a sounder and a power supply. The detector works on the photo-electric principle described in Chapter 16. It is sensitive to smoke from synthetic foam and plastics as well as to wood, hydrocarbons, oils and greases. The sounder is an intermittent vibrator giving an output of 85 dB at 3 m. The intermissions consist of 1.5 seconds on and 1 second off. It is generally accepted that an intermittent signal is more likely to arouse a sleeper than a continuous sound.

The power supply of this domestic alarm system is a 9 V alkaline battery which must be changed every year. If the battery does run down during the year an intermittent signal is given by a bleep sounder operating once every eight seconds and this bleep will continue for about three weeks before the battery is finally exhausted.

The internal circuit is so designed that a pulsating infra-red LED (light emitting diode) checks the transparency of the atmosphere every eight seconds. If smoke is present at a low level the circuit changes and the pulse rate increases to about three-second intervals. If the smoke continues to be present or increases in density the circuit will go into the alarm condition within two or three pulses. If the smoke clears after the alert stage the circuit will return to its normal beat. The life of the LED is estimated at 49 years.

A small visual indicator on the front of the unit pulses in time with the detecting impulses, giving a visual indication of a healthy state of the unit. A test button is also provided on the front of the unit which allows a functional test of the circuit to be made by simulating the prescribed smoke level limits.

Figure 7.7 is a typical example of a battery-operated optical smoke detector domestic alarm.

The installation of these self-contained units is very easy as no mains connection is needed and they are simply screwed to the ceiling and positioned in accordance with the general rules for the siting of detectors as appropriate (see Chapter 18). It is claimed that these detectors are effective if screwed to a wall but a clear distance of at least 0.25 m should be allowed below the junction of the wall and ceiling or from the corner where two walls meet. The reason for this is that in the early stages of a fire, smoke rises and 'mushrooms' on the ceiling, being gradually forced lower, but in doing so tends to miss any corners in the room until the room is filled with smoke. Generally the best place to put a detector is in the centre of the ceiling.

Given that the batteries are regularly changed each year a self-contained unit detector of this type will give excellent detection in an individual room

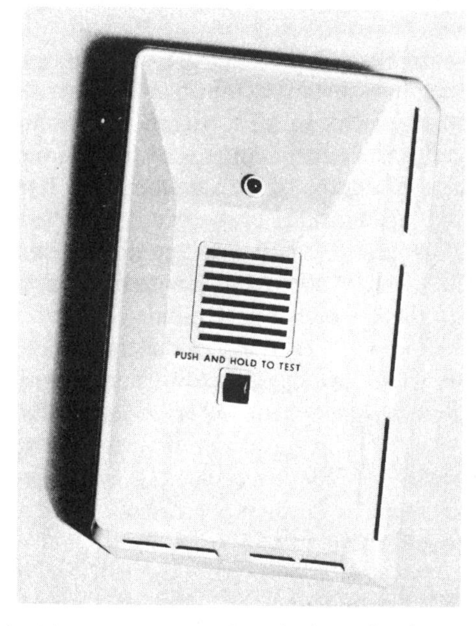

Figure 7.7 A typical battery-operated optical smoke detector. (Pyrotector)

and fair general protection if installed in a hall or short passageway that is free of draught. The top of any stairs should always be protected.

Mains-powered self-contained optical smoke warning systems

These units have the additional advantage that up to six units can be connected together and if any one of them detects smoke all six sounders will operate. Complete protection of a normal sized house can therefore be cheaply carried out. Against this must be set the possibility of a mains failure putting the entire system out of order. Having protected the whole house it is far less likely for the mains to be sufficiently damaged by fire before a warning can be given. Unlike the battery-operated system there is no sampling condition and the unit goes into the alarm condition just before smoke obscuration reaches 1 per cent. These units can also be fitted with heat detectors. In addition, a heat detector can be used to replace one or more of the smoke detectors. This is useful because a heat detector can be installed in a kitchen where it is inadvisable to install smoke detectors because of the effect of the concentration of steam associated with kitchens.

Self-contained ionization smoke detectors

In the same way that optical smoke detectors can be used in self-contained units so ionization detectors can be used. Both battery- and mains-powered units are available. As with the optical detectors, several models have facilities for interconnecting a number of units so that when one has detected the presence of smoke, they will all give a warning. These devices do contain a small radioactive source but the radiation from it has little penetration and is negligible when compared with normal background radiation.

Large numbers of these useful and reliable detectors are installed in houses in the USA, where they enjoy a good reputation. They are now readily available in the UK and are becoming widely accepted. Most of them require a 9 V alkaline battery, which again must be replaced once a year. Many of the available models provide a warning that the battery voltage is too low by sounding an intermittent bleep.

STANDARDS

Standards are available in many countries covering the equipment described in this chapter. As has already been indicated, the relevant BS code of practice is discussed in Chapter 18, because both detection and alarm systems are covered by the same document. Methods of detection and examples of practical detector design are discussed in Chapter 16.

8

Legislation

In the UK there are several hundred Acts of Parliament which contain some reference to fire but of these only three need to be discussed here. These are the Fire Precautions Act 1971, the Health and Safety at Work Act 1974 and the Fire Safety and Safety of Places of Sport Act 1987. There are all enabling Acts: relatively brief documents which give the relevant Secretary of State the power to issue detailed regulations and statutory instruments. The goverment departments responsible for the publication of these regulations and instruments will frequently issue guidance notes which explain what you should do to comply with the legislation.

The Health and Safety at Work Act replaced the earlier Factories Acts. It is unusual in that it places a duty on the employee to work safely in addition to requiring the employer to provide a safe place of work. There are parts of this Act which amended and strengthened the Fire Precautions Act, and which also resulted in the publication of the Building Regulations 1974 and 1985. These regulations include the requirement for the provision of escape routes which will remain smoke-free for a sufficient time to enable people to escape by their own unaided efforts. The regulations are discussed later in Chapter 15.

An important part of the Fire Precautions Act covers the issue of fire certificates. These are granted only when the fire authority is satisfied that the building complies with all the requirements of the Act. Fire certificates had previously been issued under the Factories Act 1961 and the Offices, Shops and Railway Premises Act 1963. The fire clauses in these Acts are now in the Fire Precautions Act, but certificates which were issued under them remain valid until they are cancelled because of alterations to the buildings, or for some other reason.

During the lengthy preparation of the Fire Safety and Safety of Places of Sport Act 1987 it had been hoped that the requirement for fire certificates would be replaced by putting a statutory duty on owners or occupiers of designated buildings to ensure the fire safety of their premises. This would have been similar to the system used to control safety at work under the Health and Safety at Work Act 1974. It would have also been in line with the move towards performance standards and codes of practice. A standard of this kind would for example say that a fire extinguisher must be capable of withstanding environmental and handling conditions, that its method of operation must be both simple and obvious, and that it must be capable of putting out a fire. Earlier compliance standards went into considerable detail about mechanical construction, operating pressures and so on, but did not specify any fire performance. It had been hoped that the new Act would simply specify the standard of safety which was required, and leave the architect and the occupier to decide how this could be achieved (with, of course, some suitable guidance). This has not been done, but some significant changes have been made to the Fire Precautions Act 1971, the Safety at Sports Grounds Act 1975, the London Government Act 1963 and the Local Government Act 1982. The effects of these changes are discussed below. It should of course be remembered that the various Acts covering fire protection enable new regulations to be written at any time and no published list of legislation can remain up-to-date for very long.

In dealing with legislation this book is only concerned with the fire clauses and this is particularly so with the Health and Safety at Work Act. No attempt is made to explain the legal side or its implications and applications. It is essential that the complete Acts and regulations should be studied together with the official guidance booklets that are issued.

In all cases affecting safety from fire, the fire authority should be consulted and they are always willing to give help and advice. In particular it is advisable to call in the fire authority at the very earliest stage when alterations or new work of any kind or even layout of plant is contemplated. This consultation should take place even before 'pen is laid to paper'. Anything that can affect the fire certificate must be notified. It is well to remember that alterations in layout once they have been drawn become increasingly expensive to change.

HEALTH AND SAFETY AT WORK ACT 1974

This Act has had a very great influence on the state of fire precautions in many premises. It brings in a new concept, that of involving employees as well as management and bringing them together to increase safety, not only with regard to fire but for many other purposes. As an enabling Act it becomes effective through designating orders, a number of which have already been brought into effect, in the form of regulations and orders made under a series of statutory instruments.

Looking at the Act itself, the first important point affecting fire precautions is found in section 2, where it is laid down that it is the duty of every employer to ensure that, so far as is reasonably practicable, the health, safety and welfare at work of all his employees. There are seven subsections to this section.

Subsection (2) para, (*c*) requires the employer to make provision for supplying information, instruction, training and supervision for this purpose and para. (*d*) requires him to maintain any place of work under his control in a condition that is safe and to provide means of access and egress that are safe and free from risks.

Subsection (3) places a duty on the employer to prepare and revise as necessary a written statement of general policy with respect to the health and safety at work of his employees and to set out the organization and arrangements in force for carrying out the policy, and to bring this and any revision to the notice of his employees. It should also be included in the Report to the Annual General Meeting of the company as information to the shareholders that the company is complying with legal requirements.

Subsection (4) gives the Secretary of State power to make regulations for recognized trade unions for the appointment of safety representatives from amongst employees. These people will represent employees in consultations with the employers.

Subsection (5) runs parallel with the previous subsection and gives the Secretary of State power to make regulations for the election by employees for representatives as above.

Subsection (6) is direct and says that it shall be the duty of every employer to consult with any such representatives to make arrangements which will enable him and his employees to cooperate fully in promoting, developing and checking the effectiveness of measures to ensure the health and safety of employees at work.

Subsection (7) says that in such cases as may be prescribed it is the duty of an employer if requested to do so by the safety representatives to establish a safety committee to carry out the functions as described above.

The whole object of these subsections is to ensure that safety from fire is not only a management duty but is also a duty of everyone at work.

It should be noted that subsections (5) and (7) were not brought in immediately but required regulations to bring them into force. These regulations were made under S.I. No. 500, known as the Safety Representatives and Safety Committee Regulations 1977.

There is a booklet available entitled *Safety Representatives and Safety Committees* published by the Health and Safety Commission in 1977 by HMSO and obtainable from them. The booklet incorporates the regulations, a code of practice, and includes guidance notes for safety representatives. The code of practice has been approved by the Health and Safety Commission and has therefore some legal standing. It is only by following the code of practice that the effectiveness of the Act is likely to be achieved. This booklet should be in the possession of every fire precautions officer and safety representative. It should be noted that safety representatives do not supplant specialist officers such as works fire or safety officers.

Section 7 of the Act places a duty on every employee to take reasonable care for the safety of himself and of other persons at work. The duty imposed on his employer, or any other person by or under any of the relevant statutory provisions, is to cooperate with him so far as may be necessary to enable that duty or requirement to be performed or complied with.

Section 8 states that no one shall intentionally or recklessly interfere with or misuse anything provided in the interests of health, safety or welfare in pursuance of any of the relevant statutory provisions.

Section 9 prohibits an employer from levying a charge on an employee in respect of anything done or provided in pursuance of any specific requirement of the relevant statutory provisions.

These three sections 7, 8 and 9 are self-explanatory but the fire precautions officer should be aware of them.

The written statement of policy

If the policy of the management is to be acceptable and successful, all employees must be persuaded to take an active interest, and the appointment of the safety representative and/or a safety committee must involve employees to the utmost.

The first step to implement the requirements of the Act following the appointment of the safety representative is a policy statement in the form of a notice which is brought to the attention of all employees. This first statement should consist of information regarding the setting up of safety representation (section 2(5), (6) and (7)) and the duties of employees (sectons 7, 8 and 9).

The next step should be an inspection of the whole premises, probably with the local authority fire officer. During this inspection notes will be made of all dangers from fire. In smaller premises this inspection will also be used for other aspects of safety and welfare. As the result of this inspection a policy notice could be set out drawing attention to the classes of defect seen rather than to individual items. Much information can be obtained from Chapters 9, 12 and 13 on this subject.

Having detected the most dangerous situations the next step should be to deal with the exceptionally dangerous ones as a matter of urgency. This should be followed up by a particular drive on one general form of danger, for example the safe storage or removal of all flammable waste. The work would be broken down by departments to small units, each of which would be responsible for reporting back to the committee what had been done. Having achieved a measure of success the committee could then turn itself into an inspecting body and comply with section 2(7) having the 'function of keeping under review the measures taken to ensure the health and safety at work of employees and such other functions as may be prescribed'.

Regulations

An important feature of the Health and Safety at Work Act is that it strengthens the Fire Precautions Act. This has been done in stages which will no doubt continue for some time. To start with the fire clauses of the Factories Act and the Offices, Shops and Railway Premises Act have been taken over by the Fire Precautions Act and there is now only one form of fire certificate.

The local authority now becomes the enforcing authority of the relevant statutory provisions for premises situated in their area, with the exception of certain premises for which the Health and Safety Executive will be responsible.

An interesting point now arises in that we have three classes of premises:

1 Those of high fire risk which are certificated by the Health and Safety Executive
2 Those of ordinary fire risk certificated by the local fire authority
3 Small premises that do not need a fire certificate but must take certain precautions against fire.

The following is a list of regulations issued under the Health and Safety at Work Act to date in the form of statutory instruments with a brief summary as a guide to their effect. This summary must not be taken in any form as an official ruling but the full text of the regulations must be studied if required.

1976 No. 2003

These regulations deal with fire certificates for special premises, i.e. the premises that are certificated by the Health and Safety Executive. Section 3 prescribes the premises for which a certificate is required and these are described in Schedule 1 of the regulations. It also lists the types of premises for which the special certificate is required. These are mainly manufacturing premises where large quantities of highly flammable or dangerous materials are used or stored and Schedule 2 sets out conditions under which a certificate may not be required. Section 4 sets out how an application for a certificate under these conditions is to be made and the steps to be taken by the Health and Safety Executive to inspect the premises before issuing the certificate. It also gives the procedure to be followed if the result of the inspection is not satisfactory. Section 5 sets out the contents of such a fire certificate and subsection (2) sets out the conditions which the Executive may impose if they consider them appropriate. Other subsections deal with matters concerning the issue of the certificate. Section 6 details matters which might affect the currency of the certificate and the action which might be taken. Section 7 is similar to section 6 except where inadequacy is caused by some person other than the responsible person. Section 8 deals with other cases where a certificate might become inadequate. Section 9 deals with cases where the responsible person wishes to vary the conditions of the certificate. Other sections deal with the operation of the regulations.

1976 No. 2004

This deals with the repeal of parts of the Factories Act.

1976 No. 2005

This deals with the repeal of parts of the Offices, Shops and Railway Premises Act.

1977 No. 294

This is an order and brings into effect a number of sections of the Health and Safety at Work Act concerned with Building Regulations. These are discussed in Chapter 15.

1977 No. 500

This is the regulation dealing with safety representatives and safety committees.

1977 No. 746

These regulations deal with matters relating to the enforcement of the Act. Subsection (3) appoints the local authority as the enforcing authority.

Subsection (4) sets out a number of exemptions of official bodies. Subsection (5) enables enforcement duties to be exchanged between the local authority and the Health and Safety Executive. Subsection (6) deals with the action to be taken where there is uncertainty as to which authority is responsible for enforcement. Schedule 1, parts 1 and 2 of the regulations lists a number of main activities which determine whether the local authority or the Health and Safety Executive will be the enforcing authority. Schedule 2 deals with repeals and modifications relating to the Agriculture (Safety, Health and Welfare Provisions) Act 1956. Schedule 3 repeals parts of the Factories Act dealing with enforcement. Schedule 4 repeals parts of the Offices, Shops and Railway Premises Act 1963 in respect of enforcement.

1976 No. 2007

In the 1971 Act, factories, offices, shops and railway premises in which no more than 2 people are employed are exempted from the requirement for a fire certificate. The SI specifies requirements in these premises for adequate means of escape and for the provision and maintenance of appropriate means for fighting fires. (See also below in Clause 16 of the 1987 Act.)

1976 No. 2008

These regulations describe the form in which an application for a certificate is to be made and revoke the regulations made in 1972 for the form of application. It will be noticed that the new form can, and is to be used, as an application whether the premises would have been certified under the Factories Act, the Offices, Shops and Railway Premises Act or any of the premises given in section 2 of the Fire Precautions Act including a new class 'If used as a place of work'. This is a very wide class created by section 78(2) of the Health and Safety at Work Act. Copies of the application form can be obtained from the local authority fire brigade.

1976 No. 2009

This order designates premises for the purpose of section 1 of the Fire Precautions Act which requires fire certificates for premises put to designated uses. In this order, these designated premises are factories, offices, shops and railway premises as described in the Acts which define them, and in each case where persons are employed to work. Section 4 of the order exempts premises where not more than 20 persons are employed to work at any one time or where not more than 10 persons are employed at any one time other than on the ground floor or if explosive or highly flammable materials are not stored or used, unless they are in such small quantities that the fire authority does not consider they materially affect the safety of the premises.

1976 No. 2010

These regulations apply to the small premises that are exempted under the previous regulation. Although they do not require a fire certificate they must observe a standard of fire safety. This standard is described generally in the next chapter.

FIRE PRECAUTIONS ACT 1971

These notes have been prepared to show the practical application of the Fire Precautions Act. the full implications should be studied by reference to the Act. If there is any doubt whether any particular premises fall within the operation of the Act or statutory instruments, the local fire authority should be consulted. Even if premises do not come within any regulations under this or any other Act of Parliament, prudent users of property will take precautions against the loss of life and property by fire. In this respect the provisions relating to the fire certificate as set out in this Act should be regarded as a minimum.

The statutory classes of usage of premises were made the subject of statutory instruments for which a fire certificate is required and without which the premises may not be so used:

1 Use as, or for any purpose involving the provision of, sleeping accommodation (this includes hotels, inns, boarding houses and in some cases apartment houses).
2 Use as, or part of, an institution providing treatment or care (this will include: hospitals, nursing homes, homes for the sick, aged and incapacitated persons, baby units and nursery schools).
3 Use for purposes of entertainment, recreation or instruction or for purposes of any club, society or association.
4 Use for purposes of teaching, training or research (this includes all schools – public, private and kindergarten).
5 Use for any purpose involving access by members of the public, whether by payment or otherwise. (This will include dance halls, bingo halls, exhibition halls and social clubs).
6 Use as a place of work. [See Health and Safety at Work Act 1974, section 78(2).]

A house which is occupied as a single private dwelling is exempted. Premises used solely or mainly as places of public religious worship were also exempted in this Act but do now require a fire certificate under the 1987 Act (see below).

The fire authority may also serve a notice on the occupier, owner or a person having management of premises, that a fire certificate is required if it appears to them that the premises consist of, or comprise a room, that is used as living accommodation if it:

- is below the ground floor level of the building;
- is two or more floors above the ground floor of the building;
- is a room of which the floor is six metres or more above the surface of the ground on any side of the building; or
- if explosive or highly flammable materials are being kept under, in or on the building. The materials will be prescribed and the regulation will not be operative unless a prescribed minimum quantity is exceeded.

There is a right of appeal against the notice.

Regulations may also be made to include vessels remaining moored or on dry land and also to include tents and similar movable structures.

Hotels and boarding houses have already been designated under the Act by 1972 S.I. No. 238. This statutory instrument requires that a fire certificate must be obtained for any premises used as a hotel or boarding house if sleeping accommodation is provided for more than six persons whether guests or staff or if there is some sleeping accommodation above first floor level or below ground level. The order extends to dining-rooms, drawing-rooms, ball-room or other accommodation for guests.

The fire authority is charged with the enforcement of the Act which it does by the appointment of fire inspectors. Officers of fire brigades may also be appointed to carry out the duties of a fire inspector.

Under certain conditions, a fire inspector has authority to enter premises at any reasonable time, require answers to questions that he may put and require a signed declaration as to the truth of those answers. If the premises are used as a dwelling-house he may not exercise this authority unless 24 hours' notice has been given to the occupier.

If requested, a fire inspector, or a fire brigade officer carrying out the duties of an inspector, must produce a document duly authenticated showing his authority.

Penalties are laid down for offences in connection with the Act and there are various safeguards by appeal to the court.

The fire certificate

It has already been mentioned that 1976 SI. No. 2009 requires that all designated premises must have a fire certificate with the exception of premises in which only a few persons work. Even then the exception does not apply if explosive or highly flammable materials are used or stored unless the quantity of these materials is so small that the fire authority determine that they do not constitute a serious risk in case of fire.

Also mention was made of 1976 S.I. No. 2010 in which premises excepted by the previous statutory instrument are required to take certain fire precautions. The effects of this statutory instrument are discussed later in this chapter.

A fire certificate is also required by hotels and boarding houses by virtue of 1972 S.I. No. 238 although certain small premises are excepted.

If there has not previously been a fire certificate for the premises or if alterations are being made which affect the fire certificate, a form of aplication for a fire certificate should be made to the chief executive of the fire authority of the district concerned, or to the Health and Safety Executive of the area where appropriate.

The form of application is laid down in 1976 S.I. No. 2008. The applicant is required to give his name, address and trading name and the name and address of the owner of the premises. Details of the premises are required in respect of the use to which the premises are to be put, the maximum number of people to be employed below, on and above ground level and an estimate of the number of persons other than staff who may be on the premises. The applicant is also required to give the number of people for whom sleeping accommodation is provided below the ground floor, above the first floor and the total for the whole of the premises. The number is to include staff, guests and other residents.

Details are required of any highly flammable or explosive materials stored or used on the premises, quantities, methods of storage and the quantity that may be exposed at any one time. The number of floors and the approximate age of the building are also required.

After receipt of the application the fire authority may demand a plan of the building to explain any of the points in the application. The next step will be a visit by the fire authority in the person of an inspector who will then examine the building. The applicant is advised to accompany the inspector and he can then discuss the points which the inspector raises and will obtain a much better idea of what is required and the best methods of achieving it.

If the inspector is satisfied that the fire precautions conform with the requirements of the Act, the fire certificate will be issued. If, however, he is not satisfied he will issue an improvement notice stating what must be done and setting a time for the work to be carried out. If the work is not completed, inspected and accepted as satisfactory by the date given, the premises may not be used for the purposes within the Act. These powers have not been strengthened by the provisions of the 1987 Act (see below).

Most fire inspectors will help to explain the requirements to the applicant, but it must be remembered that they have a duty to perform and cannot be diverted from that duty. It is in the interests of the applicant to assist the inspector in every way especially at the time of the first inspection. It is at this time that the applicant should put himself in a position to discuss in a reasonable and knowledgeable manner what steps are needed to improve the fire precautions. By such discussion alternative methods may be discovered of meeting the requirements. Generally speaking an inspector is unlikely to alter a decision once an improvement order has been made.

Before detailing methods of improving fire precautions there are certain other matters in connection with a fire certificate that must be remembered. It is the duty of the user of the premises to inform the fire authority in advance before making any alteration to the structure of the premises. Likewise the authority should be informed of any rearrangement of the premises or in the furniture or equipment, particularly of any items enumerated on the fire certificate. The reason for this is to determine whether such rearrangement will cause obstruction to the means of escape. This does not mean that the occupant must inform the authority of a simple change round of furniture but if there is any doubt at all it is simpler and better to say what is proposed. If it is proposed to make any change which involves the use or storage of highly flammable materials or explosives the authority must be informed. It may be such a small quantity that the authority will be prepared to accept it or perhaps make some simple recommendation that will make the storage acceptable.

The fire certificate, when granted, must be kept on the premises.

The contents of the certificate

The contents of the fire certificate are stated in section 6(1) of the Fire Precautions Act. Briefly these are as follows:

(a) The particular use or uses of the premises.
(b) The means of escape in case of fire which are provided.
(c) The means which are provided to ensure that the escape route can be safely and effectively used at all times. This means safety lighting, routes free of obstruction, no locked doors.
(d) The provision of firefighting equipment and the position of it. This includes extinguishers, hose reels and hydrants.
(e) The means of giving warning in case of fire.

Any of these may be described by reference to a plan. The means of escape in case of fire are discussed in Chapter 9; the position of firefighting equipment is discussed in Chapter 27; and the means of giving warning in case of fire is discussed in Chapter 7.

Additional items that may be imposed by the fire certificate

By subsection (2) of section 6 of the Fire Precautions Act, five other requirements may be imposed.

The first of these additional requirements is to secure that item (b) of subsection (1) (given above) is properly maintained and kept free from obstruction.

The second is to secure that items (c), (d) and (e) (given above) are properly maintained. All these requirements have been discussed in this book under their various subject headings.

The third is to ensure that persons employed to work in the premises are given instruction and training in what to do in case of fire and that records are maintained of the training given. Notes on this subject will be found in Chapter 11.

The fourth is to enable a limitation to be placed on the number of persons that may be in the premises at any one time. This would, of course, be decided by the fire authority who would take into consideration such items as the use of the building, the standard of construction, the escape routes and any other relevant details.

The fifth and last requirement that may be added allows the fire authority to require any other fire precautions to be taken in relation to the risk of fire to persons on the premises.

Loans to meet expenditure

Where any person proposes to make structural or other alterations to the building to comply with an order made under the Act, he may apply to the local authority for a loan. If the local authority consider that he will be able to meet his obligations and subject to the rules laid down, they may, if they think fit, enter into a contract for a loan for this purpose.

There is also an appendix to the Act which shows the effect of structural alterations in connection with the Rent Act 1968.

The following statutory instruments have also been made under the Fire Precautions Act. 1976 S.I. No. 2006 brings into operation a number of provisions of the Act which had not previously been in operation. 1976 S.I. No. 2007 amends the Act in respect of premises that are in multiple occupation and the obligations of the Act can now be placed on owners rather than on occupiers.

Premises that do not require a fire certificate

1976 S.I. No. 2010 applies to factories, offices, shops and railway premises which are exempted from obtaining a fire certificate by reason that they do not employ (i) more than 20 persons at any one time on the premises; and (ii) not more than 10 persons elsewhere than on the ground floor. It should be noted that the dividing line is the number of persons employed, and not the number of people likely to be on the premises. There is no exemption for factory premises where explosive or highly flammable materials are stored or used or in or under the premises (unless accepted by the fire authority that they do not constitute a serious additional risk).

Section 4(i) is quite straightforward. If more than 10 persons work in a room and the door from that room leads to a staircase or a corridor it must either slide or open outwards and all doors that give an access or means of exit (final exit doors) to the premises must also open outwards or be sliding doors. This eliminates revolving doors.

Section 4(ii) requires that every window, door or any way of getting out of the premises other than the normal exits must be conspicuously marked. Although for many years exit signs have been red, modern usage is to have green on a white background reserving red as a colour for danger, whereas green is for safety. Exit routes to be used only in case of fire should preferably read 'FIRE EXIT' in letters of 100 mm and if necessary arrows should be used if the route is not clear. The ordinary way out does not legally have to be marked, but it is not prohibited to be so marked.

Section 4(iii) is again quite straightforward and requires that hoistways and lift enclosures in factory premises are to be of half-hour fire-resistant construction with half-hour fire doors and if not ventilated to the open air at the top, must be enclosed by material easily broken by fire. The reason for this is obvious as it is intended that any outbreak of fire in such a construction is contained so that it does not spread laterally and endanger a route to safety.

To enclose an existing lift or hoistway use can be made of asbestos-based board or asbestos-free board that meets the requirements of the relevant parts of BS:476.

Section 4(iv) gives the fire authority power to exempt any of the requirements in subsections (i), (ii) or (iii) by notice in writing if they are satisfied that compliance with them would be inappropriate or undesirable. The occupier has no right to demand such exemption though he can ask the fire authority to grant it.

Section 5(i) requires that no doors may be locked through which an employee might have to pass to get out of the premises unless they are so fitted that they can be easily and immediately opened by him on the way out. Means of exit which comply with this regulation are described in Chapter 9.

Section 6(i) requires that appropriate means for firefighting shall be so placed as to be readily available for use and subsection (ii) reinforces this by stating categorically that 'maintained' means maintained in efficient working order and in good repair. Suitable information concerning this can be found in Chapter 27.

Section 7 contains subsections dealing with the responsibility for these requirements which are placed on owners, joint owners or occupiers.

Certain low-risk premises are exempted from the need for a fire certificate by the 1987 Act, which is discussed next.

FIRE SAFETY AND SAFETY OF PLACES OF SPORT ACT 1987

Part I of this Act makes changes to the Fire Precautions Act 1971 (discussed above). Part II amends the Safety at Sports Grounds Act 1975, which is described later. Part III provides for a system of safety certificates for sports ground stands with covered accommodation for 500 or more people. Licences for sports entertainments in indoor premises are required by amendments, in Part IV, to the Local Goverment Act 1982 and the Civic Government (Scotland) Act 1982. The more important changes to the provisions of the Fire Precautions Act are summarized below (Clauses 1 to 17) together with the Clauses (19 to 43) which are concerned with the Safety at Sports Grounds Act.

Clauses 1 and 2 enable a fire authority to exempt low-risk premises from the need to have a fire certificate, even though they are designated in the 1971 Act (and are not otherwise exempt). The authority has the power both to exempt and to withdraw the exemption, and requires the occupier to notify the authority of any proposed changes which may increase the risk in the premises. These changes include those affecting the means of escape, or the presence of explosives or flammable liquids in excess of an agreed limit, or the presence of more people than are specified in the exemption notice. The exempted premises must have adequate means of escape and suitable means of fighting a fire, and these must be maintained (see Clause 5).

This is an important change because it will leave the fire authority with more time to deal with the more significant hazards.

Clause 4 redefines the meaning of 'escape' in the 1971 Act to cover means of escape which extend beyond the premises to a place of safety.

Clause 5 imposes a duty on the occupier of exempted premises to provide and maintain means of escape and means of fighting a fire.

Clauses 6 and 7 address the fact that for some time factory inspectors have been able to issue improvement notices requiring occupiers to make changes to comply with the law covering health and safety. The same powers have now been given to the fire authorities. If the authority considers that there has been a breach of an occupier's duty to provide means of escape or means of firefighting, they can issue an improvement notice detailing the work which needs to be done. The occupier or owner must do this work unless he appeals to a Court against the notice within 21 days.

This is an important change in the law which should result in an enhancement of fire safety.

Clause 8 covers the time between the application for a fire certificate and its issue (or the decision to exempt the premises under Clause 1). During this time the occupier must ensure that existing exits can be used safely and effectively, and that people working in the premises are given instruction or training in what they should do if there is a fire.

Clause 9 constitutes another significant change which greatly strengthens the effectiveness of the fire authorities in ensuring fire safety. Under this clause an authority can issue a prohibition notice, to close down any premises where they consider that there would be an excessive risk to people if there were a fire. The occupier can appeal to a Court against the notice within 21 days. In this case however the notice remains in force from the time that it is issued until either it is withdrawn by the authority or cancelled by the Court. (Again this follows the system already in use by the factory inspectors.)

Clause 10 allows the Secretary of State to decide on the extent and frequency of inspections of designated premises.

Clause 11 concerns section 21 of the 1971 Act which restricts the discosure of information obtained during an inspection. This is now amended so as to allow information to be given to another enforcing authority which is empowered by the Health and Safety at Work Act 1974.

Clause 12 makes it clear that the amended 1971 Act does not confer a right to take civil proceedings for any breach of requirements which have been made under the Act (although existing rights are unaffected).

Clause 13 removes an unfortunate anomoly from the legislation. The 1971 Act exempted buildings used for religious worship from the need for a fire certificate. This meant that any religious organization could hold meetings of large numbers of the public in quite unsuitable premises. This clause now removes the exemption conferred by the Act from any premises appropriated to, and used solely or mainly for, public religious worship.

Clause 14 restricts the defence which can be used in the Courts in connection with a contravention of the requirements of a fire certificate.

Clause 15 changes the 1971 Act where the means for fighting fire which could be specified by the fire authority were restricted to equipment 'for use in case of fire by persons in the building'. The authority could not therefore take into account any automatic systems which had been provided. This wording has now been removed and an authority can now make allowance for the presence of automatic systems before they issue a certificate, and can also specify a system in the conditions of the certificate. This again makes the legislation more flexible and more effective.

Clause 16 tidies up the legislation by modifying the effect of the 1971 Act as it applies to some factories, offices, shops and ralway premises. This simply brings this part of the Act in line with the provisions of SI 1976 No. 2007, which has already been discussed.

Clause 17 amends the 1971 Act where it allows the Secretary of State to apply the requirements of the Act to vessels and to 'moveable structures'. This is now extended to include places of work in the open air.

Some of the remaining clauses are concerned with amendments to the Safety of Sports Grounds Act 1975. This Act gives the Secretary of State powers to designate any sports stadium which can accommodate over 10 000 spectators, which is then required to obtain a safety certificate. The certificate contains requirements for adequate means of escape and specifies the number, size and situation of exits, and the access to them. The application of these requirements is widened and strengthened by this part of the 1987 Act.

Clause 19 reflects the Popplewell Report of the enquiry into the Bradford City Football Club Fire (1985) which recommended that the distinction in the 1975 Act between sports grounds and sports stadia should be abolished. In the Act a stadium was defined as a ground wholly or substantially surrounded by accommodation for spectators, which meant that some sports grounds were exempted from the requirements even though they could hold more than 10 000 spectators.

Clause 20 removes this limitation to grounds holding 10 000 or more people. A different numerical threshold can now be substituted, or different thresholds for different classes of sports grounds.

Clause 22 changes the procedures for appeals against the requirements of a certificate, so that these now go to the Courts instead of to the Secretary of State.

Clause 23 gives the local authority the power to restrict or prohibit the admission of spectators to a certificated ground instead of having to apply to a Court for an order.

Clause 25 requires designated grounds to be inspected at least once a year.

Clause 26 allows the local authority to require safety certificates for any stands in undesignated grounds which provide covered accommodation for 500 or more spectators.

Clauses 42 and 43 deal with the licensing of indoor premises within the London area, in other parts of England and Wales, and in Scotland (this involves amendments to various London government and local government Acts).

Deaths and injury in fires

It should be remembered that the purpose of these Acts of Parliament is to protect people from the effects of fire. The legislation is not directly concerned with protecting property, although the required precautions will reduce the rate of fire spread. An indication of the effectiveness of the 1971 Act in protecting people can be obtained from the published fire statistics. It is reasonable to assume that it would take ten years for the provisions of the Act to become fully effective, and we can look at the changes during this period. In 1971 here were 89 300 fires in occupied buildings and ten years later, in 1981, there were 94 800 fires. The number of deaths due to fires in houses (to which the Act does not apply) increased from 574 in 1971 to 780 in 1981. The numbers fluctuated throughout this period, with a maximum of 865 in 1979 (due possibly to a very cold winter), but the trend showed a continuing increase. On the other hand, the fatalities due to fires in other occupied buildings, which were affected by the Act, fell from 152 in 1971 to 80 in 1981. Again the trend was continuous, but in this case downwards, although the actual numbers fluctuated from year to year.

It is reasonable to assume that if the Act did apply to houses, the number of deaths in domestic fires would also have been reduced. This emphasizes the value of even the simplest fire precautions in the home: ensuring that doors are closed at night, the installation of battery-operated detectors, and the provision of for example an emergency ladder, or even a knotted rope, to enable people to escape from upper floors.

9

Means of escape in case of fire

Under section 9A of the Fire Precautions Act (a section added by the Health and Safety at Work Act) it is laid down explicity that all premises to which the description of factories, offices, shops and railway premises apply, shall be provided with such means of escape in case of fire as many be required in the circumstances of the case. A futher subparagraph states that for the purposes of that section regard shall be paid not only to the number of employees but to the number of other people who might reasonably be expected to be on the premises.

Within the Fire Precautions Act 1971, which under the Health and Safety at Work Act 1974 becomes the regulating Act, it is possible to make two divisions: premises which are required to have a fire certificate; and those small premises which, although they are not required to have a certificate, must comply with certain requirements.

Two further divisions can be made between buildings not yet erected and buildings which already exist and have to be adapted, because of a change of use or because they have now become liable to improve their fire arrangements.

Adequate means of escape were required in the Building Regulations 1976 but there was a 'deemed-to-satisfy' clause which said that the means of escape would be acceptable if they complied with the requirements of British Standard CP3. The 1985 Regulations require that the means of escape must comply with the mandatory rules in a separate document. This document again refers to BS CP3 and also to BS 5588. However both these Regulations and the associated documents are still under review (see Chapter 5).

As the plans for new buildings have to be submitted to the enforcing authority (usually the local authority fire service), they will be brought automatically into line with the requirements but to save time and avoid other difficulties, it is advisable to study codes of practice before making any decision about plans.

This book is concerned mainly with existing buildings and what is likely to be required to bring a building up to a point that will satisfy the legal requirements. It must be pointed out that although the recommendations and advice are based on the information available, the interpretation of the regulations rests solely with the enforcing authority. It must also be admitted that while the authorities use every endeavour to see that there is a universal standard of fire protection throughout the country, there may be small variations between authorities. This is particularly noticeable if existing buildings are being brought up to modern standards when there is some discretion to individual enforcing authorities. This discretion is seen in the Fire Precautions Act where it is required that all premises shall have 'such means of escape in case of fire for the persons employed to work therein as may be required in the circumstancesof the case'. Again, where other persons may be expected to be on the premises, the means of escape must be determined by the numbers of people who may reasonably be expected to be present.

ESCAPE ROUTES

In terms of escape routes, the main points to be taken into consideration are enumerated in the following paragraphs, based mainly on BS CP3.

Use of the premises and of any materials

Under the Fire Precautions Act a certificate of means of escape in case of fire has to be obtained for premises in or under which explosives or highly flammable materials are stored or used.

A useful guide can be found in 1961 S.I. No. 917 which was made under the Factories Act 1961. As a general guide, all liquids having a flash point below 49°C (120°F) would be classed as highly flammable materials. These will include petrolem spirit, methylated spirit, acetone and carbon bisulphide. For some of these materials there are other special Acts. Also included in the list would be paraffin, white spirits, turpentine and cellulose solutions, whatever the flash point might be.

Under solids, would be included powdered metals, celluloid, cellulose nitrate and a number of similar substances; cotton, flax, hemp and jute in a loose state; paper, wood, wool and shavings when used or stored in a possibly dangerous manner.

The inherent risk in the building itself must also be considered. For example, if there is some constructional defect that cannot easily be rectified, this must also be taken into account when deciding on the exits which are required or can be approved.

Time to be allowed for evacuation

The time for the evacuation of a building is considered to be two and half minutes. The time is reckoned from the first warning until all the occupants have reached the open air. But under some circumstances it could be taken as the time for the occupants to have passed through a fire-resisting door, and to be in the comparative safety of a corridor built to resist fire for 30 minutes. The time for evacuation must be reduced for buildings and rooms where the hazard from smoke and fire is particularly great.

Number of persons involved

So far as is possible, the number of people in the buildings should be forecast accurately, and this number should be broken down room by room and floor by floor, so that it can be known how many people will use any particular exit. Where accurate figures cannot be obtained, a useful guide can be taken from the following figures:

Restaurant	0.9 m^2 (10 ft^2) per person
Shops dealing in consumer goods	1.9 m^2 (20 ft^2) per person
Light engineering or packing works, mainly female labour	4.7 m^2 (50 ft^2) per person
Factory	7.0 m^2 (75 ft^2) per person
Specialized shops, show rooms for furniture, carpets, etc.	7.0 m^2 (75 ft^2) per person
Offices	9.3 m^2 (100 ft^2) per person

Height of occupants above and below ground level

As a general rule, means of escape will be required for buildings where there are occupants more than 1.9 m (6 ft) above or below ground level. This will be dealt with more fully under Stairways.

ROOM EXITS, CORRIDORS AND HORIZONTAL EXITS

An exit begins at the point from which an occupant proceeds to safety. Doors of small individual rooms such as small offices are not referred to as exits unless they actually lead directly to the open air, but in a large room the doors are considered as exits.

The size of a room is as important as the number of people in it, because if any employee has to travel a long way across a room to reach an exit, that way out might be barred by a quick burning fire before the occupant could get to the exit.

The width of an exit is measured at its narrowest point, except that the projection of a handrail is discounted. The unit of width is 533 mm (21in), but a single unit must not be less than 762 mm (2ft 6in). The unit of width is based on the assumption that 40 persons will be able to pass through it in one minute. One double unit exit of 1 m (3ft 6in) can be accepted as passing 80 people per minute, and for normal usage 200 persons could escape in the estimated two and a half minutes.

The distance of travel to an exit is measured along the normal path of travel, either to direct access outside the building or to a door leading to a fire-resistant enclosure or fire compartment.

Where only one exit is provided the maximum distance that anyone should travel, and have to travel, until they reach the nearest exit is 12 m (40 ft). The figure can be taken to 30 m (100 ft) if the floor for at least 18 m (60 ft) of travel is of fire-resistant construction.

Where a suitable alternative exit is available, the distance of travel may be extended to 45 m (150 ft). An alternative exit must be able to pass the entire number of occupants within the estimated allowable time.

If more than one exit is required, the additional exit or exits must be placed in positions away from each other and lead to different escape routes.

An interior passageway does not become part of an exit until it is enclosed as a fire compartment with at least half an hour's fire resistance. If a smoke-free passageway can be obtained, this could rank as temporary safety.

Smoke-free passageways

These are constructed in accordance with the standard of construction for a minimum of half an hour's fire resistance grading. A smoke-free corridor must be entered and left by a fire-resisting door and, if the corridor is more than 60 m (200 ft) long, additional fire-resisting doors will be required to subdivide it into shorter sections.

Arrangements of the contents of a room

Due regard must be paid to the arrangement of the contents of a room. These fall into two parts:

Fixtures

Machinery in factories, desks in offices and counters in shops must be so arranged that there is a clear passageway through the room to both the main and any secondary means of exit. The arrangement must also be made so that employees are between the point of greatest fire hazard and the exit. For example, the operator of a lathe turning magnesium alloy should not have to go round the machine before reaching an exit. When carrying out any hazardous operation, such as the example given, a secondary exit must always be provided.

No occupant of a room should be expected to scramble over or under any obstacle to reach a place of safety. A material supply pipe in a factory is often found in a position that people have to step over it or duck beneath it to reach an exit. This is very dangerous when it is remembered what the effect would be of anyone rushing to get to an exit in the event of a sudden flare-up or in smoke or with a lighting failure.

Loose materials

Loose materials, work waiting to be done, completed work waiting for removal, stock in a large store, loose chairs in an office – all are liable to be left in the clear space which should be allowed across a room or factory floor. Because this space has always to be left clear, there is a great temptation to put anything in them 'just for a moment'. But that 'moment' quickly becomes permanent.

One of the best methods of preventing this is to paint white lines on the floor (100 mm (4 in) wide), known as 'fire lanes'. If the words can be added (by means of a stencil is a good method) it seems to have a better effect. The daily inspection then has a positive force since the exit route is clear or it is not clear. The condition of the lines will be regularly noted, and repainting carried out frequently.

Doors leading to corridors and stairways

Internal doors on the escape route must open outwards or be sliding doors. The daily inspection should be used to see that the doors open or slide easly, giving special attention to the sliding arrangements above and below the door. Oiling and greasing of runners and hinges should be carried out monthly, wiping off any surplus oil.

Nothing should be allowed in or near the doorway which could foul the operation of the door or movement through it. In offices, there is always the danger of a loose hat or coat stand. Any loose objects left just inside or outside a door can lead to an occupant stumbling or falling, and the resultant pile-up behind could lead to serious consequences.

STAIRWAYS

When there are occupants above the ground floor, suitable stairways must be provided. Nearly always, two are required. No reliance must be placed on the arrival of the fire brigade to be able to effect a rescue from a roof or other parts of the premises.

While an exit above the ground floor from one building to another can form an excellent method of escape, this can only be so if the bridge itself is safe and can be kept free of fire and smoke. The entrance into and exit from the other building must be absolutely safe and certain. Both properties should be under the control of the same occupier.

Lifts may not be used for means of escape for a number of reasons. In the first place, the number of people that can be moved by this means is very small. Second, in the event of fire, the electric power is liable to fail or to be cut off. Third, there is the danger of anyone being trapped in what might virtually be a chimney, and last, the opening of lift doors might admit smoke to parts of the building not otherwise affected.

In existing buildings it may not be possible to obtain ideal stairways, but the following notes may be of help. (New stairways are adequately described in the appropriate British Standard code of practice).

At least one stairway should be internal, and only if no other method is available should an outside stairway be tolerated for the second escape route. This is because of snow and other inclement weather conditions, as well as the danger from smoke and flames issuing from rooms below the stairway. Where an outside stairway is adopted, there should be no doorways within about 3 m (10 ft) below, and all windows within that distance should be glazed in fire-resistant construction. Lowering lines and mechanical lowering apparatus, including chutes, cannot be used with safety, and are not permitted.

Approach to stairways

The approach to the stairway should be by way of a fire-resisting door, and it is better if two doors could be provided with a distance of 1.8 m (6 ft) between them to act as an air lock.

Continuity of stairways

The stairway should be continuous except for landings, and it should lead directly to the open air. It should ideally never be necessary for escapees to cross a room to get to another staiway or a final exit. If, however, such a crossing is necessary, it should be made by an exclusive fire-resistant corridor.

Landings on stairways

Landings should be formed at intervals of not more than 15 steps, and at all turns. Spiral stairways are not normally accepted but might possibly be for one person who is accustomed to their use.

Width of stairways and handrail

Stairways should be at least 1 m (42 in) wide, that is two units' width. If the stairway is 2 m (7 ft) or more in width, a centre rail should be provided. The first and last posts of the centre rail should extend to the ceiling so that people can see that there is a division. Unless a centre rail is provided people will not use the full width of the stairway. The centre handrail should be a double rail at a height of 1 m (3ft 6in) above the stair tread. If one side of the stairway is open, there should be at least one intermediate rail unless other protection is provided,.

Angle of stairway

The angle of a stairway should not be greater than 45 degrees. Vertical ladders are not permitted otherwise than for one or two men, and provided that they are accustomed to using them. Any such ladder should project 1 m (3 ft) above the upper floor level unless a platform is provided giving suitable stepping off facilities.

Treads and risers

Treads should be 250 mm (10 in) in width and risers should be 190 mm (7½ in) in height. The tread should be solid so that even the smallest heel of a shoe cannot be caught.

There should be not less than three nor more than 16 steps without a landing or platform. There should be no doors within 6 m (20 ft) of an outside stairway, and all windows within that distance should be of wired glass. It should not be possible to open them.

Ventilation of stairways

There is no doubt that the safest stairway is one built away from the main building with connecting bridges to various floors. The bridge should be entered and left by fire-resisting doors and the bridges should be enclosed in fire-resisting material. Particular attention should be paid to preventing the entry of smoke. This independent stairway should be ventilated by open windows at each level or by permanent ventilation at the top.

Stairways to basements

Basements wherever possible should be served by a separate stairway leading direct from the ground floor to their own final exit. Where this is not possible, there should be a fire-resistant lobby at the ground floor.

Rooms giving on to stairways

With the possible exception of toilets, no rooms should open directly on to a stairway but should be approached via a corridor and fire-resistant door. Cupboards, particularly cleaners' cupboards, should not open directly on to a stairway and where cupboards are situated in corridors they should be lined with asbestos-based board and kept locked. A notice to this effect should be permanently attached to the door.

EXIT DOORS

No doors may be locked or fastened in such a manner that they cannot be immediately and easily opened from the inside. To secure doors against unlawful or unwanted entrants, panic latches or panic bolts can be fitted on the inside. If there is a possibility that such doors might be opened for wrongful purposes, a simple intruder detection device can be attached to the door which will give an audible warning at an attendant's position and also switch on a light over the door.

Panic latches

A panic latch consists of a bar which when pressed releases the door catch. Figure 9.1a illustrates such a latch. A panic bolt consists of a bar which shoots a bolt to top and bottom of the door, as shown in Figure 9.1b. As with the panic latch, panic bolts can be provided with locks against entrance from the outside. If the door is to be used only in an emergency, no handle

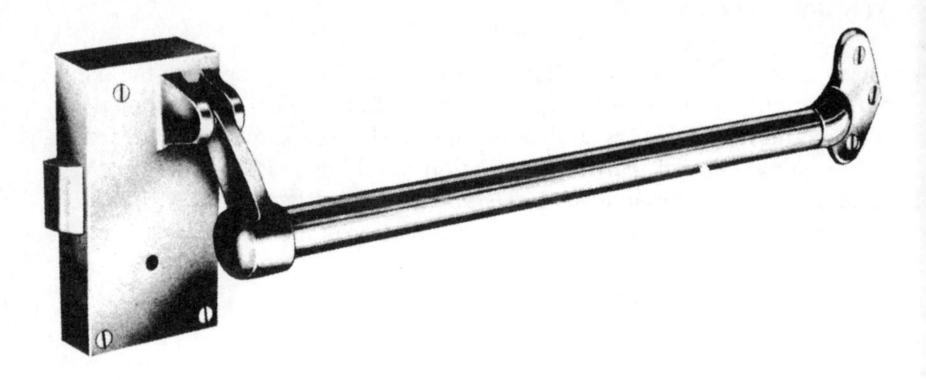

Figure 9.1a Panic latch.
(William Newman & Sons Ltd)

need be fitted to the outside, and the door is always secure against intruders. If entrance is required from the other side, and if it need not be restricted, a door handle can be fitted on the outside. If entrance may be required from the outside, but on a restricted basis, a lock can be fitted as illustrated in Figure 9.2a and b. The lock may be either a barrel lock or a lever lock, either of which can be keyed to a master key suite. No matter whether the key is turned or not, the operation of the bar on the inside will always open the door.

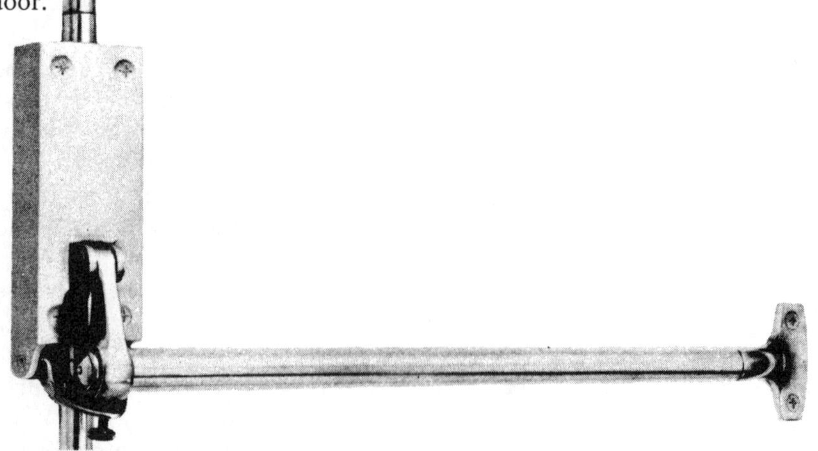

Figure 9.1b Panic bolt
(William Newman & Son Ltd)

Figure 9.2a Barrel lock for panic bolt or panic latch (William Newman & Sons Ltd)

Figure 9.2b Lever lock for panic bolt or panic latch (William Newman & Sons Ltd)

The bar on a panic bolt has a small catch, illustrated in Figure 9.3, which, when the bar has been pushed, retains the bar and so holds the bolts in the withdrawn position. This prevents the bolts from dragging or even jamming on the floor. To close the door it is only necessary to pull the door shut, and then release the catch. A further pull on the bar then shoots the bolts.

A particular watch must be kept on panic bolts during the daily inspection to see that dirt and rubbish do not accummulate in the bottom hole. This may happen accidentally or by carelessness, but it may be deliberately put there to beat the bolts. Sometimes it will be found that a short section has been sawn off the bottom and/or the top bolt for the same purpose. It is very difficult to detect, so that, when inspecting a door held by a panic bolt, the door itself must be tested to ensure that the bolts are holding as well as releasing. The correct method to test a panic bolt for release is by turning the back to it and leaning on the bar. It should not be necessary to jerk on a panic bar to make the bolts or lock release.

When panic bolts are fitted with locks, special provision has to be made to retain the bolts in the withdrawn position while the door is being shut. To be able to do this the catch has to be dispensed with, and an extra handle or grip put on the inside of the door. The procedure for closing the door on the inside is to push against the bar to hold the bolts withdrawn, and to pull the door shut with the extra handle. When the door is shut the bar can be released, and a sharp pull will shoot the bolts. To close the door from the outside the handle must be unlocked and turned to hold the bolts withdrawn

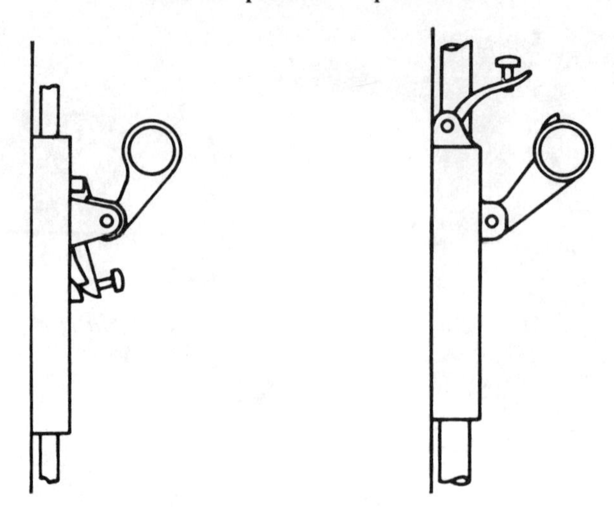

Figure 9.3 Catches for holding the bar of a panic bolt in withdrawn position.
(William Newman & Sons Ltd)

while the door is closed. The bolts can then be shot by turning the handle, and the door locked normally.

A much better method, however, is to install a self-locking top catch as illustrated in Figure 9.4.

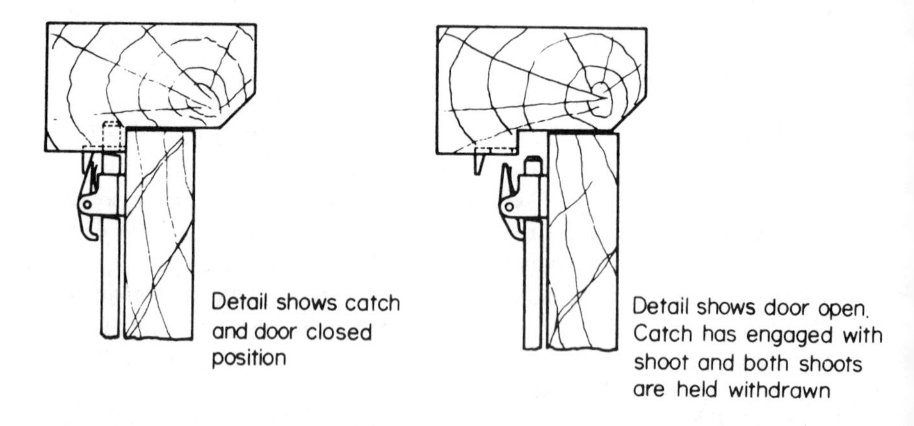

Figure 9.4 Self-locking top catch for panic bolt.
(William Newman & Sons Ltd)

Panic latch with cylinder

A panic latch is available for single rebated doors with push bar action giving immediate withdrawal of the latch tongue. It is fitted with a cylinder lock with key to open from the outside. It is very useful as a staff entrance lock and acts as an ordinary cylinder lock would (see Figure 9.5).

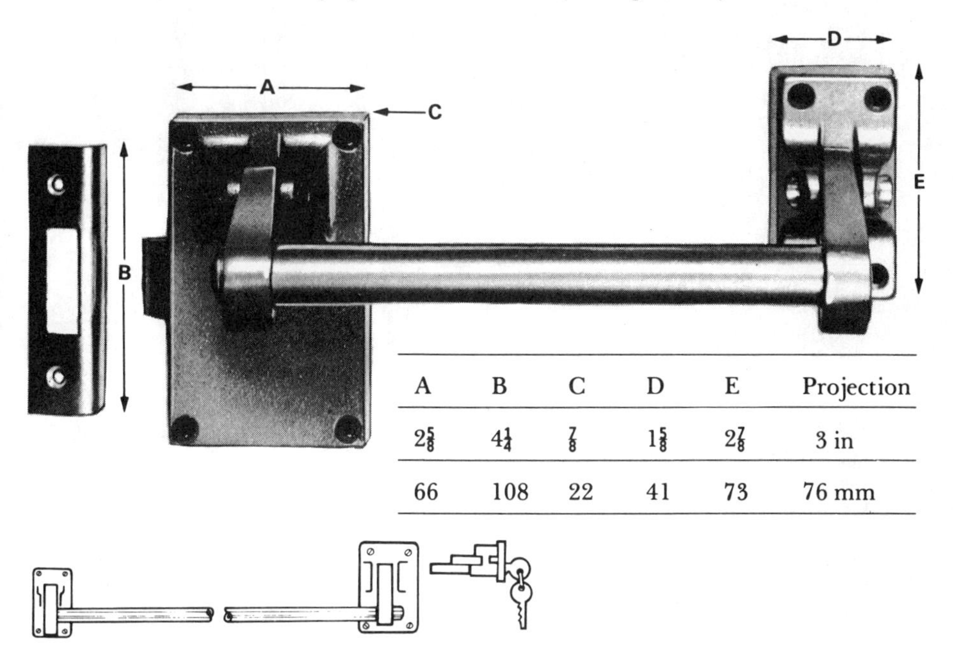

A	B	C	D	E	Projection
$2\frac{5}{8}$	$4\frac{1}{4}$	$\frac{7}{8}$	$1\frac{5}{8}$	$2\frac{7}{8}$	3 in
66	108	22	41	73	76 mm

Figure 9.5 Panic latch with cylinder lock on the outside.
(W & R Leggot Ltd)

Wellington bolts

Where entry from outside is required, the Wellington bolt (see Figure 9.6) has many advantages. The design is different from other types of emergency bolts that offer re-entry facilities, in this case the key operates the striking plate and not the bolt.

The Wellington bolt can be obtained in a double sided version: when the glass is broken on either side, powerful springs immediately withdraw the bolt. The door cannot be rebolted until the glass panel has been replaced.

Figure 9.6 The Wellington emergency bolt with key-operated striking plate. (Albert Marston & Co Ltd)

Redlam bolts

Where entry from outside is never required, a cheap and more effective method of securing a door to comply with the requirements of the law is by means of a Redlam bolt. This is illustrated in Figure 9.7.

The bolt is held under tension so that as soon as the glass tube is broken the catch is released. The Redlam bolt can be fitted to existing doors.

Glass key safes

Glass-protected keys should no longer be installed but where a glass key safe is permitted there is an excellent type that can be screwed directly on to the door. The frame is first screwed into position, then the key is hung on its hook and finally the glass panel is inserted. The glass panel is protected in a frame, and the frame is so fitted that once it has been put into the main frame, the key cannot be removed without breaking the glass.

Whenever key safes of any description are used, the key should be firmly attached to a chain of sufficient length to enable the key to be inserted into the lock. The object of this is to ensure that if the key is dropped it can be found instantly. The key safe should always be attached to the door so that the chain cannot foul the door opening and so obstruct travel through the door.

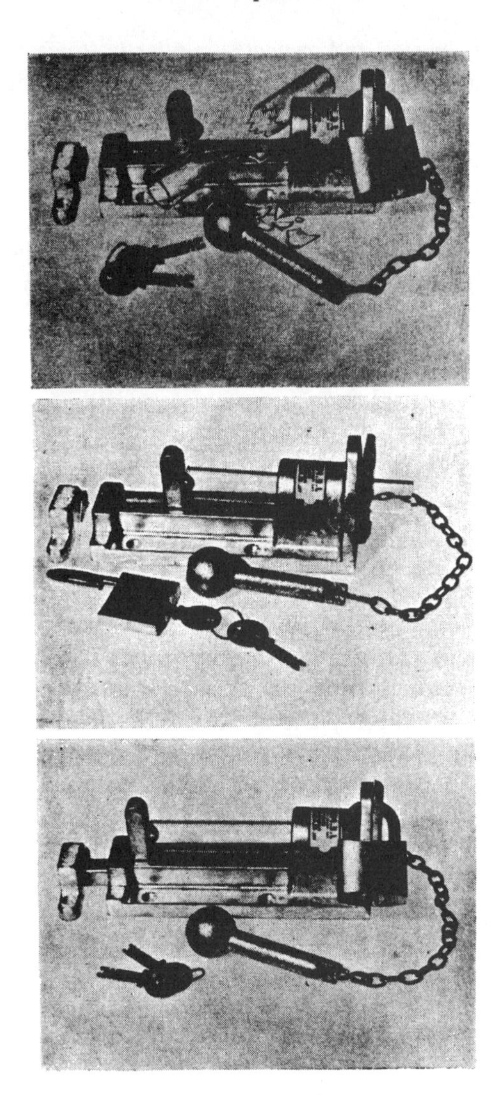

Figure 9.7 Redlam panic and security bolt.
(J E Mercer)
(a) Emergency operation, break glass cylinder and the bolt slides back automatically with quick positive action.
(b) To reset, slide in replacement glass cylinder.
(c) Bolt set in locked position.

Key safes should be listed and inspected regularly. At least once a year the key should be removed from its hook and the lock operated and the door opened.

FINAL EXIT DOORS

All doors affording a means of exit from the building must open outwards. This precludes the use of sliding and revolving doors as final exits from a building. Revolving doors can often be by-passed by a suitable door or doors placed on either side of them.

Outside landings

Final exit doors must not give immediately on to a step or steps, but a suitable landing must be extended from the door for a distance of not less than 1 m (3 ft). Guard rails should be provided on two sides of the landing.

Outside clearance

Final clearance should lead completely away from the building, and must not end in an enclosed yard. It is somtimes permitted for gates leading from a yard to the street to be kept locked in the open position when the premises are in use. Any passageway into which a final clearance leads should be kept clear as far as possible, and occupants should not be considered to be safe until they are sufficiently far away from the building to be safe from falling debris.

EXIT ROUTES TO BE SIGNPOSTED

Suitable signs must be placed to show the exit routes. Signs do not have to be exhibited along the normal route used by occupants, but they must be incorporated on secondary routes.

Once having commenced to signpost a route, it should be followed through with a sign and an arrow at every turn until the final exit to the open air.

Lettering should preferably be white on a dark background, or light green on a black background in letters of about 125 mm (5 in) minimum size. Where a facing wall can be obtained the sign should be placed flat on the wall, but where there is a turn partway along a corridor, a projecting sign is required.

ALTERNATIVE MEANS OF ESCAPE

There is a number of reasons why alternative means of escape are required, for example, the height of the building, the presence of readily ignitable materials and any other reasons that might lead to a rapid spread of fire or indeed anything that might render the main escape route impassable. Even a slight amount of smoke can be frightening and even dangerous.

A second stairway is the usual means of providing an alternative exit. To be effective a second stairway must be in another direction from the first and at the opposite end of the building. The second stairway must have all the safety features of the principal escape route and should lead to a final exit on the opposite side of the building.

Where an effective stairway is available the distance of travel across a floor can be increased.

SURFACE FINISHES ON WALLS AND CEILINGS

The heating of surface finishes in the early stages of a fire can produce considerable quantities of smoke and may also spread fire and flame across the surface of walls and ceilings. It is often extremely difficult to discover what surface finishes have been applied once they have been seasoned. If necessary the surface finish should be removed and a known material resistant to the spread of flame and fire applied. Expanded polystyrene tiles or surfaces which have been coated with gloss paint should definitely be removed.

Advice on many groups of materials and the situations (stairs, corridors and lobbies) listed where they can be used will be found in the *Guides to the Fire Precautions Act*, obtainable from HMSO.

10

Emergency lighting

Emergency lighting, which is often referred to as safety lighting, can be divided into two parts: escape lighting and stand-by lighting.

Escape lighting is the illumination required to enable people to escape from the building with safety and to enable fire precautions to be operable under fire conditions.

Stand-by illumination is the illumination required for people to be able to carry on their ordinary work under circumstances such as the rupturing of a fuse when there is no need to evacuate the building.

Frequently, emergency lighting will consist of a mixture of both systems but where this occurs the requirements of escape lighting must be met in full irrespective of the stand-by lighting. This section deals only with the requirements of escape lighting.

The lighting points on the emergency lighting system are known as 'luminaires', the term embracing all the apparatus within the case and the case itself. An internally illuminated sign can be considered as a special type of luminaire.

REQUIREMENTS OF ESCAPE LIGHTING

Escape lighting must provide sufficient illumination to meet the following requirements.

Visible exit directional signs

Signposting has been described in the previous chapter and it is essential that the signs can be readily seen at all times. Under ordinary circumstances primary lighting will be sufficient for normal purposes although where signs do not stand out particularly well, additional lighting from the emergency lighting system may be advisable.

To make a sign particularly outstanding it can be made self-contained in a box. The box must be made from a self-extinguishing polycarbonate with an opalescent front panel. The wording can be silk screened and should be in green lettering.

Other than being self-contained in a box, exit route signs must be adequately illuminated and light must fall on the sign at all times. When designing the emergency lighting due attention must be paid to this point. Such lighting must not constitute a glare to anyone approaching it and if necessary the lighting point must be shaded to achieve this purpose.

Safe movement along the exit route

The level of illumination is generally required to be of not less than 0.2 lux but a rather higher value is preferable and this can be obtained by using a rough guide from the following formula:

$$\text{Wattage required} = \frac{\text{Area in square metres}}{2}$$

Using this formula, the luminaires should be not more than 3.5 m and not less than 2 m from floor level.

It is not sufficient, however, to provide only enough luminaires to meet this formula and they should be so placed as to illuminate any special hazards. Principally these will be:

- At any intersection of corridors
- At each change of direction
- Over each flight of stairs and so arranged as to illuminate and emphasize the actual steps
- Above the final exit door on the inside of the building
- If necessary the immediate route from the final exit.

In addition to the foregoing positions for luminaires, there is a number of other points of danger that require illuminating under fire conditions such as:

- Manual (break-glass) fire alarm points
- Any telephone that may be used to call the fire brigade
- Fire equipment points
- Kitchens and places where hot plates or moving machinery may be found. It is possible for the power supply to this equipment to remain on after the lights have failed. In particularly dangerous situations of this kind an interlink is sometimes madc so that the failure of the lighting circuit automatically cuts out the power circuits. However, where there is any overrun, e.g. the machinery does not stop instantly, it is essential to use emergency lighting of a type that has instant change-over.
- To allow for fire brigade requirements in the early stages of a fire.

MODES OF OPERATION

Three modes of operation are recognized:

1 Non-maintained
2 Maintained
  (a) floating battery
  (b) change-over relay
3 Sustained

The non-maintained mode of operation

In the non-maintained mode of operation the emergency lamps are not energized until there is a falure of mains current. Figure 10.1 shows the outline circuit of a non-maintained mode.

It will be seen that while current is flowing in the relay the switch operating the lamps is held off but on failure of the power supply the emergency lighting is switched on.

Various refinements may be added such as an indicator light to show that the battery charging equipment is healthy, and a test switch to break the relay circuit which will be the equivalent of a mains failure and should cause the emergency lamps to be energized from the battery.

The maintained mode of operation

Floating battery

In the maintained mode of operation the lamps are energized at all material times. This may be accomplished by floating the battery across the mains

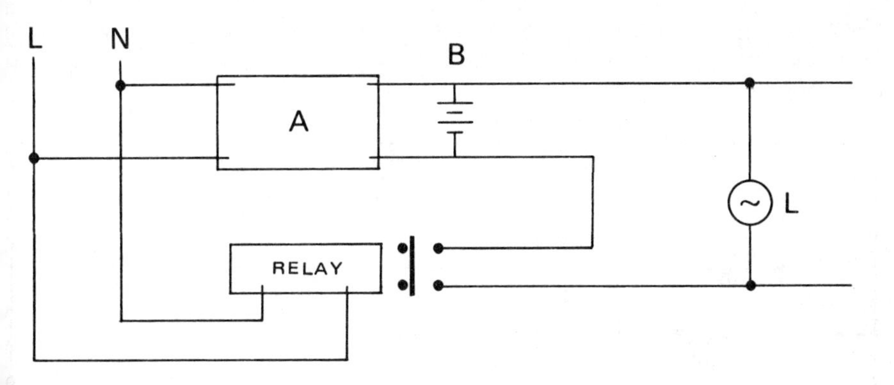

Figure 10.1 Non-maintained mode of operation – (mains alive).
A Charging equipment B Battery L Lamp(s)

via the charging unit and directly across the emergency lamp (or lamps). Figure 10.2 is an outline drawing of such a circuit.

This system is good insofar as there are no working parts to effect the change from normal to emergency working and the change is instantaneous. There is usually a slight falling off of light output when the battery takes over. The chief objection to the system is the difficulty of recharging the battery after a period of use as the normal battery charging equipment is for a trickle charge.

In the alternative system of a change-over relay a boosting charge can be given to the battery to bring it back to a fully charged condition.

Change-over relay

In the maintained mode of working with a change-over relay, the battery is isolated from the lamps, which are directly energized from the mains. Figures 10.3a and 10.3b show the circuit when (a) the mains are normal and (b) the mains have failed.

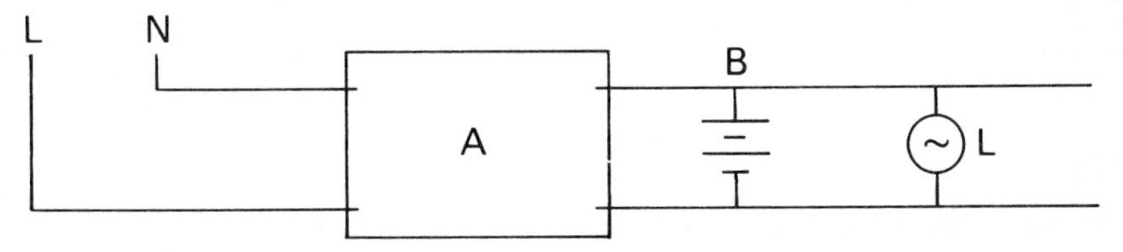

Figure 10.2 Maintained mode of operation – floating battery
A Trickle charging equipment B Battery L Lamp(s)

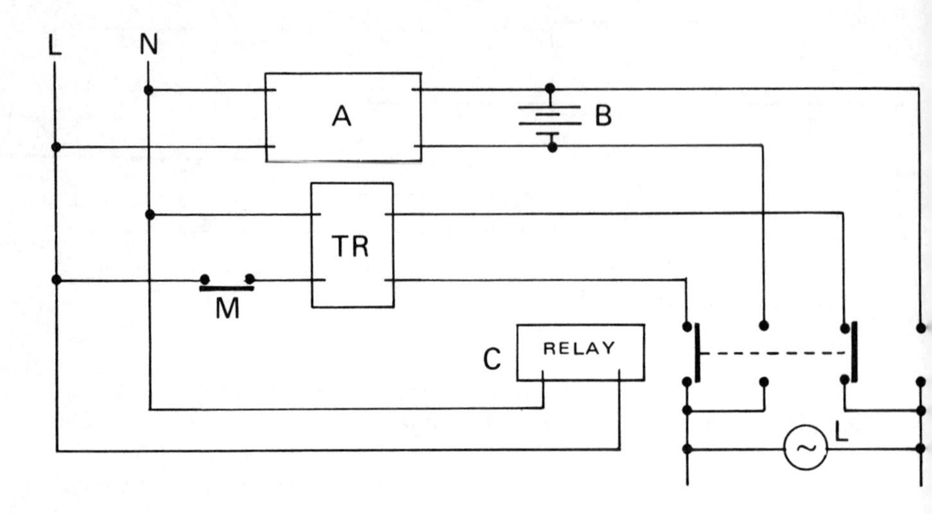

Figure 10.3a Maintained mode of operation – change-over relay (normal)

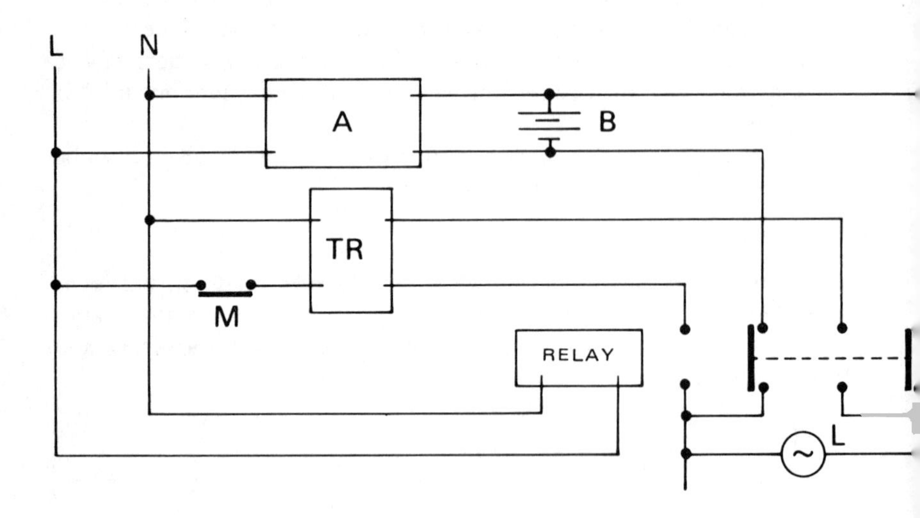

Figure 10.3b Maintained mode of operation – change-over relay (with power failure).

A Battery-charing equipment
B Battery
C Relay (with two coupled armatures)
M Manual switch for normal lighting control
L Lamp(s)
TR Transformer

In both diagrams a switch is shown (at M) which can be used to turn off the lights during normal daylight hours but it will be seen that emergency lighting circuit will bc cnergized on mains failure irrespective of the operation of this switch. This was a useful addition when tungsten lamp bulbs were used as continuous operation rapidly burns out these lamps whereas the modern fluorescent units are not so badly affected.

The sustained mode of operation

The sustained mode of operation has two separate circuits in each luminaire. One circuit is operated by the mains and one by the battery. It is difficult to see why this is known as a separate mode of operation as it is in practice a non-maintained mode with a mains-operated bulb brought into the luminaire casing instead of being on the outside. Figure 10.4 is a simple outline circuit of the position.

The system has a considerable disadvantage in that the heat of the mains bulb tends to diminish the life of the battery.

POWER SUPPLY SYSTEMS

Power supply systems may consist of:

1 Self-contained batteries
2 Central batteries
3 Batteries for emergency lighting
4 Stand-by generators
5 Static inverters

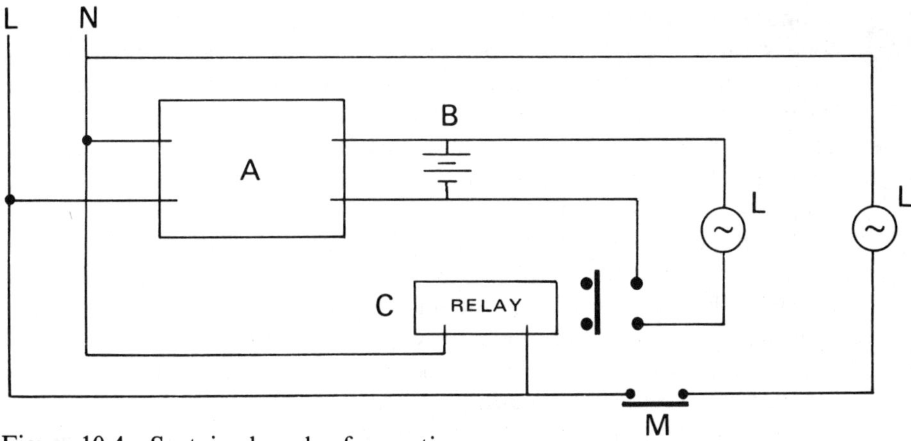

Figure 10.4 Sustained mode of operation.

A Battery-charing equipment
B Battery
C Relay
M Manual switch
L Lamp(s)

Self-contained systems

These are wired into the live side of the lighting subcircuit which supplies the normal lamps in the vicinity of the emergency light. In this way the emergency lamps will light up if there is any form of lighting failure whether due to fire or any form of wiring breakdown such as a blown fuse.

The connection to the mains can and should be in the form of ordinary PVC-covered cable so that damage to the wiring will cause the emergency lamps to energize. If an emergency lamp lights up for no apparent reason it should be taken as a warning that there may be a breakdown of the wiring insulation in a concealed place. Such a breakdown might be an indication of an incipient fire.

When connecting a non-maintained system to the mains electric supply, particular care must be taken to ensure that supply voltage is correct with the charging circuit requirements of the luminaire. The reason for this is that if the charging circuit is designed for a 240 VAC supply and only 220 volts are supplied to it, charging up from a flat battery could take considerably longer, possibly even twice as long. This could be a serious problem as the recharge time is usually specified by the fire authority.

Care must also be taken to see that the battery is not subjected to undue heat. The maximum temperature to which the small sealed batteries should be exposed is 30°C; above this temperature the battery life will be quickly reduced.

The British Standard relating to emergency lighting for premises other than cinemas and certain other places of entertainment is 5266 which recommends that self-contained luminaires should be energized from the battery for a short period at least once a month by simulating a power failure. This test is to ensure that the apparatus is functioning correctly. The luminaire should also be energized for at least one hour every six months and if it is required to satisfy a condition that the battery will maintain full emergency lighting for longer than one hour, it should be energized for its full period every three years.

The provision of a simple non-locking press switch is now usually standard with the circuitry of the luminaire but it is as well to check that the switch can be operated without too much labour. It should not be necessary to dismantle any part of the luminaire to get to the switch.

The longer test and three-hour tests are more difficult to arrange. At the moment it is generally necessary to remove the subcircuit fuse. If the premises are in use for 24 hours a day this may not be possible. The use of a lock-on switch has the grave drawback that it may be forgotten and not returned to normal at the end of the test. It must also be in such a position that it is not subject to vandals and yet not be too expensive in labour for easy operation. The removal of a subfuse where possible is also subject to the same grave consequences of not being replaced at the end of the test, although this may partly be offset by the fact that the normal lighting in the area will not be available. The most satisfactory method would seem to be some form of time-delay switch, incorporated in the luminaire circuit. This would consist of a press-button switch which would turn on the emergency lamp immediately and would continue in the on position until automatically reset by the time delay operation. During this time each luminaire could be checked just before the three-hour test completes itself.

Central battery systems

These systems consist of a large battery supplying a number of luminaires. It used to be considered that this was the cheapest method of supplying power to emergency lighting but it has many disadvantages and the individual non-maintained system is now more commonly installed.

From an operational or fire service point of view probably its greatest disadvantage is that should the main distribution cable become damaged the whole of the system is out of order. Similarly, unless some form of subcircuit failure is brought back from individual subcircuits, the emergency lighting does not operate until the main electric power supply breaks down.

From an installation point of view, wiring costs militate against the central battery system. Only in very large systems is the emergency supply likely to be other than a direct current (DC) system (and the system then would probably be not a central battery but a static inverter system). If the supply is DC then the wiring must be sufficiently heavy to take the load and avoid voltage drop. Thus a 12-volt central battery could supply very few luminaires and the selected voltage would have to be at least 24 VDC and probably 50 VDC or higher. Unless the wiring is sufficiently heavy to ensure no voltage drop in the length of run, the lamps may be 'under-run', that is there would not be sufficient voltage at the lamp for efficient working. Even a slight voltage loss in tungsten lamps means a high loss in light output and a fluorescent tube might not even strike.

The output wiring must also take into account that the lamps nearer to the supply will be at a higher voltage unless suitable calculations for wiring size have been made. If this has not been done it may be found that in operation there is a variation in light output.

Some authorities specify the type of wiring to be used. MICC is often quoted as is PVC in metal conduit; both are more expensive than ordinary PVC and both are more expensive in labour to install. However, this protection is needed for a central battery supply system and both lose the value of energizing the emergency lamps if fire damages the cable. It is interesting to note that PVC in plastic tubing is gaining in popularity as damage to the tube does not result in a short circuit. In any event wiring for central battery systems should not be brought through areas liable to damage by fire.

A further disadvantage of the central battery system is that special precautions must be taken in housing the batteries. The reason for this is that one of the advantages of a central battery system is that lead-acid batteries can be used, but lead-acid batteries have to be ventilated to the open air because of the highly flammable and explosive gas hydrogen that is produced during charging operations. It is essential that the battery room should be a fire-resistant compartment.

Batteries for emergency lighting

Sealed nickel cadmium cells

Sealed nickel cadmium cells are specially recommended for use in self-contained luminaïres because they stand considerable variations in discharge and recharging conditions. The chief form of abuse is using the battery at high temperature, and cells should not be subjected to temperatures exceeding 40°C, which means that the ambient temperature should not exceed 30°C.

The discharge rate for standard nickel cadmium cells used in self-contained luminaires is about 1 ampere, which will mean that a 4 ampere hour cell is required to ensure a three-hour discharge. Each cell when fully charged develops a fraction over 1.2 volts so that a 6-volt battery will consist of 5 cells.

The normal charge rate needed to achieve 80 per cent cell capacity within 14 hours is about 0.125 of the ampere hour capacity of the battery at a voltage of about 1.2 volts per cell. After the 80 per cent capacity is achieved, the rate of acceptance of charge of the battery is reduced and much of the charging current is dispersed as heat. The higher the ambient temperature, the slower will be the acceptance of the charge.

If unused and fully charged, nickel cadmium cells will retain 60 per cent of their charge for about a year at 20°C. If stored at 0°C they will retain about 90 per cent charge for over a year but if stored at a temperature of 40°C they will be flat in about six months. In use, the life expectancy of sealed nickel cadmium cells is between two and five years but much depends upon the temperature in which they are used.

Vented nickel cadmium batteries

Vented nickel cadmium batteries are often used in central battery systems. They have a life expectancy of some twenty-five years. Maintenance is very simple and only consists of topping up the cells at infrequent intervals. They are resistant to damage from overcharging and high discharge rates and they will not deteriorate so rapidly as other types of batteries if left in a low state of charge.

The use of translucent cases for the cells enables the liquid level to be checked at any time and a special low electrolyte level indicator can be provided as an additional safeguard.

Lead acid batteries

Pasted plate lead-acid cells specially made for emergency lighting will give reliability for a life of about five years if well maintained. Charging should be by means of a fully automatic self-regulating constant potential. If correctly designed and adjusted only occasional topping up should be required.

Flat plate cells have a life of about ten to twelve years. Their chief advantage is their low cost to life ratio.

Plante cells have a life expectancy of about twenty-five years and maintain their voltage well to the end of discharge period.

Car batteries should not be used for emergency lighting. The design of cells for car batteries is for intermittent use, whereas those for emergency lighting are required for constant use.

Static inverters

Apparatus such as iron lung machines, computers and other special systems require a constant AC supply of power to be available, so use can be made of static inverters for an emergency supply of power. These are supplied by batteries having a DC input to the converter of 24 V, 50 V, 110 V or 220 V and the standard AC output is 240 V single phase 50 Hz. Power ratings vary between 100 VA and 1500 VA or more if needed.

The inverter contains a thyristor together with control circuits. Protection has not only to meet overload conditions but must prevent current reversal.

Three systems of mains change-over are available:

1 The first system is a stand-by passive system with the load normally carried by the AC mains and the battery on charge. On mains failure the battery is switched to the converter and it returns automatically to the mains on restoration of power. The change-over may take a few seconds.

2 The second system is an active stand-by system in which the inverter is permanently connected to the mains. Because the inverter is already activated there is practically no delay when the mains fail and the inverter takes over.
3 The third system is when the battery and inverter are supplying the load. There is then no break in the supply of power; it simply means that the inverter draws the full load from the battery.

In all cases, a battery charger will be required which will be of a standard type.

Stand-by generators

These are generators operated usually by a diesel engine. They can usefully be compared with a diesel engine employed to drive a booster or stand-by pump for a sprinkler system. The precautions as for maintaining such a motor are fully described in Chapter 20.

There is always some delay before the diesel motor can reach full power output. As much as 10 seconds' delay can sometimes occur, and such a delay would not be acceptable for emergency lighting. The delay, however, can sometimes be bridged by the use of a battery.

LIGHT SOURCES FOR EMERGENCY LIGHTING

For self-contained luminaires there is a choice between:

1 Incandescent lamps
2 Fluorescent tubes
3 Cold cathode tubes.

Incandescent lamps

The incandescent lamps used for emergency lighting in self-contained luminaires usually operate on 6 or 12 volts. Special Krypton filled lamps are available to give a better light but in general they are not so efficient as fluorescent tubes. It is also possible to obtain 6 V, 10 W tungsten halogen lamps which give a very high light output. However, even this is not so great as an 8 W fluorescent tube. The life of these small bulbs may be as low as 100 hours. They are not suitable for maintained operation.

Fluorescent tubes

Fluorescent tubes used in self-contained luminaires are usually 150 mm to 525 mm in length and 16 mm in diameter. Two short tubes are often used in place of one long one. They are rated at 4, 6 and 8 watts and give 100, 110 and 160 lumens output respectively. These small fluorescent tubes often have a life of a year.

Cold cathode tube

Cold cathode tubes are operated by an electronic inverter which develops a high frequency alternating current and a much lower voltage with a low safe current. The cold cathode tube has a much longer life than a conventional fluorescent tube and is better able to withstand a falling off of battery power below the recommended voltage. It can be used in maintained and non-maintained circuitry. It is claimed that because of the high frequency and low voltage it gives greater illumination.

LIGHT METERS

Expensive analogue light meters are available which will accurately measure low light values in order to assess whether an emergency lighting system adequately meets the legislative requirements laid down by the various authorizing bodies. These light meters are necessarily delicate and are therefore easily damaged both mechanically and by light overloading.

There is available, however, a simple hand-held battery-powered meter device which will give readings from 0.1 lux to 0.5 lux in stages of 0.1 lux (see Figure 10.5). There are on the face of the machine five red LED signals and the light level is shown by the number that lights up.

A big advantage of this piece of apparatus is the fact that even in low ambient light levels the number of LEDs signalling can clearly be seen whereas the indicator on a dial cannot easily be read. It is a robust electronic piece of equipment incorporating inbuilt protection against light overload.

PHOTOLUMINESCENT MATERIALS

If an emergency lighting system is well designed and properly maintained, the probability that it will operate correctly when needed is high. However, it must be accepted that failure of the complete system, or some part of the system, is possible. People trying to escape could then be left in complete darkness, possibly in an unfamiliar area. In this situation a very useful back-

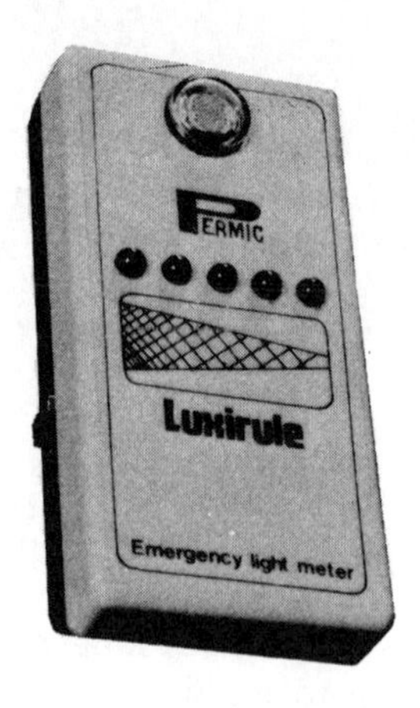

Figure 10.5 Battery-operated low level light meter. (Permic Emergency Lighting Ltd)

up can be provided by photoluminescent materials. These glow in the dark in a similar way to the luminous paint used on watches and clocks.

The range of products of Jalite Ltd include standard safety signs to BS 8578, together with tapes, arrows, sheet materials and paints. The photoluminescent material is not radioactive but is capable of absorbing energy when it is exposed to light, and of releasing this energy as visible light. Immediately after activation (by a brief exposure to daylight or other light) the luminous intensity can be in excess of 90 mCd/m^2 (millicandlelight per square meter). This intensity falls continuously to around 8 mCd/m^2 in 30 minutes. The manufacturers point out that the average person's vision becomes dark-adapted at a rate which roughly matches this decrease in intensity, and the signs can remain visible for several hours.

In addition to marking out an escape route, the material can be used to identify fire extinguishers, manual call points, telephones, and panic bolts. A particularly useful development is the self-adhesive arrows which can be attached to the floor along an escape route. If smoke penetrates into this route while people are trying to escape, the emergency lighting will be obscured but, with their heads beneath the smoke layer, the people will be able to see the arrows.

It is important to note that the photoluminescent material must receive light at a sufficiently high intensity (although only for a brief period) before it can function. It is unsuitable for use in a dimly lighted position.

11

What to do in case of fire

If a fire should break out, the first consideration is to see that all occupants of that fire compartment can reach a place of safety.

A fire compartment is considered to be a building or section of a building so constructed that a fire will not spread into or out of it for some considerable time. For small buildings, all occupants will be evacuated at once. But where large numbers of people are employed in buildings such as multi-storey office blocks, which are constructed in fire compartments, it is better not to evacuate everyone at once, but only those people in the section affected. Immediately they are clear, it might be advisable to evacuate the next adjoining sections. Separate warning signals for this purpose will be given by means of the signal distribution panels or by some similar means.

It is a legal requirement that every employee should be familiar with the exit routes from a building, and the only satisfactory way to ensure this knowledge is to prepare an evacuation scheme leading to regular evacuation practices of both the primary and secondary ways out of the building. The scheme suggested here should be considered for model purposes only. It will need modification to suit the size and nature of the premises, but the general principles of the scheme should be followed.

ORGANIZING AN EVACUATION SCHEME

Formulating committee

The preliminary work for an evacuation has already been discussed: the means of giving warning in the case of fire, and the escape routes and doors through which the occupants must pass. The next step is to appoint a general organizer and a small formulating committee. The general organizer will be responsible for putting into effect the decisions of the committee, and in the event of a fire will be available to the incident controller. The general organizer should not be the fire precautions manager nor the security officer. The committee must work in close conjunction with these officers, as well as with the personnel officer, union officials and the company executive responsible for fire. If the local fire officer is not a member of the formulating committee, he should be kept fully informed of the proposals, and the scheme must be submitted to him before it is brought into effect.

Check points

The check points are the places where employees report that they are clear of the building, and also where the wardens finally report that the section for which they are responsible is clear of people. The duties of checkers and wardens are described later.

The committee will decide where the check points are to be situated. They must take into consideration that one of the instructions issued to employees will be that, when they hear the evacuation signal, they must immediatey leave by the nearest exit and may not under any circumstances go to a cloakroom. Employees who sense no immediate danger (although unknowingly they may be in great danger) will not go out into inclement weather without some form of protection. As it is very important to obtain the fullest cooperation from everyone, it is a good idea when giving notice of a practice to state that it will be cancelled if the weather is inclement. Cancellation of a practice is quite simple – the evacuation signals are just not operated. The whole purpose of the practice is to give people confidence in the signal, and to ensure that their reaction is correct.

Having decided on the location of the check points, the committee will arrange for suitable signs to be prepared and erected. The most satisfactory signs consist of one letter. The sign should be very large, and plainly visible over a long distance. It should be fixed about 3 m (9 ft) from the ground.

Checklist

For each check point, there should be a checklist of names of people who are expected to report at that point. This list should be prepared with the help of the personnel department, and kept up to date. In one factory having a fairly high fire hazard, the list is brought up to date by the checker as his first job, morning and afternoon.

The list should be kept near the exit leading to the check point, and its position known to several people so that in the absence of the checker someone else will pick it up and carry out his duties. A good rule is that the first person out of the room takes the list. The list should be typed with a box beside each name, so that all the checker has to do is to tick the names as people report. No time should be lost by the checker having to stop and write in a name, nor should he have to make a roll call. It should be impressed on employees that they must give their names, and not wait to be asked.

Assembly point

The assembly point is where people gather in safety until they can be returned to the premises or dismissed to their homes. If possible, it should be a place well protected from the weather. A canteen clear of the building is ideal, but very often it has to be the same place as the check point.

It must be understood that an employee once having passed through the check point may under no circmstances return to the building until the incident is officially closed or when he has obtained the permission of the fire brigade given through the incident controller.

PRINTED INSTRUCTIONS

In many companies, a book of the rules of the company is given to every employee on joining the firm. The rules usually include instructions on what to do in case of fire. But the book may be very long, and seldom read because of its length, and in any case the employee is more interested in the benefits he may obtain than the rules of safety. It is therefore advisable to give every employee a personal card, not larger than a postcard, giving on one side short concise instructions on what to do in case of fire, and on the other side some useful instructions about fire precautions. It should be noted that the card gives the check point for that particular individual.

Every member of the company, from the chairman to the most junior member receives a card, and a note should be recorded on the personal folder of the employee to the effect that the card has been so issued so that no one can plead ignorance of what to do.

Large forms printed on waterproof card should also be prepared. The wording of these can be much the same as the front of the card, but these forms should be numbered and fixed permanently to suitable places throughout the building, including the inside face of every office door. A record should be prepared showing the position of each form by its number and the notices should be inspected during the daily inspection of the premises. It should be specially noted that this large notice describes the nearest exit and the alternative exit from the point of exhibition. It should be emphasized that employees not in their own department when the warning is given leave by the nearest exit from wherever they are. Then, having reached the open air, they must make their way back to their own check point without going into or through the building.

ALLOCATION OF DUTIES

According to circumstances certain duties are required following an evacuation signal. The following lists some of these duties but other special duties such as the closing of fire doors may also have to be allocated. It must also be remembered that the working hours of staff and employees may not be the same, and also that the organization must be able to take care of night shift and out-of-hours workers.

Incident controller

If not already appointed, an incident controller must be nominated. The incident controller should be the senior executive officer of the company likely to be present. His duty is to be the central point of information for everything connected with the incident. He must be in a position to give orders and instructions on behalf of the company. His position will be near to the entrance by which the fire brigade will enter the premises and, should the incident be large enough for the fire brigade to set up a control point or a mobile control unit, he would be established in close proximity. The exact arrangements should be made with the fire brigade which will be only too pleased to know where to find an executive official of the company and to be able to refer to him for information and liaison.

Incident officers

On the staff of the incident controller there will be two or more incident officers. In the absence of the incident controller an incident official will take over his duties. Incident officers will receive the reports from the check points, and it is to be hoped that by the time the fire brigade arrive information can be given them that everyone is clear of the buildings – or, if not, in what area people may be trapped. It may not be possible to achieve absolute reliability of the reports, but every effort should be made to account for everybody.

Wardens

A note should be circulated to heads of departments asking them to select wardens and make arrangements for someone to do the duty in the absence of the warden. Section wardens will generally be senior members of the staff, and foremen or forewomen on the shop floor. Their duties will be to see that all people in their section promptly leave the building, and that no one is left behind in cloakrooms or toilets. They will generally be the last to leave the section and, if time permits, will see that doors are closed and machinery shut down. Should they at any period of the incident know that people are trapped, they will bypass all reporting channels and send a report direct to the incident controller, or use any other means of notifying the fire brigade immediately.

Checkers

Checkers will be employees who can arrive early at the check points and can then record that people on their nominal lists can be accounted for. They cannot be expected to check more than 25 to 30 persons. Despite protests that it is not possible, fairly accurate information can be obtained of the whereabouts of all employees if the cooperation of everyone is obtained.

As soon as the checklist is complete, and as quickly as possible, a report must be taken to the incident oficer who is receiving reports from that particular section. The incident officers will report to the controller as soon as they know the state of their locations. Even if the list is not complete, an immediate report should be sent to the incident controller, and nearest fire officer, if the checker has information of any persons being trapped.

Special service marshals

Under certain conditions, and provided that danger is not imminent, it may be possible to permit certain key people to continue to work after the evacuation signal has been given. Mention of these people is made in the next section. If smoke is seen, evacuation is compulsory. In order that these people may be looked after, special service marshals are appointed. They are senior officers who have power to order complete evacuation if they consider there is real danger, or if instructed to do so by the incident controller or senior fire officer. Special service marshals must be in close contact with the incident controller. They will not necessarily be with the employees remaining at work, but must be able to communicate with them very quickly.

INSTRUCTIONS FOR EMPLOYEES HAVING SPECIAL DUTIES

Ambulance staff

The ambulance rooms should, if possible, be situated outside any fire risk area, in which case the staff would remain on duty. If this is not possible, they should report to a position from which they can easily be called by the incident controllers. A portable first aid kit should be kept ready assembled so that the ambulance staff have equipment ready to convey to any emergency.

Boilerman

Appropriate instructions in the event of fire should be posted up in the form of a large notice in the boiler room. The engineer will draw up these instructions as necessary.

Canteen staff

Consultation should be carried out with the canteen manager so that suitable instructions can be posted up and given to the staff. It would be bad policy to ruin the meals of everyone because a fault occurred in the alarm system. According to the position of the canteen, it might be possible to leave at least one cook behind under supervision by a special services marshal.

Cashiers and wages clerks

If conditions are suitable, the wages clerks and cashiers might reman behind to secure any exposed money. They would be under control of a special services marshal, but they would leave immediately that security of cash had been attained. Under no circumstances would they be allowed to continue normal work.

Computer staff

One member of the computer staff can remain behind if conditions are suitable, until he has completed the run, so that valuable time and records may not be lost unnecessarily. Should danger become imminent, as by smoke being seen, or should a special marshal order evacuation, the computer must be switched off and left immediately. If necessary, a large notice giving instructions on what to do in case of fire should be prepared and exhibited. Under no circumstances may the operator continue his normal work.

Directors and their secretaries

During the evacuation of the building, the directors and their secretaries will go through a check point. But arrangements should be made so that in the event of a serious fire they can go to an assembly point and meet at another office to make such decisions as may be necessary to keep the business running or to restore it to running order. It is possible that reciprocal arrangements could be made with a neighbouring company.

It is most important that directors set a good example and evacuate the building, treating the evacuation seriously, even for practice. If at any time, directors or senior management take advantage of their position, it will be almost impossible to persuade employees to carry out their instructions correctly.

Maintenance engineer and staff

Part or all of the maintenance staff should be organized as a salvage unit. The remainder, if any, should be allocated a special assembly point where the engineer can quickly obtain the technicians he requires.

The maintenance engineer must at all times be in contact with the incident controller.

Fire teams

Instructions to fire teams and any private fire brigade will contain the methods by which they will assemble or report directly to the incident. In medium sized premises an instruction is sometimes included that members of the fire team will take a fire extinguisher or other equipment with them when attending directly at an incident. This brings considerable additional firefighting power to a small incident.

Messengers

A number of boys, or if boys are not available young girls, should be detailed as messengers. The number will depend on the size of the premises, but some must be attached to the incident controller. They should have message pads, and if the size of the premises warrants it, they should be provided with a cycle.

Process workers

Process workers are a special class, as the sudden shut-down of plant in itself might cause an incident. Very careful instructions should be prepared by the process engineer to govern the shut-down of plant in the event of various contingencies. But it must be clearly understood that if there is any threat of danger, particularly from smoke, process workers must evacuate the building immediately. A special service warden will be allocated for their protection.

On some plants it may of course be possible to arrange a completely automatic shut-down.

Security staff

Security staff must not be involved in other duties. They will see that traffic is diverted and the roads kept clear for the fire brigade. At all times they will bear in mind the possibility that a fire has been caused to create a diversion.

Telephonists

Instructions to the telephonists must be prepared according to local conditions. The essential feature is a very large notice that can be read by every telephonist with ease, and without moving from her position. This notice will contain in bold letters on a distinctive background the address of the premises as agreed with the fire brigade, generally in the form of a cross reference, giving the name and address of the firm, with an additional landmark. For example: *Smith & Son, 25 Side Road, off Town Road.*

The notice should also contain an alternative number for the fire brigade in case of difficulty in dialling 999. No other information should be displayed on this board. Any other emergency telephone numbers or information must be given on another board and in another colour.

In addition to this notice, detailed written instructions should be given to each telephonist. These should also be pinned up on a notice board near the switchboard, and should be reviewed and renewed each year. The essential points in these written instructions are to see that the telephonist does not have to ask permission before sounding the evacuation signals if they are not automatically sounded. Neither should she have to obtain permission before calling the fire brigade or confirming that they have received the call if it is automatically transmitted. Having made this call successfuly, she is then at full liberty to go immediately to safety if there is any sign of smoke or imminent dangcr.

After the evacuation signal has been given, the fire brigade contacted and there is no sign of danger, calls may be made to other people such as the engineer, manager or commissionaire.

Alternative means of calling the fire brigade must also be arranged. These should consist of details of telephones outside the premises to which access can be gained in the event of a complete breakdown of telephonic communication within the premises.

If the switchboard is readily accessible to the open air, it might be possible to keep one telephonist on duty under the protection of a special service warden. If immediate escape to the open air is not possible, the telephonist should throw the night switches and leave the board.

If the board is kept manned, the telephonist must firmly refuse all incoming calls, and accept outgoing calls only from fire brigade officers or on the instruction of the incident controller.

Telephonists should be instructed not to give any information to the press, nor may they connect the press to the incident controller or any other officer, but simply say that an official statement will be issued as soon as it is available. They should also be warned not to give their own name or that of any other officer of the company.

Vehicles and drivers: transport

Vehicles must not move on the private roads of the premises while employees are moving in the roads although, as soon as possible, vehicles will be cleared for fire brigade appliances. Drivers will report to a special check point, and the checker should be a dispatch clerk who knows which vehicles and drivers are likely to be on the premises. After reporting, drivers should return to their vehicles, provided that they are not in jeopardy.

Visiting vehicles: transport

A printed slip should be part of any order involving incoming transport, to the effect that the driver will report to a particular check point, and carry out any instructions given to him by the checker, should the evacuation warning be given while he is on the premises. This is an essential instruction, because visiting drivers are likely to say that they do not belong to the premises and can do as they please. If should be part of the general security instructions to ensure that visiting drivers are under the same obligations as the employees of the company.

Fork-lift truck drivers

Fork-lift trucks can be of great value during and immediately after an incident. They should be instructed to wait until employees have cleared the building and then bring their vehicles out to an allocated standing place within reach of easy communication by the incident controller. If the standing place is not a check point, they should next report to their checker and then return to stand by their vehicles. If danger is imminent they will leave their vehicles at any time and proceed to safety.

Visitors

Visitors to the premises should be escorted to a check point by the officer they are visiting, or by someone delegated to do so by him if he himself has other duties to perform. The checklist for visitors will be provided through the security section by reference so the visitors' book. If possible, it is better that visitors are checked through reception, but the main point is to endeavour to account for them.

Outside contractors

A printed instruction should accompany all orders placed with outside contractors that, while on the premises, they will comply with the instructions issued to employees of the company. In particular, if an evacuation signal is given, they will proceed to a given check point. At the check point, their foreman or senior man will ascertain that they are all accounted for, and will see that a report to this effect is given to the chercker at the nearest check point.

Technical officers

In large works it is advisable to see that technical officers assemble at a point within easy communicaton of the incident controller after passing through their check point. Such officers would be production controllers, chemists and others whose specialist knowledge might be required to advise the fire brigade or to assist in restarting production.

When considering the evacuation scheme, it is useful to remember that an emergency scheme has also to be established for incidents which occur out of working hours.

EVACUATION SCHEME

First draft

Before finalizing the evacuation scheme, heads of departments should be given an opportunity of commenting on it to the committee. They should also be allowed to ask for items to be added in respect of their own department. For example, they might request that employees should carry out with them valuable documents, or that certain working trays should be put into a fire-resistant safe. Provided that the action does not endanger life, it should be agreed to, because it is of the utmost importance that the business should be able to survive after a major incident.

Stand down

The final point in the consideration of the evacuation scheme is to decide on the method by which employees are told that it is safe to return to work, or to dismiss them to their homes.

For this purpose a loud hailer may be necessary if the assembly area is very large. Notices held up may be sufficient or, in very small companies, a message passed by word of mouth might suffice. It should be unnecessary to add that the incident controller should never return people to work under actual incident conditions unless the senior fire officer present gives his authority. It might be well to incorporate this when practices take place.

When the final scheme is prepared, every copy should be numbered and a record kept of the number issued, and to whom it is issued. Every year old copies should be called in, and a new copy issued, with amendments bringing it up to date.

Evacuation practice

As soon as an evacuation scheme has been drawn up, regular evacuation practices should take place. If possible, use should be made of outside observers stationed at various points. The local fire brigade should always be told before a practice is to take place, and they will often provide observers. The local fire brigade will also indicate the extent to which they will be able to cooperate and whether they want a practice message sent through and, if so, in what form. At times they will provide a token turn-out, and this may provide information as to the access to the premises, particularly under conditions where employees are moving to safety. The local police should also be informed of any practice, in case outsiders send in a message. The cooperation of the police is very necessary if people are being sent on to public streets and roads in large numbers. In this case, the police will appreciate 24 hours' notice.

Having decided on the date and time, this information must be fully circulated, so that everybody is warned. They will, of course, check up what they should do, but this is surely part of the exercise. The exercise is not required to prove that people do not know what to do, but to train them to do the right thing.

The first reaction to the notice may be a spate of managers claiming to be excused for some reason or another. A firm stand must be taken and a full practice must take place. The directors and senior management must not only give their support fully, but must show by their example the seriousness with which they regard losses both of life and property from fire.

When the time of the actual alarm is due, the general organizer should go to any employee, and say something to this effect: 'I want you to imagine that you can see a fire. Please take the action that is laid down in your fire instructions'. The starting point should be varied for each practice.

The fire compartment under practice should be cleared in about two and half minutes, and observers posted to time the clearance should note the length of time taken for the last person to reach safety. It will take several minutes for all the checkers' reports to come in, and a note should be made of any that fail to report.

In small companies it is often possible to carry out a demonstration of the use of fire extinguishers before the practice is called off. But it is much better if these demonstrations are carried out with only 25 to 30 people at a time.

When the general organizer is satisfied that the practice is complete, he should report to the incident controller. He, in turn, will initiate the action to return people to work, first seeking permission from the senior fire officer present as a matter of form. Immediately after the practice, all the officers taking part should assemble, and any weaknesses and improvements should be brought to light. If necessary, points may have to be referred back to the committe. After this, a short notice should be drawn up and issued, giving the results of the practice. This notice should be widey circulated.

GENERAL INSTRUCTIONS IN FIRE PRECAUTIONS

To maintain a high degree of safety from fire it is necessary to take every opportunity of making people fire-conscious. To this end, regular instruction should be given by fire officers. Every person joining the company should be given a short preliminary course of instruction; about twenty minutes is sufficient. This should be given on their first day if possible, but should not be delayed beyond the first week New employees must not be allowed to start work until they have been taken through the nearest exit and on to their assembly point. They must also be taken through the secondary exit. This first course of instruction might be given by a training officer, but it achieves far greater importance if given by a fire officer, particlarly if he is in uniform.

SUGGESTED LECTURE FOR INSTRUCTION OF NEW ENTRANTS

Using the personal instruction card as a basis for a lecture, the instruction should consist of the following information.

Discovery of a fire or suspected fire

The important point is to sound the alarm immediately to enable everone to get out in safety, and so that help in the form of the professional fire brigade is started on its way. If there is any doubt whether it is a fire or not, time must not be lost in trying to find out. There must be no investigation for this purpose, but the alarm must be given immediately. The fire brigade would prefer any number of false alarms given with good intent rather than have even one second of delay on a genuine fire. Under no circumstaces should a door be opened, not even a cupboard door, to see whether there is a fire behind it. A fire that has only been smouldering could suddenly obtain the air it has been lacking, and burst into flame with almost explosive violence.

Attacking a small fire

After the alarm has been raised and no danger threatens *and* there is an immediate exit to safety, a small fire may be tackled with the nearest fire extinguisher, provided that the operator is between the fire and the exit. This is necessary because if the fire is not promptly extinguished the operator can then escape in safety.

Frequently, there are arguments as to whether a small fire should be tackled first or the alarm raised first. There should be no argument. The alarm must come first because there can be no definition of what constitutes a small fire or the conditons under which it may occur. A fire that can be fought safely by one person could not be fought by another. A fire occurring in one place might be comparatively safe and easy to fight, whereas a similar fire elsewhere might be dangerous. The most important point is to enable people to start moving out of danger.

In practice, the question seldom arises because in the majority of cases more than one person is present, and both operations can take place simultaneously. But if there is only one person present, it is very hazardous for that one person to attempt to fight a fire single-handed.

Emphasis must be placed on an instruction that under no circumstances must a blazing object be moved either with the intention of carrying it out into the open air or throwing it out of a window. Practically everyone who does so is more or less seriously injured by thc flames being driven back on to them, and the fire is spread all along the route. This point must be stressed as heavily as possible. Inform the class that the same rule applies to a frying pan on fire. In this case it is only necessary to cover it with a saucepan lid or a damp towel. The personal home reference is always appreciated and is good publicity for fire prevention generally.

On hearing the evacuation signal

The evacuation warning must always be treated as an indication of danger and, so that there may never be any reason to doubt the possibility of danger, evacuation signals will never be sounded for practice unless previous warning has been given, stating the time that a practice will take place. The point to be emphasized is that the building must be evacuated immediately, by the nearest route.

Having cleared the building, the employee must report to the checker, and make sure that he is recorded as having reached safety. It is not sufficient to go to the check point and rely on being seen. The checker may be looking at his check sheet. It is the duty of the employee to state his name and see that it has been recorded, and so save the fire brigade from having to divert valuable men and equipment to search for missing persons.

No one must start to run, because it tends to cause confusion and interferes with other people. If anyone trips and falls other people may be brought down, so causing danger and further confusion. If everybody leaves immediately they hear the warning, there is ample time for everyone to reach safety. On no account should anyone go in the reverse direction to the general run of traffic. Neither may they go to the cloakrooms. By doing so they may be putting themselves in danger, and certainly will delay the warden from reaching safety himself. Any particular personal belongings should always be kept with the employee because, if an attempt is made to go back for them, not only will the employee delay his own escape but he may also be going against the stream of traffic.

If there is no apparent danger, shut down machines and switch off all gas and electricity other than electric lights. These may be needed to help the fire brigade. Any sign of smoke is an indication of danger. The last person out of the room should always close the door. Any door will tend to check the spread of fire, and it will certainly prevent draughts from feeding a fire. This is also good advice at home at night. It is a fallacy that if a bedroom door is left open, a person will smell smoke before there is any danger.

Things an employee should know

The lecturer should instruct the new entrant on the importance of hearing and recognizing the evacuation signal, and further that for his own safety he should inform the proper authority if he cannot hear it. He should not only inform the proper authority, but should see that something is done to rectify such a dangerous state, even if it is only to provide a temporary method of receiving the signal. The proper authority for him to inform should be told to him.

Similarly the new entrant should be instructed how to report that an exit door is locked, or an exit route blocked or obstructed. A particular warning should be given about loose materials left in an evacuation route. Loose materials and obstructions left in an emergency route lead to people being tripped up. While it is appreciated that draughts can be most uncomfortable, the use of curtains or hangings is very dangerous, and alternative methods should be found that cannot endanger the free movement of people.

Although not so important for safety of life, a report should be made through the proper channels if any fire extinguisher is missing, obstructed, damaged or apparently out of order. A report should also be made if an extinguisher is used or partly used, or if it is knocked over or knocked off its rest or hook. Similarly, a report should be made if a fire alarm point is obstructed.

The lecturer should then briefly talk about the common causes of fire and how they can be avoided. It is sound practice not only to talk about fires at work, but also to mention safety measures in the home. All such information will help to make employees fire-conscious.

Conducted tour

Every new employee on arriving at the place where he is to work should be taken through the evacuation procedure, actually walking over the escape route and through the fire exit doors right to the check point. The same procedure should be adopted for the alternative means of escape. He should also be shown the nearest fire extinguisher to where he is working, and requested to read out aloud the instructions printed on the label. His attention should be particularly drawn to the means of starting the extinguisher.

Every new employee should be shown the nearest fire alarm point. If no hammers are provided for breaking the glass in a manual alarm point, ladies should be told that the best way of breaking the glass is to slip a shoe off and strike the glass a really smart blow with the heel. Men should be told to use the point of the elbow, pausing a second before removing the elbow to allow the glass to fall clear. The ringing of the evacuation bells will disclose that the alarm has been signalled.

Films and videos

One series of instruction is not enough; efforts should be made periodically to keep interest in fire prevention alive. This can be done in many ways. One of the best is by the occasional showing of one of the excellent and interesting films and videos that are available, many of which can be obtained on free loan.

Posters

The occasional use of posters is another way in which specific dangers can be emphasized, and a list of very useful literature can be obtained from the Fire Protection Association.

FURTHER READING, PART TWO

Canter, D., *Fires and human behaviour*. John Wiley & Sons, Chichester.

Best, R. L., *Reconstruction of a tragedy: the Beverly Hills Supper Club fire*. NFPA, Boston.

Everton, A. R., *Fire and the Law*. Butterworths, London.

Everton, A. R., Holyoak, J. and Allen, D., *Fire, safety and the Law*. Victor Green Publications Ltd, London.

Marchant, E. W., *A complete guide to fire and buildings*. Medical and Technical Publishing Co., Lancaster.

PART THREE

FIRE PREVENTION

12

Causes of fire

It was Percy Bugbee, when he was General Manager of NFPA, who said that there are three main causes of fire: men, women and children. This is very true because fires are much more likely to result from human activity than from equipment failure. Fires are particularly likely to happen for example during the construction of buildings and plant, and also during the annual shut-down of an existing facility. The activities of contractors in factories or on production sites are notorious for causing fires. This is not necessarily due to carelessness. It may simply happen that the contractor's employees are not familiar with the hazards on the site, nor with the established work procedures. There is also unfortunately the temptation to cut corners because the sooner the work is completed, the more profitable the contract. Training is always an important part of fire prevention: it ensures that people are aware of the hazards on the site and that they understand the reasons for the working methods which they have to adopt. This applies equally to everyone working on the site, whether they are full-time employees or contractors' men. Training should of course go beyond this to include means of escape and methods of first aid firefighting.

Fire statistics often provide analyses of the sources of ignition, the materials first involved, and the location of fires. This information is valuable because it enables fire prevention activities to be concentrated on the more vulnerable areas, and hence to be cost-effective. There is however an unfortunate tendency for the information media to regard the source of ignition as the only significant cause of a fire. If for example an investigation shows that people died in a fire because there were too many flammables, too many people, and not enough exits the press will seize on the supposed ignition source and announce that the fire was caused by a cigarette, or whatever. The fire may indeed have been started by a cigarette, although it is difficult to ignite many fuels in this way, but to blame the fire on smoking is to divert attention from the real causes of the tragedy. A headline which appeared very soon after the London Underground fire (1987) said 'King's Cross fire "caused by cigarette"' (*Daily Telegraph*). This does nothing to explain the very rapid spread which caused the 31 deaths, and does tend to give the impression that the blame for the tragedy rests with a careless smoker.

The two main sources of information on fire are the fire brigades, who prepare a report on every fire that they attend, and the insurance companies. There is also a largely untapped source in the records maintained by industrial fire brigades: on many sites less that 10 per cent of the fires are attended by the public brigade. Fire brigade and insurance statistics are usually published annually and the data are frequently reviewed by fire research establishments to identify significant trends and relationships.

One useful application of statistical data is in assessing the value of fire protection methods. For example a county fire brigade in the UK provided an advisory service to householders over a period of 11 years: firemen made regular visits to every house. At the end of this period the number of house fires in the country had fallen to two-thirds of the national average. Studies of this kind make it possible to assess the cost-effectiveness of both fire prevention and fire protection methods.

Statistics are also useful in indicating the relative risks in different situations, and hence in identifying areas in which fire precautions need to be improved. A study by the UK Fire Research Station published in 1970 showed that on average there were 3.8 deaths in hotel fires for every 100 million exposure hours. The corresponding figure for hospitals was 0.35, and for houses it was 0.19 (100 million hours is about 200 lifetimes). Another FRS study showed that in half of all building fires the fire spread beyond the immediate source and also that half of the fires in buildings were attended by the fire brigade. The chance of fire spreading beyond the room of origin varied from 11 per cent for flats and maisonettes to 42 per cent for a multi-story warehouse. However the chance increased to over 50 per cent for fires caused by arson. Yet another FRS report contains the surprising information that one fire is attributed to smoking for every 10 million cigarettes which are sold.

This chapter is concerned particularly with the causes of fire as reported by fire brigades and insurers. The figures in Table 12.1 were derived from an analysis of fires costing in excess of £250 000 in the year ending September

Table 12.1
Causes of fires costing more than £250 000

	(%)
Malicious ignition	32.7
Electrical equipment	21.1
Friction, heat and sparks	8.2
Smoking materials	7.6
Rubbish burning	3.5
Spontaneous combustion	2.3
LGP equipment	1.8
Acetylene equipment	1.8
Oil and oil-fired equipment	1.8
Other known causes	1.2
Unknown	18.0
Total	100

1987. This was published by the UK Fire Offices' Committee and is typical of many lists prepared by them, and by other insurance organizations, in recent years. The fact that these were all very costly fires may mean that the sample is biased, but in effect we are clearly dealing here with the most serious and possibly the most hazardous fires. The proportion due to malicious ignition (arson) may be higher than it would be had we accounted for all reported fires. This is simply because of the greater probability of these fires spreading from the room of origin and hence becoming larger fires. Table 12.2 also derived from the FOC data, shows the places in which these fires originated. Here, the majority of the large number of fires which occurred in educational establishments may well have been due to malicious ignition.

If we examine Table 12.1 it is not difficult to accept that over half of the fires can be attributed to human actions (both deliberate and accidental), which supports Percy Bugbee's analysis. Clearly education can play an important role in fire prevention, by increasing the awareness of hazards, by reducing carelessness, and by changing the attitudes which may lead to arson. The fire protection associations maintain a continuous flow of well presented information, some of which is directed towards children. It is unfortunate that fire prevention is not included regularly in the education of most children in the UK, although fire brigades and other organizations are very willing to participate. Other countries do devote much more time and effort to education and training. This is particularly true in Japan, where the public attitude towards fire is very different from our own: a fire in the neighbourhood is regarded with shame and concern.

Table 12.2
Location of fires costing over £250 000

	(%)
Engineering factories	12.6
Educational establishments	10.8
Warehouses, wholesalers	10.8
Food and drink factories	8.1
Shops	6.5
Timber, furniture and upholstery factories	5.4
Textile and clothing factories	5.0
Chemical and plastic works	5.0
Clubs, restaurants, pubs	4.5
Unoccupied buildings	4.1
Multiple occupancies	3.2
Paper factories	2.3
Dwellings	2.3
Hotels and boarding houses	2.3
Others	16.8
Total	100

In the UK and many other countries the increase in crimes of violence has been paralleled by an increase in arson. In Table 12.1 this is the single largest cause of high-cost fires. Security must therefore now be considered as an important part of fire protection. Detailed security precautions are beyond the scope of this book, but one possible approach is worth mentioning. The demand for economies in industry and particularly in the supposedly non-productive areas has resulted in combined fire/safety/security departments in many of the larger firms. A useful development in this situation has been the establishment of combined fire prevention and security patrols using Land Rover or similar vehicles equipped for firefighting. The men are trained to cover all three aspects of work: fire prevention, fire fighting and security. This is discussed again in Chapter 9.

The second largest cause of fire in the table is shown as electrical equipment. Some of the hazards associated with electricity, and some of the other possible sources listed in the table are discussed in the remainder of this chapter. Because of the potential fire hazards inherent in all electrical installations there are many standards, rules and codes of practice covering their design, construction and operation. In the UK very comprehensive regulations are published by the Institution of Electrical Engineers and there is a number of British Standards covering the design, testing and installation of equipment (including equipment for use in explosive atmospheres, which is discussed below). In the USA the *National Electrical Code* is published by NFPA. NFPA70 is the main code, and 70A covers one- and two-family houses.

It was mentioned above that fire brigade records are a useful source of statistical data. In the UK the Chief Inspector of Fire Services issues an annual report which contains an analysis of the fires attended by all brigades, although the report is not published for some time after the year which it covers. The report for 1986 showed that there had been 957 fire deaths, of which 615 were due to the inhalation of toxic gas or smoke. Three hundred deaths were attributed to burns and the remainder to other causes like physical injuries. The total number of deaths in the previous year had been higher, at 978, but this figure included the deaths at the Bradford City Football Ground and the Manchester Airport fires. The number of fire injuries in 1986 was the highest recorded, 12 700, and there were 8450 rescues.

A total of 387 000 fires were attended by the brigades. Of the many analyses in the report perhaps the most surprising showed that the number of road vehicle fires had nearly doubled in the 10 years to 1986, with a total of 47 680. However this does not necessarily indicate that cars and other vehicles had become more hazardous. There had been an increase in the total number on the road and, what is probably more significant, the number of fires which were not thought to have been started deliberately increased in the 10-year period from 1800 to 14580. This again illustrates the disturbing increase in arson. The most frequent causes of accidental fires in houses were cooking appliances (22 500 fires), smoking materials other than matches (6500) and electrical appliances other than cookers and space heaters (6400). The statistics for accidental fires in occupied buildings other than houses were closer to those in Table 12.1. The greatest single cause was electrical equipment but this was followed by smoking materials and matches. Over the 10 years to 1986 there had been a significant increase in the number of fires in houses, rising to the highest recorded total of 63 500. Despite this, the total number of fires in all occupied buildings showed only a slight increase over the previous year's figure.

ELECTRICITY AS A CAUSE OF FIRE

An electric circuit is composed of conductors carrying the electric current from its source via its controlling equipment to the final load. These conductors are normally covered with an insulating material to prevent the current from passing directly between the conductors or to prevent a person touching them. Electric current is measured in ampères or 'amps', as a common abbreviation, and the pressure behind this current is expressed in volts. The amount of power so conveyed is given by the product of these two figures – that is, volts × amps, which is expressed in watts.

All conductors offer some resistance to the passage of an electric current through them and the passage of the current through this resistance creates heat. The amount of heat created depends upon the amperage and the resistance of the conductors themselves. Provided that the cables selected for any particular application are large enough to carry the current concerned, any heat produced in the cables will be so small that it will be harmlessly dissipated. If, however, the rating of the cables (the current which they can carry without producing excessive heat) is exceeded, there is a very serious fire risk. Consequently, any proposed increase of loading of an electrical circuit must take into consideration the ratings of the cable to see if the cables themselves are capable of withstanding the extra load or whether they should be increased in size.

The thickness and type of insulation covering the cable is governed by the voltage to be applied to that conductor. The ability of insulating materials to withstand the voltage or pressure applied will be lowered by mechanical damage, age or heating.

Mechanical damage is usually caused by abrasion or cutting through the insulation. The effects of age usually show up in rubber cables perishing or the insulation becoming hard and dropping away from the conductors. Damage by heating can be caused internally by excessive loads or externally by heat from the appliances they are feeding or from some other source. In either case, and again as in ageing, the insulating material tends to become hard and brittle and may eventually break away from the conductor. The plastic materials now being used for insulation may melt and so expose conductors.

The risk of an outbreak of fire within an installation and connected appliances can be due either to overcurrents; short circuits; or earth leakage faults. These three differing kinds of faults are illustrated in Figure 12.1.

This figure shows the simplest form of distribution circuit. Power is taken along the lines marked 'live' and 'neutral' from a transformer to a load protected by a fuse. The equipment being protected is contained within a metal case connected to earth. Although supply transformers are normally three-phase, only one phase is shown in use on this diagram for the sake of simplicity. A three-phase transformer will have three secondary (output) windings instead of the single winding shown in the figure. One of the ends of each of these windings are connected together at the 'star point', which is connected to earth.

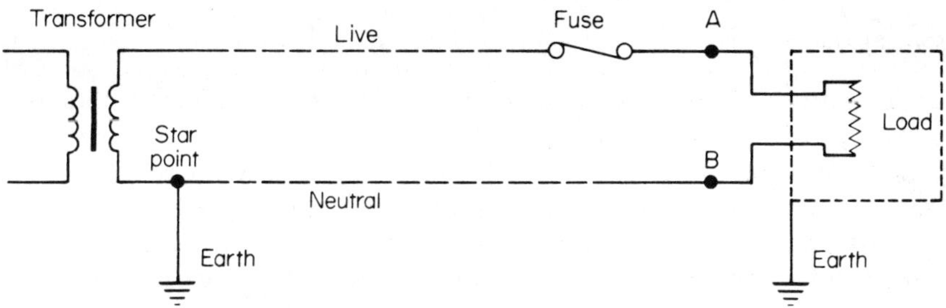

Figure 12.1 Simplified form of distribution circuit.
(J A Crabtree & Co Ltd)

Electrical faults

Overload

The transformer and cables shown in Figure 12.1 are each capable of carrying a certain amount of current. This current is their full load rated value. If this current is exceeded, the cables and transformers will run hot and if this is allowed to persist it would constitute a fire risk due to overloading. If these conductors are allowed to run hot continuously, it could be detrimental to their insulation and could also cause heat to be transferred to surrounding material. If slow heating takes place, any combustible material in the immediate vicinity will eventually give off flammable gases and, when the gases or material reach ignition temperature, they will automatically take fire (see Chapter 5).

Short circuit

If a connection is made between points A and B of Figure 12.1, giving a resistance-free path for current to flow along, an extremely high current will pass between these points. Such a current is limited only by the resistance of the transformer and cables and this is known as a short circuit. This current could easily be 20 to 30 times normal rating of the cables and could under some instances rise to a much higher value. Such short circuits are caused by failure of insulation for example by nails driven through cables or by live conductors touching each other. These short circuits constitute a fire hazard not only by virtue of the higher current they carry but by the fact that sparking can be produced if the points A and B contact each other. If sufficient protection is not given, the arcing produced could be hot enough to heat local material to a sufficient degree that it will give off flammable vapours and be hot enough to ignite these vapours as soon as they reach their flash point. (For the moment, we are ignoring the presence of the fuse in the figure. This is discussed below).

Earth fault

Under normal circumstances, current is constrained to flow within the transformer, conductors and load only. If, however, a fault develops, permitting current to flow from a live conductor to earth, it will return to the star point via the earth connection shown in Figure 12.1. This circuit comprises an earth fault, and gives rise to a fire risk if that current is being caused to flow through paths that were never intended to carry an electric current.

PROTECTING DEVICES IN ELECTRIC CIRCUITS

Electric circuits must contain elements to give protection against these three types of fault. The rating of the protective device to be used in a circuit will be determined by the size and types of conductors in use. It is very important that although the protective equipment will allow the circuit to carry its full rated load continuously without operating, it should operate quickly if there is either a sustained overload, a short circuit or an earth fault.

It is not practical to fit circuits with over current devices that will switch off the very moment the rated current of the circuit is exceeded. For example, if a circuit rated at 10 amp is feeding a motor taking a 5 amp full load current, it will be impossible to start the motor. When first switched on such a motor will have an initial starting surge of up to six times full load current for a brief period and this would cause such overload protection to operate immediately. Provided that this six times full load current is not allowed to persist, the cable will handle it safely. It is therefore necessary to provide protective equipment that will allow a circuit to carry its rated current continuously, to operate within a short period of time on a modest overload and to operate on a shorter period of time with a more severe overload. Such a characteristic is known as an 'inverse time characteristic'. The higher the sustained overload, the more quickly the device must operate because the heating effect of an electric current is proportional to the square of the current – that is, twice the current will have four times the heating effect and be a much more serious fire hazard.

Short circuits are only extreme cases of overload; consequently an overload device which becomes instantaneous in operation above six times full load current will give protection against overloads and short circuits. Earth leakage protection is a much more complex subject, consideration of which will be given later in this chapter.

Overload and short circuit protection devices

There are two types of overload and short circuit devices: fuses and circuit breakers.

Fuses

The rewirable fuse was the earliest form of fuse used and is still in common use today. A short length of wire composed of a metal that will melt readily when heated is carried within a fuse carrier. This carrier is then used to make connection between the supply and the cables feeding the load such that the full load current passes through the short length of wire. The gauge of the wire is selected according to the current at which the cable is rated so that it will carry the current continuously, but will melt if the current is exceeded for any appreciable length of time. Such rewirable fuses are being replaced by cartridge fuses which are in use on the heavier fault current circuits. Such cartridge fuses carry the short length of wire within a ceramic tube and the wire may be surrounded within the tube by silica sand.

When these fuses are subjected to really heavy faults (short circuits), the arc produced by the wire melting is extinguished by agitation in the sand. Consequently, these fuses, which are also known as high rupturing capacity fuses (HRC fuses), are capable of handling much heavier fault currents than a rewirable fuse. Cartridge fuses have many other advantages over the rewirable type. Their fuse holders are designed so that only a certain maximum size of fuse can be placed in the holder. There is no such restriction on a rewirable fuse and as a result of this, almost any gauge of wire can be fitted or even other materials can be used in place of fuse wire. In this way, rewirable fuses are open to abuse and many fires have occurred through nails or hairpins being used in place of the correct wire. Cartridge fuses are not so easily subjected to this form of abuse.

The use of electricity is growing rapidly and is putting increasingly heavy demands on all types of installations. As full load currents increase, the short circuit current available also increases. HRC fuses are extremely suitable to handle heavy fault currents in view of the arc quenching effect given by the silica sand. Rewirable fuses contain no such facilities and are not safe when subjected to very heavy fault currents. Thus, where there is a choice between cartridge and rewirable fuses, cartridge fuses should be chosen for the following reasons.

- they are not likely to do any damage when they 'blow';
- the arcing is contained within the ceramic tube;
- they are more likely to be of the correct current size as they are not so easy to misuse;

- they act faster and are less likely to deteriorate through age or by harmless overloading.

The only disadvantage of the cartridge fuse as against the rewirable fuse is that it is more expensive, but if the use of cartridge fuses can save the cost of the premises being damaged by fire, then it is well justified extra expense.

Circuit breakers

Recent years have seen a considerable increase in the use of miniature circuit breakers for the protection of final circuits. A miniature circuit breaker is essentially an automatic switch containing overload and short circuit protection. Miniature circuit breakers complying with the appropriate British Standard specification are tamper-proof devices because they are set to a specific rating by the manufacturer and this rating cannot be changed by the user.

A well designed circuit breaker cannot be held in against a fault although it could be safely closed while there is still a fault on the circuit. In the case of a fused circuit, it would be very dangerous to attempt to replace a fuse unless the controlling switch had already been opened. Miniature circuit breakers also include an arc-quenching device equivalent in action to the sand incorporated in an HRC fuse and are therefore capable of breaking heavy fault currents.

From the purely maintenance point of view, once a miniature circuit breaker has operated because of a fault, it can be reclosed once the fault has been cleared and can continue in service. Conversely, once a fuse has operated it is of no further use and must be replaced by a new fuse cartridge or be rewired. The controlling switch must then be opened, resulting in other circuits being shut down (even though it is only a momentary shut-down, it can cause considerable inconvenience) while the fuse carrier or cartridge is replaced. Added to this, the right size of fuse wire may not be readily available which leads to the temptation to use anything that will complete the circuit or, in the case of an HRC fuse, the circuit must remain open until the correct fuse is available.

Inverse time characteristic

It has been shown earlier that a protective device must have some form of inverse time characteristic operation. This is achieved in a fuse by the time taken to obtain sufficient heat to melt the wire, which is dependent upon the amount of current flowing. In a miniature circuit breaker, this inverse time characteristic is obtained either by the use of a bimetallic strip which has

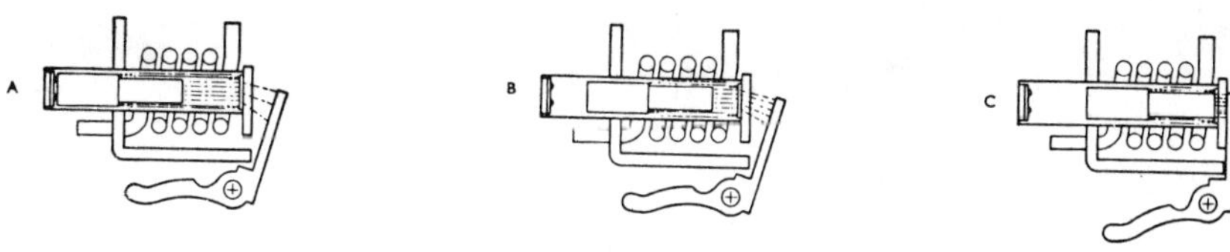

Figure 12.2 Typical magnetic-hydraulic tripping mechanism (J A Crabtree & Co Ltd)

similar thermal properties to a fuse or by the use of a magnetic-hydraulic trip mechanism. The magnetic-electric trip mechanism is to be preferred for final circuit protection, and it is shown in Figure 12.2.

This device contains a ferrous slug mounted within a non-ferrous tube. The slug is normally held at one end by means of a restoring spring (condition A). The tube is completely filled with a silicone fluid and hermetically sealed. Surrounding the tube is a trip coil and the remainder of the magnetic tripping circuit. When full load current flows in the trip coil, there is no reaction because the solenoid coil will not have sufficient power to move the slug. Under overload conditions, the magnetic pull exerted by the solenoid will exceed the restraining spring and the slug will start to move towards the other end of the tube, the time taken for the slug to reach the end of the tube being decided by the magnitude of the overload and the viscosity of the fluid (condition B).

As the slug moves through the tube, the total air gap in the magnetic circuit is reduced and the resultant pull on the armature is increased. When the slug reaches the end of the tube, the pull is sufficient to operate the trip lever and so trip out the circuit breaker (condition C). This method of construction automatically gives the required time delay characteristic in small and moderate overload conditions due to the time effect of the slug moving. The same construction also ensures almost instantaneous operation on heavy loads or short circuit conditions. This automatic instantaneous operation occurs because the larger short circuit current flowing in the trip coil will produce sufficient pull on the trip leaver even though the slug is at the far end of the coil. In smaller moderate overload conditions, the pull is insufficient to operate the trip lever until the slug has moved, but at about six to ten times full load current the initial pull becomes large enough to operate the breaker even without the slug moving from the rest position.

Sustained overloading

From Figure 12.3 it will be seen that a miniature circuit breaker will give faster protection than an HRC fuse, which in turn is giving faster protection than a rewirable fuse. These remarks apply up to four or five times full load

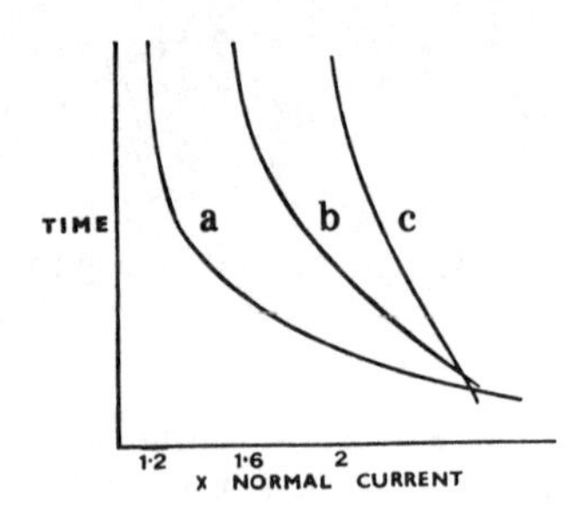

Figure 12.3 Typical tripping characteristic curves
(J A Crabtree & Co Ltd)
(*a*) Miniature circuit breakers
(*b*) HRC cartridge fuses
(*c*) Rewirable fuses

current, above which the characteristics are transposed. The British Standards specification states that a miniature circuit breaker must trip on a sustained 35 per cent overload. The magnetic-hydraulic mechanism described previously can hold this figure to 25 per cent on all ratings, whereas an HRC fuse is in need of a 50 per cent overload to blow and a rewirable fuse could need twice full load current before it would blow.

Short circuit

From Figure 12.3 it appears that, under short circuit conditions or extremely heavy fault condition, the rewirable fuse is the fastest and the miniature circuit breaker the slowest. These deductions are purely theoretical as there are other considerations that become effective. Such deductions would suggest that for rapid clearance of a short circuit a rewirable fuse is best. This is dangerous thinking as rewirable fuses are not designed for use on heavy short circuits. A miniature circuit breaker, although the slowest, will still operate in less than 10 to 15 milliseconds – that is, half a cycle – and will give more than adequate protection for the circuits in which they are installed. Instances can arise, however, where the level of short circuit current is in excess of the value that the miniature circuit breaker can handle safely. There are techniques for overcoming this difficulty, which only rarely occurs in practice. The manufacturers of miniature circuit breakers will always be pleased to give advice on individual cases as they arise.

Other advantages of circuit breakers

Table 12.3 shows the advantages of circuit breakers in comparison with fuses. There is no doubt that if miniature circuit breakers were used instead of fuses, the risk of fire, particularly from sustained overloading, would be considerably reduced. Furthermore, their tamper-proof construction ensures that the protection originally inbuilt into an installation would remain throughout the life of the installation, irrespective of the misuses that a final circuit user often perpetrates.

Earth leakage protection

The thirteenth edition of the IEE wiring regulations introduced the concept of the earth fault loop path, which is the path taken by earth fault currents. This is illustrated in Figure 12.4, which shows a similar supply transformer, conductors and load to Figure 12.1, but also shows an earth fault whereby current is flowing from the load to the metalwork of the premises and then back to the supply transformer through earth. The first problem in applying earth leakage protection is to recognize when an earth fault exists. An overload is easy to detect in that a certain current through the line conductor is exceeded. An earth fault, however, may be much smaller than a load current and more difficult to detect, especially as it is flowing in paths not intended to carry current. It would be very convenient if the fuse or other overload device shown in Figure 12.4 could also be used to clear an earth fault. This is possible if the resistance of the path taken by the earth fault current is low enough to permit the flow of a current sufficiently high to blow this fuse. If the resistance in this path, known as the 'earth loop impedance' is too high other means of protection must be used.

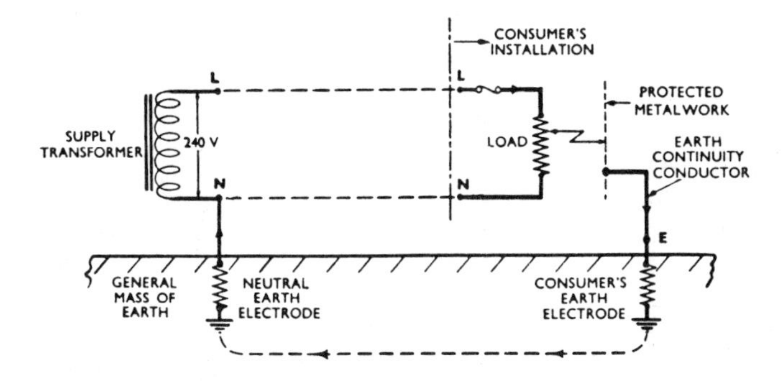

Figure 12.4 Flow of fault current
(J A Crabtree & Co Ltd)

Table 12.3
Advantages of circuit breakers compared with fuses
(J A Crabtree & Co Ltd)

The circuit breaker	Fuses
May be used as a switch to disconnect a circuit.	Must be withdrawn by hand to isolate a circuit.
The circuit may be restored immediately a fault is cleared by simply switching on again.	Must be repaired and master switch controlling all subcircuits must be switched off before fuse can be replaced.
Individual subcircuits do not need to be switched off.	Master switch must be turned off to remove and replace a fuse.
Affected circuit can be identified immediately even in complete darkness.	Search has to be made for the blown fuse with the master switch turned off.
There is no danger when switching on again even if fault persists, neither can it be held in against a fault.	Danger in replacing fuse if the fault persists especially if there has been neglect in turning off the master switch.
Does not age.	Steadily deteriorates with time.
Is resistant to tampering.	Can be abused by over-rating fuses or substituting wrong material.
Large sizes of circuit breaker can be fitted with a signalling contact so that their status may be known at a remote control at any time.	No indication of a blown fuse can be given with most designs.
Large sizes of circuit breaker can be fitted with a device so that they can be tripped by remote control.	
When used with a three-phase motor, a fault on one phase will break the circuit of all three phases.	A fault in one phase may not break all three phases, leading to a dangerous running condition.

Although protection by fuses and circuit breakers as above is usually acceptable, under certain conditions even closer protection will be given by the use of earth leakage circuit breakers. The passage of a current to earth through a fault constitutes a very serious fire risk as this current could be passing in a path never intended to carry a current and which will therefore produce local heating. A further important consideration, which very often is not appreciated, is that even a small leakage current if allowed to persist can within an hour or so render what was a reasonably good earth connection completely valueless. This is because the earth leakage current can dry out the soil surrounding an electrode and hence render useless the protection expected from fuses or overload circuit breakers. The use of an earth leakage circuit breaker of the voltage type would overcome this.

OTHER MEANS OF PROTECTION

It has just been stated that where certain conditions relating to the impedance of the earth loop path cannot be met, other means of protection must be used. These fall into two categories, earth leakage circuit breakers or a distribution system known as 'protective multiple earthing' or 'PME'.

Earth leakage circuit breakers

There are two principal forms of earth leakage circuit breakers: voltage-operated or current-operated types.

Voltage-operated earth leakage circuit breakers

This form of breaker is primarily used for protection against the dangers of electric shock when the earth loop impedance is very high and it will function when earthing conditions are extremely bad. A voltage-operated breaker comprises a set of contacts operated by a switch dolly on the front of the mechanism which contains a trip coil. This trip coil has one lead connected to the metalwork that is being protected and the other end of the coil is connected to an earth electrode distinct from any other earthed metal on the installation. Figure 12.5 shows the basic circuit of a voltage-operated earth leakage circuit breaker.

For a voltage-operated breaker to give fire protection, the metalwork that it is protecting must be completely isolated from earth. This is an ideal situation which is almost impossible to attain in practice. On most installations, the metalwork involved will almost inevitably have a second path to earth known as a 'parallel earth'. This second path will be caused either by connection with structural metalwork which is in contact with the

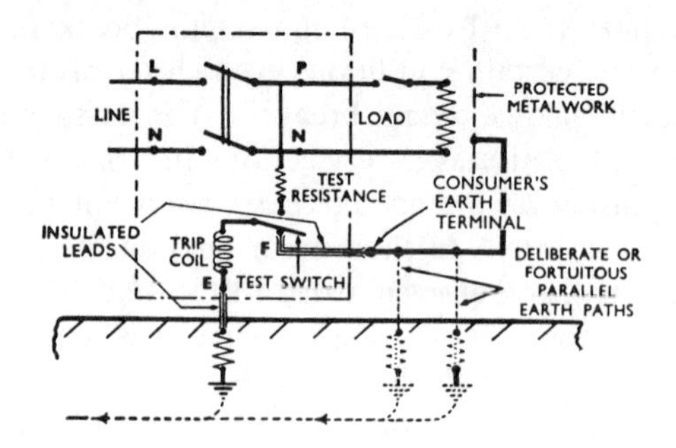

Figure 12.5 Installation of a voltage-operated earth leakage trip (J A Crabtree & Co Ltd)

ground or through any water pipes in the installation which must also be bonded onto the the protected metalwork. The lower the resistance of this second path, the less effective will be the fire risk protection given by a voltage-operated earth leakage circuit breaker. If a parallel path is present, the protection against fire is given by the overload protection alone. If there is no parallel path the impedance of the trip coil of the breaker will limit the maximum earth fault current which can flow to about half an amp. As the breaker will be capable of detecting small leakage currents of 50 mA or above, it will often detect the fault before it developes into a more serious one. This actually constitutes a very high level of fire risk protection.

Current-operated earth leakage circuit breakers

The problems introduced by parallel earth paths have resulted in the development of the current-operated earth leakage circuit breaker. The basic principle of operation of this type of breaker is illustrated in Figure 12.6. It will be seen that the load current of the appliance is fed through two equal and opposing coils wound on a common transformer core. When the line and neutral currents are balanced, as they would be on a healthy circuit, they present equal and opposing fluxes in the transformer core and there is, therefore, no resultant voltage in the fault detector coil. If, however, more current flows in the line side than in the neutral side, an out of balance flux will be produced which will induce a current in the search coil and so trip the breaker. Normally, the reason for more current flowing in the line than in neutral is that some line current has returned to the supply through an earth fault; hence this method may be regarded as a means of directly detecting earth fault currents and giving fire risk protection. Such units have normally a sensitivity of half an amp and can only be used where

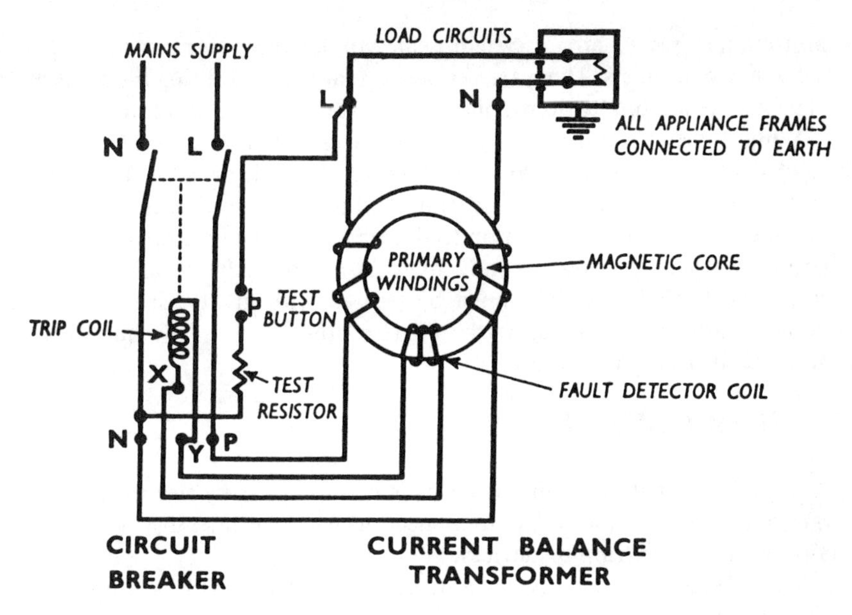

Figure 12.6 Installation of a current-operated earth circuit breaker (J A Crabtree & Co Ltd)

the earth loop impedance will permit this amount of current to flow before a dangerous voltage rise can occur on the protected metalwork under fault conditions.

Of the two types of earth leakage breaker referred to above, the current-operated unit is that primarily designed to give protection against a risk of fire, as it gives protection against current flowing in an earth fault. It cannot, however, be used on installations where the earth loop impedance is high in value, but under these conditions the prime risk is one of shock and not fire and a voltage-operated breaker will be more suitable.

Earth leakage circuit breakers also provide valuable protection to people using portable electric equipment like power drills and lawnmowers. Serious injury or death could be caused by the passage of a current which is less than the minimum needed to actuate a fuse or a normal circuit breaker (see below).

Protective multiple earthing (PME)

This method of protection requires the metalwork of the premises to be bonded to the neutral conductor, so that any earth fault to the metalwork becomes a short circuit to the neutral and so operates the overload protection. This is not a system that can be introduced casually or by individual consumers. There are various regulations concerning the application of protective multiple earthing, but these are beyond the scope of this book.

While a PME system correctly installed gives extremely sensitive protection against shock and fire risk by earth faults, any fault or error in the system can create fire hazards considerably more dangerous than those it is designed to protect. It can indeed introduce such hazards into other premises where the system may have been correctly installed. For this reason, a PME installation must be treated with great respect and must never be altered or introduced by anyone other than a fully qualified and authorized electrical engineer.

ELECTRIC SHOCK

Although not strictly within the terms of causes of fire by electrical fault, the question of shock does arise in connection with the earthing of appliances and earth leakage tripping devices.

The effects on the body of contact with different voltage supplies are shown in Table 12.4.

However, this table does not show the full story because the effects on the body depend on the current which is flowing and this, by Ohm's Law, must depend on the total resistance. The highest body resistance is in the skin and for dry hands this can be as high as 3600 ohm. If the hands are wet the resistance can fall to 500 ohm. This means that at for example 15 volt the current flowing through a contact with dry hands will be about 4 mA and this might give a slight tingling as shown in the table. With wet hands the current could be 30 mA, which could cause quite a painful shock. The effects of current flow in the body are shown in Table 12.5. This was obtained from a different source to Table 12.4, but the two tables relate

Table 12.4
Physical effects of electricity (voltage)

Supply voltage	Effect
15	Can be felt as a tingle
20–25	Can be felt as pain
25 or more	Muscular contraction, burning
70 or more	Can cause death
220–240 or more	Can cause serious burns, shock, death from heart or breathing failure

Table 12.5
Physiological effects of electricity (current)

Current, mA	Effect
5.2	Slight tingling
9.0	Shock: not painful and muscular control not lost
62.0	Painful shock, but muscular control not lost
76.0	Painful shock: release current threshold
90.0	Painful and severe shock, muscular contractions, breathing difficult

quite well if we assume that the results in Table 12.4 were obtained with a skin resistance of 2800 ohm, which is quite a typical value.

The significance of the release current threshold is that at current flows about this value (76 mA) it may not be possible for a person to let go of a live conductor because of muscular contraction (this effect is more serious with DC than with AC cicuits). The total effect on the body may depend on the time during which the current is flowing. For example, the heart may stop beating after 1300 mA has been flowing for only 0.03 s, but at 500 mA it may take 3 s.

The values quoted in Table 12.5 are surprisingly low and it is fortunate for us that the skin does have a high resistance. If it is penetrated then death can be caused by only a few volts. The importance message is that severe shock, injury and death can result from contact with· normal electrical supplies. Because people are used to the safe operation of electrical equipment they tend to forget the dangers. Constant vigilance is needed if the unpleasant consequences are to be avoided.

Portable electric tools

Where a flexible cable is used to convey electric power to a tool, any one of the conductorss can break through constant flexing. If the break is in the live or neutral conductor the user will immediately become aware of the fact because the tool will not work. If, however, the earth conductor breaks, the shock risk protection is lost and the operator will be unaware of this because the appliance will go on working.

There are two main means of increasing the safety factor: one is by monitoring the earth lead and the other by increasing the sensitivity of the earth leakage protection applied.

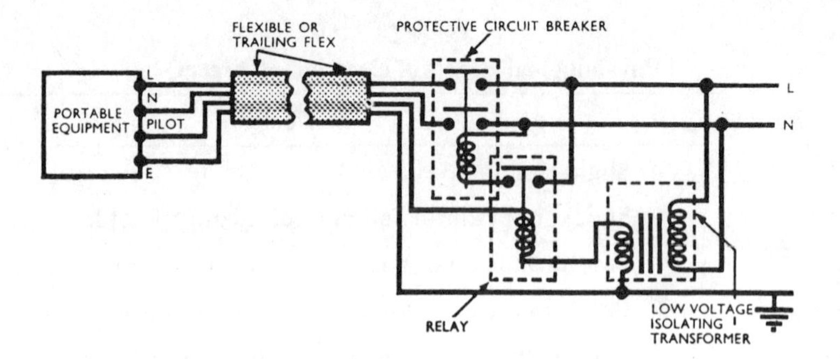

Figure 12.7 An earth monitoring circuit
(J A Crabtree & Co Ltd)

Monitoring the earth lead

The basic circuit of an earth monitoring device is shown in Figure 12.7. It consists of a low voltage transformer (usually less than 12 volts output), a relay and a protective circuit breaker. There are two earth leads to the appliance: the primary earth continuity conductor which is bonded to earth in the normal manner, and a pilot conductor.

The low voltage output from the isolating transformer is used to drive a circulating current (usually less than 3 amps), round a loop consisting of the primary earth continuity conductor, a section of the metallic housing of the appliance, the pilot conductor and the relay operating coil. The circulating current holds in the relay which in turn energizes the hold-on coil of the circuit breaker. If this circulating current is interrupted, as it would be if the primary earth continuity conductor or the pilot conductor failed, the relay armature would drop off. This in turn would interrupt the hold-on coil circuit and so trip the protecting circuit breaker and isolate the equipment.

It will be noted that earth monitoring does not in itself form any earth leakage protection, and devices of this type usually incorporate earth leakage protection.

Ultra-sensitive earth leakage trips

There are two methods of increasing the sensitivity of current-operated earth leakage circuit breakers: one is by means of a solid-state amplifier and the other is by means of breaking down a magnetic field.

In the first method, the amplifier is used to increase the output from the search coil so that it has enough power to trip the circuit breaker. This has

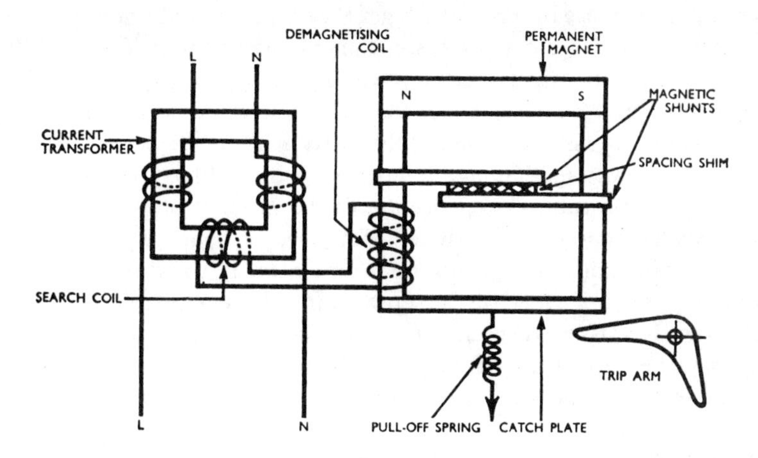

Figure 12.8 Ultra-sensitive magnetic earth trip
(J A Crabtree & Co Ltd)

the advantage of providing sufficient power to ensure the positive action of the earth leakage circuit breaker. The other method, which is illustrated in Figure 12.8, is to use current from the search coil to demagnetize a special trip mechanism.

The main path for the magnetic flux from the permanent magnet is through the magnetic shunts, the thickness of the spacing shim being so adjusted that only a weak stray magnetic field persists in the remainder of the magnetic circuit, this field having sufficient strength to retain the catch plate against the force of the pull-off spring. By applying output from the search coil to the demagnetizing coil, a weak signal from the search coil can be arranged to cancel out this already weak magnetic field and so release the catch plate and trip the circuit breaker. The power available is obviously very limited and the design has to be good enough to ensure that the device works satisfactorily. It has the advantage that it does not rely on the presence of mains voltage for its operation.

FIRES CAUSED BY ELECTRICAL APPLIANCES

Many fires started by electrical appliances are associated with lamps and mainly rise from a lack of appreciatioan of the heat developed by filament lamps. It is very significant that the incidence of fires started by lighting failures is exceptionally high in commercial premises, that is shops and offices as compared with factories and workshops. For some years, there has been an increase in the use of higher wattage lamps, but even a 100 watt lamp will ignite tissue paper. It is not uncommon to find 200 and 250 watt bulbs in lamp standards and fittings with shades that were never intended to endure such heat. Bulbs of this power will ignite woodwork and a bare bulb laid on a packing case can easily be the cause of a fire.

When high wattage lamps are suspended by ordinary lighting flex, the heat of the lamp causes deterioration of the installation and it is strongly recommended that such flex wires should be of heat-resistant insulation. The weight of shades should not be taken on the flex and the weight of the bulb should not be carried on the screw connections in the ceiling rose.

Danger also arises when electric lamps are placed in show cases or shop windows where the heat is confined. It is especially dangerous when highly flammable liquids such as scents and perfumes are also present.

Portable lamps

Portable lamps are a frequent source of trouble, the common causes being as follows:

- Lead wires damaged
- Lamp taken into an atmosphere where there is an explosive dust, gas or vapour
- Lamp placed on flammable material (even with a guard)
- Lamp used in a damp or wet situation (this may also cause a risk of shock)
- Bulb loose in socket and causing arcing
- Bulb easily broken

In addition to the precautions of installing voltage-reducing transformers and earth leakage circuit breakers, as described in the previous section, all wandering leads and portable lamps and tools should be kept in a locked store under the control of a storekeeper and only issued by him against a requisition signed by a responsible person. The storekeeper should be responsible that they are issued in good condition and that they are promptly returned.

Where frequent use is made of portable lamps and tools beneath cars and lorries and in workshop pits, special precautions should be taken to prevent damage to the leads, lamps and tools. It is much better to have ample fixed lighting equipment in the form of bulkhead fittings and suitable socket outlets.

Electrical heaters

There are five types of electrical heater in common use: open element heaters, convector heaters, fan heaters, oil-filled units and storage units. Each is capable of being a source of ignition. This may happen after a considerable delay, which is of particular significance when premises are unoccupied.

Open elements are run at a temperature well above the ignition temperature for most materials. A piece of paper or cloth falling in contact with the element may ignite almost at once. There is an additional hazard because of heat which can be transferred to a fuel surface by radiation, and here it may take several minutes or even hours before the fuel is heated to its spontaneous ignition temperature. If the fuel is sufficiently close it may only be necessary to heat its surface to the fire point, when flammable vapours will be released which can ignite on the element. A flame will then flash back to the fuel surface. The gradual increase in surface temperature due to radiation can be readily appreciated if a hand is held close to the element. It may feel pleasantly warm at first, but eventually the skin temperature will be intolerable. In the same way clothing which is placed near to an electric fire to dry or air may be thought to be receiving a safe amount of heat, but it will eventually ignite. Free-standing heaters of this type are particularly dangerous in an office or work area where paper or other readily ignited materials can be left close to them or may fall onto them. Ignition may occur after a delay and when the room is unoccupied.

A convector heater is so designed that a convective flow of air over a heated element keeps it at a relatively low temperature. At the same time the air passing through the unit is not heated to a temperature at which it could become an ignition source. These heaters are therefore less hazardous than the open element type, but they are still capable of causing ignition. If for example clothing is placed over the heater so that it partially or completely blocks the airflow, the temperature of both the heating element and the air will increase. The clothing may then be ignited. Heaters of this kind usually have a thermal cutout which is designed to switch them off if the temperature rises above a predetermined value. Even so, they are still capable of causing ignition, particularly if the airflow is only partly restricted by say a piece of paper. Again, free-standing heaters of this kind are unsafe in an office or work area. They can be acceptable if they are secured to a wall and protected in such a way as to prevent the accidental reduction of the air flow. Exposed element heaters may also be acceptable if they are attached to the wall or suspended from the ceiling at a high level.

In an oil-filled unit the heating element is completely immersed in oil which is contained in a sealed panel or in tubes. The surface temperature is relatively low and most of the heat is lost by convection. It is still possible for ignition to occur if for example clothing is left over one of these heaters to dry. Wall-mounted heaters can be made safe if they are guarded in such a way that combustible materials cannot come into contact with the hot surface.

Night storage heaters contain a mass of solid material of high specific heat in a well insulated box. The material is heated to a high temperature overnight, using cheap off-peak electricity. The heat is released slowly by convection when a slow flow of air passes through the units. Like the other heaters they can become an ignition source if this air flow is obstructed, and similar precautions are necessary. However these units are not usually free-standing, being too heavy to move.

The unauthorized use of electrical heaters can be a serious hazard. Constant vigilance is needed to ensure that people are not hiding them under desks and benches (where they are particularly likely to start a fire). If people are working late, or overnight, there may also be a temptation to bring in toasters and other food-heating equipment.

Temporary wiring

The provision of temporary wiring by people who are not electrical engineers can be potentially hazardous. For example, someone working with desk-top computers may wish to add accessories, printers, display units and so on. This can conveniently be done by using a multi-socket extension: a strip of four or more three-pin sockets connected by a lead to a three-pin plug. The arrangement can be extended, apparently indefinitely, by plugging in further extensions in series, forming a 'daisy chain'. Each of the plugs will contain a fuse which may be rated at 13 amp, so this is the maximum current which can be flowing in the circuit. It may therefore be argued that the leads to the extensions are designed for this current and should be safe. Unfortunately in an arrangement of this kind it is usual to find all of the wires jumbled together, and often covered by papers and cardboard boxes. The result is that the heat generated in the wires cannot escape, and the temperature increases. Heat may also be generated at a poor contact within a plug or socket, or at the points of contact between the two. Fires can be started in this way, particularly if the equipment is left running at night. The proliferation of electronic equipment should be carefully controlled, and daisy chaining should be prohibited. Ideally each work station should have an adequate supply of sockets, perhaps located at the back of the working surface. Only one appropriately fused plug should be supplied by each socket and there should be no surplus wire between the plug to the unit. Wires should not be permitted to trail across the floor. The electronic equipment itself can become an ignition source despite the provision of fuses within the equipment and in the connecting plugs. A faulty component may draw a current which is insufficient to blow the fuse but which causes the gradual overheating and ignition of insulating materials.

People engaged in research are particularly tempted to by-pass the system and do their own temporary 'hook-ups'. Again constant vigilance and proper instruction are necessary to eliminate potential ignition sources.

If premises are patrolled at night, one of the first checks should be for the presence of unauthorized heating equipment and temporary wiring. The patrol should be provided with details and locations of all permitted heaters, and should also be informed about any electrical equipment which is left on at night. Pilot lights on sockets are particularly helpful in enabling a patrolman to check whether things like soldering irons and hot plates have been left on. Strict management procedures are necessary to control the use of electric and electronic equipment, and the provision of temporary wiring.

Electric motors

Danger of ignition from electric motors is high unless proper care and maintenance are exercised. The following points should be carefully examined.

The location of the motor which, unless it has been specially designed for use in wet situations, should not be in a damp position or one in which corrosive or flammable vapours are present.

Dust and dirt should not be allowed to accumulate on or near the motor. Open motors should be cleaned by blowing air through them at least once a week.

Mention has already been made of circuit breakers and these should be of the correct type and setting. If there is any doubt, an electrical engineer should be consulted. A warning should be given as to the difference between an electrician who supplies and fits electric wiring and an electrical engineer who is properly qualified and who is responsible for the correct protection of electrical equipment.

The starting equipment should also be carefully checked to see that it is properly installed and protected against overloading, heating or arcing. The same cleanliness should be maintained in and round the starter as for the motor. These points can be particularly dangerous in an old installation.

Cooking apparatus

Cooking apparatus, both domestic and industrial, causes a large number of outbreaks of fire. The majority of these fires are caused by fat and oils boiling over or 'spitting' because of water contamination.

Chip pan fires are particularly prevalent. If the fat is overheated it will eventually reach its flash point, when the vapour can ignite on the source of heat. If it cannot be ignited in this way, heating may continue until the auto ignition temperature is reached. Large deep-frying equipment will be fitted with thermostats to prevent overheating, but they can become faulty. Kitchen staff should be trained regularly so that they can cope safely with this emergency. On no account should any attempt be made to carry a pan of burning fat. Instead, the source of heat should be turned off and then the flames extinguished by smothering. A large fryer should be provided with a metal cover which can be lowered remotely. Failing this, and for small pans, air should be excluded by placing a damp towel, or a fire blanket, over the pan. This can be done in such a way that the person's face and hands are protected. The fat should then be left undisturbed until it is cool. If water is poured into burning fat it will immediately flash into steam and the contents of the pan will be ejected violently, spreading the fire and causing serious injury. Very clear instruction should be given about this danger, preferably by means of a demonstration during a training exercise. The contents of a pan will also erupt if Halon 1211 is directed into the fat. It is possible to extinguish the flames by a carefully applied Hallon 1211 or powder extinguisher but this should not be attempted by anyone other than a fireman (and then with great care).

If the contents of an oven are on fire it is safe to use a powder extinguisher, having first turned off the source of heat. Halon 1211 would also be effective but a quantity of breakdown products would be released in a small volume. When the oven door is opened to attack the fire, it should be used as a shield so that the person is not facing directly into the oven.

It is of course possible to use a foam extinguisher successfully against both these fires. However the violent flashing which occurs for the first few seconds can be frightening and an unskilled person may well abandon the attempt.

ADDITIONAL PRECAUTIONS

Whenever possible, main power and main lighting switches should be situated close to the principal exit and the last person to leave the building at night should turn off the main switches. Where a number of buildings are under constant surveillance either by a watchman or automatic watching service, it is good practice to use a status signal to show that the switch has in fact been turned off. An inspection of electrical equipment should form part of a regular (preferably daily) fire prevention survey. The inspection should include all defects such as frayed flexible cords, loose or defective switches, open fuse or junction boxes, loose or trailing leads, unofficial alterations or repairs, dangerous positions of electrical appliances, the heating of any cables or fittings and the deterioration of the insulation of cables. Particular attention must be paid to authorized temporary circuits and also to equipment which has been left on overnight, again with proper authorization (it should be ensured that the permitted time limit has not been exceeded).

TESTING THE ELECTRICAL INSTALLATION

The electrical installation of any premises should be under the supervision of a properly qualified electrical engineer. In the case of a small company, the electrical engineer should be an outside firm of electrical contractors, an arrangement that is much preferable to the employment of an unqualified engineer.

Every electrical installation should be completely tested before being taken over by the company, after which a complete test should again be carried out within five years. After the first five years, testing should take place every two years until the installation is 20 years old. The tests should then be made every year and rewiring should be carried out after 25 years or earlier if there is any appreciable deterioration. Any alterations, additions or extensions should also be tested before being handed over.

ELECTRICITY AND FLAMMABLE ATMOSPHERES

It it is possible for a flammable gas or a liquid at a temperature above its flash point to escape into an enclosure, an explosive atmosphere can be formed. The minimum ignition energy for these mixtures is very low and most normal electrical equipment can cause ignition. Special precautions must therefore be taken if electrical equipment is to be used safely in these enclosures. Ideally the electrical equipment should not be there at all. For example, lighting could be outside the enclosure shining through a transparent panel, or an electric motor could also be outside with its shaft passing through a seal in the wall. However this can rarely be done and it is necessary to use specially designed equipment which has been tested and approved for use in flammable atmospheres.

It is usually arranged that the standard of protection which is provided in hazardous areas depends on the probability that a flammable mixture will be present. Different methods of assessing this probability are described in various codes and standards. The area classification accepted in Europe is perhaps the simplest, although still typical:

Zone 0: in which a flammable atmosphere is continuously present or present for long periods.
Zone 1: in which a flammable atmosphere is likely to occur in normal operation.
Zone 2: in which a flammable atmosphere is not likely to occur in normal operation and if it does occur it will exist for only a short time (not more than 10 hours a year).

Only certain types of 'safe' electrical equipment are allowed in the different zones. An explosion will occur only when a flammable atmosphere is present at the same time that the electrical equipment fails. The probability of failure must therefore be extremely low for Zone 0 electrical equipment but the requirements can be less rigid for Zones 1 and 2.

There is a number of different ways in which electrical equipment can be rendered safe for use in flammable atmospheres, and these are described below. The choice of a particular type must depend on the area classification (the zone number) and on the gases which can be present in the enclosure. Gases are placed in various groups depending on their ease of ignition and on the minimum quenching distance for a premixed flame. The testing and approval of the equipment must be undertaken by a specialized laboratory. In the UK this work is done by BASEEFA (British Approval Service for Electrical Equipment in Flammable Atmospheres).

Explosion-proof equipment

This term is used to describe any equipment which is sufficiently strong to withstand an internal explosion.

Flameproof equipment

In designing this equipment it is assumed that electrical ignition sources can occur and that the equipment must therefore be enclosed. It is also assumed that the flammable atmosphere can penetrate this enclosure and that explosions can therefore occur within it. The enclosure must be capable of withstanding the effects of the explosion and must not allow the transmission of a flame to the flammable atmosphere outside. This is achieved by having a sufficiently strong (explosion-proof) case and by ensuring that any gaps in the case are sufficiently small to prevent the passage of a flame. It is usual to design for at least 1.5 times the maximum explosion pressure. The maximum gap between flanges and joints must depend on the quenching distance for the particular gas/air mixture. For joints not less than 26 mm long the maximum permissible gap varies between 0.5 mm for methane and 0.2 mm for gases like ethylene and coal gas. It is not practicable to manufacture equipment with gaps which are safe for hydrogen, acetylene and carbon disulphide, all of which have very small quenching distances. Flameproof equipment is suitable for use in Zone 1 (and Zone 2) areas, but not when one of the three gases just mentioned is present. The equipment is not considered suitable for Zone 0 areas.

Intrinsically safe equipment

For every flammable gas/air mixture there is a minimum amount of energy which is necessary for its ignition (in Chapter 5 the need was discussed of raising a certain minimum volume of mixture to a critical temperature). If the energy available from the equipment either under normal working conditions or reasonable fault conditions is less than the minimum ignition energy, the equipment is intrinsically safe. The minimum ignition energies are very small amounts of energy indeed but some electronic devices operate at surprisingly low voltage and current values. Intrinsically safe equipment is suitable for Zone 0 areas.

Pressurized and purged equipment

The penetration of a flammable atmosphere into electrical equipment can be prevented either by maintaining a positive pressure in the equipment or by having it continuously purged. Some automatic fail-safe mechanism is needed to ensure that the equipment does not continue to operate if the pressure or the purge is lost. This method of protection can be applied to an entire control room. This is frequently done offshore where all the process modules are maintained at a pressure slightly below atmospheric, and the control rooms (and the living accommodation) are at a slightly positive pressure. However this kind of protection is not sufficiently safe for use in Zone 0 areas.

Oil immersion

Transformers and switchgear are sometimes immersed in oil to ensure that there is no exposed arcing if the circuit is broken. This method cannot be used on DC circuits and is suitable for Zone 2 areas only.

Surface temperature restriction

Even when precautions have been taken to ensure that ingnition cannot spread from the inside of equipment to the surrounding atmosphere, ignition will still be possible if the surface of the equipment is sufficiently hot. A certificate can be obtained from a testing authority giving the maximum surface temperature of the equipment. It should be remembered that some gases can be ignited at quite low temperatures (Chapter 5). Equipment to be used in coal mines and on which coal dust can collect should not have a surface temperature greater than 150°C.

STATIC ELECTRICITY

It is probable that a great many fires which are caused by a static discharge are reported as being of unknown origin. This is mainly because many people are still not aware of the conditions under which static electricity can be generated, stored and discharged.

The generation of static charges was discussed in Chapter 5. In general, a charge is produced whenever two materials which have been in contact are separated, particularly if one or both are non-conducting. The materials can be solids, liquids or gases in any combination other than gas/gas. Flow conditions constitute continuous separation, which can result in the generation of high charges. Examples of situations in which hazardous charges can be generated are given in Table 12.6, grouped according to the two materials which are being separated.

Almost all human activity and probably the majority of industrial processes develop static charges. Most of these are too small to cause any appreciable effects. For example the simple process of writing on paper must be continuously generating charges but these are lost rapidly and harmlessly. We can sometimes detect discharges which result from our own activities. In a darkened room the sparks produced by pulling of a woolly can be seen and heard. If you stroke a cat too vigorously and then hold a finger near to its ear you can feel the shock of the discharge, as indeed can the cat. Thus we are continuously surrounded by static charges, most of which are leaking quietly away. We are concerned here with the conditions

Table 12.6
Generation of static charges

Liquid/solid	Flow of liquids through pipelines Passage of liquids through filters Splash filling of tanks
Liquid/liquid	Settling of drops of one liquid through another Stirring of immiscible liquids
Gas/solid	Filling of containers with powder Pneumatic conveying of solids Dust filters Fluidized beds Discharge of carbon dioxide
Gas/liquid	Liquid sprays, including water Discharge of wet steam
Solid/solid	Reeling or unreeling of paper or plastic sheets Conveyor belts Belt drives Movement of human body

under which discharges can occur which are sufficiently energetic to ignite a flammable atmosphere. This usually results from the charging of a surface to a high voltage over a period of time, and then the discharge of all or part of this energy as a spark. Some of the situations in which this can happen are shown in the table, and a few cases histories are summarized below to indicate the wide range of this problem.

A wig maker was cleaning human hair in a bucket of petrol. The raising and lowering of the hair and possibly the splashing developed a sufficiently high charge for a spark to ignite the petrol vapour. Two people were killed in the subsequent fire.

A fatal accident occurred in an operating theatre when a spark from a static discharge ignited ether vapour which was being used as an anaesthetic. A number of similar accidents led to the conclusion that static charges were developed in several ways including the use of nylon underwear and the movement of rubber tyres on trolleys over non-conducting flooring.

A large underground petrol store was built and was protected by a carbon dioxide total flooding system. At the opening ceremony it was decided to demonstrate the system by discharging the carbon dioxide. This resulted in a static discharge and the ignition of petrol vapour. A number of people were killed by the explosion.

It was decided to cut open a tanker vessel which had been wrecked. The vessel had been carrying naphtha (light distillate) so it was necessary to create an inert atmosphere in the tanks before starting to cut. This was attempted by discharging a number of carbon dioxide extinguishers. Again there was an explosion which killed a number of people.

A man was washing out an open container with toluene. As he swirled the liquid round a static spark ignited the vapour.

In a hairdressing establishment the static charge created by the brushing of one client's hair caused the ignition of the vapour of a flammable hair spray which was being used on another client. The results were fatal.

Human beings and static electricity

Human beings can be involved in hazardous static discharges in three ways.

First, they can generate a charge. This can be done by body movements and particularly when clothing is worn which contains synthetic fibres. If the charge cannot leak away because of non-conducting footwear or an insulated floor, then a charge can be built up which, if it is discharged as a spark, can ignite flammable vapour/air mixtures. A charge can also be generated by walking on a synthetic carpet in an air-conditioned hotel. In the last case mentioned above, the hairdresser acquired a charge simply by brushing his client's hair. People with naturally dry skin are particularly prone to develop charges.

Second, they can accept a charge. If a man approaches very close to a charged objected then part of the charge can be transferred to him by a flow of current. The current will depend on his capacity and on the voltage, and may be sufficiently high to cause a spark. Alternatively his mere presence near to a charged object can cause him to acquire an induced charge by the movement of electrons on the surface of his body. This charge can then be released as a spark when he approaches say a metal object which is earthed.

Third, they can earth a charge. A man can accept a charge as described above only if there is no conducting path to earth (usually through his shoes). If such a path does exist he can drain a charge which has accumulated on an insulated object. Again if the voltage and current are sufficiently high, an incendive spark can pass between the object and the man. It is possible to experience quite a painful shock by touching a vehicle which has been driven for some distance on a very dry day. The charge which has been generated by the movement through the air, and possibly also by the separation of tyre contact on the road, is discharged through the body before it has had time to leak away through the high resistance of the tyres.

It might appear from this description that a man can create a static hazard whatever happens. Where people are working in areas where flammable vapour/air mixtures can form, they are provided with conducting footwear, or with a strip of conducting material which can be tucked around their shoes. They cannot then build up a charge (although they can continue to generate static electricity). At the same time the equipment on which they are working is adequately bonded and earthed so that it too is incapable of acquiring a hazardous charge (see below).

Detection of static charges

The presence of a static charge can often be demonstrated by using a simple neon tube of the kind used to indicate the presence of mains voltage, or to test the ignition circuit on a car. This is not a sensitive method of checking and the absence of a glow in the neon tube does not necessarily show that there is no static charge. Instruments are available which are much more sensitive but instruction is needed in their use if reliable information is to be obtained.

Protection against the dangers of static electricity

Hazardous static charges can be avoided in two ways: by preventing the generation of static electricity or by draining it away so that a charge cannot be built up. In hazardous areas it is usual to find both methods in use, as was described above in the discussion on human beings and static. However it would be wrong to think that the generation of static electricity can be completely eliminated. What can be done is to reduce the rate of generation and increase the rate of leakage so that it is not possible for a hazardous charge to accumulate (a hazardous charge being one in which the voltage is sufficiently high to cause a spark and the released energy is sufficient to ignite a flammable mixture).

The transfer of hydrocarbon liquids through pipelines and into tanks can be hazardous because of the high resistance of these liquids, which reduces the rate at which a charge can leak away. In discussing this problem it is usual to refer to the conductivity of the liquid rather than its resistance. Conductivity is measured in reciprocal ohms and the unit used to be called a mho. The SI name for the unit is now the Siemen, and the conductivity is 1 S when the application of I V between the points causes 1 A to flow. The dissipation of a charge is called relaxation and the time taken for a hydrocarbon liquid to lose about two-thirds of its charge is the relaxation time. (It is not possible to give a time for the complete loss of charge because the voltage, and therefore also the current flowing from the charge, decreases continuously. It will always take the same time to lose two-thirds of the remaining charge and indeed in theory the charge can never be completely lost). Hydrocarbon liquids which generate hazardous charges have conductivities between 0.1 and 10 pS/m (picosiemens per meter). The corresponding relaxation times are 180 and 1.8 seconds.

The rate of generation of charges when a liquid is flowing through a pipeline can be reduced by lowering the velocity of the liquid. If this is done by using a pipeline of larger diameter, the surface to volume ratio will also be reduced but this does not have a significant effect. The presence of even small amounts of free water, or any other immiscible liquid, increases the rate of charge generation. Water droplets will act as charge collectors and their presence can increase the charge in the receiving vessel by as much as 50 times. If a contaminated liquid like this is being transferred, the maximum velocity in the pipeline should be 1 m/s. If the hydrocarbon is completely dry, higher velocities can be used, perhaps 2 m/s at the lowest conductivity to no more than 7 m/s at the higher conductivities.

If the liquid passes through a pump, and particularly if it goes through a micro-filter, charging is considerably increased. A length of earthed conductive pipeline should be provided between these and the tank to allow the excess charge to leak away. The residence time in this pipeline should be about three times the relaxation time of the liquid, but not less than 30 seconds (which would be sufficient for any liquid with a conductivitiy of 2 pS/m or more). The residence time can become inconveniently long for liquids with very low conductivities but the required time can be reduced by the use of an antistatic additive. The ones used for aviation fuels will typically increase the conductivity to greater than 50 pS/m.

The presence of water in the bottom of a tank can cause problems when it is mixed into the incoming hydrocarbon. For this reason water should be drained regularly from tanks. Also, if there is a flammable atmosphere in the tank, at least 30 minutes should be allowed after filling has stopped for the charge to relax, before any manual dipping or sampling is allowed.

Splash filling will generate charges as well as releasing flammable vapour and spray. The filling velocity should be restricted to 1 m/s until the outlet of the fill line is covered to a depth of at least 0.5 m. This velocity applies also to the filling of road and rail tankers although the velocity may be increased once the outlet is covered.

A hazardous discharge happens when the surface of one object is at a considerably higher voltage (because of a static charge) than that of a neighbouring object. No discharge will happen if they are both at the same potential and this can usually be ensured by bonding the two together electrically and connecting them to earth. It is of course quite usual for a non-conducting surface to acquire a charge and clearly there would be little prupose in bonding this to its neighbour. Fortunately the discharge current from the surface would necessarily be low and would probably not be hazardous. The real danger is from the discharge of a large metal surface, from which the whole of the charge can flow almost instantaneously. The main hazards on a plant in which potentially dangerous static charges are being generated can be eliminated by bonding and earthing all the metal parts. In a road tanker filling bay for example it would be ensured that there was good electrical contact between all the pipeline flanges, and both the structure and the tanker would be connected together and to earth before any liquid was pumped. Any flexible hoses would also be made conducting by a metal thread in their construction.

Dust collecting bags in filter units can also be made with metallic threads in the fabric. Rubber and plastics can be made sufficiently conductive by the addition of carbon black, and conducting rubber is used in hospitals for floors and trolley tyres in operating theatres. Bonding and earthing are particularly important when rotating machinery is mounted on wood or concrete. It is sometimes necessary to collect a charge from shafts and belting by means of brushes, but care must be taken that these do not themselves cause a spark. Belting can be made conducting by the use of a suitable dressing.

The resistance of a connection to earth can be quite high and it will still be effective because of the high voltages and low currents which are involved. However the practice of using a length of chain on a concrete floor as an earth is not safe. It is wise to have a really good earth connection so that it will provide protection against electrical faults and against lightning, which is discussed below.

Hazardous static discharges are less likely to occur if the conductivity of the air is increased. This is because the charge escapes as a steady flow of current through the air instead of jumping suddenly as a spark. Air becomes sufficiently conductive to produce a significant reduction of the static hazard when the relative humidity is 50 per cent or more. The conductivity of dry air can also be increased by the use of alpha radiation from radioactive material which causes the air to be ionized. Suitable sources are now incorporated in many production lines in the process industries.

LIGHTNING

The voltage needed to cause a static discharge from a cloud to earth is in excess of 100 MV. The current which flows in the lightning stroke can be between 3 and 200 kA. The current is so high that quite appreciable currents can be induced in conductors which are parallel to a lightning conductor. Also even the very low resistance of a properly designed conductor can result in sufficiently high potentials at the top to cause arcing to other metallic earth routes. Surprisingly the increase in temperature of the conduct is only about 1°C. This is because of the very short duration of the lightning stroke: perhaps only 0.1 ms. The temperature increase can be very much greater in a high resistance conductor, for example a tree. The water within the trunk can flash into steam, generating sufficient pressure to shatter the tree. The smaller fragments of wood may then be ignited. The temperature of the air which is conducting a lightning stroke can be as high as 30 000°C, corresponding to a peak power of about 100 MW/m. The rapid expansion of the air results in a shock wave which is capable, for example, of dislodging tiles from a roof.

Clearly a lightning stroke itself and the associated arcing when it hits a structure can be powerful ignition sourses. Considerable protection from these effects can be provided by a properly designed lightning conductor. However we need to decide whether the cost of installing and maintaining lightning protection can be justified by the probability of being hit by lightning. In countries which are within about 20° latitude north or south of the equator there can be up to 180 thunderstorm days a year. The southern states of North America have these storms on between 40 and 80 days a year. In the UK the number goes down to between 5 and 15. These numbers are averaged over about 11 years, which corresponds to the sunspot cycle: the variation during this period can be as high as two to one. It is more convenient to use the values obtained in a decade and most published statistics are based on this period. In the UK about a million flashes strike the ground each decade but there are more strikes in the east than the west, and more in the south than the north. For example there is only about one stike per km^2 per decade in the north of Scotland and the west of Wales, compared with as many as 7 per km^2 per decade in the south-east of England. Using values like these, obtained from meteorological records, the probability of a particular building being struck by lightning can be calculated. This is done by working out the 'effective collection area' of the building: the plan of the building extended in each direction by the height of the building, and then the corners formed by suitable radii. For example Figure 12.9 shows a simple rectangular building 20 m × 10 m and 5 m high. The effective collection area is made up of a number of rectangles plus four quadrants of a circle with a radius of 5 m. The total area is about 580 m^2. If this building is located in Suffolk, where the probability of a strike is 7 per km^2 per decade, the building should be hit once in about 2500 years. (Note

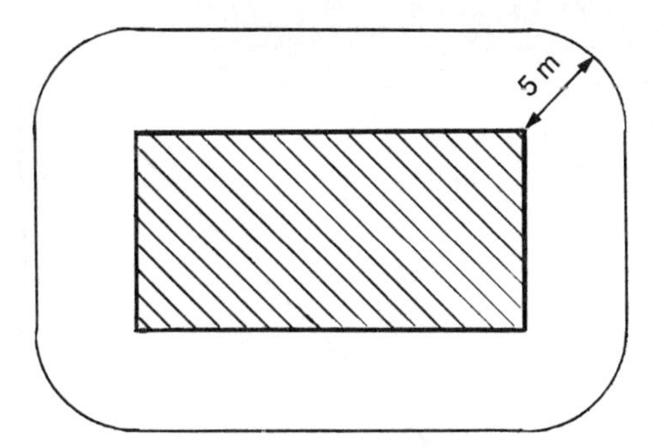

Figure 12.9 Effective collection area of a building 5m high

however that this is the probability of being hit: it could still happen tomorrow.)

The probability that a building will be hit and damaged is greater if it is tall, isolated, on high ground, and does not have a steel frame. The consequences of a strike will be greater if the building contains explosives or flammable liquids, or if it is occupied by a lot of people. All these factors, and others, are taken into account in a method of calculating a risk factor in the British Standard on lightning protection (BS 6652: Code of practice for the protection of structures against lightning). This and similar standards contain a great deal of very detailed information on the design and construction of various protective systems, the materials of construction, and the procedures for inspection and maintenance. Only a few very general points are reproduced here.

The protection of explosives factories and magazines in the UK is a requirement of Orders in Council made under the Explosives Act 1875. This is sometimes achieved by means of suspended air terminations, in which a network of horizontal cables is suspended above the buildings from a number of masts. The more usual method is to provide one or more vertical conductors. The volume protected by a conductor is usually assumed to be a cone with its base on the ground and its apex at the tip of the conductor. The angle between the conductor and the side of the cone is 45°, that is, the total solid angle is 90°. Since the angle between the side of the cone and the ground must also be 45°, the triangle formed by these two and the conductor must be isosceles. The radius of the base of the cone must therefore be equal to the height of the conductor. This is shown in Figure 12.10 where it can be seen that the building shown in Figure 12.9 can be protected by a conductor which is 27 m high. This method should not be applied if the building to be

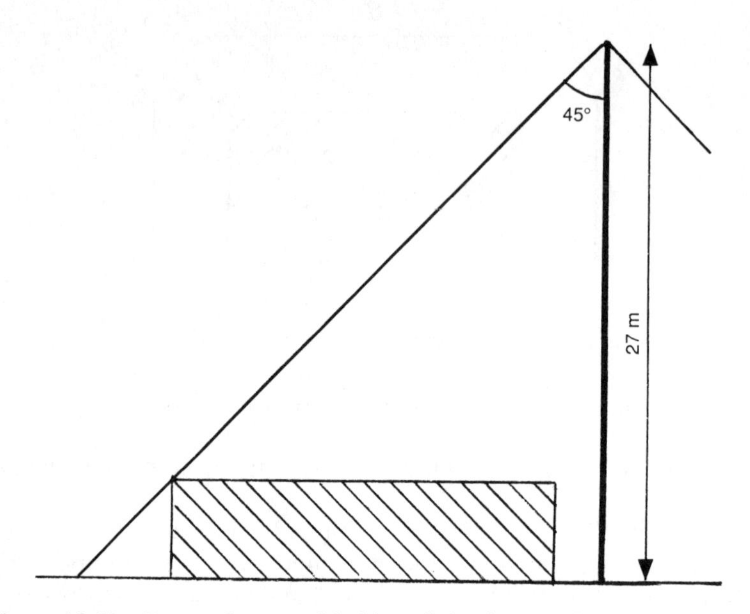

Figure 12.10 Protection provided by a lightning conductor

protected is more than 20 m high because strikes could occur on the sides of the building. The British Standard describes methods of assessing the protection needed for buildings which are tall or of complicated shapes. It also recommends basing the protection of buildings with explosive or highly flammable contents on a cone angle of 30° instead of 45°. This is because the 45° angle, although based on experience, is not absolute. Certainly the probability that the protected building will be struck decreases as the protective angle is reduced. Suspended conductors and masts should be at least 2 m away from hazardous buildings to prevent side flashing. The earth electrodes should be suited to the physical nature of the soil and can be a closed loop not less than 20 m in length, a vertical rod not less than 9 m in length, or radial conductors not less than 20 m in length. The British Standard requires that the resistance to earth must not be greater than 10 ohms. (Lower values, eg, 7 ohms, are given in other standards.)

If a steel tank containing a flammable liquid is struck by lightning, the current should flow harmlessly to earth through the walls of the tank. It is necessary for the tank to be properly earthed and this should be done even when the tank is standing on the ground. There may not be a sufficiently low resistance through the foundations, particularly where a bitument/sand mixture, or concrete, has been used. The passage of a heavy current through the tank wall may induce a current in a metal structure within the tank. For this reason and for the prevention of static charges all the metal within the tank should be bonded to the main structure. In floating roof tanks it is particularly important that an adequate low resistance connection is made between the roof and the shell. Metal structures used to support process plant should also be bonded and earthed.

Electrical storms

In a thunderstorm the static charge in a cloud gradually increases until the voltage is sufficiently high to cause the discharge through a lightning stroke. During the time before the lightning passes, the charge in the cloud will induce an equal and opposite charge on the ground (and buildings) beneath it. This charge will persist for as long as the charge cloud is present. If the cloud discharges to another cloud, or to a point beneath its extreme edge, then the sudden release of the charges on tanks and structures beneath the cloud may cause sparking if they are not adequately bonded. It is because of the existence of this opposite charge on the ground beneath a cloud that the lightning may sometimes travel upwards rather than downwards. There may be a series of strokes of which the first, called the leader, is from ground to cloud.

If the atmospheric conditions are right, some of the induced charge on structures may be dissipated in a corona discharge. This is seen as a glow, rather like that in a neon tube, sometimes accompanied by a hissing sound. The corona is usually attached to the highest point and may typically be seen on vent stacks. These corona discharges are capable of igniting a flammable mixture.

SMOKING

Until recently smoking and smokers' materials were considered to be the greatest cause of ignition, and the cause of an outbreak of fire was attributed to this if no obvious reason was at hand. Although it does not now seem to be quite so frequently given, there is no doubt that smoking is a major cause of the outbreak of fire.

Prohibition of smoking, unless it can be absolutely enforced, is not sufficient. Even in mines containing explosive gases and in which the danger of naked flames is well known to miners, men have still been known to smoke. Prevention rests considerably on the education of people to the dangers of smoking and to the simple measures of prevention that should be taken.

Smoking must be prohibited in certain places and under certain conditions. Thus, where an absolute prohibition is necessary, safe areas must be designated, and time allowed for people to visit them. Even in petrol depots it is possible to install small glass-sided structures in which a man may smoke in safety. Such smoking islands should be equipped with a chair and table, and ample ash trays of safe construction. They should be situated reasonably close to where the man is working and preferably reached under cover. Lighters should be provided so that matches are not used. The chair or chairs should not be of a type in which a cigarette end can lodge and smoulder. Before setting up smoking islands, the insurance company should be consulted.

Opposition to smoking islands sometimes occurs because employees feel that they are under constant observation. It is not intended that the user should be under observation, but only that a possible fire can be immediately seen. Although plain wired glass is to be preferred, a figured glass could be used rather than have opposition to the proposal.

Where smoking is permitted in offices and similar places of comparative safety, ample ash trays should be provided. The ash trays should be either of the self-extinguishing type or of heavy glass or porcelain, with facilities for resting a cigarette. But it should be so arranged that, if it is forgotten, it cannot burn down to a point where it will overbalance and fall outwards on to a table.

Where smoking is permitted, it should be confined to rooms or offices and not allowed in corridors or open spaces, so that lighted matches or cigarettes will not be dropped. A rule should also be imposed that smoking must cease 30 minutes before the office or room is vacated for the night.

The emptying of ash trays is very often a source of trouble and special arrangements should be made for all ash trays to be emptied into a metal bin, and not into sacks containing waste paper.The best method to accomplish this is by means of a special patrol which not only ensures that ash trays are emptied in safety, but also starts the first fire patrol.

WELDING OR CUTTING OPERATIONS AND BLOW LAMPS

Blow lamps and gas and electric welding and cutting operations are a very frequent cause of fire in industrial premises.

The Fire Offices Committee issues an excellent leaflet setting out the precautions that should be adopted for this work. In cotton mills, the insurance company must be consulted before any of this type of work is carried out.

The best method of applying these rules is by means of a permit to work. This form should be originated by the engineering department, and should go in the first place for counter-signature by the manager of the department and then to the fire precautions manager who will have the site of the work visited.

Examination of the site

The fire precautions manager will see that all flammable materials are removed from the site and will also check the opposite side of any walls or partitions, ensuring that there are no flammable materials in danger from heat transmitted through the walls. Special attention should be given to the possibility of heat being transmitted by steel girders or steel plates. If necessary, portable fire-resistant metal screens should be ordere to the site.

The floor should be inspected to see that there are no cracks through which sparks or hot metal could pass or be lodged. If the floor is of combustible material it should be wetted and kept wet or a layer of sand spread over it. If no other means of protection can be used, the floor should be covered with fire blankets which are well overlapped. If the floor is soaked with flammable material – and even a concrete floor can become so – special protection may be necessary. A good practice is to cover the floor with a carpet of foam immediately before work commences. The practice of laying down sheets of metal is not advisable because sparks can easily be blown underneath.

Having carried out the inspection, the fire precautions officer should countersign the request and insert the name of a trained person who will be present during the work as an observer. This trained person, probably a fireman or a patrolman, must not take any part in the work so that he is not distracted from his prime duty of observing any danger. If he considers there is any danger, he should have authority to stop the work until a more senior officer can attend.

Equipment for the observer

The observer should have water available, together with other suitable portable extinguishers. If there are flammable liquids within the general area, foam extinguishers should be available and, if the flammable materials are solvents that will break down standard foam such as alcohols, acetones, ketones, aldehydes and so on, special foam extinguishers will be required.

Wet sacks and a fire blanket can be very useful, and a metal receptable for the stub ends of welding rods is necessary. Where much of this work is carried out, a small trolley can be made up to carry these items. It can then be kept ready for use at any time. A trolley such as this should always be available in a welding shop.

Instructing the observer

The observer should *not* be equipped with tinted glasses, but must be warned of the severe injury and danger to eyesight of looking at a welding or cutting flame and particularly of looking at an electric arc. He should be instructed to take up a position where the operator will screen him from the working point, but where he will be in a position to see and appreciate the fall of sparks. He should for preference be at least 10 m away from the site of operations.

Site duty

On arriving on site, the observer will check for himself that precautions have been taken and that his equipment is in order. He will also look at the welding or cutting apparatus that is to be used, ensuring to the best of his ability that the gas hoses are in good condition and that electrical cables are not worn or frayed. If there is any sign of damage to electric cables, either on the welding or supply side of the welding machine, the entire cable must be replaced.

In the case of electric welding, the earth return cable must be securely clamped to the work and to the machine. Use should not be made of any metalwork or the steel frame of the building for earthing purposes. If blow lamps or torches are to be used, the observer will ensure that they are filled to capacity and lit in the open, that refilling is done in the open and that no refilling is done while the lamp is hot. Gas-operated lamps must be extinguished and allowed to cool before changing the cylinder. All blow lamps should be extinguished as soon as the work is completed, and on no account should a lighted blow lamp be left unattended even for a few minutes.

The observer will not leave the site for 30 minutes after work has finished, and the whole area including adjacent rooms above and below the working site should be inspected after about one hour. Where heavy steel girders have been heated, danger may exist up to six hours and the danger extends the whole length of any girder even where it has passed through a wall.

Where a flame is maintained in a welding shop, it should be in a fixed position and not on a flexible tube. There must be no woodwork or other combustible materials near to it.

A special warning must be given on the use of metal drums as welding tables. It is very common to see work placed between two drums which make an excellent table. But, whatever the drums may have contained, there is always the danger of heating any residue inside. Even if this was heavy oil or some chemicals, the effect of local heat might cause a serious explosion.

REPAIR OF TANKS AND DRUMS

The application of heat in any form for the repair of tanks or drums that have contained an explosive or flammable substance or liquid is not only dangerous, but is contrary to section 31(4) of the Factories Act, until adequate measures have been taken to ensure that the material and any vapours have been removed or rendered non-flammable or non-explosive. The Act also forbids the admission of any explosive or flammable material until the metal has cooled sufficiently to prevent ignition.

Materials at risk

The most common risk is the repair of car and lorry petrol tanks. It is not necessary for a spark to be present to cause an explosion, as the ignition temperature of petrol is so low that the heat of a soldering iron is sufficient to cause an explosion.

Many other materials which are often contained in tanks and drums such as diesel oil, vaporizing oil, linseed oil and even solid waxes have low ignition temperatures.

These materials, and many others, give off flammable vapours when heated, and a very small volume of liquid is sufficient to produce an explosive mixture. Small quantities, such as might be held in a seam, produce the maximum explosive mixtures, whereas a heavier concentration might be too rich to explode although it might burn with explosive violence.

Repairs without heat

Emergency repairs can sometimes be effected without the use of heat. One suggestion is the use of cold plastic metal. A number of such materials are available on the market, some of which are very strong and could almost be accepted as a permanent repair under ideal conditions. Similar remarks could also apply to cold curing synthetic resins. The manufacturers should be consulted as to the suitability of the materials.

It might be possible to use a patch with a suitable packing material held on by self-tapping screws. Great care would have to be taken in drilling the tapping holes and only a sharp new drill used, as it must not be allowed to run hot.

Removing flammable materials: how not to do it

Before outlining measures that can be taken to remove flammable material, mention must be made of methods by which flammable material cannot be removed. It may be thought that containers left for some considerable time, with the filler cap removed and in the open air, will eventually become safe. They do not become safe, and many serious explosions have been caused by children playing with fire near the open petrol tanks of derelict motor cars. Another suggestion is that blowing out with compressed air will eventually render the tank safe. This might be so in a few select cases (where there is no danger that material could be trapped in seams or joints) but it is a method that cannot be recommended. There is also some danger from the development of static electricity when blowing compressed air into an explosive atmosphere. The danger results mainly from entrapped materials such as drops of water. Cold water, even with the addition of caustic, is not effective.

How to remove flammable materials

There are two main methods of removing flammable material from tanks and drums: steaming and boiling out with a strong alkaline or detergent solution.

Steaming out

For steaming out, low pressure steam is used and precautions must be taken against the development of static electricity. A well earthed armoured hose with a metallic nozzle should be used, and the hose and nozzle bonded to the tank. The tank itself should also be earthed.

There must be ample opening for the steam to escape and there should be an opening at the bottom of the tank to allow any distillate to escape, together with any sludge that may be solved out. If there is only one opening, as is the case of some drums, the hole should be at the bottom.

Steaming for two or three hours will probably be required, longer for some types of containers. To check when the tank is ready, the inside of the tank can be examined by means of a small mirror and a flameproof torch. This should be backed up by examining the distillate. Preferably this test should be made by a chemist. For large tanks, an explosimeter can be used, though the results might be misleading if the contamination is by materials that are not very volatile.

Boiling out

As much as possible of the material that the tank has held should be removed and then, with all filler caps, plugs and the like removed, the tank or drum should be immersed in and filled with the boiling solution which should be either a strong alkaline or detergent solution. Because alkaline solutions attack aluminium and aluminium alloys as well as some other materials which may be used to line tanks and drums, the advice of a manufacturer of drums should be obtained before using a boiling alkaline solution on any material other than iron or steel.

Boiling should be continuous until the container is clean. This will take at least half an hour and much longer for some substances. The same testing routine should be carried out as is described for steaming.

Rendering ignition impossible

After thoroughly cleansing the container, the atmosphere inside it should be rendered inert before attempting any process involving heat.

Where steaming has been carried out, it may be possible to complete the repair work while the steam continues to flow, thus maintaining an atmosphere almost free of air. Sometimes it is possible to use nitrogen gas. Carbon dioxide in the form of a solid block of 'dry ice' can be used and at least 2.75 kg per m^3 is required. This should be crushed and dropped in the container. The container is allowed to stand for a short while and then sealed up, leaving a small vent to relieve any build-up of pressure.

Halons

A small concentration of Halon vapour would certainly inhibit the combustion of a flammable mixture in a tank. Its use is not recommended because the Halon liquid is miscible with many liquid fuels and some of the discharge could be lost in this way. Also, the vapour would decompose in contact with hot metal and sparks, producing a potentially corrosive atmosphere.

Foam

Large tanks can be filled with high expansion foam which will displace the atmosphere in the tank. The mass of bubbles in HiEx contain air and it is possible that some of the fuel vapour could eventually diffuse into the bubbles, forming small pockets of flammable mixture. There is no record of premixed flames propagating through HiEx in a tank and it seems unlikely that this would happen. The possibility can be averted by blowing the HiEx with nitrogen instead of air. This has the advantage of retaining the nitrogen in the tank. On a windy day there could be a considerable loss of free nitrogen through openings in the top of a tank.

Before commencing any heating operation

Make sure that one of the reliable forms of preparation has been carried out successfully. Unless this is certain any work involving heating must not be started. Even with all other precautions taken, work should be undertaken in the open air if possible.

It should be remembered that the inert atmosphere in a tank can be lethal. It may be tempting for a man to put his head inside the vessel to observe the effects of his work. If he is breathing pure nitrogen for only a brief time he may become unconscious and fall into the vessel.

SPARKS FROM POWER UNITS

There is a growing use of mobile cranes, fork-lift trucks and small power-driven trucks. These are powered by various means, particularly by electric batteries, petrol, diesel and liquid petroleum gases, and sometimes by means of trailing or overhead power cables. Each system has its own fire hazard, as well as that from the load which it is carrying and the use of the premises in which it is operating.

Special precautions are necessary where these machines are lifting and carrying flammable liquids, and this applies not only to manufacturing concerns but also to general storage warehouses, where a container of flammable liquid could fall from a pallet. The damage to the container caused by the fall might release flammable liquid, the vapour of which could then be ignited by the prime mover.

Battery power units

Electric battery power units carry heavy batteries that make them very suitable as cranes and fork-lift trucks, because the batteries provide a counterbalance weight. When used where there is any possibility of flammable liquids being present, the motor and controls should be flameproof.

Charging the batteries must not be carried out in the warehouse or premises, but in a separate fire-resistant building. If this is not possible, charging should be carried out in a fire-resistant compartment, each machine being separated by a non-combustible partition. The building must be well ventilated at top and bottom, because hydrogen gas will be developed from most batteries while being charged. Hydrogen, being lighter than air, will rise and it must not be possible for it to be trapped under roofs or canopies. Fresh air must be able to enter at the bottom of the charging compartment, so that a free draught will carry the gas away. Charging must be carried out under weatherproof conditions, and all leads should be carefully inspected for signs of wear or damage, especially at any point where they could chafe. Damaged leads should be replaced, and no attempt made to repair them.

Patrolmen should watch for materials being left on charging apparatus. If the materials are flammable any defect in the charging plant could start a fire, but whether the materials are flammable or not they would interfere with the proper ventilation of the apparatus. This, in turn, could cause overheating and lead to trouble.

Although the charging plant is designed to cut out when the batteries are charged, or after a given period, it is advisable for a time limit to be posted up for the patrolman to check. If the automatic device fails to cut out he is then able to take such steps as his instructions provide.

Regulations of the Institution of Electrical Engineers should be adhered to for all batteries.

Petrol and diesel power units

If petrol or diesel-powered machines are to be used where there is any combustible material, whether it is highly flammable or not, flame arrestors should be fitted both to the induction and exhaust systems. Refuelling should be carried out at least 6 m away from any building or storage of materials and the driver should not be seated on the machine. The laws and bylaws regarding petroleum spirit both in regard to storage and use must be strictly observed. Spare cans of fuel should not be carried on the machine.

If electric starting or lighting batteries are used, a master switch should be incorporated at the battery terminals to isolate the battery in case of emergency.

No self-propelled vehicle should be allowed to remain in a factory or warehouse overnight or unattended unless it is in a properly constructed and enclosed fire-resistant compartment.

Each machine should be equipped with a suitable fire extinguisher. Halon 1211 would be the best choice. Powder could be effective, but would be very difficult to clean out of the moving parts of the machine. Carbon dioxide could also be used but it is less effective than the Halon and does have a limited throw.

If there has been a release of flammable gas or vapour into the area where a petrol or diesel unit is operating, this will be drawn into the engine with the air. The engine will begin to race because of the additional fuel. If its own fuel is then cut off, the engine may continue to run if the concentration of gas or vapour in the air is sufficiently high. A petrol engine can of course be stopped by switching off the ignition circuit, but a diesel engine will continue to run. If the mixture is rich the engine will race and overheat. It will then probably release sparks and flames from its exhaust, and these will ignite the flammable cloud which surrounds it. Incidents like this have happened, resulting in the ignition of released vapours in an area which was otherwise free of ignition sources. A runaway diesel can be stopped only by shutting off the air supply to the engine, or by injecting an extinguishant. Devices are available which will do this and diesel units which are used in hazardous areas should be provided with this protection.

Power units using LPG

In addition ot the precautions given under petrol and diesel power units, the following precautions should also be taken with machines using liquefied petroleum gases:

1 The connection from the gas cylinder to the fuel pipe should be a quick acting, self-sealing coupling.
2 There should be a fuel shut-off valve at a point before the carburettor, so designed that the valve closes automatically when the engine stalls or is switched off.
3 There should be a pressure relief valve between the cylinder and the automatic shut-off valve as well as on the cylinder itself. Neither relief valve should discharge near to any source of ignition.
4 The cylinders of LPG, whether full or empty, should be stored outside the buildings, preferably in a small separate compound. No part of the compartment should be below ground level because LPG vapours are heavier than air and will 'flow' along trenches, drains and culverts into cellars and into hollows, where they will lie as an invisible pool. Care must be taken in siting the compartment away from these areas.
5 Ample ventilation to and from the open air should be provided at the top and bottom of the compartment and all cylinders should be stored upright.

Trailing cables

Machines with trailing electic cables should not be used inside a factory or warehouse where combustible or flammable materials are stored or used.

NON-ACCIDENTAL CAUSES OF FIRES

The term 'non-accidental' causes has been coined to cover a number of causes of outbreaks of fire which are either malicious, deliberate or of such gross carelessness as to be equally culpable.

Bonfires and rubbish fires

These fires, technically said to be 'burning under control', do from time to time get out of control. The site for rubbish fires should be carefully selected, at least about 20 m away from doors or windows that might be open. Properly enclosed incinerators should be used and material fed into them slowly, so that no burning embers are blown into the air. Material waiting to be burnt should be so kept that it is impossible for a spark to come from the incinerator and set fire to it.

Rubbish fires, incinerators and bonfires should never be left unattended and, if not completely burnt out and cold at the end of the day, they should be doused with buckets of water while the ashes are being turned over.

Paraffin should not be thrown on to a fire to revive it, and petrol will almost always flash back to the container with serious results, probably burning the operator. It it is necessary to assist a fire to burn or to revive a dying fire, a piece of rag soaked in paraffin will be more successful and much safer.

Children playing with matches

Matches as a form of ignition provide many non-accidental fires, and in the hands of children and young persons they are a positive menace. The only safeguard is to keep unaccompanied children under careful observation or, if they have no right to be on the premises, they should be promptly removed. Good security is the only answer to this trouble.

Adequate safeguards should be taken with pavement lights and similar places where children can amuse themselves by dropping lighted paper or matches through a grating.

Arson

Arson is the crime of deliberately setting fire to property but it is not described as such until a court has found a case proved. The correct expression is a fire of 'doubtful origin', whereas the fire the cause of which cannot be determined is said to be of 'unknown origin'.

Arson is generally taken to be due to a desire to obtain money from insurance companies and there is no doubt that this is very often the case – more often, in fact, than can be proved. But there are many other forms, and reasons, for arson. A very frequent cause is that of employees with a grudge against their employer, even against a foreman, or another employee. It is often used by thieves and other criminals to cover up the evidence of their crime.

Another very frequent cause of arson is that of persons who try to arrange a fire so that they can obtain praise or credit as being the person to discover it or put it out (in which very often they to not succeed). In the same category are firemen who get bored and want to attend a fire, and people who are mentally disturbed and who obtain satisfaction from seeing an uncontrolled fire. There is an unfortunate additional cause: the actions of terrorists, mobs and demonstrators.

The prevention of fires of this kind depends very much on security, which must therefore be regarded as an essential part of fire protection. The amalgamation of fire, safety and security activities in many companies has practical as well as financial advantages.

Also under the term 'non-accidental fires' can be listed such items which could be described as wilful neglect. This is neglect to maintain fire precautions deliberately so that, if a fire does occur, it will be serious. With this goes the wilful neglect to make the best use of established fire precautions and firefighting facilities, for example by omitting to call the fire brigade. This neglect stems from the feeling that after all a fire might be a good thing to tide over a period of bad business.

Sometimes carelessness can be taken to such an extent by the neglect of even the most elementary precautions against an outbreak of fire, or the spread of an incipient fire, that it can be considered as criminal negligence, as in smoking in a fiery pit or a petrol store.

IGNITION BY CHEMICAL ACTION

Spontaneous ignition

This has already been discussed in Chapter 5. It is often the result of oxidation which has taken place in a mass of particulate or fibrous material. If the conditions are such that heat is generated by the reactions faster than it is lost to the surroundings, the temperature will increase. Smouldering combustion may then be established. If this progresses to the outside of the heap of material, flames may develop which will spread fire to other combustible materials.

Spontaneous ignition is more likely to occur in still, warm conditions than for example outdoors at night. Some materials are more likely to heat up if they are damp, as is the case with hay. Here the initial heating is due to the activities of living organisms, mainly pyrophilic bacteria. Similar metabolic processes can cause the ignition of manure heaps.

A surprising number of liquid and solid fuels are capable of spontaneous ignition when the ambient conditions are right. A few of the more common substances are discussed below.

Sawdust

Sawdust is not usually recognized as being liable to spontaneous ignition, but nevertheless when green timber is being cut up spontaneous ignition is liable to take place. It was observed during a hot summer weekend when rain had penetrated into the sawpit that the top surface of the sawdust had caked over and heating had taken place underneath. When found, it had been untouched and unobserved for about 30 hours, and was then glowing red with the wooden supports of the walls smouldering.

It is possible that the self-heating of green sawdust may also be the result of microbiological action. Once smouldering has been established in a large heap of sawdust it can take several days or even weeks to penetrate to the surface.

Oily materials

The most common of these are linseed oil and boiled oil, including colours in oil. Tung oil is less liable to spontaneous ignition than is olive oil, while castor oil and turpentine have little tendency to heat.

Vegetable oils and animal oils generally are more likely to give trouble than mineral oils. Lubricating oils are not liable to spontaneous ignition unless they have been degraded previously by heat.

The reaction is more likely to occur if the oil is soaked into cotton waste, rags, fibrous or granular material. Well known examples are when spillages are mopped up and the contaminated material is placed into sacks for disposal. A typical case occurred when an undertaker spilled linseed oil with which a coffin was being polished. It was a hot day, and the oil was mopped up with cotton waste which was later swept into a sack containing sawdust and waste paper. The fire was discovered some nine hours later when the sack burst into flame.

Most paints will behave in the same way when they are absorbed by cleaning rags and the rags are collected in a heap.

Although mineral oils are not likely to fire spontaneously, it is always possible that there may be some contaminating material also present. So it should be a rule that all oil-soaked material in workshops, garages and in factories should be placed in a metal bin that is kept closed with a metal lid, preferably hinged so that it may not be detached from the bin and so that it cannot be left open. The bins should be stood outside the building during non-working hours.

Waste paper

There is no report of dry paper having ignited spontaneously, but this has happened with soggy heaps of waste paper. It is necessary for the paper to be just sufficiently damp and presumably there is a critical size of heap. Again it is possible that the initial rise in temperature is due to micro-organisms.

Coal and charcoal

Rather surprisingly the surface of coal can be oxidized at room temperature. This is particularly true of freshly created surfaces. The heat released by this reaction can easily be dissipated from an isolated lump of coal, but not if the lump is buried in a heap. Because of this, the temperature within a heap may continue to rise until smouldering begins. This can then spread thoughout the heap. The conditions under which a heap can ignite depend on the type of coal (soft coals have a greater affinity for oxygen), the sizes of the lumps, the moisture content, the ambient temperature, and the time since the lumps were broken up. The last effect is important because of the greater reactivity of recently formed surfaces and it is for this reason that coal should not be dropped from a height on to the storage area. The presence of pyrites in the coal also seems to increase the ease of ignition. In general, lumps larger that 25 to 75 mm give little trouble and stacks of less than 200 tonnes and less than 2.5 m high have not been known to ignite. Stacks up to 4.5 m high should be safe if they do not contain more than 50 tonnes. It sometimes happens that fine coal which has begun to heat up in a storage heap is delivered to a hopper which is feeding a pulverized fuel (PF) plant. Even though a relatively small amount is kept in thc hopper, it may begin to smoulder. Surprisingly, smouldering coal has been fed into the ball mill of a PF plant without causing a dust explosion. This may be because the dust concentration was above the upper limit of flammability, but clearly this situation should be avoided.

It is good practice to insert a number of vertical metal pipes into a large heap of coal so that a thermometer can be lowered down them to check the temperature within the heap. Any temperature above 27°C is hazardous, and a fire will almost certainly occur if the temperature rises above 55°C, unless the heap is broken up and cooled. Men who work on coal storage sites can usually detect the presence of hot spots by the smell which is released at the surface of the heap.

Charcoal is used in many industries, and is always present after a fire. Water seems to play an important part in the initiation of the reaction and wet materials should be especially suspect. They should be removed and spread to cool in the open air. Pyrophoric carbon is formed when wood and textile materials are heated to a comparatively low temperature over very long periods of time, even as along as many years. It is particularly found where timber has been enclosed in stone or brickwork chimneys. Pyrophoric carbon is liable to spontaneous ignition on being exposed to air at normal temperatures.

Fibres, fabrics, rags and sacks

These materials, which also include wool, straw, hemp, jute, sisal and the like, do not themselves ignite spontaneously but only where contamination with other substances occurs. The danger is increased if the materials are also wet, and even more so if the materials have been damaged by fire.

Rags and empty sacks must always be regarded as hazardous and stored under cover, dry and away from other buildings.

Pyrophoric metals

Quite a number of metals are subject to spontaneous heating, especially when in the form of finely divided powder; fresh turnings of iron are a common example. In this case preheating has taken place during working operations, but heating conditions are continued if the turnings are put into sacks. Subject to other conditions, such as the presence of combustible material and the inability of heat to disperse, ignition may take place.

REACTIVE CHEMICALS

Combustion is itself a chemical reaction in which sufficient heat is released that the reactants and their products are incandescent. These reactions are between oxygen and a fuel but there are other reactions which can be equally energetic. For example, the traditional way of manufacturing pure hydrochloric acid (HCl) is to burn a diffusion flame of hydrogen in an atmosphere of chlorine. High temperature flames can also be produced by the reaction of fluorine with a wide range of fuels. Indeed the most energetic of all normal chemical reactions is that between fluorine and hydrogen to form HF, with a flame temperature of 3600°C. Some of these halogen mixtures will ignite spontaneously, as will several mixtures of fuels with oxidizing agents. Examples of spontaneous ignition of reactive liquids can be found in the many self-igniting (hypergolic) bipropellant rocket systems. The earliest of these was the T-stoff/C-stoff combination in the motor of the Messerschmitt Me 163B interceptor fighter developed during the Second World War. T-stoff was a mixture of hydrazine (N_2H_4) and methanol, and C stoff was high strength hydrogen peroxide (about 83 per cent H_2O_2). A catalyst was added to the fuel to reduce the ignition delay to less than 60 ms. Other hypergolic systems include nitric acid (either the red fuming acid or 100 per cent acid) with aniline, furfuryl alcohol and butyl ether. Some liquids release sufficient energy when they decompose to be used as rocket monopropellants, and the reaction proceeds at temperatures near to those in flames. Hydrazine and high strength hydrogen peroxide can both be used in this way.

Clearly the accidental mixing of these very reactive materials can cause a fire. There is a wide range of possible reactions which could result either in an immediate ignition or a rapid rise in temperature. Whenever chemicals are being stored or processed a check should be made of the consequences of any possible accidental combinations.

The most widely encountered group of hazardous chemicals is that of the oxidizing agents. These are used in many industries (fertilizer production, heat treatment plants, electroplating and mirror silvering plants, water treatment, and plastics fabrication are a few examples). The oxidizing agents are stored and transported in large quantities. Some are sufficiently unstable to be hazardous in their own right, but most of them create hazardous conditions by the release of oxygen (particularly when they are heated). The presence of oxygen ensures that fuels are ignited very readily and that they then burn very fiercely. Mixtures of some oxidizing agents with powdered fuels, and even with ordinary dust, are sensitive and may be ignited by friction. Examples of the more commonly encountered oxidizing materials are listed below.

Bromates

These are similar to chlorates (see below) but rather less reactive.

Chlorates

These are powerful oxidizing agents which can form sensitive mixtures with materials like sulphur, sugar and charcoal. Sodium chlorate spilled on to an old wooden floor can form mixtures with the powdered wood and dust which will ignite when they are walked over.

Chlorites

These behave like the hypochlorites (see below) when heated, but higher temperatures are required. Like the hypochlorites they are used for bleaching.

Chromates

These are used in paints. They are not powerful oxidizers and are not usually hazardous.

Hypochlorites

These release chlorine and some heat in contact with water and are used as bleaching agents. When heated they release oxygen. A number of fires have been attributed to the storage of hypochlorites in damp conditions.

Nitrates

These release oxygen on heating and also form sensitive mixtures with solid fuels. Wood which has been impregnated with a nitrate solution may subsequently ignite by friction.

Ammonium nitrate, either on its own or mixed with other chemicals, is used in large quantities as a fertilizer. It can form dangerous mixtures with fuels and a mixture with diesel fuel is used as an explosive in quarrying. A number of disastrous fires and exlosions have resulted from the storage of very large amounts of ammonium nitrate. Confinement and local pockets of high pressure cause a violent exothermic decomposition.

Sodium nitrate is the oxidizing agent in black powder (gunpowder).

Nitrites

These contain less oxygen than the nitrates but are almost equally hazardous.

Perborates

These are used as disinfectants and bleaching agents (for example, in denture cleaners). Sodium perborate is used in commercial quantities. In decomposes at 48°C to release oxygen but its own water of crystallization prevents the development of high temperatures.

Perchlorates

These are powerful oxidizing agents but are less hazardous than the chlorates, for which they are sometimes substituted.

Permanganates

These can form sensitive mixtures with powdered organic materials.

Peroxides

Solid peroxides form very dangerous mixtures with organic materials. Sodium peroxide decomposes very readily with the release of oxygen when it is heated or when it is in contract with water. A mixture of sodium peroxide and a solid fuel can therefore be ignited either by friction or by the presence of moisture.

Many organic peroxides, like benzoyl peroxide, decompose explosively and must be handled with great care. Special precautions are needed for their safe storage and transportation.

High strength hydrogen peroxide will cause the ignition of organic materials on to which it is spilled. It will also decompose violently. The dilute solutions usually found in industry are less hazardous but are a readily available source of oxygen.

Persulphates

These are used as bleaches and disinfectants. They are strong oxidizing agents.

Dangerous chemicals

Acids

Strong acids such as pure sulphuric acid, nitric acid and hydrochloric acid, when in contact with combustible material, may create sufficient heat to start a fire.

All such materials are usually kept in glass jars or carboys. They should be stored in a separate building from the main premises, on concrete floors and narrow concrete shelves. The reason for the narrow shelves is to prevent bottles being stored more than one deep, so that in reaching over for a bottle at the back others are not likely to be knocked over.

There should be separate rooms or compartments for acids and alkalis, well ventilated and bunded so that spilt liquids are contained. No combustible materials should be allowed in the rooms.

The chemical reaction between most common acids and metals produces large quantities of hydrogen gas which can easily form an explosive mixture.

Metallic sodium and potassium

Metallic sodium and metallic potassium are both found in most laboratories. They take fire immediately on contact with water and are generally stored in paraffin or some similar light oil. It is essential that firefighters should know the position of any containers and that as far as possible they should be segregated from other materials. They should also be guarded against accidental damage, bearing in mind surrounding conditions.

Phosphorus

Again, this chemical substance is nearly always found in laboratories. It takes fire on contact with air, and is normally stored under water. Great care must be taken to see that the water cannot evaporate. Storage should be such that the containers are not likely to be knocked over or damaged so that the phosphorus is exposed to the air.

It is very good practice to persuade the head chemist to go right through his stock of dangerous chemicals as least once every year, and remove any that are no longer immediately required.

Aluminium alkyls

These are being increasingly employed in industry. Their danger lies in the fact that they take fire spontaneously when exposed to the atmosphere. Any slight leak will therefore cause a fire immediately. The danger is even greater because they react violently with water and Halons.

Extra precautions should be taken in buildings where these alkyls are stored and used. They should be of fire-resistant construction throughout, with additional exits, door cills raised and all floor openings sealed. Suitable extinguishers are dry powder and CO_2. Carbon dioxide is only effective while it can be continuously applied. The powder should be allowed to fall gently on the burning material in the ratio of two of powder to one of alkyl. Fibreglass blankets should be readily available in case of splashes on clothing.

Very dry sand can be used to mop up spillages. Water or any form of foam must not be used as it will produce violent reactions.

13

Initial spread of fire

The means by which fires can spread in buildings has already been discussed in Chapter 3. Here we are concerned with the more practical aspects and with the precautions which can be taken to slow down or prevent fire spread. These should reduce the probability that an incipient fire will grow until it involves the contents of a room (probably resulting in a flashover). The prevention of fire spread from room to room depends very much on structural design and this is discussed in Part Four. In this chapter we deal first with some of the simplest and most cost-effective of fire precautions before looking at some of the methods of dealing with flammable gases and liquids.

GOOD HOUSEKEEPING

The term 'good housekeeping' used in connection with fire prevention refers to the control of combustible and flammable materials which tend to accumulate in any place of work. Good discipline and constant vigilance are needed to ensure that there is no unnecessary increase in the potential for fire spread due to normal working activities. The application of good housekeeping in a number of typical industrial situations is discussed below.

Packing department

The most common danger in a packing department is pieces of packing materials cut off and dropped loose on the floor. Arrangements should be made for these to be continually collected and placed in covered metal bins. Wood wool is a particularly hazardous material being light, very combustible and creating large quantities of smoke. Wood wool in bulk should always be kept in large metal bins each not exceeding about 1.5 m^3 content. The lid should be held open by counterbalanced weights and so arranged that without the weights it would fall to the closed position. The weight could also be held by fusible links and quick release couplings. So important is the closing of these bins that, in commercial premises such as large stores, it would be good practice for the weights to be released automatically with the alarm signal.

It is absolutely essential that no loose wood wool should be allowed to remain on packing benches or accumulate on the floor, all surplus material being immediately returned to the bin and no parcels containing wood wool allowed to remain uncovered.

Similar arrangements should be made for straw and sawdust; the latter could be shovelled from the bottom of the bin without opening the lid except for refilling. Only a minimum amount of these materials should be brought inside the building at any one time, never more than would be used during one session of work.

Corrugated paper has an additional danger to ordinary packing paper because it has a central space which brings a much greater surface of paper in contact with air and which can also act as a form of flue to increase the draught and carry fire through the roll. As little as possible should be brought into the building and kept in roll form rather than loose sheets.

Goods inward department

Here again is the danger of loose packing materials. In this case, the problem is one of temporary storage pending removal outside the building. This should be achieved by the use of metal containers. Here the lids should remain closed and opened by foot operation, power-assisted if necessary, so that waste packing can be collected continually and held in storage. Regular collection is essential and nothing should be left in the buildings or bins overnight.

The office

Dangers from smoking have already been separately discussed; the main dangers in an office are from waste paper baskets and loose electric leads. Loose electric leads as a source of ignition have already been mentioned. The main hazard is loose paper on the floor dropped outside the waste paper basket. Waste paper baskets should be made of metal as also should be letter trays and as far as possible all office furniture, on the principle that the less material there is present to burn, the less likely it is that a fire can start or continue to burn.

Special care is needed in the printing and duplicating departments. Here extra large waste paperbins are required, but there will also be additional hazards from cleaning materials. Many of the cleaning fluids are highly flammable and some even come within the scope of the Petroleum Acts. In any case, all flammable cleaning liquids should be kept in metal bins with a top opening lid, so designed that the lid cannot be left open. All tins or bottles of fluid should be immediately returned to the bin and not left lying on desks or work tables. Cleaning rags should also be kept in metal bins and, essentially, all used cleaning rags should be immediately disposed of in a separate metal bin from that containing waste paper. There is an excellent system where special wipers are provided by a firm who collect, clean and re-issue them. This avoids small torn off bits of rag being dropped about and picked up with waste paper. It encourages the use of a fresh wiper for each job.

Offices also provide a danger from the temptation of abusing heating systems. This relates especially to the drying of clothes in front of or on top of radiators, and to leaving papers and similar materials on heaters. This is especially dangerous in the case of night storage heaters which may seem perfectly cool and safe in the late afternoon but may prove extremely dangerous when the current comes on automatically during the night hours.

Old office equipment

In almost every office building there will be a store of old office equipment, tables, desks, chairs and the like. This junk store is nearly always in an inaccessible and dangerous place – the cellar or the attic. Combined with it will be a store of old newspapers, magazines, periodicals, old records and even old clothing, overalls, etc. Documents and records are mentioned separately in the next section.

A regular inspection of these places should be included in the daily inspection of the premises and periodically they must be thoroughly cleaned out. A very good method of carrying this out is to have everything removed from the room and only essential items returned to it.

Documents and records

The storage of papers is often haphazard and is often a cause of the spread of fire. An instruction should be given to the head of each department requiring that all documents are to be classified under one of five headings:

- Irreplaceable. Deeds of property, etc.
- Replaceable but only after considerable research and expenditure
- Of vital importance. Incoming orders
- Useful but pending destruction. Copies of goods dispatched
- Papers that are of no further value. Out-of-date acknowledgements, etc.

This exercise should be carried out conscientiously and if possible all stored documents and papers should be moved and checked. The exercise should be repeated annually. The first result is nearly always a large quantity of paperwork that nobody claims. This should be checked by a senior executive and either allocated to a department or destroyed. In issuing the instruction for this 'spring cleaning', the basis of the classification should be taken as the effect on the business of the total destruction of the documents at that particular moment irrespective of how they are stored.

Having classified the papers in a general manner, the next step is to classify them according to how they will age. Title deeds, for example, will always be irreplaceable, but incoming orders and similar prime documents will only be irreplaceable until they have been recorded. As soon as this has been done they can be re-graded as being of vital importance until the order has been completed and paid for by the customer, when they can be destroyed. Many documents will only be required until the next audit has been completed, although some may have to be kept for a period covering the Statute of Limitations. Some otherwise useless documents should be kept for historical reasons. A most frequent source of useless papers which are overdue for destruction will be found in the correspondence files. A company may not only reduce its fire risk but gain filing space if only copies of routine acknowledgements of incoming and outgoing letters and orders are destroyed as soon as they are of no further value.

All papers that are of no further value should be destroyed immediately. Papers that are pending and almost ready for destruction should be tied up in convenient parcels marked with the date on which they are to be destroyed, the essential point being that they are neatly parcelled up and not allowed to lie about loose.

In relation to papers that are considered of vital importance, it should be possible to have many of these recorded and the originals destroyed. Some documents considered to be replaceable only after considerable research and expenditure might also be treated in a like manner. With computer and recording data methods, a great deal of information can be recorded on tape and similar machines. The protection of these tapes and devices must then be treated in the same way as the originals should have been treated.

Irreplaceable documents should always be kept in vaults as described later, and where the company does not have its own vaults, arrangements should be made to deposit such documents with a bank or other concern that has fire-resistant vaults. In some cases of documents of vital importance, duplicates might be stored with an associate company or on some other form of mutual assistance basis.

Protection of documents and records

Vaults

Vaults can give six-hour protection if they are built to the following specification, but this takes no account of resistance to the collapse of the building during a fire and extra strength might be necessary for protection against this possibility. After a fire, high temperatures will be maintained for several days and no attempt should be made to open any storage until the contents have cooled down to ambient temperature.

1. *Situation.* On solid ground above water level. On ground floor if sufficient strength can be obtained; alternatively in basement.
2. *Walls.* Solid brick or reinforced concrete not less than 230 mm thick.
3. *Roof and floor.* Reinforced concrete slab not less than 175 mm thick.
4. *Doors.* Double steel or metal-covered doors separated by the thickness of the wall.
5. *Windows.* No windows.
6. *Lighting.* Electric, in metal conduit, and preferably no conduit inside the vault. No flexible leads, and fittings to be of the bulkhead type.
7. *Racks.* To be of incombustible material.
8. *General conditions.* Combustible materials other than stored documents to be kept to a minimum. No highly flammable liquids to be kept in the vault. No smoking permitted. Storage to be above possible flood level.
9. *Containers.* The documents themselves should be contained in boxes so that no readily ignited edges are exposed.

Filing rooms

Filing rooms are usually larger than vaults and a lower standard of fire protection can generally be accepted. The standard of construction should be much the same as for vaults, but wired glass windows would be acceptable if they do not overook a fire hazard.

Safes

Safes are difficult to assess but generally if they are old they are suspect. Reputable manufacturers will be able to give some idea as to the present condition of a safe if the serial number is quoted. Safes that are no longer considered suitable for cash are rarely suitable for the protection of documents from fire. Frequently one sees a first-class safe containing the petty cash, and documents of untold value in an old safe or even an ordinary filing cabinet.

Fire-resistant steel filing cabinets

Specially designed modern steel filing cabinets can be obtained from reputable manufacturers that will give considerable resistance against fire conditions. These cabinets are not only resistant to fire, heat and smoke but also to the impact of fire brigade jets of cold water and of falling from a height as they might well do in a fire.

Ordinary steel furniture

Ordinary metal office furniture including filing cabinets and cupboards have no resistance to the effects of fire, heat, smoke or water. They are only to be preferred above wooden furniture because they make no contribution of combustible material to a fire.

Filing cabinets on wheels.

Prime documents should be exposed to the hazard of fire for as short a time as possible and the use of wheeled metal cabinets that can be placed beside a machine is to be strongly recommended. Unless they can be wheeled quickly into a vault or filing room, a fire-resistant type should be used. With this type it is only necessary to drop the lid and the papers will be safe under the conditions of a normal office fire. Part of the evacuation procedure should be the detailing of staff to close cabinets, safes and vaults and, in some cases where adequate protection cannot otherwise be obtained, staff should carry out a tray of papers with which they are working.

Salvage

If the worst happens and valuable documents are damaged by fire and water it may still be possible to salvage them and to recover the information. A specialist firm should be contacted immediately and their instructions closely followed. Although the documents may be capable of a quite remarkable recovery from extensive fire damage, they can be ruined very quickly by fungal attack. If the documents are wet, as they probably will be, then it may be necessary to put them into deep freeze storage until the specialist firm can deal with them.

Office cleaners

Many fires start in cleaning cupboards. Special attention should be given to see that no loose material is left on the floor, strict control must be kept over flammable cleaning materials, waxes, etc., and under no account should beeswax be heated over gas rings or other open flames. Surreptitious smoking is one of the grave dangers and procedures should be laid down for proper smoking facilities.

The work of the cleaners in regard to emptying ash trays has already been mentioned, but it is again emphasized that precautions should be taken to ensure that ash trays are not emptied into containers with waste paper.

For some reason cleaners are the worst offenders in respect of leaving fire doors open and propping smoke stop doors open. It seems impossible to make them understand that by so doing they are not only endangering their own lives but those of other people. There is only one real cure and that is to fit automatic door release gear so that the doors can remain open at all times other than when there is an alarm (described later). Failing this, constant vigilance is necessary and an attempt must be made to educate people in the dangers of smoke and draughts.

When an outside firm of cleaning contractors is employed, it should be part of their contract that they will ensure that no fire or smoke stop doors are fixed open and it would also be sensible to include a clause that they will make a separate collection of the contents of ash trays in a metal container.

Factory waste

The removal of all industrial waste, especially of a flammable nature, should be a continuous process and should not just be left to the end of the day. The removal of waste should be a two-stage operation: the immediate deposit of material into temporary storage bins with lids and then the general removal of the bins outside the premises. Collections of the outside bins could be left to the end of the day under suitable conditions of safety, but it is preferable that this also should be a continuous process with the collecting trolley leaving an empty bin for each full one collected.

The modern method of putting all flammable waste into large metal enclosed containers which are picked up periodically by lorry is an excellent system. If the collection is made by contract, it is an added advantage because people will normally demand that a contract is fulfilled where they would not be so particular in performing the same supervision over their own work. The whole advantage of the system can be lost if the top of the container is left open or if rubbish is allowed to accumulate round the container instead of being put right inside and the lid closed down.

Dust and dirt should not be allowed to accumulate round machines and, where oil leaks cannot be prevented, drip trays should be provided to prevent oil leaking on to and soaking into floors. Even a concrete floor will absorb oil which can subsequently be ignited. Electric motors in little used rooms or concealed spaces can be a source of danger and open electric motors should be cleaned weekly, where necessary by blowing a jet of air through them.

Boiler room

It has often been said that the state of the boiler room will depict the whole standard of good housekeeping. Certainly, a clean well kept boiler room is a very good sign. In the boiler room, check the fusible link device which shuts off the oil feed in case of fire. Ensure that the fusible links have not been painted or tampered with. Sometimes they will be found to have been bypassed by tying a wire across the link or even replacing the link with a piece of ordinary wire. Check the wire over the pulleys making sure that they all run freely and finally check the weighted drop-arm. This should be fenced off to prevent anything from being placed beneath it and so stopping the fall of the arm.

EXTERNAL HOUSEKEEPING

Cleanliness outside the premises is as important as inside. No flammable material should be allowed to accumulate against the walls of the building nor should any be allowed to collect round outside storage areas. Waste that is removed from the buildings should be properly safeguarded until final removal to a dump or incinerator.

Grass and weeds should not be allowed to grow near buildings or stores. Grass fires start very easily; even an old broken glass bottle has been known to concentrate the rays of the sun sufficiently to start a fire. The danger is particularly bad in the UK during February and March, when all growth has ceased and droughts and frosts are prevalent. As soon as the sap starts to rise and new growth takes place the danger becomes less.

Weeds and grass should be killed before they can obtain any long growth by the use of a selective weed killer. On no account should weed killers containing sodium or potassium chlorate be used, because both of these materials are oxidizing agents.

Incinerators

Properly designed incinerators, installed on open ground at least 18 m away from buildings, are an excellent way of disposing of waste paper and similar materials. Maintenance and operation should be strictly in accordance with the manufacturer's instructions. Material waiting for destruction should be contained in such a way that it cannot blow about or be fired from the incinerator. The incinerator and its compound should never be left unattended while there is any fire present. Fire extinguishers should be readily available. (Remember to protect water extinguishers against frost if they are in the open.)

Open rubbish fires should not be tolerated unless they are so far removed from buildings that there is no possibility of a flying brand reaching the building. Flying brands can travel several hundred yards.

It should be noted that charred material is more likely to take fire than unburnt material; it is thought possible for it to fire spontaneously. Ashes from an incinerator should be spread thinly to cool.

Where rubbish dumps are used they should be well removed from buildings and covered over after each day's additions. There are a number of reasons why rubbish dumps take fire and when they do so they become a nuisance from smell as well as a danger from fire. They are difficult to extinguish and usually have to be turned over and damped down completely. A bulldozer is unsuitable for this work as it tends to bury the fire, which then breaks out again. The best method of extinguishing rubbish dump fires is to turn the material over by the use of a mechanical shovel and to damp down each shovelful as it is turned.

EXPLOSIVE DUSTS

Dust explosions were described in Chapter 1. A suspension of the dust of almost any combustible solid will explode under the right conditions. The main factors affecting the explosibility of a dust are the particle size, the concentration (which must be within the limits of flammability), the moisture content, and the nature of the ignition source. Several laboratories have facilities for testing dusts. They are a number of standard pieces of apparatus in which various concentratons of the dust are exposed to different ignition sources. Measurements are made of the maximum pressure and of the rate of rise of pressure. This latter measurement is useful when precautions against the effects of an explosion are being considered, particularly the use of relief vents and of explosion suppression systems. The UK Fire Research Station has been testing dusts for many years in addition to conducting research on dust explosions and developing new testing methods. They publish comprehensive data on materials which have been tested, listing both explosible and non-explosible dusts.

The minimum explosible concentration (the lower limit of flammability) varies quite widely with different materials and is also dependent on the particle size. Some of the lowest concentrations have been recorded for benzoic acid, epoxy resin, polyethylene and zirconium dusts, all at 10 g/m^3. At the other end of the scale are instant coffee at 150, milk sugar at 300, dried gelatin at 500 and ferrochromium at 2000 g/m^3. Even the low concentrations cause appreciable obscuration and it is not possible to see very far through a hazardous dust cloud.

The particle size and the concentration of dusts also affect the maximum explosion pressure. Values recorded for some of the materials mentioned above are benzoic acid 6.3 bar, epoxy resin 5.4 bar, instant coffee 4.4 bar. Pressures as high as 9.6 bar have been recorded for milled barley, and 10.0 bar for corn starch. These values are of course far in excess of the pressures needed to wreck normal process plant and buildings.

Some of the materials most frequently involved in dust explosions include aluminium, sulphur, starch, malt, flour, sugar, rubber, synthetic resins, cork, wool, flour, coal and sawdust. Some of the plant in which explosions have happened include cyclones, filters, conveyors, ball mills, milling and grinding machines, dryers, blenders, bucket elevators, storage bins and silos. In many of the incidents a relatively minor explosion in the plant caused a pressure pulse which dislodged layers of dust from surfaces throughout the building. The secondary explosion propagating through the resultant dust cloud then caused severe damage to the building and the remaining plant. Some examples of dust explosions are described below.

A bucket elevator ran through all 13 storeys of a sugar refinery. A fault caused this to run out of alignment and friction caused a rise in temperature which was sufficient to ignite a sugar dust suspension in the elevator shaft. Severe damage was caused to the building and two people were injured.

A screw conveyor was used to transfer milk powder from a hopper to a blender. Deformation of the conveyor caused the screw to grind against the metal casing. Sufficient heat, and possibly sparks, were produced to ignite a suspension of dust in the conveyor. The explosion burst the casing and a secondary explosion completely destroyed the building, killing three employees.

An inspection lamp which had been left inside a grain silo ignited the dust which was released when grain was charged into the silo. A reinforced concrete wall 0.5 m thick was blown 12 m from the silo and smaller items travelled up to 100 m.

Starch which had been left in a dryer over the weekend had been smouldering. When the plant, used to manufacture confectionery, was started up, a primary explosion occurred. This caused a dust cloud in the building and the secondary explosion destroyed brick walls on the third and fourth storeys. Fifteen employees were killed.

A grain silo complex comprised 73 reinforced concrete silos each 35 m high and with diameters between 8 and 10 m. An explosion which occurred in one of them in the early morning may have been ignited by a static discharge. The head house, 85 m high and containing the elevating system, was damaged by the explosion. The top 35 m of this building collapsed on top of the office buildings beneath it and completely destroyed them. Throughout the morning explosions spread from one silo to another until all of them had been involved. More than half of the silos were totally destroyed and the rest were damaged or displaced. Thirty-six people were killed.

The prevention of damage by dust explosions requires a carefully planned and constructed plant, properly supervised procedures, and adequate venting or suppressing techniques. A great deal of guidance is available, which can be summarized only briefly here. In the UK Section 31 of the Factories Act (1961) requires that 'all practicable steps' must be taken to prevent and to restrict the spread and effects of dust explosions. The Health and Safety at Work Act (1974) does not mention dust specifically but is quite clear on the responsibilities of both employers and employees in dealing with hazardous processes. Legislation, codes of practice, and the recommendations of particular industries, are available in many other countries.

A dust explosion cannot occur unless a combustible dust is of the right size and has been suspended in the right concentration. There must also be a sufficient concentration of oxygen, and an adequate ignition source. If dust suspensions can be eliminated, explosions cannot happen. For this reason, screw conveyors should be so designed that they are always full of powder. It is not usually practicable to eliminate completely the loss of dust from the plant but clearly this should be reduced to an absolute minimum by good design and careful maintenance. Within the building containing the plant surfaces on which dust can collect should be minimized. This applies particularly to window ledges, light fittings, pipe and cable runs, beams and girders. Wherever possible there should be no horizontal surfaces and any unavoidable projection should have an inclined angle of 60°. Pipes and cables should be embedded and doors should be flush. Good housekeeping is still essential: the regular and thorough cleaning of the entire building.

If dust is generated by a process which cannot be contained, for example, by the work done on machines in the woodworking industry, adequate dust extraction and collection must be used. Although these systems remove a hazard from the working area they may themselves contain explosive suspensions of dust and must be designed accordingly. The accumulation of dust deposits within the ducts must be avoided, if necessary by regular cleaning through suitable hatches.

The elimination of ignition sorces has already been discussed and is clearly of vital importance in the prevention of dust explosions. One possible source results from the accidental introduction of iron or steel objects into a grinder or ball mill. This can be avoided by the use of electromagnets in the duct feeding the machine. The magnets should be so connected that any failure in the supply to them, or their removal from the duct, will shut down the process. Flameproof electrical equipment is essential and strict permit to work procedures must be applied to control cutting and welding operations and work on electrical equipment. Precautions must also be taken against the accumulation of hazardous static charges.

The effects of a dust explosion can be mitigated by the use of properly designed vents and, in plant, by the installation of an explosion suppression system (described later). The explosion suppression technique depends on the detection of a small rise of pressure in the early stages of an explosion. A similar system is available in which an infra red detector can sense the presence of a spark and trigger the discharge of an extinguishant before an explosion can develop.

FLAMMABLE LIQUIDS

If large quantities of flammable liquids are to be used in process plant, careful planning is needed at the design stage to reduce the probability of a fire. Fire precautions should begin at the research phase of a new project: it may be found that by altering the process flammable liquids can be eliminated, or at least replaced by less hazardous materials. When it has been accepted that hazardous materials, reactions or processes are unavoidable, a detailed assessment of all possible failure modes must be made before the design of the plant is finalized. A HAZOP study, discussed in Chapter 31 may be the best way of doing this. Some of the basic precautions which should be taken in the storage and processing of flammable liquids are summarized below.

Briefly, a fire can happen if there is a loss of containment in the presence of an ignition source. The size of the fire can be controlled by limiting the area occupied by the released liquid. Fire spread by radiation can be reduced by cooling plant and structures which are exposed. An explosion can happen if there is a release of a flammable gas, or a liquid at a temperature above its flash point, and ignition is sufficiently delayed. The severity of an explosion and of its effects can be reduced by ensuring that there is good ventilation.

If a flammable liquid is stored in a fixed roof tank at a temperature above its flash point, over a particular temperature range there will be an explosive mixture in the vapour space of the tank. With volatile liquids it is possible for the mixture in the vapour space to be above the upper limit of flammability: it is too rich to burn. This happens in petrol storage tanks at temperatures down to about –30°C. However, air will be drawn into the tank through the vent when liquid is taken out, and flammable mixtures can then persist for some time until equilibrium is re-estalished. To prevent the ignition of these mixtures it is usual to have a flame arrestor in the vent. The formation of flammable mixtures in a fixed roof tank can be eliminated by nitrogen blanketing. If a closed circuit is used, with a variable volume nitrogen tank (like an old coal gas holder), the system may be able to pay for itself by preventing the loss of expensive vapour. The problem of explosive atmospheres in the storage of low flash liquids can also be solved by using a floating roof tank. The roof, which has a sliding seal around its edge, floats on a pontoon on the surface of liquid. There is thus no vapour space at all. In general, fixed roof tanks are used for liquids with high flash points (above 32°C) and liquids with low flash points (less than 23°C) are kept in floating roof tanks.

Tanks should be in bunds which will contain any spillages and will also limit the area of a fire resulting from a large spill. The bunds are usually designed to hold 110 per cent of the contents of the largest tank, thus allowing sufficient space for the application of foam. The bund wall should be sufficiently high to provide some protection to firefighters but should not be so high as to hinder firefighting or attempts to escape from the bund. About 1.5 m seems to be a reasonable compromise. Even quite small tanks of things like fuel oil should be bunded.

Fixed systems can be provided through which foam can be applied to the fuel surface in a burning tank, having usually been pumped in at the bund wall. A fire can be established in a fixed roof tank only after the roof has been blown off by an explosion in the vapour space, which is likely also to damage the foam pourers. For the protection of fixed roof tanks it is better therefore to rely on the use of suitable foam monitors to project foam into the tanks from a position outside the bund. Good access to all of the tanks from any direction is essential and of course adequate supplies of water and foam concentrate. On a very large installation it may be possible to provide a foam concentrate main in addition to the fire main. The problem of damaged foam lines can also be overcome by base injection: the use of a product line to pump foam into the bottom of a tank. The foam then rises through the fuel and floats on the surface. It is necessary to use special concentrates, like fluoroprotein, otherwise the foam may pick up sufficient fuel in its passage through the tank to become flammable. Special equipment is also needed so that the foam can be generated and pumped at a pressure which is sufficient to overcome the static head of the fuel in the tank.

If one of a group of tanks is burning it will be necessary to cool the adjacent tanks to prevent them from becoming involved. This could happen if the surface of the fuel, or a film of fuel on the wall in a neighbouring tank is heated to a temperature above its flash point. An explosion would result when vapour issuing from the vent was ignited. Ignition of the contents can be delayed by the presence of a flame arrestor but will eventually happen once the arrestor is hot. It is also possible for the unwetted wall of a tank to be heated to the AIT of the fuel. Cooling can be provided by a fixed installation, which is likely to remain undamaged on the tanks which require to be cooled (that is, all the tanks with the exception of the one on fire). Again pumping-in points can be provided at the bund wall through which water can be applied. Alternatively water can be provided at a sufficiently high pressure from a suitable main and if necessary the whole system can then be operated automatically by detectors. The same application rate is used to protect tanks, steelwork and process plant (including offshore installations): 0.17 litres/m^2/sec (this is a direct conversion from the original 0.2 UK gal/ft^2/min). It would of course be possible for a fire brigade to apply cooling water from monitors or jets. The advantage of a fixed system is that little (if any) time is needed to turn it on and it will then look after itself, leaving the firemen to get on with other work.

Small amounts of flammable liquids are often kept in drums. These should be stored in secure compounds (because of the possibility of arson). The compounds should have concrete aprons surrounded by a low sill and preferably sloped so that any spilled liquid will drain away and can be disposed of safely.

If a liquid is processed at a temperature above its flash point, any loss of containment must result in the formation of flammable mixtures. Ideally the plant should be in the open air. This will ensure that vapour/air mixtures are rapidly diluted to below the lower limit of flammability, and also that if there is an ignition the explosion will not be confined. If it is necessary for the plant to be inside a building then the more ventilation which can be provided, the safer will be the process. If large areas of the walls can be replaced by louvres, not only will there be natural ventilation, but the effects of an explosion will be minimized by pressure relief. The probability of a loss of containment can be reduced by the use of all-welded pipelines and by careful consideration of the mechanical strength, and the liability to damage, of all the plant. Where it is necessary to break into the containment, for example to clean a filter or to remove a pump for maintenance, adequate precautions should be taken against a loss of liquid. The closing of a single valve between the affected equipment and the rest of the plant is not an adequate means of isolation. The valve may leak, particularly if it is a gate valve and has been closed on to a solid deposit (which could result from corrosion). Also, the valve may be opened inadvertently. Slip plates (or 'spades') will ensure that leakage cannot occur. These are blank metal plates which can be bolted between two flanges, or at the end of the line, downstream of the isolating valve, and after this section of line has been drained. Their use is illustrated in Figure 13.1a and b. The short length of line which has been removed in b is sometimes called a bobbin piece. A spectacle plate, which comprises a blank and an open plate connected together (Figure 13.1c) provides a positive identification of the status of an isolation. A superior method of isolation which should be employed whenever highly flammable liquids or high pressures are involved is illustrated in Figure 13.1d. This is double block and bleed: the two large valves are closed and the smaller one is open to drain away the trapped liquid and any liquid which may subsequently pass through the left-hand valve. Without this bleed, any increase in temperature would result in the generation of high pressure in the trapped liquid, and this could cause a leak. The bleed should be connected to a safe drain or to a vent stack. All the valves in the figure should be locked shut (or open in the case of the bleed valve) and should have notices attached which say 'Do not open' (or 'close'). In addition the equipment which is to be removed (perhaps a pump or a filter) should be identified by a numbered tag and the same number should appear on the permit to work.

The elimination of ignition sources, already discussed in Chapter 5, is essential. In particular flameproof electrical equipment is required, and a rigidly enforced permit to work system. Automatic detection of both fires and flammable gas may be necessary, as well as automatic fire and explosion suppression systems. These are described in Chapters 16, 17, 25 and 26. Any structural steelwork which can be exposed to a fire may need to be protected by thermal insulation or by water deluging. Plant may also need to be deluged if it is not adequately spaced from other plant (for this reason all offshore plant is deluged).

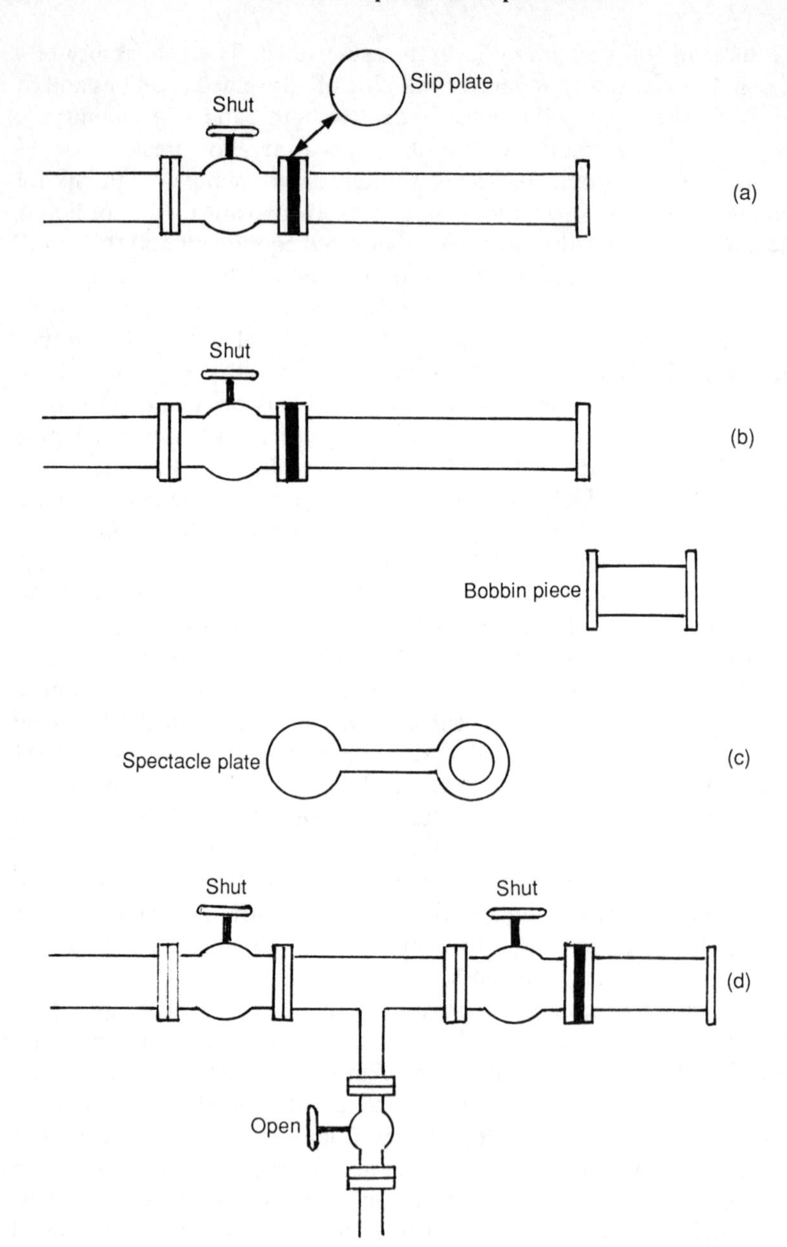

Figure 13.1 Methods of isolation: (a) and (b) slip plates, (c) spectacle plate, (d) double block and bleed.

FURTHER READING, Part Three

NFPA, *Fire protection handbook*, NFPA, Boston.

Klinkenberg, A. and Minne, J. L., *Electrostatics in the petroleum industry*. Elsevier, Amsterdam.

Palmer, K. N., *Dust explosions and fires*. Chapman and Hall, London.

Bartknecht, W., *The course of gas and dust explosions and their control*. Elsevier, Amsterdam.

Meidl, J. H., *Flammable hazardous materials*. Glencoe Press, N.Y.

Meidl, J. H., *Explosive and toxic hazardous materials*. Glencoe Press, N.Y.

Bretherick, L., *Handbook of reactive chemical hazards*. Butterworth, London.

Lees, F. P., *Loss prevention in the process industries*. Butterworth, London.

Hughes, J. R., *Storage and handling of petroleum liquids*. Griffin, London.

Vervalin, C. H., *Fire protection manual for hydrocarbon processing plants*. Gulf Publishing Co., Houston.

Magison, E. C., *Electrical instruments in hazardous locations*. Instrument Society of America, New York.

PART FOUR

STRUCTURAL FIRE PRECAUTIONS

14

Limitation of fire spread

COMPARTMENTATION

The main objective of some of the earliest UK legislation was to prevent the spread of fire from one building to the next. This resulted from a number of disastrous fires in congested cities. The houses contained a great deal of timber and frequently had thatched roofs. They were very close together, and the upper stories of many were built out over the streets, to be within a short distance of the houses on the other side. In 1136 large numbers of houses were destroyed in London, Bath and York; and London suffered again in 1212. This latter fire led to the introduction of a law restricting certain trades, and specifying the materials which could be used for new or restored roofs. Despite this, 80 per cent of London was destroyed by fire in 1666. Royal commissioners, including Sir Christopher Wren, were appointed to investigate the causes of the fire. Their work, and that of newly appointed city surveyors resulted in the earliest building control regulations embodied in the Rebuilding Act of 1667. The provisions of this Act were strengthened by a new Rebuilding Act of 1774. This laid down minimum thickness for walls, required at least 8 inches of bricks between two pieces of structural timber, and between the timber and a chimney, and specified the design of chimneys and hearths. The materials which could be used for walls and roofs were specified, and again dangerous trades, like turpentine distilling, were restricted.

The vast amount of experience, and more recently research, which has been accumulated since then has led to different concepts of fire protection, but the basic provisions of the 1774 Act have been largely retained. Modern legislation is concerned with reducing the incidence of fires, providing adequate means of escape, and with controlling the spread of fire within buildings. Thus, in addition to controlling the spread of fire from one building to another, the new legislation is concerned with reducing the probability of spread from one part of a building to another. This embodies the concept of compartmentation: the construction of separate cells within a building each of which is capable of completely containing a fire.

When modern methods for the construction of large multi-storey buildings began to develop at the turn of the century, many 'fireproof' buildings were erected. A surprising number of these were destroyed by fire. This happened because although the architects had used non-combustible materials, they had not anticipated that a sufficiently high fire load (timber flooring, partitions, panelling, staircases, combustible furniture and finishings) could cause their destruction. Also the resistance of the buildings was often drastically reduced by the presence of lift shafts, ducts and staircases. It became obvious that it was necessary to relate the fire load in a building to the fire resistance of both the elements of construction and the fire stops provided at openings in the structure. Both sides of the equation needed to be quantified: fire load and fire resistance.

FIRE SEVERITY

The destructive potential of a fire can be related to the total heat output and the rate of burning. Fires in compartments are usually air-controlled, so that the rate of burning is dependent on the ventilation. The thermal properties of the compartment itself also have a slight effect because of the heat conducted away, or radiated back into the compartment by the walls. However, the assessment of fire resistance would be unnecessarily complicated if account was taken in each situation of the probable ventilation. In any case this would vary with the opening of windows, or with their failure during the fire. Instead, standard fire conditions are used in the measurement of fire resistance, based on the averaged results of a large number of practical tests. The standard fire is characterized by a temperature/time curve, and this is conveniently related to other fire conditions by the equal-area concept. If the area beneath the temperature/time curve of an experimental fire is measured and found to be equal to the area under the standard curve (plotted on the same scale), the severity of the two fires is about the same. Thus, a fire which burns for a shorter time, but reaches a higher maximum temperature, will have the same damaging effect as a slow-burning fire which achieves a lower maximum temperature. This will happen when the total heat output from the two fires is the same. Thus, for the purposes of assessing fire resistance, the severity of a fire can be related directly to thc firc load.

The fire load in a compartment could be measured by checking the weight of each individual item (furniture, cupboards, shelving, carpets, curtains and so on) and then determining the heat of combustion of the materials. However it has been found that the average heat of combustion of materials like these is usually between 16 000 and 19 000 kJ/kg. The approximate fire load can therefore be expressed simply as a weight per unit area.

This series of approximations may at first sight be difficult to justify. They are however based on long experience and on a vast amount of data from experimental and actual fires.

The result is sufficiently accurate to enable a reliable decision to be made about the degree of fire resistance which is required, within the range of half an hour, one, two or four hours. For most practical situations the fire hazard in buildings can be placed in one of three classifications, which have their origins in the Ministry of Works Post-war Building Studies. These are shown in Table 14.1, together with the corresponding equivalent standard fire test.

In determining the hazard classificaton for a particular building, factors other than simply the potential fire severity may need to be considered. These could include the potential hazards to people in the building, the value of the contents of a compartment (a computer, for example), and difficulties which could be experienced in fighting the fire. In a normal situation, however, if a compartment has a one-hour fire resistance and the fire load is no greater than 49 kg/m^2, it should be possible for the entire contents to be consumed without the fire spreading to the next compartment. This is quite a high fire load for an ordinary occupancy: in a large open-plan office for example it is unlikely that the load will exceed more than about a quarter of this value.

Table 14.1
Fire load related to standard fire tests

Hazard	Maximum fire load kg/m^2	kJ/m^2	Equivalent standard test condition hours
Low	49	186 800	1
Medium	100	372 160	2
High	100+	372 160+	4

FIRE RESISTANCE

The fire resistance of various elements of construction – walls, floors, columns, roofs, beams, doors, glazing, and so on – is determined by exposing them in a furnace in which the temperature follows the standard temperature/time curve. In the UK Standard, BS 476, the curve is defined mathematically, and the temperature control on the furnace is programmed to follow the curve. Very roughly, the temperature reaches 940°C after one hour, 1055°C after two hours, and 1160°C after four hours. The curve was estalished in the very early days of fire testing (in the 1930s) but has proved to be reasonably satisfactory in meeting current developments. The basic requirements were changed only slightly when the original Part 1 of BS 476 was replaced in 1953 by Part 8, which is equivalent to the corresponding ISO Standard. The curve itself is assumed to start at 20°C in Part 8, and this makes a difference of perhaps 30°C in some parts of the curve, or a difference of 10 minutes in the time taken to reach a particular temperature. Also in the original standard the furnace was at a slight negative pressure to ensure that smoke was removed up the chimney. The latest version requires a slight positive pressure and this has a significant effect on for example the time taken for fire to penetrate a door.

The various parts of the standard are under continuous review, the major concerns being the recognition of new developments, and the need for international parity in order to facilitate the removal of trade barriers. Also, new parts are added as new methods of testing are developed. Part 31 for example is concerned with the leakage of smoke around doors and shutter assemblies, of which Section 31:1 covers leakage at ambient temperatures. However, at this point we are concerned with the exposure of elements of construction to furnace tests: other tests defined in the standard are discussed under the next heading.

Fire resistance, as measured by the standard tests, is defined in three different ways: stability, integrity, and insulation. Stability refers to the ability of an element of structure to perform its normal function for a certain time in the fire situation. For example a column or a beam must continue to support the load of the floor above. The floor itself is not load-bearing but it should not collapse during the period of the test: it might, for example,be part of an escape route. For a load-bearing element to pass the test it must support the load during the heating period, and again when it is re-loaded 24-hours later. If the element fails during the heating period it is rated at 80 per cent of the time to collapse. Floors and beams are regarded as having failed at the time at which they have deflected by more than one-thirtieth of the total span.

A wall may survive a test and continue to support a load, but it may allow the passage of fire through it because of cracks and other openings. The integrity of the wall is the time during which it resists the passage of fire. This is determined by the observation of a flame on the unexposed side, or by the ignition of a cotton pad held close to any cracks or other openings which have appeared. The integrity of a column is not checked because it is heated on all sides during the test. The critieria may not be appropriate for a beam either.

Even if a wall, or other element of construction, remains intact during the test, it may still fail to prevent the passage of fire to the next compartment. This will happen if the unexposed surface reachs a sufficiently high temperature to cause the ignition of materials which are near to it. Insulation is expressed as the time during which the mean temperature on the unexposed side has increased by no more than 140°C, or the temperature at any point on the surface by no more than 180°C.

Thus, the measurement of fire resistance has three components, stability, integrity, and insulation, all of which are expressed as a time. Building regulations may not necessarily require an element of construction to achieve the same time for each component. Thus a floor in an ordinary house may be required to have a resistance of 30 minutes for stability and 15 minutes each for integrity and insulation, and a door would not normally be expected to have any insulation characteristic at all.

Many countries have standards for the measurement of fire resistance on which their own building regulations are based. Although there is general acceptance of the ISO temperature/time curve, the actual fire exposure in different countries may not be the same. Heat transfer to the specimen by convection may be different because the geometry of the furnace will affect the amount of turbulence. Heat transfer by radiation will be affected by the thermal properties of the furnace wall and the nature of the fuel: gas flames radiate less than oil flames. Unfortunately the heat transfer to the specimen is not measured, only the temperature of the furnace. Even the temperature may vary by as much as 200°C because of the tolerances allowed on the temperature/time relatonship (±15 per cent for the first 10 minutes, ±10 per cent for the first 30, and ±5 per cent after that). A new generation of furnaces has been produced which will provide much closer agreement, and much of BS 476: Part 8, which is concerned with these tests, is being revised. These changes include the introduction of a new term, imperiousness: the ability of the specimen to prevent the egress of hot gases from the unexposed surface from exceeding specified limits.

The fire resistance of roofs is covered in Part 3 of the standard. The roof is exposed to radiation, but not in a furnace. Four radiant panels are used to create a radiation intensity of 14.6 kW/m^2. This is the calculated intensity resulting from a typical house fire located 13.5 m away from the roof. Tests are conducted to determine the ignitability of the roof surface, and the resistance of the roof to fire penetration.

OTHER FIRE TESTS

Materials used in building construction can affect the progress of a fire not only because of their resistance, but also because of the contribution that they make to a fire. Tests are available in parts of BS 476 which measure non-combustibility, ignitability, flame propagation, surface spread of flame and (in preparation) the development of smoke and toxic gases.

In the non-combustibility test (Part 4) a small sample is heated in a furnace to 750°C for 20 minutes. The material passes the test if the increase in temperature due to its presence is no more than 50°C, and if any flames which are seen do not persist for longer than 10 seconds. This is a severe test which is failed for example by plasterboard, although this material can make a useful contribution to fire resistance.

Part 5 describes a simple ignitability test in which a flame is applied to a 225 mm square specimen for 10 seconds. During this time flame should not spread to the edge of the specimen. Flames should not persist for more than 10 s after removal of the flame, nor should they spread to the edge. This part is to be replaced by a new test in which a range of ignition sources will be used. Part 13, published in 1987, describes a test which simulates heat transfer by radiation, as might occur during the growth stage of a room fire. Five different levels of radiation are applied to the specimen in the presence of a pilot flame.

The fire propagation test in Part 6 of the standard checks the amount of heat which is released by a material, and the rate at which it is released. In effect, it produces a temperature/time curve which is compared with that of a standard material.

In the surface spread of flame test (Part 7) the specimen is mounted at right angles to the edge of a radiant panel. A pilot flame is applied at the end of the sample for the first one-and-a-half minutes of the test, and then the rate of propagation of flame over the surface is measured for a total of 10 minutes. The material is placed into one of four classes depending on the distance travelled in one-and-a-half minutes and 10 minutes. It should be noted that the test does not measure vertical spread of flame, which is of great practical importance. Fire spread would be faster than that measured in the test if ignition started in the corner of a room, but would be slower if the material was mounted on a backing which acted as a heat sink.

The results of the last two tests can be combined to provide a Class 0 rating for materials which produce sufficiently low scores in both tests. However a more useful test is being developed in ISO and will eventually become a new Part 15 of the BS.

15

Building regulations

Separate Building Regulations apply to England and Wales, to Scotland, and to Northern Ireland. There is a need for harmonization not only within the UK but, under the Single European Act, throughout all European countries.

The Building Regulations 1976 (for England and Wales) comprise a large and complicated document. Their application to the design of a complex building was extremely difficult, and particularly in determining the level of fire protection. The requirements of the 1985 Regulations are much simpler to understand. This is because detailed specifications have been replaced by functional requirements. Also a number of 'trade-offs' have been allowed. One such relaxation was available in the earlier Regulations: the increase in the allowable volume of a shop if a sprinkler system was installed (see below).

This very welcome move towards functional requirements rather than detailed specifications is happening both in legislation and standardization. It is comparatively recently for example that fire extinguisher standards have required that the extinguisher should be able to put out a fire. Reliance had been placed before on detailed dimensional and mechanical specifications. In essence the standards now say that the unit must withstand the effects of its environment and function correctly when needed.

The new regulations allow a much more reasonable approach to be made to the problems of fire protection. The purpose and the objectives of these regulations is summarized in one of the Guidance Notes issued by the DOE as early as 1975: they are 'designed to set down basic requirements for a building structure that will not itself be unduly hazardous in the event of a fire and which will, when coupled with a satisfactory means of escape, ensure within reasonable limits the safety of all persons who might become involved'.

However, the practical fire engineer is more likely to be concerned with the ways in which fire protection has been achieved in existing buildings (most of which were built before 1985) than with the specification of new buildings. Also, in any involvement in discussions on new buildings, the details of the 1976 Regulations provide a practical guide to methods of achieving the functional requirements. For these reasons the outline of the earlier regulations has been retained in this edition of the book. It should again be emphasized that the 1985 Regulations and the associated documents will be under active review until complete European harmonization has been achieved.

It should be remembered that planning for fire protection may need to extend far beyond the requirements of the building regulations. It is necessary to check for example the manning, equipment and attendance time of the local fire brigade. Difficulties which could be experienced in attending the site, and in using water supplies, should also be checked. The availability of water should be discussed with the brigade and it is also necessary to check what will happen to water used for firefighting. The threat posed by other premises in the area should be considered. The siting of storage tanks of flammable liquids is of particular concern. They should not for example be on a hillside above the proposed site.

THE BUILDING REGULATIONS 1976

The building regulations apply only to:

1 The erection of any building
2 The alteration or extension of any building
3 The execution of any works and fittings
4 Material change of use (if any).

These regulations mainly concern the architect and builder and are enforced by the local authority. The help and advice of the appropriate authority should be sought at a very early stage. The regulations do not apply to an existing building until the status is changed.

In studying the regulations it must be remembered that they apply to all buidings including private houses as well as commercial and industrial premises.

It is not suggested that a fire precautions officer should be an expert on the Building Regulations but he should be familiar with the fundamental requirements and where to find them. In the following pages will be found brief notes on the various sections. These notes are not intended in any way as an interpretation of the regulations.

The regulations are divided into a number of parts. The three parts of major interest to the fire precautions officer are parts E, J and L.

PART E – SAFETY IN FIRE

Section I of Part E deals with structural fire precautions.

Section E1

Subsection (1) gives a list of the definitions used in the section.

Subsections (2) and (3) make exceptions to the rule and subsection (4) relates to storeys and galleries.

Subsection (5) is the actuating clause which requires that an element of structure, e.g. a door or other part of a building, should have a specified period of fire resistance. This period of fire resistance is acceptable if a specimen of the same construction satisfied the requirements of stability, integrity and insulation in BS 476 Part 8 for not less than the specified period of time set out in Table 1 of Regulation E1.

Subsection (5)(*b*) is a deemed-to-satisfy clause for walls, columns and floors. Within its context it refers to Schedule 8 of the Regulations. The Schedule is in eight parts relating to the elements of structure. Each part is in the form of tables giving details of construction and materials set out against the thickness of the element and the degree of fire resistance that is deemed to satisfy the provisions of the Regulation. Table 1 of this Regulation relates various parts of a building to the methods of test and the minimum period as to stability, integrity and insulation given in BS 476 Part 8.

Subsection (6) applies the requirements to a roof. This has to be constructed similarly to a specimen that has passed the relevant provisions of test of BS 476 Part 3. There is a deemed-to-satisfy clause if the roof conforms with one of the specifications set out in Schedule 9.

Subsection (7) refers to the use of plastic materials and relates the types given in BS 2782 with the criteria given in Table 2 of the Regulation which specifies type, description of material and method of test with these criteria.

Section E2

This section lists the purpose groups into which every building or compartment is divided. The grouping is based on the hazards of the purpose and use to which the premises or compartments are put.

Where more than one purpose exists, only the main purpose is taken into account. The following is a list of these purpose groups:

Group I	Small residential
II	Institutional; hospitals, homes, schools, etc.
III	Other residential
IV	Office
V	Shops
VI	Factories, within the meaning of the Factories Act
VII	Other places of assembly
VIII	Storage and general

Section E3

The danger from fire increases with the height above ground, the area and the cubic capacity of the building. This section defines the rules of measurement of these component hazards.

Section E4

This section requires that beyond certain limits of height, area and cubic capacity for each purpose group, the building is to be divided into compartments by means of walls or floors or both. The limits are laid down in Table 1 of the regulation. The section does not apply to single-storey buildings with a floor area of less than $3000m^2$.

There is a special provision in the case of purpose group V (shops) which permits the tabulated area and cubic capacity to be doubled provided that an automatic sprinkler system is installed and which complies with BS 53061 Part 2.

Section E5

This section deals with the fire resistance of the elements of structure. It takes each purpose group and in the form of tables defines the maximum dimensions in floor area and cubic capacity against the height to show the fire resistance of the elements above ground and below ground in terms of hours of fire resistance.

Although the tables only set out the compartment size and purpose, the height is the height of the building. There are two tables, one for buildings having more than one storey above ground level and the other for buildings having only one storey above ground level.

Section E6

This deals with the fire resistance of floors in conjunction with suspended ceilings and in a table to the regulation relates the height of the building with the type of floor, the required resistance in hours of the floor and a description of the suspended ceiling. In the description reference is made to Class 0 and Class 1 surfaces. Definitions of these terms are given in E15(1)(*e*) and (*f*) of the regulations.

Section E7

This rather complex section relates to the fire resistance of external walls. With some exceptions walls within 1 m of the relevant boundary must meet the required fire resistance (as laid down in the table to Regulation E5) from both inside and outside. Walls more than 1 m from the relevant boundary must meet this from inside only, with the provision that the insulation criteria of BS 476 Part 8 have to be met for 15 minutes only. There are certain additional requirements for buildings of purpose group VII. Tables in Schedule 10 deal with unprotected areas and distances from relevant boundary.

Section E8

This section relates to the fire resistance of separating walls. (The term 'party wall' is not used as it has legal significations which are not relevant to building construction.) Subject to a number of conclusions,the Regulations require that separating walls shall be imperforate and be carried through above the roof, but there are a number of exceptions and safeguards for the exceptions. For example provided that pipes comply with section E12, they may be allowed to pass through a separating wall, as may doors complying with section E11. There are also conditions governing certain types of roof construction.

Section E9

This section deals with compartment walls and floors. It is somewhat similar to the previous section in that it requires any compartment wall or floor to be imperforate. But there are a number of exceptions, each with its own conditions.

1 Provides for an opening if fitted with a fire-resistant door

2 Permits a protected shaft
3 Permits a ventilation duct, provided that the space surrounding it is fire-stopped and that it is fitted with a fire-resistant shutter
4 An opening for a pipe as in the previous regulation
5 A chimney or flue pipe provided that it complies with the regulations concerning flues.
6 This is the operative clause which requires (subject to exceptions) that any compartment wall or compartment floor shall have a fire resistance of one hour or more.

Similar requirements apply to fire stopping of the junctions between compartment walls and floors and external walls.

Section E10

This section deals with protected shafts. A protected shaft is a stairway, lift shaft, escalator, chute or similar structure. The section ensures that they are fire-resisting and that any openings are protected as for fire compartments.

Section E11

The terms 'fireproof door' and 'fire-resisting door' may both be encountered by the fire engineer.

The term 'fireproof door' is the term used by the Fire Offices Committee. While the term is technically incorrect, it is acceptable by common usage. Fireproof doors are usually of two- or four-hour resistance to fire and of a construction acceptable for fire insurance.

The term 'fire-resisting door' is used in the building regulations and by the fire service. These doors are generally described as being of half-hour or one-hour construction through occasionally a two-hour fire-resisting door may be required. The term 'fire check door' and 'modified half-hour' fire door no longer appear in the regulations. Where fire-resisting doors are referred to in the regulations it is taken to mean a door of similar construction to a specimen that has been exposed to a fire resistance test and which has passed the test for stability and integrity laid down in BS 478 Part 8. Fire doors are not called upon to withstand the test for heat insulation.

The report of a test gives the actual time achieved during the test before falure occurs so that a particular make can be selected to meet the requirements of the regulations.

Three types of fire doors are recognized in the building regulations. These appear in Table 1 to Regulation E1 as items numbered (14), (15 and (16).

(14) Doors which are required to withstand fire test conditions that may be imposed for a specified purpose. These are referred to in BS Code of Practice CP3 Chapter 4 Part 1 as Type 1 doors.
(15) Doors which are required to withstand the test conditions for stability, for not less than 30 minutes and the test for integrity for not less than 20 minutes. They are referred to as Type 2 doors in the code of practice but are more commonly known as 30/20 fire doors.
(16) Doors which are required to withstand the test conditions for stability for not less than 30 minutes and the test for integrity for 30 minutes. They are referred to as Type 3 doors in the Code of Practice but are more commonly known as 30/30 fire doors.

There is a Type 4 door recognized by the Code of Practice CP3 Chapter 4 Part 1, which is concerned with flats and maisonettes over two storeys. It is for use in a corridor within an escape route and differs from a 30/20 door in that it may swing in one or both directions and the frame may have either no rebate or rebate of an unspecified depth.

In the case of the 30/30 door the test has to be made in any frame but in the other cases the doors have to take the test in their own frame.

While on the subject of frames it must be remembered that the expression 'fire door' in the regulations means not only the door but also the frame. It is therefore better to speak of a 'fire door assembly' rather than a 'fire door'. This will be more noticeable when the construction of a fire door assembly is discussed. Fire doors are necessary to enable occupants to pass through a fire compartment. It is therefore axiomatic that a fire door must be as resistant to fire as the structural element in which it is installed, and that all structural elements in the same compartment shall be of the same resistance to fire. Table 15.1 shows the position of door openings with door and frame requirements.

Section E12

This section deals with the penetration of the structure by pipes; it includes pipe fittings but excludes flue pipes, and ventilating pipes. There is a table to the section giving specifications of pipes and the maximum size of those pipes which may penetrate a structure. Two types of pipe are recognized:

(a) Pipes made of non-combustible material which, if exposed to a temperature of 800°C, will not soften or fracture to such an extent as to permit the passage of flames or hot gases through the wall of the pipe. The size of the pipe in this case must not exceed an internal diameter of 150 mm.

Table 15.1
Position of door openings with door and frame requirements
(Capc Boards and Panels Ltd)

Doors and frames, requirements				
Position of door opening	Fire-resisting requirements of door and frame: to BS 476: Part 8: 1972			Closer
	stability	integrity	Type of swing (see key at foot)	
Penetrating protect shaft				
Above ground level in purpose group III, IV or VII, or where protected shaft has 1 hour fire resistance or less	30 min	30 min	No requirement	Door closer required
As above, but opening to hall, lobby or corridor with ½ hour fire resistance	30 min*	30 min*	A,B,C,D,E	Door closer required
Purpose group I: access from internal stairway to habitable room or kitchen in building of 3 or more storeys	30 min	20 min	A or C	Door closer required or rising butts
Any other case	½ period required for protecting structure		No requirement	Door closer required
Penetrating separating wall				
Permitted only where necessary as means of escape	Full period required for wall		No requirement	Door closer required
Penetrating compartment wall				
Access to flat or maisonette from space in common use	30 min	20 min	A or C	Door closer required or rising butts
Any other case	Full period required for wall		No requirement	Door closer required
Access from attached small garage to house				
Garage of 40 m^2 max floor area, door sill 100 mm above garage floor	30 min	20 min	A or C	Door closer required or rising butts
Access to refuse storage chamber				
Sole access for removal of refuse must be a flush door situated in an external wall	30 min	20 min	No requirement	Door closer required

* In this instance only, if the door meets the 30-minute test requirements when fitted in any rebated frame, it is deemed to provide 30 minutes fire resistance when fitted in any frame where gaps at the perimeter of the leaves are as small as possible. In all other instances the frame must be to the same specifications as that test.

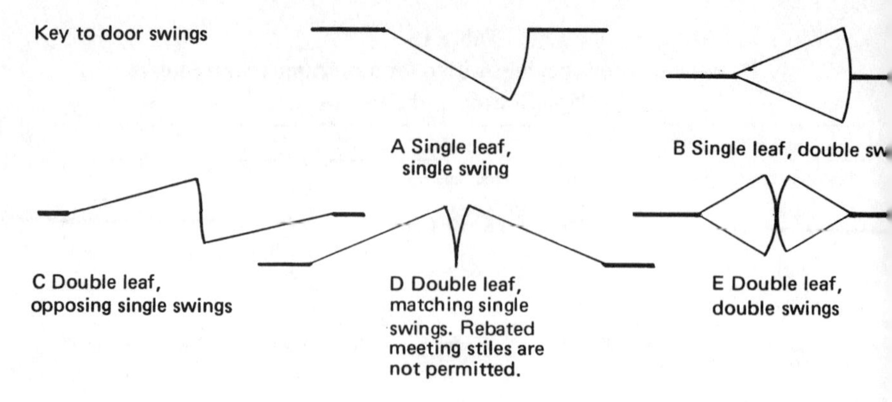

Notes

GENERAL

Any doorway in a fire-resisting construction must be protected by a fire-resisting door. The door may be a shutter, cover or other form of protection comprising one or more leaves and including any necessary frame.

The general requirements for periods of fire resistance and constraints upon the types of swing permitted are shown in the table.

METHOD OF INSTALLATION

Hinges. Any fire door must be hung upon hinges which are made entirely of non-combustible material with a melting point not lower than 800°C, i.e. steel or brass hinges.

Closers. Any fire door must be fitted with an automatic self-closing device: rising butts may be used to meet this requirement only where shown in the table. There is an excepton in the case of a doorway penetrating a protected shaft containing exclusively lifts, see below.

Hold-open devices. The only devices which may be used to hold open a fire door are as follows:

1 For doors other than escape doors, a fusible link.
2 For an escape door and for any other door which can readily be opened manually, an 'electro-magnetic or electro-mechanical device susceptible to smoke'. Such a device must allow the door to close automatically upon:
   (a) Detection of smoke
   (b) Manual operation of a switch
   (c) Failure of electricity supply
   (d) Operation of a fire alarm.

EXCEPTIONS

Fire door used as means of escape. Any fire door which may be used as a means of escape must readily be openable manually, and may be held open only by an electro-magnetic or electro-mechanical device susceptible to smoke.

However, the same opening may include also a door which cannot readily be opened manually provided:

1 It is fitted with an automatic self-closing device and held open by a fusible link.
2 The manually openable door has a fire resistance not less than 30 minutes.
3 The necessary fire resistance is achieved by the two doors together.

Protected shaft containing excusively lifts. A door without a closer may be used under either of the conditions below:

1 The opening does not penetrate a compartment wall, it is the only doorway at that level, and the door provides fire resistance for the full period required for the wall.
2 The door provides 30 minutes' fire resistance, and a second door is installed in the same opening. The second door must be fitted with a closer, held open by a fusible link, and provide fire resistance for the full period required for the wall.

(b) Pipes made of lead, aluminium or alloy, asbestos cement or unplasticized polyvinyl. These pipes may not exceed 100 mm if they penetrate a structure (but a separating wall is excluded), enclosing a protected shaft not regularly used for the passage of people but may not exceed 38 mm in all other cases. Any opening must be as small as possible and is to be fire-stopped around the pipe.

Section E13

This section deals with stairways and landings and requires that they shall be constructed of non-combustible materials.

Subsection (1)(*a*) makes exceptions mainly concerned with maisonettes and shops.

Subsections (2) and (3) make special requirements for buildings of purpose group I (private dwelling-houses), which have three or more storeys.

Section E14

This section deals with the provision and construction of cavity barriers and fire stops.

This is a very important section because the lack of cavity barriers and fire stops in many buildings of comparatively recent date has often accounted for the rapid spread of fire throughout the building. It is one of the points which should be borne in mind when inspecting existing buildings. Every effort should be made to improve any such conditions that may be found. The subject is closely linked with the provision of fire compartmentation (see Chapter 14).

Subsection (2)(*b*) is especially important and requires that where an element of structure containing a cavity meets other such element, they are to be so closed that they do not communicate which each other. A small opening of even 1 mm can permit the passage of smoke and flame, which will rapidly enlarge the opening and allow a fire to spread to an adjacent compartment, where additional fuel would be available, thus increasing the intensity of the fire and so leading to far greater destruction and danger to occupants.

There are a number of subsections which require study bearing in mind the principle behind the section. When applying this section it is of vital importance to achieve an absolute seal because of the possible shrinkage of materials or settlement of the building as well as the effects of fire and heat. The use of mineral wool or a gun applied intumescent mastic can assist in achieving this.

Section E15

This deals with the restriction of the spread of flame over walls and ceilings. A Class 0 material is defined as one that is non-combustible or one that meets a certain performance standard when tested for fire propagation to BS 476 Part 6, and having a Class 1 surface. Class 0 is not a spread of flame classification, but in terms of surface class it is to be regarded as better than Class 1. A spread of flame classification is given to a material depending on its performance when tested to BS 476 Part 7. Class 1 is the best followed in descending order by Classes 2, 3 and 4.

A table to this regulation relates the maximum floor area of each purpose group to the class of surface for walls and ceilings. In general, this table shows that except in small residential houses having not more than two storeys, the restriction to the spread of flame over the surface of circulation spaces and protected shafts should be of the order of Class 0. Rooms other than small rooms should have Class 1 protection (except purpose group II Institutional which should have Class 0 protection for walls). Small rooms – defined in the table according to the purpose group – should have Class 3 protection except again in purpose group II which should have Class 1 protection.

Section E16

This section deals with exceptions permitting the use of plastics.

Section E17

This section deals with roofs and in particular the danger of fire spreading from one building to another by way of the roof. It takes into consideration the construction of the roof, the size of the building (which governs the ferocity of the fire), and the distance the building is from the boundary fence.

It is very important to know the roof designations as given in BS 476 Part 3 and the details for the test procedure.

By this British standard roofs are described by two capital letters. The first of these refers to the resistance to penetration by fire and the second to the spread of fire. Four letters are used in each group. Thus, the first letter refers to penetration as follows:

A Roofs which can withstand penetration by fire for one hour under test conditions.

B Those that withstand penetration between 30 minutes and one hour.
C Those which are penetrated in less than 30 minutes.
D Those which fail a preliminary test.

The second letter refers to the spread of flame under test conditions and is as follows:

A Roofs on which there is no spread of flame.
B Roofs on which there is not more than 500 mm spread of flame.
C Roofs which have more than 500 mm spread of flame.
D Roofs which continue to burn for more than five minutes after withdrawal of the test flame or if the spread of flame is more than 375 mm in a preliminary test period.

Thus BC would indicate a roof which could withstand penetration and on which flame would not spread for more than 500 mm under test conditions.

Subsection (1) states that no part of the roof of the building, which has a cubic capacity exceeding 1500 m^3 and is wholly or partly of purpose group VII or VIII or is a house in a continuous terrace of more than two houses, may be so constructed as to be designated by any one of the nine worst categories from BD to DD, nor may it be covered with thatch or wood shingles.

Section E18

This deals with small garages. A small garage in this context is one not exceeding 40 m^2 of floor space. If such a garage is attached to a house the wall between the garage and the house must have fire resistance of not less than half an hour. Any opening between them must be not less than 100 mm above the level of the garage floor and any door, shutter or cover between them must be a 30/20 door and in other respects comply with the regulation E13.

Section E19

This deals with small open car ports. These are designated as having a floor area not exceeding 40 m^2.

Section II of part E comprises E20, 21 and 22 which deal with means of escape in case of fire.

Section E20

This section applies to all three following sections for buildings consisting of flats or maisonettes and to offices and shops with only one stairway and limitation as to height or floor area.

Section E21

Subsection (1) interprets the expressions used in the section and subsection (2) states that it must not be prejudicial to the requirements under the Offices, Shops and Railway Premises Act 1963 (this will now of course be under the Fire Precautions Act 1971). Subsection (3) states that any structural fire precautions under the next section are additional to other requirements of the Building Regulations.

Section E22

This section deals with the provision of means of escape and requires that sufficient and adequate means of escape are to be provided as may reasonably be required to enable occupants to reach a place of safety in the event of fire and that any other work shall be provided to ensure that the means of escape can be safely and effectively used at all material times. This section does not set out to detail how the means of escape are to be achieved.

Section E23

This is a deemed-to-satisfy clause in which it is accepted that the provision of means of escape will be satisfied if the building complies with parts of BS CP3 Chapter IV Part 1 1971 (Flats and maisonettes); CP3 Chapter IV Part 2 1968 (Shops and department stores); CP3 Chapter IV Part 3 1968 (Office buildings).

PART J – REFUSE DISPOSAL

Section J1

This Regulation applies to refuse storage containers constructed in buildings comprising more than one dwelling. In general it applies to a chamber constructed to accommodate refuse containers. The chamber has to be constructed as a one-hour fire-resistant compartment with an opening consisting of a half-hour fire-resisting flush door and a refuse chute or a hopper for depositing refuse into it.

Section J2

This describes the construction of a refuse chute. It is to be constructed of non-combustible material so arranged as to prevent ignition of any part of the building.

Section J3

This deals with pipes or shafts for ventilating refuse containers and chutes.

Section J4

This deals with loading hoppers. These are also to be constructed of non-combustible materials. No hopper may be situated inside a building.

PART L – CHIMNEYS, FLUE PIPES, HEARTHS AND FIREPLACE RECESSES

Section L1

This gives definitions and interpretations of the expressions used in the section.

Attention is drawn to the definitions of Class I and Class II appliances. A Class I appliance means (a) a solid fuel appliance or oil-burning appliance having, in either case, an output rating not exceeding 45 kW; or (b) an incinerator having a refuse combustion chamber exceeding 0.03 m^3 but not exceeding 0.08 m^3 in capacity. A Class II appliance means (a) a gas appliance having an input not exceeding 45 kW; or (b) an incinerator having a refuse combustion chamber not exceeding 0.03 m^3 in capacity.

A high rating appliance means (a) a solid fuel appliance or oil-burning appliance having, in either case, an output rating exceeding 45 kW; or (b) a gas appliance having an input exceeding 45 kW; or (c) an incinerator having a refuse combustion chamber exceeding 0.08 m^3 in capacity.

A constructional hearth is one that forms part of the building and a superimposed hearth is one that does not form part of the structure of the building.

Section L2

This deals with general structural requirements.

Subsection (1) requires that subject to certain exceptions, any chimney, flue pipe, constructional or fireplace recess must be constructed of non-combustible materials of such a nature, quality and thickness as not to be unduly affected by the heat of the products of combustion and so placed or shielded as to prevent ignition of any part of the building.

Subsection (5) requires that if provision is made for a solid fuel fire to burn directly on a hearth, provision shall also be made for the secure anchoring of a fire guard in the surrounding structure.

Section L3

This deals with fireplace recesses for Class I appliances.

Subsection (2) deals with fireplace recesses in general construction of which has to be 200 mm thick on each side and at the back.

Section L4

This deals with constructional hearths for Class I appliances and gives details of constructional hearths for Class I appliances. In general the hearth must be not less than 125 mm thick. If it is constructed in conjunction with a fireplace recess:

(a) it must extend to the back and jambs of the recess;
(b) it must project 500 mm in the front of the jambs; and
(c) it must extend 150 mm beyond each side of the opening between the jambs.

Other subsections deal with the distance that combustible material must be kept from the hearth.

Section L13

This governs the distance the outlet of a flue in a chimney must be above the building.

Section L14

This contains requirements for Class II appliances similar to those for Class I appliances.

FIRE COMPARTMENTS

A firc compartment is a space bounded by the elements of construction into which it is difficult for hot gases (smoke), flame and fire to enter or leave. The time that it takes for this resistance to break down is determined by tests made under standard conditions.

Recent studies have shown that in the past insufficient attention has been given to the effective construction of fire compartments and even today we are only just beginning to realize the importance of absolute sealing against penetration by smoke and hot gases.

A fire compartment has three functions:

1. By limiting the size of a compartment the amount of fuel is restricted and the amount of available oxygen is restricted and so the severity of a fire is limited.
2. By creating a compartment that will be constructed so that it will reduce the time before smoke and fire can enter into it and so occupants are given more time to escape.
3. By the same token it is easier for occupants to escape if it takes a fire longer to break away from the point of origin.

External walls

External walls should be constructed in accordance with Schedule 7 of the Building Regulations. It is a deemed-to-satisfy provision.

Internal walls

Internal walls can be either loadbearing or non-loadbearing. Schedule 8 of the Building Regulations (which is a deemed-to-satisfy provision) refers to notional periods of fire resistance. It is set out in the form of a table showing the construction and materials for loadbearing or non-loadbearing walls with various covers as against the minimum thickness of the material to give a defined period of fire resistance. The periods of resistance are four, two, one-and-a-half, one, and half an hour. For example under masonry loadbearing walls of brick, of clay, concrete or sand lime:

(a) unplastered 200 mm thickness will give four hours' protection and 100 mm thickness will give protection for any shorter length of time. If the wall is non-loadbearing four hours can be obtained with 170 mm thickness, two hours and one-and-a-half hours' protection requires only 75 mm.

Against this

(b) a similar wall plastered with 12.5 mm of vermiculite-gypsum plaster requires only 100 mm of thickness for both loadbearing and non-loadbearing walls to achieve four, two and one-and-a-half hours of resistance and only 75 mm of thickness to achieve one hour or half-hour resistance for a non-loadbearing wall.

The tables are very extensive and show many other materials that can be used to cover brickwork and increase fire resistance. Several other constructions are given including materials to be added to framed and composite construction.

For escape route purposes half an hour fire resistance is the usual requirement, although a fire authority inspector may ask for higher resistance under special conditions as when the risk of fire spreading quickly could be a feature of the building or for any factor which makes escape difficult and the conditions are otherwise difficult to rectify.

Detail	Specification	Fire resistance
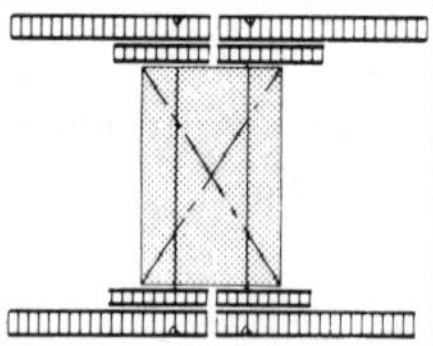	Minimum 75 mm × 50 mm softwood studding at 610 mm maximum centres, 9 mm Asbestolux facing fixed to both sides through 6 mm Asbestolux fillets using 60 mm × 11 swg nails or 50 mm × no.8 woodscrews at 400 mm maximum centres. Fillets to be 25 mm wider than stud.	½ hour
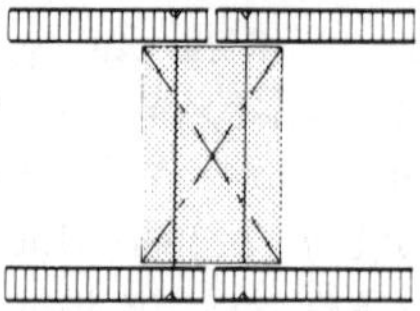	Minimum 75 mm × 50 mm softwood studding at 610 mm maximum centres, 12 mm Asbestolux facing fixed to both sides using 60 mm × 11 swg nails or 50 mm × no.8 woodscrews at 400 mm maximum centres.	½ hour
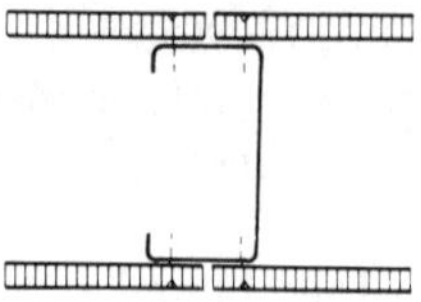	Minimum 75 mm × 40 mm × 20 swg steel channel framing at 610 mm maximum centres, 9 mm Asbestolux facing fixed to both sides using 20 mm × no.8 self-tapping screws at 400 mm maximum centres.	½ hour

Figure 15.1 International non-loadbearing walls. (Cape Boards and Panels Ltd.)

In some cases improvement can be made by the use of 12.5 mm plasterboard or asbestos-based insulation board added to both sides of the wall or partition.

Figure 15.1 shows the detail of the construction of three partitions which will give half an hour fire protection. Note that Superlux asbestos free board may be used in place of Asbestolux.

Where a higher degree of fire resistance is required, a construction using a 25 mm thick mineral wool mat stapled to the inner of one of the facings will give up to one hour of fire resistance. Up to two hours' protection can be obtained by using the channel method together with 9 mm fillets of asbestos-based insulation board at the joints and the 25 mm mineral wool mat as described above.

Surface finishes

It is essential that throughout all escape routes the surface finish of the compartment shall not permit the passage of flame across that surface. The standard for the surface spread of flame is BS 476 Part 7.

Floors and ceilings

Floors and ceilings are usually considered together and the building regulations in Schedule 8 and section E6 and the table to that regulation gives details of the construction for new buildings.To construct fire-resistant compartments it is first necessary to assess the existing elements of construction and the fire resistance obtained from them. This can best be done by a comparison with part VII of Schedule 8 of the Building Regulations and part VIII if the floor is of reinforced concrete.

In part VII, which deals with construction and material for timber floors, as would be expected, tongue and groove timber floors achieve a much greater period of resistance than do plain edged boards. It is not considered that plain edge boarding can achieve more than half an hour fire resistance.

Tables are given in part VII of Schedule 8 comparing the material for the ceiling beneath the timber floors against three periods of time – one hour, half an hour and a modified half-hour. This table is the only place in the building regulations where the 'modified half-hour' requirement is found. It refers to item 10 of table 1 to Regulation E1 and concerns the floor of the upper storey in a building of Purpose Group I which deals with small residential buildings. (These are private dwelling-houses excluding flats or maisonettes.)

The tables are in three parts:

A Plain edge boarding on timber joists not less than 38 mm wide with different covering materials. The thickness of the covering material determines the period of resistance. For example item VII of the tables, which is two layers of plasterboard with a total thickness of 25 mm, gives half an hour resistance; if the total thickness is 19 mm it would satisfy a modified half-hour resistance.

B Tongued and grooved boarding or chipboard of not less than 16 mm (finished) thickness on timber joists not less than 38 mm wide with a ceiling of, for example (item i) timber lath and plaster with 16 mm thickness of plaster, will give a modified half hour of protection but if the covering (item ii) is plasterboard 9.5 mm thick it achieves half an hour fire resistance.

C Tongued and grooved boarding or chipboard not less than 21 mm (finished) thickness on timber joists not less than 175 mm deep by 50 mm wide then (item iii) with a ceiling of metal laths and sprayed asbestos (in accordance with BS 3590) to a thickness of 19 mm will give one hour fire protection or if the thickness exceeds 12.5 mm half an hour protection is achieved.

One of the easiest methods to bring an existing floor of plain edge boarding up to half an hour protection for an escape route can often be achieved by adding a 3 mm thick hardboard to the floor and plasterboard or asbestos-based insulation board of suitable thickness either to form a ceiling or to reinforce the ceiling. Manufacturers of such materials will give advice on the necessary thickness if full details are submitted to them.

If a higher degree of fire resistance is required it can often be obtained but the advice of the manufacturers should be sought.

Cavities

Cavities are any spaces enclosed by the elements of a building or contained within an element other than a room, cupboard, circulating space, protected shaft or the space within a flue, chute, duct, pipe or conduit. The space between the floor and ceiling of a building is a cavity and by reason of their elements of construction they must be fire compartments.

CAVITY BARRIERS

Cavity barriers are constructions within a cavity designed to restrict the movement of smoke and flame within the cavity. Figure 15.2 shows positions for cavity barriers between fire compartments.

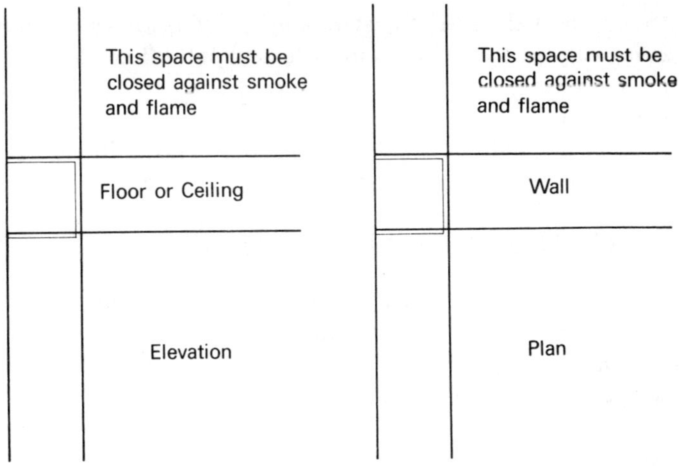

Figure 15.2 Positions where cavity barriers are required.

Cavity barriers are also required to limit the size of cavities, as for example to prevent a fire which has originated on the ground floor of a building bypassing higher storeys behind panelling. In this instance the fire could travel upwards to the roof where it might 'mushroom' out and spread across the top of other fire compartments undetected.

The requirements of a cavity barrier are four-fold:

1. Smoke and hot gases must be prevented from passing from one compartment to another, for example by passing through the point where the wall meets the ceiling.
2. Smoke and hot gases must be prevented from bypassing a compartment, for example if a separating partition (a non-loadbearing wall) is not carried up through the ceiling to the roof, fire could travel along the roof cavity or ceiling void appearing in another compartment having the same construction.
3. Smoke and flame must be prevented from travelling long distances. An example of this would be a fire on a ground floor getting into a cavity wall and being able to travel up to a roof void or other temporary check (where it could 'mushroom' out). In this case the cavity could also act as a gigantic flue.
4. A cavity barrier must also maintain its integrity throughout the life of the building. This means that:

   (a) the materials that form the barrier must not age so that they will no longer contain smoke and hot gases;

(b) the barrier must still maintain its ability to seal completely the compartment even with some settlement of the building;
(c) the barrier must retain its properties despite any minor thermal changes due to the weather or domestic heating;
(d) it must retain its stability even under fire conditions for at least as long a period as the compartment is designed for.

OPENINGS THROUGH FIRE COMPARTMENTS

Having dealt with the constructional details of a fire compartment, it is now necessary to examine the openings that pass through the compartment to ensure that its integrity is maintained.

These openings will consist of:

- Doors
- Columns and beams
- Pipes, cable ducts and cables.

Doors

These are dealt with below. Again emphasis is laid on the point that it is the fire door assembly and the junction between the assembly and the compartment wall that are very important features.

Columns and beams

Fire tests for columns are established in BS 476 Part 8 section 4 and the tests for beams are established in section 5.

The danger of columns collapsing and beams sagging under fire conditions has been recognized for many years and various methods have been devised for their protection.

The constructional details for new reinforced concrete columns are given in Schedule 8 of the Building Regulations Part II, the details for reinforced concrete beams is given in Part III and those for prestressed concrete beams with post-tensioned steel are given in Part IV. Part V, sections A and B give various methods of protecting steel columns and beams for up to four hours. Materials mentioned range from plasterboard with a gypsum plaster coat to asbestos insulation board, but many other methods can be used. These include a calcium silicate-based vermiculite reinforced non-combustible board which can be screwed or stapled together, and a speedily erected prefabricated galvanized or PVC coated sheet steel with a bonded lining of asbestos-based or asbestos-free insulation board.

Figure 15.3a Typical four-sided casing to a column.
(Cape Boards and Panels Ltd.)

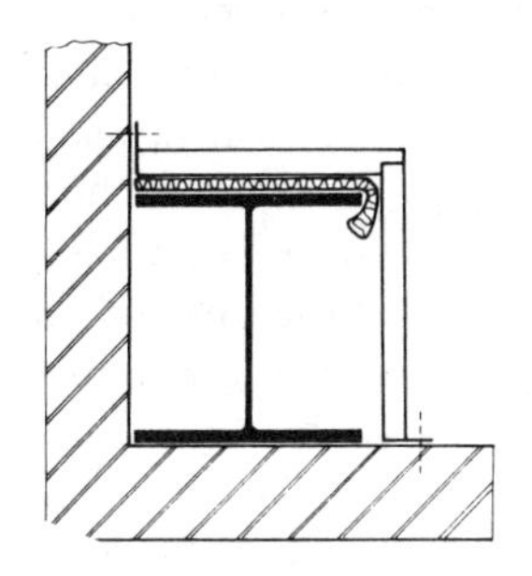

Figure 15.3b Typical two-sided casing to a beam adjoining a wall and ceiling.
(Cape Boards and Panels Ltd.)

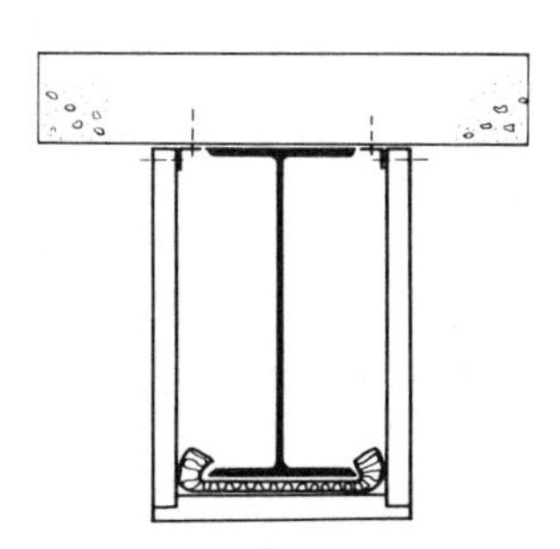

Figure 15.3c Typical three-sided casing to a beam.
(Cape Boards and Panels Ltd.)

To obtain protection for existing columns and beams there is a very suitable method of fire-casing them. This is illustrated in Figure 15.3a, b and c and two- or three-sided beam casings and two, three or four-sided column casings can be supplied.

The casings are made to site measurements and although the maximum standard length is 3.048 m, longer casings can be made in sections with unobtrusive joints. This type of construction can achieve up to four hours' fire resistance, depending on the thickness of the lining.

Pipes, cable ducts and cables

The Building Regulations, section E12 lays down the sizes of pipes that may penetrate a compartment. The size is related to the material of which the pipe is composed. Thus a pipe made of non-combustible material which, if exposed to a temperature of 800°C, will not soften or fracture to such an extent as to permit the passage of flames or hot gases to pass through the wall of the pipe, may be of a size up to 150 mm internal diameter.

Where other materials are used the internal diameter of the pipe may be as low as 38 mm. The general rules given in the regulations should be studied as there are rules which can be followed if the pipe or duct is encased in fire-resistant casing. Such a casing has to comply with the requirements as to the surface treatment to secure a Class 0 specification.

FIRE STOPS

Fire stops are required to seal imperfections between essential openings in fire compartments and cavity barriers and also to ensure a smoke- and flame-proof seal between fire compartments and the openings that are permitted through them.

A number of materials may be used to make fire stops, of which the following is a selection:

- Brick, stone and cement
- Cement as a filling material
- Asbestos insulation board
- Asbestos-free non-combustible fire-resistant insulation board
- Intumescent material
- Woven asbestos caulking material.

Some of these materials will be suitable for where a small movement due to thermal changes has to be accommodated; others will be more suitable where no movement is expected.

Where a beam has to pass through a fire compartment it can generally be fire-stopped in a similar manner to pipes. Where there is a complicated system of trusses it may be necessary to wrap each individual truss with mineral wool held in place with wire netting.

FIRE DOORS

There are no regulations laid down as to how a fire door is to be constructed. The criteria are those of constructing a door which will meet the requirements of BS 476 Part 8.

There is no doubt that if all doors, especially those in private houses, were constructed as fire doors and kept closed, there would be a considerable saving in fire damage and many lives would be saved. The difference in cost between an ordinary door and a fire-resisting door is not very great but the saving of life and property would be well worthwhile.

When a fire starts in a room smoke and hot gases are immediately given off. The expansion of these gases causes them to rise to the ceiling where they will 'mushroom' and spread outwards to the walls. Due to the expansion of the hot gases light pressure will build up forcing smoke and hot gases out through any gaps between the door and the frame. It should be noted that the same effect would occur if there were any gap between the frame and the wall. If any smoke and hot gases find a gap and can pass through it, flames will quickly follow and the complete assembly will soon be destroyed.

The damage now comes first from the shrinkage of the wood and finally the actual charring until a gap is created. The shrinkage of the wood results in distortion of the door itself. If the side of the door that is exposed to the fire is the door stop side, the distortion will be restrained by the door stop and disintegration will be delayed. The final collapse of the door is usually preceded by the penetration of fire through the edges of the door and the fixing of any glazing, particularly through a glazing bead if not properly constructed.

Failure usually starts at the top of a door, obviously where the temperature is higher. Wired glass 6 mm thick will itself withstand the temperature developed under test conditions for over an hour because most of the heat falling on the glass will be transmitted through it and then dissipated on the cooler side. Details of glazing for fire resistance are given later in this chapter.

Factory-made doors are nearly always to be preferred to local made doors. The safest way to obtain good fire-resistant doors is to purchase them from a reliable manufacturer obtaining from him a copy of the test certificate of the prototype, together with a statement that the particular door purchased is guaranteed to conform with the specification for the particular regulation.

Local construction or conversion

Before setting out any details of local construction or conversion, it should be emphasized that it is not considered possible to convert an existing door to one-hour fire resistance. However, upgrading to 30/20 standard is possible with panelled doors and the more substantial flush doors by facing with suitable fire-resistant materials, in the manner recommended by the manufacturer which should be based on fire tests or assessment by a recognized testing laboratory. Doors less than 35 mm thick are not suitable for upgrading.

Even before the consideraton of any fire door, thought should be given to the fixing of the door jamb into the wall. There must be no gap and the fire resistance must be carried right up to the jamb. If necessary an intumescent strip can be inserted between the wall and the door jamb (see Figure 15.4). When inspecting a new or converted door care should be taken to see that cover strips are not used to hide defective workmanship in this respect.

The door jambs and head should be made of well seasoned wood morticed together and not just nailed. The door jambs should be 60 mm by 48 mm (57 mm for one-hour doors). The door stops should be 25 mm deep cut from solid timber. In the case of half-hour doors the stop may be built up to 25 mm by means of a non-combustible fire-resisting strip glued and screwed with rust-proof 57 mm no. 8 screws placed not more than 300 mm apart.

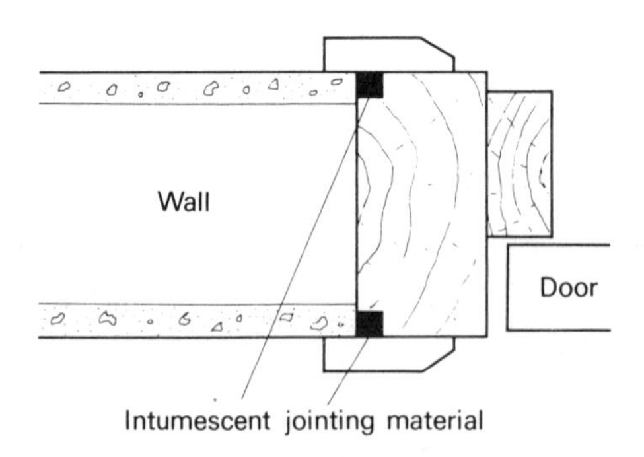

Figure 15.4 Sealing a gap between wall and door jamb.

In the case of one-hour doors, the jambs and head should be pressure-impregnated to a depth of not less than 13 mm. Care must be taken to ensure that there is no distortion and it is preferable to machine the wood to size after impregnation.

The door must be close fitting into the door frame with a gap of not more than 2 mm all round or if an intumescent strip is incorporated in the door or frame the gap may be 3 mm.

Intumescent strips

These are now an essential part of fire-resisting doors and when combined with a suitable draught excluder greatly improve the fire-resisting qualities of a door.

An intumescent strip consists of a material that when heated to about 200°C swells and is converted into a crust-like material which is highly resistant to fire and heat. There are a number of these substances all with slightly differing characteristics. In general they are inserted in the door and/or jamb as shown in Figure 15.4.

To achieve one-hour fire resistance two intumescent strips not less than 10 mm wide and 3 mm deep will be needed, one strip in the edge of the door and one in the frame. Half-hour fire resistance can be obtained with only one intumescent strip. Protection must be continuous over hinges and door furniture. Special hinges and door furniture can be obtained for this purpose.

The weakness of intumescent strips is that cold smoke can pass the door as the seal is not made until sufficient heat is available to raise the temperature of the strip to effect its conversion and expansion. To overcome this, fire doors should have a draught-excluding strip. Ordinary draught excluders will give protection for a time but deteriorate over the years or may be rendered useless by painting. The only successful draught excluders are the metallic spring clip or neoprene wiper type. A combination draught excluder and an intumescent strip can be obtained. Figure 15.5 illustrates such a type for a two-leaf double swinging door. Another type is available which is suitable for single doors. Special intumescent strips can be obtained to provide protection against heat transfer at hinges and locks.

Construction of fire doors

An existing panelled door having a thickness of not less than 45 mm with stiles and rails may be brought up to half-hour fire check (30/20) standard if the face of the panels only on both sides is covered with insulating board not less than 6 mm thick (not asbestos cement board), the door stops being of

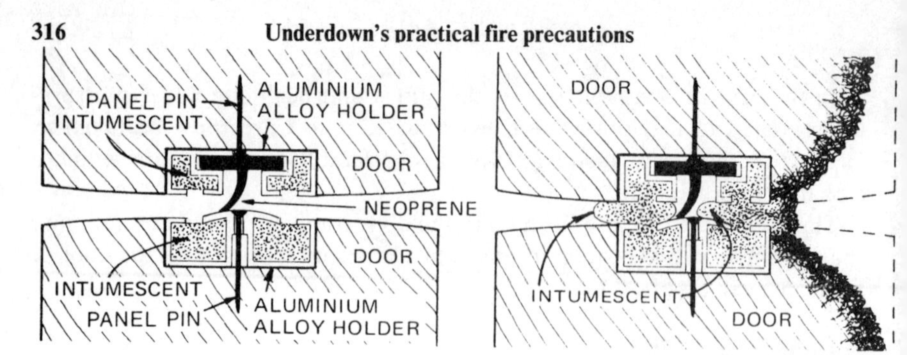

Figure 15.5 Combined intumescent strip and draught excluder fixed in the meeting stiles of a double swing double leaf door (Sealmaster of Cambridge).

25 mm depth. A door having stiles and rails over 35 mm thick may be similarly upgraded from one side by completely facing that side with 9 mm asbestos insulating board (AIB) (or asbestos-free equivalent) and infilling the panel voids, or may be treated with 6 mm thick AIB on both sides in which case panel voids do not have to be infilled. Solid laminated doors 45 mm thick are also acceptable.

Glass panels in fire-resisting doors

Where possible it is advisable to have at least a small glass panel in a fire-resisting door because of the danger to occupants using the door on any occasion.

Wood frames for fire-resisting glazing

The essential feature of glazing in wooden fire doors is to ensure that sufficient wood remains for the period of test to hold the glass in place.

One method is to use a frame construction with the rails and stiles of the door but not part of them and with the opening limited to the size of permitted area of glass.

For either 30/20 or 30/30 doors the frames should be cut from 56 mm by 44 mm material with a solid rebate not less than 13 mm. For the protection of wood beading, which should also be not less than 13 mm wide, it is possible to use intumescent paint or a metal protection.

Probably the best method, however, is to use a 44 mm wide asbestos-based channel developed by the Timber Research and Development Association and with the joint to the framing covered with a hardwood strip. The channel should be fixed to the framework of the door with 38 mm by no. 6 countersunk screws at 200 mm centres, each side of the glass.

For glazed panels in a one-hour fire door, 50 mm wide 'trada' channel should be used.

Door furniture

It has already been said that a fire-resisting door usually breaks down first where the door meets the door frame and the weakest part of the gap is where the furniture is fitted. The reason for this is fairly obvious as the metal conducts heat and so increases the area and depth of charring.

Three hinges are to be preferred and though brass hinges can be used for half-hour fire resisting doors steel is necessary for one-hour doors. No part of the hinge on which any fire door is hung may be made of combustible material or material having a melting point less than 800°C (Building Regulations E.11(2)). Despite the extra metal, the screws must be long enough to ensure a good grip even after severe charring. To achieve this it may be necessary to provide hinges with extended flaps. In cases where rising butt hinges are permitted, the stop at the top of the door frame must be extra deep to compensate for the amount that has to be cut away from the door to enable it to open.

On the other side of the door some form of spring catch is essential; otherwise the door may be drawn open by the pressure of the heated gases. The lock also may reduce the effect of distortion of the door by the heat generated by a fire. Where a mortice lock is used, an absolute minimum of wood should be cut away. The lock will however still present a zone of weakness; this can be reduced to some extent by the liberal use of intumescent paste to fill in any gaps.

The latch itself should be of steel and should engage the latch plate by at least 13 mm. The latch plate can offer a weakness but this again can be reduced by the use of intumescent paste. Letter plates should not be introduced into fire-resisting doors but where absolutely essential should be as low down as possible. There is at present no satisfactory method of making letter openings fire resistant.

Spring closure

Building regulations require that every fire-resisting door shall be fitted with an automatic self-closing device. Spring closures are frequently fitted to the top of the door but if this is on the exposed side of the door a sudden burst of heat might easily warp or damage the closing device. On no account may any part of a fire-resisting door or its jamb be cut away to accommodate a spring closing device.

Some devices are fitted to the centre rail and the spring is protected within the door jamb. It is certainly an improvement but it also makes for a weakness in the door jamb. The best position is beneath the sill of the door, although this can be very expensive if alterations are required to fit an underfloor spring closer. It must not be possible for the door to overrun the spring and remain open.

One of the best systems is to house the spring closure in a box beneath floor level controlled by a hydraulic hold-open device. In this way not only is the operating system at the coolest point but it is far less likely to be tampered with by people wishing to put it out of action or adjust it. Figure 15.6 shows such a box the size of which is 290 mm by 195 mm and only 60 mm in depth.

Loss of electric power or a pull on the open door opens an electromagnetically controlled valve which releases the door. Loss of electric power can be initiated by an automatic fire detector or by the evacuation signals.

In the system illustrated in Figure 15.7 the hold-open angle of the door can be selected at any point between 85° and 180°. This angle does not have to be predetermined as it is only necessary to open the door to between the two angles, and it remains at that point.

The electric control current is 12 or 24 VDC and approximately 2.8 watts. The manual disengagement force is 60 kN/m^2.

The spring must be powerful enough to close the door from any degree of opening and even be sufficiently powerful to overcome the spring of the door catch. It is particularly important in the case of two-leaf double swinging doors that the springs hold both doors in the exact closed position.

Figure 15.6 Housing of an underfloor spring closer which includes a hold-open device (Dorma Door Closers).

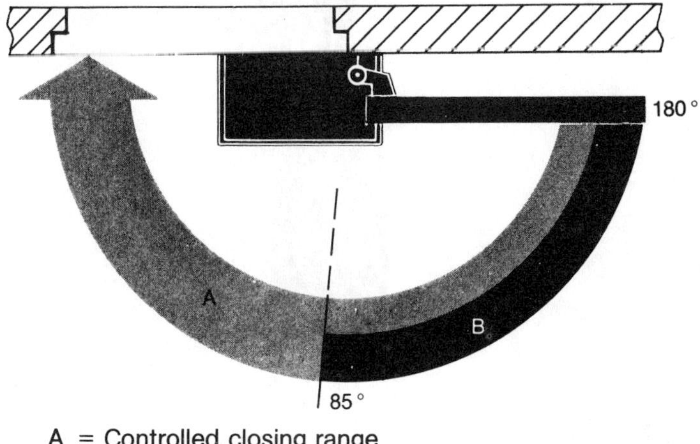

Figure 15.7 Controlled range of the Dorma electro-magnetic/hydraulic underfloor door closer (Dorma Door Closers).

Electro-magnetic door holders

Where it is necessary to hold open a fire door against the door closing device, it is permitted to use an electro-magnetic door holder. There are several ways of holding a door by these means and Figure 15.8 shows a suitable device which can be used with any type of spring closure. It consists essentially of a bolt running through a metallic case. The bottom of the bolt has a friction cap which is held against the floor by a powerful spring. The bolt is held in tension by a detent against the spring. The detent is controlled by a closed circuit solenoid in a 12 or 24 VDC circuit of about 2.4 watts.

When the solenoid circuit is broken the detent is withdrawn releasing the bolt which is then drawn away from the floor by the spring. The door will then be closed by the action of the spring closer on the door. The solenoid circuit can be broken by the foot switch on the bolt casing, by a switch beside the door, by a release action from the fire alarm system or by a smoke detector preferably one on each side of the door.

To set the door holder, the solenoid circuit is switched off and the door opened to the degree required. The solenoid circuit is then restored and the bolt pushed down. The bolt will be caught by the detent and held in position until the solenoid is de-activated. Figure 15.9 shows the angle through which an electro-magnetic door holder can operate.

Figure 15.8 Electro-magnetic door holder (Dorma Door Closers).

A similar type can be used for other purposes where it is required to release a spring-held bolt.

Some difficulty arises for old and infirm people when opening a door against a powerful spring. At the moment there does not seem to be an answer to this problem other than to take advantage of the relaxation of the regulations where electro-magnetic or electro-mechanical devices are fitted. In this case the door remains permanently open. A separate door would then have to be used for the occupants' comfort from draughts. This is no real answer because if the door is released before all the occupants have

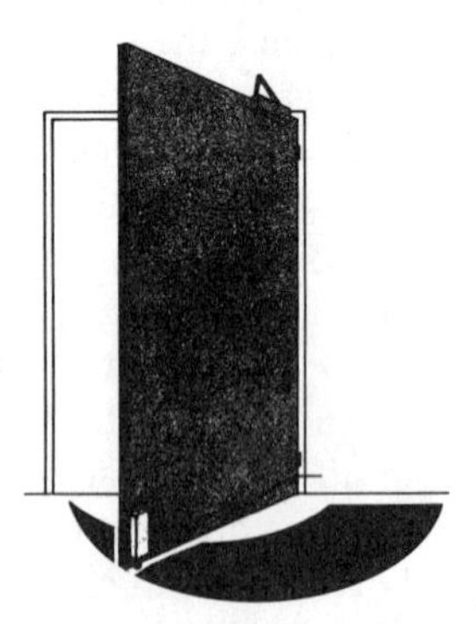

Figure 15.9 Door showing the angle through which an electro-magnetic door holder can operate (Dorma Door Closers).

(white lettering on blue background)

Figure 15.10 The sign of a fire-resisting door (as described in BS 5499 Part 1: 1978).

escaped, as it would if a fire alarm is connected to the release device, the problem is still there.

It is possible that the answer may be found with some form of opening controlled by a proximity device whereby anyone approaching the door causes it to open.

All fire-resisting doors should exhibit the sign 'FIRE DOOR – KEEP SHUT' as shown in Figure 15.10.

FURTHER READING, PART FOUR

Shields, T. J. and Silcock, G. W. H., *Buildings and fire*. Longman, Harlow.

Malhotra, H. L., *Design of fire-resisting structures.* Surrey University Press, Glasgow.

Butcher, E. G. and Parnell, A C., *Designing for fire safety*. John Wiley, London.

Marchant, E. W., *Complete guide to fire and buildings.* Medical and Technical Publishing Co., Lancaster.

Read, R. E. H. and Morris, W. A., *Aspects of fire precautions in buildings.* HMSO, London.

Home Office, *Manual of firemanship, Book 8: Building construction and structural fire protection.* HMSO, London.

Lie, T. T., *Fire and buildings.* Applied Science Publishers, London.

PART FIVE

FIRE AND GAS DETECTION

16

Fire detection

A fire can be detected by sensing the presence of one of a number of emissions from the flames: heat, radiation, smoke, and other combustion products. The heat is usually that transferred to the plume of hot gases rising from the fire, and is detected at ceiling height. Occasionally the detector may be immersed in the flames and there will be a direct transfer of heat by conduction. Flame radiation is not only in the visible region of the spectrum, but also in the ultra violet region and well into the infra red. This is of course all the same kind of radiation, but at different wavelengths: it just happens that our eyes are sensitive to only a narrow range somewhere in the middle. The wavelengths of the radiation from the flames range from about 0.1 micron (UV) to 30 microns (IR). The visible range is between about 0.35 and 0.7 microns. X-rays, gamma rays and cosmic rays all have wavelengths shorter than those classed as ultra violet, and wavelengths longer than the infra red are used for radio transmissions. Flames will radiate over a very wide range of wavelengths, but the strongest and most readily detected radiation is in the range indicated here.

The solid particles in the smoke can be detected by their ability both to obscure and to reflect light. Invisible combustion products, which comprise particles too small to be seen, can be detected by their ability to attract ionized particles. Methods of using these effects in practical detectors are described below.

Most detectors respond only to the conditions existing at the place where they are located: these are therefore point detectors. A line detector may be several metres long and will respond to an increase in temperature at any place along its length. A similar response can be obtained from a beam of light which can be deflected by thermal effects. Surveillance (or volume) detectors are sensitive to radiation and will respond to a flame anywhere within a volume.

When detectors were first developed they were required to give simply a yes/no response: either there was a fire or there was not. Modern systems make use of the additional information provided by detectors that give a continuous response, and which varies with the magnitude of the particular effect to which they are sensitive. These intelligent systems are discussed in the next chapter.

The great value of human fire detectors should never be ignored. People respond very quickly to seeing a flame and to both seeing and smelling smoke. They are also capable of rapid and effective action in suppressing the fire. The means by which their information is conveyed to a detector system is the manual call point, which is a vital component of most systems.

A large number of physical and chemical properties change with increasing temperature. Many of them have been exploited at one time or another in the design of fire detectors. Expansion provides a reliable and usually reversible effect, although the actual movement of a metal sensing element may be very small and will need to be magnified. In an early design, the May-Oatway detector, the sensitive element was a piece of copper wire 2.3 m long. This was held under slight tension so that it sagged in the middle. The expansion due to an increase in temperature caused the sag to increase, and this tripped a mercury switch, one end of which was attached to the centre of the wire.

The expansion of a liquid in the glass bulb of a sprinkler causes the bulb to shatter, releasing the water. The temperature at which this happens depends on the composition of the liquid and on the size of a small bubble which is trapped in the bulb. If a liquid or a gas is trapped in a length of tubing, the expansion can be readily detected by the movement of a diaphragm or bellows, which can be arranged to close, or open, electrical contacts. Such a device can of course be a line detector, and tubes up to 230 m long have been used in this way.

The melting points of metal alloys can be precisely controlled, and low melting point solders are used as the sensitive element in many sprinkler heads. Devices using these solders are simple and reliable, but cannot be re-set. The melting of plastic materials has also been used as a method of detection. A simple line detector comprises two spring wires, insulated by a plastic coating, and twisted together. When the plastic melts, the two wires make contact and complete a circuit.

Differential expansion of metals can be used to provide a larger movement than that of simple linear expansion, and there is then no need for any mechanical amplification of the movement. If thin sheets of two metals with different coefficients of expansion are fixed together (by compression welding for example) they will form a bimctallic strip. When this is heated, one metal will expand more than the other and the strip will curl, with the metal having the greatest expansion on the outside of the curve. This effect is exploited in many simple thermostats, the movement of the strip being used to open and close an electrical contact. One way in which bimetallic strips have been used for fire detection is described below. It is possible also to make a dished bimetallic disc which at a certain temperature will snap over so that it is dished in the opposite direction. This is another device which has been used in sprinklers (the on–off head described in Chapter 23).

Changes in electrical properties have found application in detectors. A thermocouple, comprising a junction of two dissimilar metals, will generate a current when it is heated and can be used as a precise method of temperature measurement. However very little use has been made of this effect, possibly because the current is small and would need to be amplified. Changes in electrical resistance with temperature can be quite large. This provides a useful effect, particularly with certain materials (like glass for example) which show a sudden drop in resistance at a particular temperature. The change in electrical capacitance with temperature has also been exploited, and both effects are used in line detectors which are described below.

The curie point of a ferromagnetic material is the temperature at which it is no longer attracted by a magnet. Detectors have been made in which this effect is employed. Another unusual method of detection depends on the occlusion of gases in solids. This term is used to describe the ability of certain solids, especially the platinum group of metals, to absorb large volumes of gas, and particularly hydrogen. Palladium is the most active of these metals and when it is heated in an atmosphere of hydrogen it will occlude neary 1000 times its own volume of gas. When the metal is cooled, the gas is retained, but it is released if the metal is heated to a certain critical temperature. A line detector can therefore be made from a small diameter tube of palladium, or a palladium alloy, which contains occluded hydrogen. The release of the gas can be detected by the increase in pressure, or by other means.

An obvious temperature-dependent chemical change is the ignition of pyrotechnic compositions. If two spring-loaded electrical contacts are kept apart by a piece of matchhead composition, the circuit will be made when the composition is ignited. A detector based on this design has been used to actuate systems protecting aircraft engines. One of the early sprinkler heads was opened by the combustion of a gunpowder charge, and this was ignited by a length of fuse which was looped up during the day but could be lowered at night.

A practical detector must be robust, reliable and sensitive. It should also have a long life, should not give false alarms, and should be easily tested and readily replaced. It must also be compatible with the system by which the signals are processed. In a modern system this may mean that it provides a continuous indication of local conditions and that it is capable of being interrogated.

Manufacturers have learned to rely on certain basic designs which have been modified and improved over the years. The descriptions below cover examples of most of the available types. Each has its own advantages and disadvantages: these are discussed. In general, the heat detectors are the least expensive and also the least sensitive. A fire of 100 kW or more may be needed to operate a ceiling-mounted heat detector but a smoke detector in the same situation would respond to 10 kW or less. A flame detector could be triggerd by a single small flame, but in practice they could be expected to pick up a 3 kW fire at 20 m.

HEAT DETECTORS

A heat detector will normally be located in the plume of hot gases which is rising from the fire. Heat transfer to it will be mainly by convection. Its sensitivity will depend on the rate at which heat is being transferred to it and the quantity of heat needed to cause actuation. This is usually expressed as

$$\text{Sensitivity} = \frac{C}{HA}$$

where C is the heat capacity of the detector (the specific heat multiplied by the weight, of the sensitive element), H is the heat transfer coefficient (the rate at which heat can be transferred through the surface of the element), and A is the area of the sensitive element. Sensitivity can be expressed as a time constant, the time to actuation in a specified situation. Typical times for heat detectors can be as low as 20 to 40 seconds, and for conventional sprinklers, about two minutes. The important point is that a sensitive heat detector will have a large surface area and a small thermal mass. In practice its response time will be greatly affected by its height above the fire. Research on the properties of a rising plume has shown that the heat output of a fire which is needed to produce the same temperature at different heights above the fire is proportional to $(\text{height})^{2.5}$. This means for example that if we measure the heat output from the smallest fire which will operate a detector 4 m above it, and we then increase the height to 8 m, the heat output needed to operate the detector will be 5.7 times greater.

If a detector has been designed to operate at say 58°C (one of the standard ratings for sprinklers), then if there is a very slow increase in temperature the device will actuate as soon as this temperature is reached. On the other hand, if there is a rapidly developing fire the temperature of the gases surrounding the detector may be several hundred degrees at the moment when it operates. In this case a detector which responded to the rate of rise of temperature rather than to a static value would give a more rapid actuation. This can be achieved by comparing the response of two elements, one of which heats up more slowly than the other either because it has a greater thermal mass, or a lower conductivity, or is thermally insulated. Some typical designs are described below. A device like this might however fail to detect the presence of a smouldering fire, because the rate of rise of temperature would be too low. Because of this it is usual to arrange for the detector to respond to a fixed temperature no matter how slowly it is achieved. This provision is a requirement of UK and US standards, but pure rate of rise detectors are used in France.

Even the simple May-Oatway detector described above is more sensitive to a rapid increase in temperature. This is because the steel beam which supports the copper wire has a greater thermal mass and a lower coefficient of expansion.

The advantages of the heat detectors are their reliability and cheapness. The main disadvantage is the lack of sensitivity: a fire has to be quite large

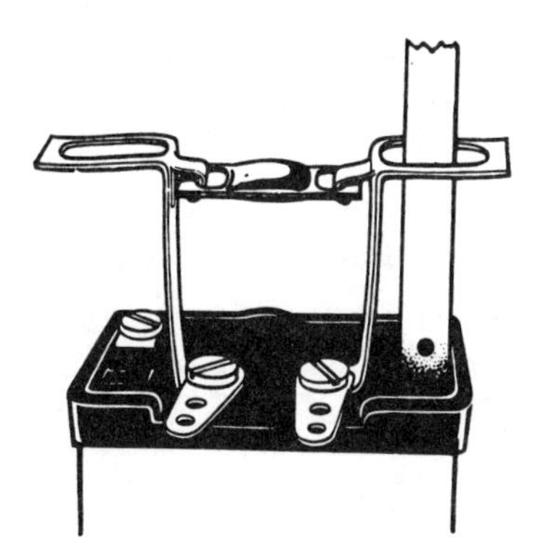

Figure 16.1 Fixed temperature fusible type detector. (Sound Diffusion Ltd.)

before it can be detected. Also, with the exception of the last one described below, they are simple yes/no devices and do not have a continuous output.

Fixed temperature fusible detectors

A typical design is illustrated in Figure 16.1. The fusible link comprises two strips of metal held together by solder (usually with a melting point of 68°C). This is held under tension by the two vertical phosphor-bronze springs. When the solder melts, these spring apart and break the circuit. Alternatively, a circuit can be made instead of being broken, by connecting it to the vertical contact, which has a knife edge, and the right-hand spring. Devices like this do not re-set after operation but in some the fusible link can be replaced. In one design the spring-loaded contacts are made of thin wire and are held together by a very small blob of solder. This will clearly have a low thermal mass and therefore a fast response. In another device the melted solder falls away from two electrodes to break a circuit, but drops down on to other contacts to make a circuit.

Detection by expansion

A very robust design is illustrated in Figure 16.2. The tubular outer case is the sensitive element and is in direct contact with the hot gases. When it expands the contacts on the bow mechanism are pulled together and the circuit is completed. This again is more sensitive to a rapid increase in temperature because there is less time for heat to penetrate to the two bows and to cause them to expand.

A bimetallic device is illustrated in Figure 16.3: a dished bimetallic disc, again exposed directly to the atmosphere, is designed to snap over to its

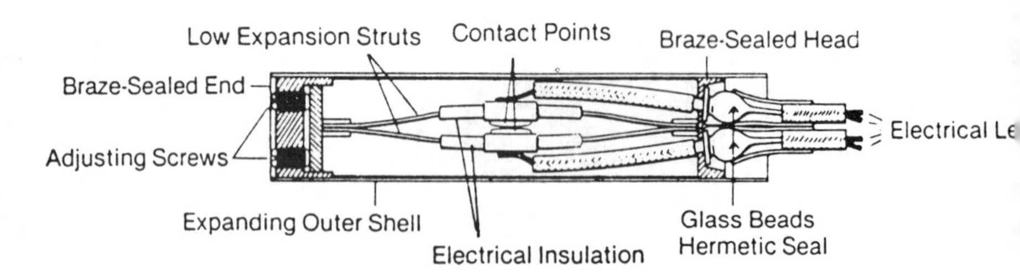

Figure 16.2 Cartridge type heat detector. (Fenwal International)

Figure 16.3 Bimetallic disc heat detector.
(Chloride Pyrotector Ltd.)

alternative position at a fixed temperature (in this design either 57° or 93°C). The movement can either make or break a circuit depending on the model. The disc will revert to its original condition as it cools and can be used repeatedly. It can also be tested quite readily. A hair dryer is often used for this purpose, a note being made of the time taken for the disc to snap over.

A bimetallic rate of rise detector

One of these devices is illustrated in Figures 16.4a and b. In the cut-away drawing in (b) it will be seen that two bimetallic strips have been formed into an incomplete circle. They are both cut from the same sheet so that their characteristics are the same. The metal with the highest coefficient of expansion is on the outer surface, so that as the temperature increases the free ends of the two strips will move inwards. It will be seen that they both carry electrical contacts, and that there is a slight gap between the contacts. If both strips move together, contact will not be made. It will be seen however that the lower strip is covered by a thin aluminium cup but the upper one is surrounded by a plastic moulding (of very low thermal conductivity). If there is a very gradual increase in temperature both strips will receive the same amount of heat and will move together.

Figure 16.4a External appearance of rate of rise heat detector. (Chubb Fire Security Ltd.)

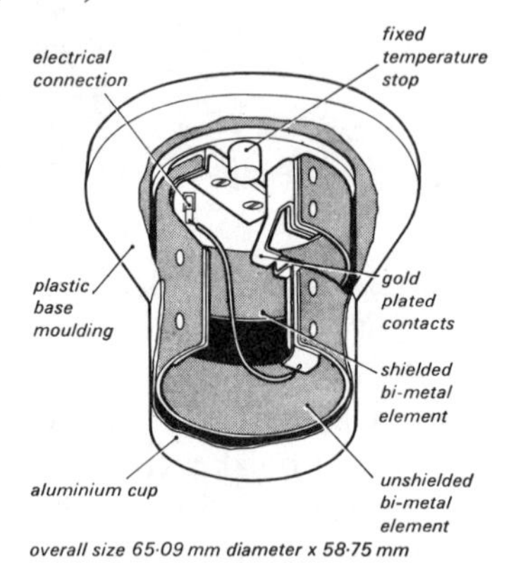

Figure 16.4b Internal layout of rate of rise heat detector. (Chubb Fire Security Ltd.)

If there is a rapid rise, the lower strip will be heated more rapidly than the upper one; it will bend more and the contacts will be closed. The fixed temperature response is provided by the stop shown in the figure. If both strips move together in response to a slow rise in temperature, the upper one will eventually come against the stop. The lower one will continue to move and the contacts will be closed.

Thermistor rate of rise detector

A thermistor is a semiconductor with a negative temperature coefficient: its electrical resistance decreases with increasing temperature. The effect can be quite marked, with the resistance changing from about 100 k ohm at 20°C

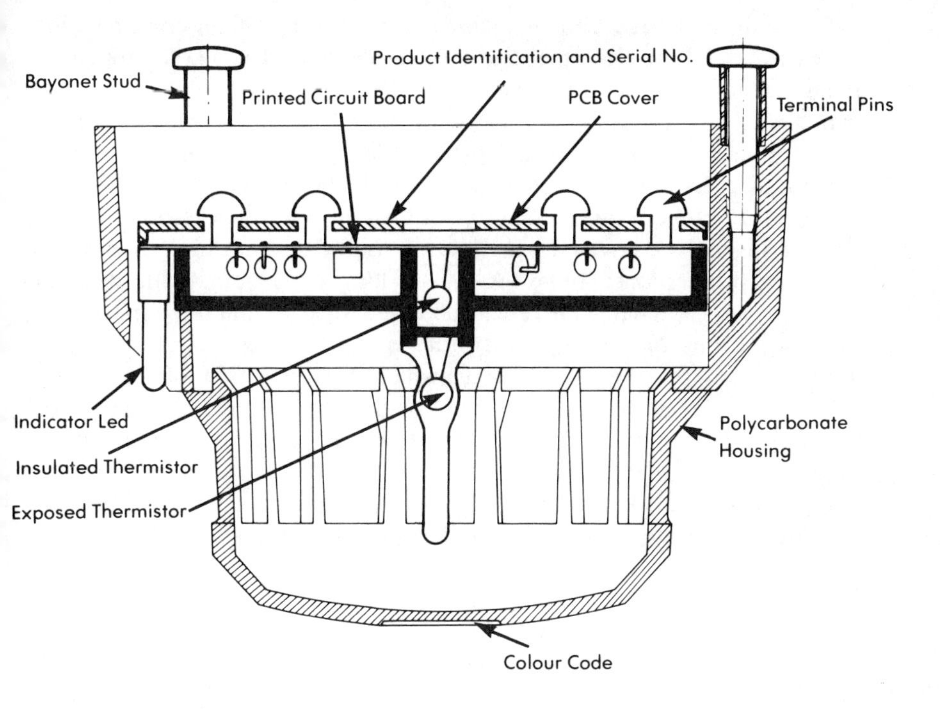

Figure 16.5 Thermistor rate of rise heat detector.
(Apollo Ltd.)

to only 100 ohm at 100°C. In Figure 16.5 a matched pair of thermistors is used to provide a rate of rise sensitivity. As in the previously described detector, one is exposed to the atmosphere and the other is thermally insulated. The exposed one is protected from contamination by a PVC sheath which is sealed to the plastic bases. The air can circulate freely around this thermistor because of apertures in the housing. The circuit within the detector is designed to alarm when the difference in the resistances of the two thermistors exceeds a present value. A fixed temperature setting, for the detection of fires which cause only a very slow rate of rise, is provided by a resistance which is in series with the insulated thermistor. It has no significant effect at low temperatures, when the resistance of the two thermistors is high, but as the temperature increases a point is reached at which the difference between the resistance of the exposed thermistor and the combined resistance of the insulated one and its added resistor is sufficiently high to trigger the alarm.

The indicator LED shown in the figure is a red, light-emitting diode which is turned on when the detector has alarmed. The whole unit can be plugged into a bayonet fitting base in which the terminal pins make contact with the system circuitry. The range of detectors of this particular design include three which conform to the three grades of BS 5445 Part 5 (discussed later) and two which respond to higher temperatures, the actual temperature range being 60°C to 95°C. One of two which respond at 60°C has the highest sensitivity to the rate of rise of temperature. At a rate of 1°C/min the response time can be as long as 32 min, with detection occurring at 58°C. When the rate is increased to 30°C/min the response time drops to 1 min 20 s, at a temperature of 65°C. These detectors can be checked easily by using a hair dryer which delivers a flow of warm air at 40°C to 50°C. If this is 2 or 3 cm from the detector with the high rate of rise sensitivity it should respond in 5 to 10 seconds and re-set in about one minute.

Detectors of this kind can be used to give a continuous signal: they will provide an accurate measurement of the ambient temperature.

Line detector: liquid expansion

Figure 16.6 shows a liquid-filled line detector. The metal tube containing the liquid has a bore of about 2 mm diameter and can be up to 230 m long. Each end is connected to a diaphragm, but the route to the left-hand one in the figure is through a small orifice. A slow increase in temperature causes both diaphragms to move together and contact is not made between the two electrodes. A rapid rise causes the right-hand diaphragm to expand more quickly and contact is made. Again with this device there is a stop which ensures that there is an alarm at a fixed temperature if a smouldering fire produces only a slow rate of rise. The three-way cock, and the cylinder on the right of the diagram, are used to pressurize the tube in order to check the mechanism.

Line detector: pressurized tube

This detector comprises a length of plastic tubing containing a gas under pressure. If any part of the tube is heated sufficiently it will burst and release the gas. The loss of pressure is sensed by a pressure switch.

Line detector: electrical

'Firewire' (a Graviner Ltd trademark) was developed originally for use in aircraft engine systems and is therefore both robust and reliable. It

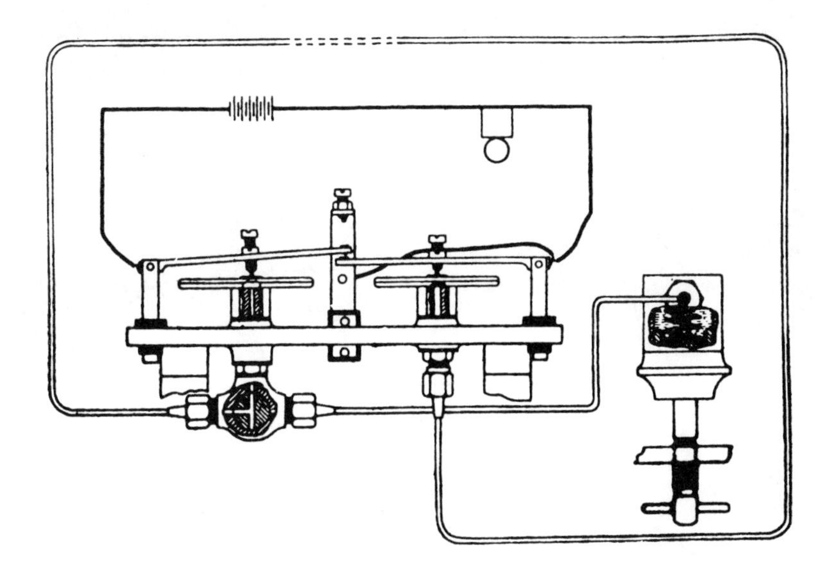

Figure 16.6 Liquid-filled line detector.
(A F A Minerva Ltd.)

comprises a thin stainless steel tube containing an axial wire. The gap between the wire and the tube is filled with a semiconductor material which is a glass. There is a gradual drop in electrical resistance as the element is heated, and a large reduction as the melting point of the glass is approached. However, it was found in practice that there was also a change in capacity with temperature and that this could be more readily detected than the resistance change. Both ends of the detecting element are connected to a control box, so that detection is still possible if the 'Firewire' is cut. Indeed because of the method of manufacture, the element will continue to detect even when part of it has been hammered flat.

A line detector of rather simple construction is illustrated in Figure 16.7. Two spirally wound conductors are separated by the temperature-sensitive material, and the cable is protected by an outer protective insulation. The total diameter is 2.74 mm and the construction ensures good flexibility. This detector is normally supplied in lengths of up 200 m. The temperature-sensitive material again has a negative coefficient, but this is linear over a wide range of temperature and is such that a 25°C change in temperature will result in a decade change in resistance.

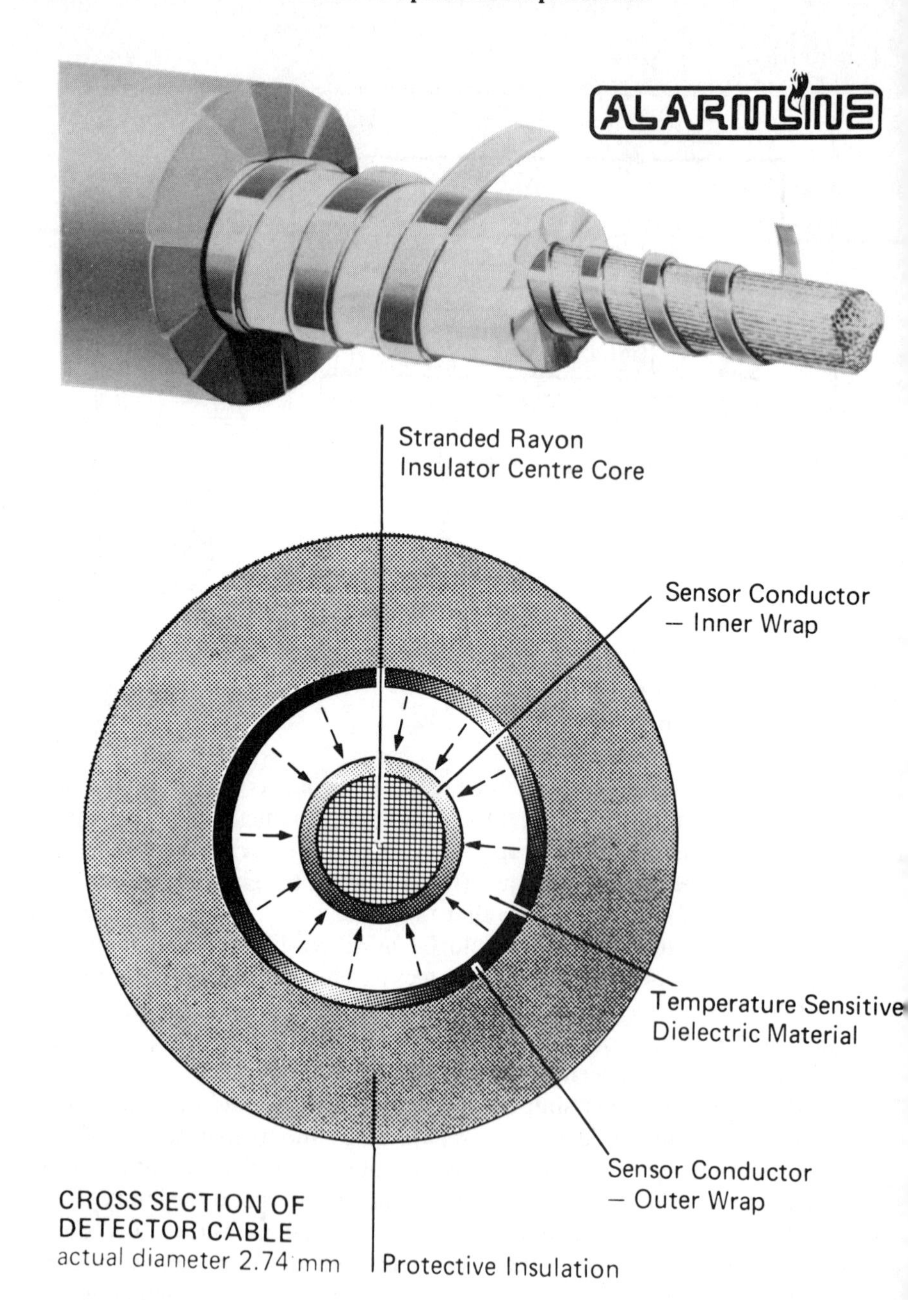
ALARMLINE
Stranded Rayon Insulator Centre Core
Sensor Conductor – Inner Wrap
Temperature Sensitive Dielectric Material
Sensor Conductor – Outer Wrap
Protective Insulation
CROSS SECTION OF DETECTOR CABLE
actual diameter 2.74 mm

Line detectors: general

The pressurized tube device described above will respond only when some part of it is heated to a critical temperature. The other devices are integrating: they will provide the same reponse if a long section is heated to a relatively low temperature or a short length is heated to a high temperature. They can therefore be expected to detect both smouldering combustion (which will cause a gradual increase in temperature in the protected enclosure), and the direct contact of a flame. This effect should be remembered if a line detector of this kind is installed in an area in which there could be a substantial increase in ambient temperature. For example, an extinguishing system protecting a furnace room was actuated by a line detector on the day when the air-conditioning system was shut down.

The two tubular devices provide only a yes/no signal but the electrical detectors give a continuous output. They can be used for temperature control in process plant.

Optical beam detectors

The plume of hot air rising above a fire has a different refractive index from that of the colder air surrounding it. The presence of fluctuations in the refraction of transmitted light (or schlieren) produces a shimmering appearance of objects viewed through the plume. This effect can be used to detect the presence of a fire and is included in this section because it is yet another of the physical changes produced by heat. If a beam of light is projected from one end of a room to a photoelectric device at the other end, the presence of schlieren along its path will be detected as a flicker or as a complete displacement of the beam. Visible light is not used because the detector would be swamped by the normal illumination of the room, and infra red radiation is usually employed instead. This can conveniently be a laser beam, which overcomes the difficulty of focusing normal radiation. The device will also respond to the presence of smoke. It is sometimes convenient to have the radiation source and the detector in the same place rather than at opposite ends of the enclosure, and this can be done by using a reflector. With either arrangement there can be problems if the walls to which the devices are attached are likely to vibrate. False alarms can also result from the presence of intruders (even a bird), but the device does have the advantage of covering a considerable volume with a single unit.

SMOKE AND COMBUSTION PRODUCTS DETECTORS

Combustion processes release enormous numbers of solid and liquid particles, most of which appear as smoke. The particles range in size from about 0.5 millimicron to 10 microns. Those which are less than 0.3 micron are too small to scatter light and are therefore invisible. Particles which are released at temperatures below the fire point of a fuel (that is, in the pre-ignition heating) are generally in the range of 0.5 to 1.0 millimicron. Those released by smouldering combustion range from about 0.1 to 1.0 micron. These ranges are important because a detector which operates by optical processes will fail to detect either the pre-ignition particles or the invisible part of the smouldering emissions. It will also be relatively insensitive to the visible particles from smouldering because its sensitivity increases with particle size (because the size is nearer to the wavelength of the light which is being used). Fortunately ionization detectors are most sensitive to the smaller particles, so it is possible to cover a wide range of combustion products with the different types of detector. However, the smallest particles do tend to coalesce into larger clumps with time, so that 'old' smoke which has followed a tortuous route from a smouldering fire may have particles in the visible range. Conversely, large droplets will evaporate and become smaller with time.

At least one manufacturer supplies a detector which contains both an optical and an ionization device, thus being sensitive to different types of smoke. However, neither of these detection methods is suitable for fires of fuels which burn without producing smoke. Alcohol is an example of such a fuel and ionization detectors have failed to respond to whisky fires. It is normal practice in hazardous areas (offshore for example) to install pairs of detectors, one of which responds to smoke and the other to a rate of rise of temperature.

The sensitivity of the smoke detectors is adjustable and in general the cleaner the atmosphere the higher the sensitivity to which they can be set. Cigarette smoke has little effect on ionization detectors because the particles have time to coalesce into large clumps. This smoke will however be readily detected by optical devices. There may be problems with both detectors in areas where food is being cooked.

Optical detectors

Two kinds of optical smoke detector are available: those which measure obscuration and those which measure light scatter. Obscuration refers to the ability of solid objects in a beam of light to reduce the amount of light which arrives on a target (in this case a photoelectric device). For large objects the proportion of light which is lost (that is, the amount of obscuration) depends simply on the total cross-sectional area of the objects. As we have seen the situation is more complicated when the solid material comprises particles less than 10 microns because their ability to obscure light decreases as the size is reduced. In a practical detector design, a beam of light is emitted from one side of a light-tight box and is focused on to a photocell on the opposite side. Smoke-laden air entering through a labyrinth obscures some of the light and when the obscuration reaches a pre-set value the detector alarms. Early designs became unreliable after a few years because the sensitivity of the photocell diminished with continuous use. Also, dirt deposited on either the light source or the cell could cause false alarms. The useful life of the detector was also limited because filament bulbs were used as the light source.

If a beam of light is directed into a cloud of particles some of the light will be scattered sideways to the beam. Particles which had previously been invisible may be revealed in this way, an effect first noted by Tyndall. The amount of scatter is not the same in all directions and in practice there is an optimum angle from which the maximum amount of scattered light can be observed. In a smoke detector the photocell is in a position where it receives no light from the source, but the maximum of scattered light from smoke particles. The cell is therefore used only when smoke is present and its sensitivity is unlikely to deteriorate in the normal life time of a detector. Also in modern designs the filament bulb, with a life of perhaps three years, is replaced by a light-emitting diode, with a life of at least 100 years. Its life can be extended even further if it is used to produce a series of flashes. A flash tube can also be used in this way. The accumulation of dirt is still a problem and regular cleaning is a necessary part of a maintenance programme.

Figure 16.8 shows a modern light scatter detector. The complete device is shown in (a) and the optical system in (b). It will be seen from (b) that the light source is an LED and that the lens system produces a hollow cone of light. The silicon photocell is positioned in the dark area at the centre of the cone where it receives light only when smoke enters the detector through the labyrinth. The LED is driven to produce short pulses of light at about 3 kHz. The alarm is initiated when the photocell has received light from three pulses. The unusual shapes which have been moulded into the base of the housing,and can also be seen in (a), are to reduce the internal reflection. From (a) it can be seen that this detector also has an external red LED to indicate when it has alarmed. There is a very convenient arrangement for testing this detector. It will be seen that there is a test plunger in the drawing.

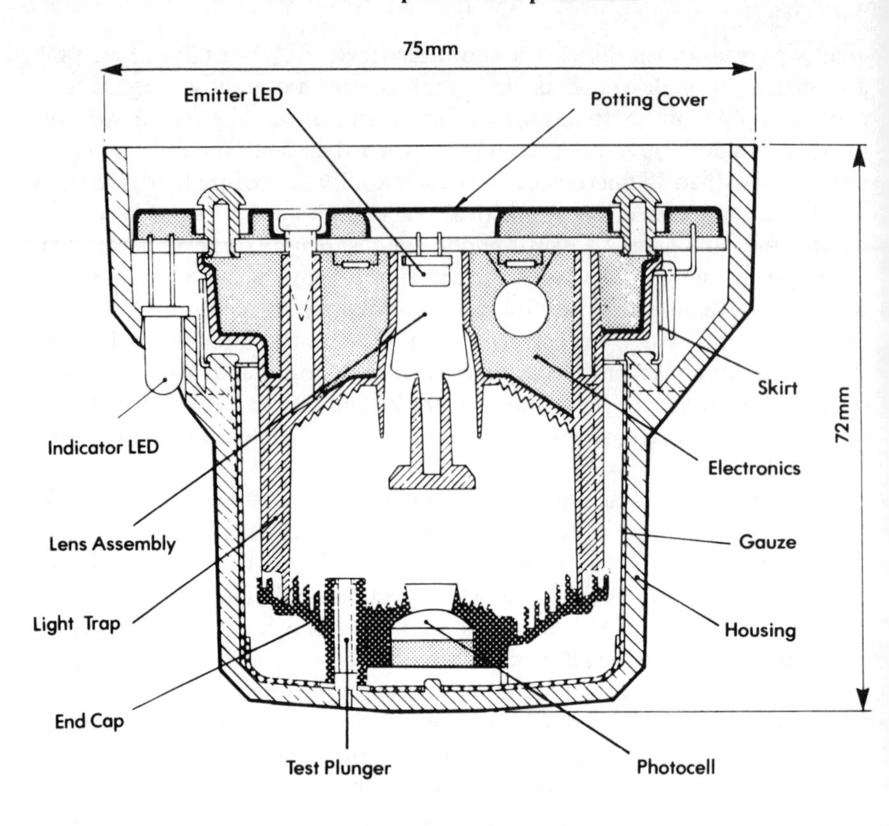

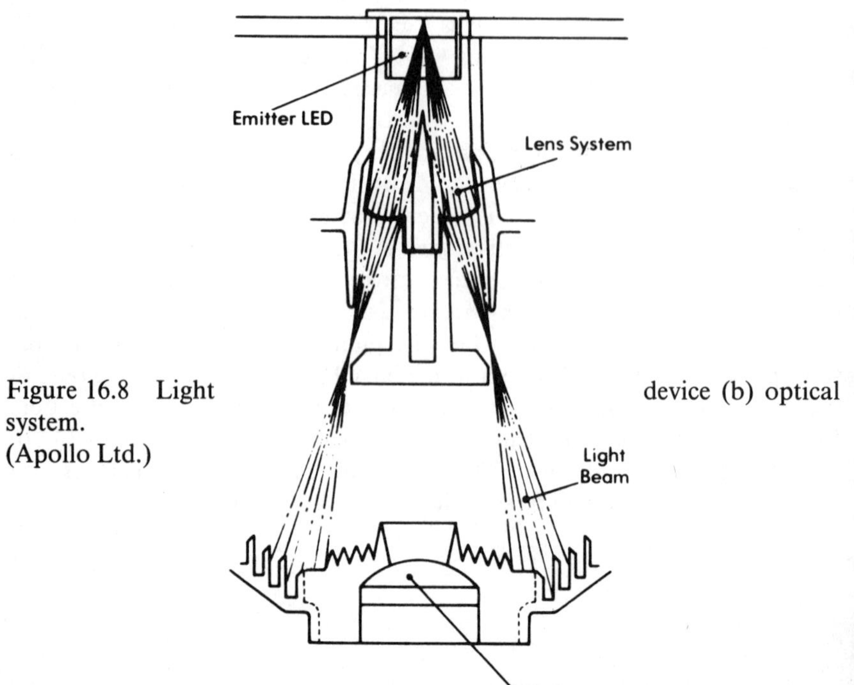

Figure 16.8 Light device (b) optical system.
(Apollo Ltd.)

When this is pushed in by a special tool sufficient light is scattered to activate the photocell.

Detectors like this are more sensitive to grey smoke (produced by smouldering fires) than to black smoke. This particular design will respond to grey smoke with an obscuration of 3 per cent/m. (There are a number of ways of expressing obscuration, but this is self-explanatory: the loss of 3 per cent of the light in its passage through 1 m of smoke.)

Ionization detectors

This is really a misnomer: these devices do not detect ionization. They are sometimes called 'combustion gas detectors' and this is not correct either. As we have seen, they are sensitive to invisible particles which are released particularly in the initial stages of a fire. Their operation depends on the ionization of a small volume of air in a chamber by a radioactive source. This is an alpha emitter, usually americium or radium. Ionization splits molecules in the air into particles (ions) which are either positively or negatively charged. If a potential is applied between two electrodes in the chamber, the positive ions will be attracted to the negative electrode and the negative ions to the positive one. This movement of ions constitutes an electric current, which can be measured. When smoke particles enter the chamber, ions are captured on the surface of the particles and the current flow is reduced. (The specific surface of small particles is greater than that of an equal weight of large particles, hence the detector's greater sensitivity to the former.) The actual current flows are extremely small, between about 1 and 100 picoamps. The development of the system became possible only when Meili, a Swiss physicist, had devised a cold cathode tube which was capable of amplifying these currents. Modern detectors use solid state circuitry but the currents being detected are still very small. It is for this reason that false alarms can be caused by radio transmissions. Detector systems in public places have been triggered by people using portable cellular telephones. The detectors can also give a false alarm if the radiation from the source is reduced, and this can happen if dirt or moisture is deposited on its surface. The ionized air can be displaced from the chamber if the device is located in a current of air, and this will also cause a false alarm.

Because of the low current consumption of these detectors they can be operated for long periods by a small battery. This makes them ideal for domestic use and many reliable but relatively inexpensive models are now freely available. Their use in houses is compulsory in some states in America.

The detector shown in Figure 16.9 is intended for use in a system from which it obtains its power supply,. It will be seen that the radioactive source is double sided and that it serves two chambers. The lower one is open to the atmosphere and is the sensing chamber. Air flow to the upper reference chamber is restricted and its purpose is to compensate for changes in atmospheric conditions. The detector alarms only when the current flow in the sensing chamber is less than that in the reference chamber. The electrodes may not be immediately obvious but in the sensing chamber one is supporting the radioactive source and the other forms the shell of the chamber. A red LED is provided in the plastic housing. The sensitivity of these detectors can be checked either by an aerosol spray or by a smoke generator in which an oil is pyrolysed on a hot wire.

The radioactive source in this detector is a foil disc 10 mm in diameter containing two thin layers of americium 241, each protected by a thin layer of gold. The gold seals in the radioactive material but still permits the release of radiation. The alpha radiation cannot penetrate the housing of the detector (and as noted above it is also absorbed by a thin layer of dirt or moisture). Some low energy gamma rays do pass through the housing but at 100 mm away the radiation is only a fraction of the background radiation and well below the permitted levels. Ionization detectors are exempted from the requirements of the UK Radioactive Substances Act 1960, the Health and Safety at Work Act 1974, and the Radioactive Substances (carriage by road) Regulations 1974 (which are based on the International Atomic Energy Regulations). The detectors do not constitute a health hazard in normal use, but it would be inadvisable for example to store large numbers of them within a metre or so of a normally occupied area. They are exempted from the sections of the Radioactive Substances Act which regulate the disposal of radioactive waste, but sensible precautions should still be taken. They should be sent for disposal either to the manufacturer or to someone who is authorized under the Act to dispose of americium 241 waste. A local authority can also dispose of them with other refuse provided that there is not more than one detector in any 0.1 m^3 of refuse.

FLAME DETECTORS

The radiation from a flame will travel to a detector at the speed of light. Flame detectors therefore have the fastest possible response time. They are also very sensitive and will respond to quite small flames. However it is necessary for the detector to see the flame (although a certain amount of reflection is possible with infra red devices). Flame detectors cannot respond to smouldering combustion, unless a sufficiently large glowing surface is

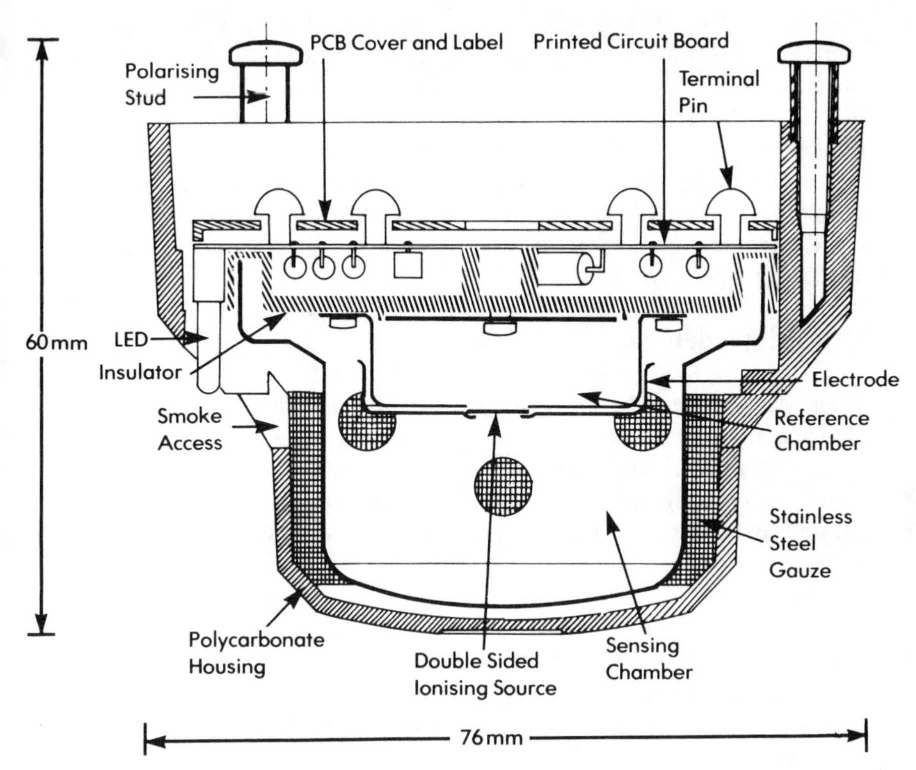

Figure 16.9 Ionization smoke detector
(Apollo Ltd.)

exposed to an infra red device. False alarms can be caused by radiation which has not originated in a fire, burning and welding operatons being a frequent cause. The main advantages of these detectors are their almost immediate response, and the fact that they are surveillance devices: a single unit can protect a large volume.

Sunlight contains the complete spectrum of radiation from the far ultra violet to very long infra red, and it is necessary for flame detectors to ignore this radiation while still being able to respond to fire. Fortunately some solar radiation is absorbed as it passes through the atmosphere and this makes sunlight sufficiently different from flame radiation for some detectors to discriminate between them. Some manufacturers prefer to use a different characteristic of flames, that of flicker. The detector ignores continuous radiation but responds to a flicker at the right frequency. The frequency varies from about 15 Hz for a pool fire with an area of 10 cm^2, to 5 Hz at

100 cm^2, and to around 2 Hz for 1000 cm^2 and above. With large flames there are also smaller fluctuations due to local areas of turbulence around the edges. The detectors are normally set to respond to frequencies between 5 and 15 Hz. It is however possible for false alarms to be caused by reflections from rotating machinery. Detectors mounted outdoors have responded to stellar scintillation, to the movement of trees and to car headlights reflected from railings. It is possible to overcome these effects to some extent by blanking out from the view of the detector those areas which are likely to cause trouble.

Ultra violet detectors

These detectors can be made solar-blind by being operated in a region of almost complete atmospheric absorption of ultra violet radiation. The cut-off is at 0.29 micron and the detectors are sensitive only to radiation at wavelengths less than this. Unfortunately only a small fraction of the energy emitted by flames is in this region and it is necessary to use a very sensitive photocell. A Geiger-Muller tube is usually employed. This comprises a glass envelope (transparent to the desired radiation) containing two electrodes, and filled with a suitable gas. The electrodes are close together and a voltage is applied between them. The metal of the electrodes is selected so that when a photon of radiation at the right wavelength is absorbed, an electron is released. As this travels to the opposite electrode it releases other electrons by colliding with the molecules of the gas. This produces an avalanche effect, and the amplification can be as high as 10^{12}.

When a small Geiger-Muller tube is mounted behind a UV window in a suitable housing it can cover a conical volume with a solid angle of 90° or more. Typically it will detect a 3 kW fire at 20 m. Radiation from lightning strokes and other transients can be ignored by incorporating a time delay in the circuit. It is however necessary to mask or disconnect the detector before any welding is attempted in its vicinity. The device will also respond to radioactive sources or X-rays used for inspecting welds. It is important that the window of the detector is kept clean because UV radiation can be absorbed by a thin film of oil (which is why oil is used to prevent sunburn). The sensitivity of the detector can be checked by incorporating a UV source and arranging for some part of the housing to reflect this radiation back through the window. The source can be conveniently actuated from the control panel. UV radiation is also absorbed by smoke and fog, and the response of a detector will be affected by their presence. Some solvent vapours will also absorb UV radiation.

Infra red detectors

It is possible to enable an IR detector to discriminate between solar and flame radiation by comparing the relative intensity in two extremes of the IR range: a band between 0.6 and 0.9 microns and another between 7 and 30 microns. Sunlight contains much more radiation in the first band than in the second; electric heaters and other hot surfaces radiate mainly in the second band, and flames radiate about equally in both.

Infra red radiation from the sun is absorbed strongly in the atmosphere by carbon dioxide molecules and by water vapour molecules. The wavelengths affected are between 1.5 and 1.8 microns and between 4.2 and 4.5 microns respectively. These are also the wavelengths at which the molecules will emit radiation when they are heated. Since these are the main products of combustion and will always be released at flame temperature, photocells sensitive to these wavelengths are the obvious choice for an IR detector. However such detectors would then respond to the gases released from car exhausts and chimneys. It is for this reason that most IR detectors depend on flicker frequency to identify flame emissions.

The detector shown in Figure 16.10 has a 90° field of view and responds to flicker between 1 and 10 Hz. The photocell is behind the upper window and an LED is behind the lower one. This particular unit is approved by BASEEFA for use in explosive atmospheres (see Chapter 3). In some tests in which a detector was mounted 4 m from the ground it responded to a petrol fire of 0.05 m^2 at 15 m, 0.19 m^2 at 30 m, and 0.37 m^2 at 40 m. The response time is set at 3 seconds so that transient effects are ignored.

A detector like this will continue to function when the window is contaminated with oil and dirt and will also detect a fire in the presence of smoke or fog. It will not respond to the 'invisible' flames of alcohol, ammonia and hydrogen.

In another design of IR detector the photocell is mounted above a rotating reflector, which enables the device to scan a large volume. The presence of an IR source causes the device to lock on to its location and this setting is held for a few seconds before the alarm is given. These detectors are often mounted on a pole outdoors, where there is the possibility of spurious signals. Areas where these are likely to originate can be masked on the protective transparent cover of the device.

OTHER DETECTION METHODS

Research on new methods of detection continues in many countries. Improvements to existing methods are frequently made, but the development

Figure 16.10 Infra red flame detector.
(A F A Minerva Ltd.)

of an entirely new method is a difficult and expensive undertaking. One problem is that a manufacturer will be reluctant to introduce a new detector which will be competitive with his current range, although there might be no objection to one which would be complementary to the range. Of the methods outlined below, only the first two are commercially available.

Sampling methods

In some circumstances there is an advantage in taking samples from the atmosphere in a risk area and checking them for smoke or combustion products. A single detector can be used to check a number of points in turn by using a rotating multi-way valve connected to say 12 sampling lines. Air is drawn through all of them by a vacuum pump and they are connected sequentially to the detector. This method can be used where for example vibration levels are very high, or where the environment is hot (the samples then being cooled before they are passed through the detector). In designing such a system it must be remembered that conventional ionization detectors may false alarm in an air stream. One method of detection which has been used in systems of this kind is a nephelometer. This is a delicate instrument which measures the turbidity of fluids by the scattering of light. This is exactly the same principle as that employed in a conventional light scatter smoke detector, but a very intense light source is used (the discharge of a xenon flash tube) and a very sensitive receiver.

A system based on this method is commercially available. This is the VESDA detector (very early smoke detecting apparatus). The sample is drawn through a light-tight chamber and is subjected to the intense radiation of a xenon flash tube. The scattered light is detected by a very sensitive solid state receiver. The device can detect smoke with an obscuration of 0.1 per cent/m. A single detector can cover an area of up to 2000 m^2.

Where the entire atmosphere in an enclosure is passed through an air-conditioning system it is possible to monitor the air in the exhaust ducting. In this case however the smoke may be diluted before it reaches the detector, and the sensitivity of the system will be reduced.

Carbon monoxide detectors

Carbon monoxide is produced in most fires and is also an initial product of surface oxidation. The presence of both fires and smouldering combustion in coal mines can therefore be detected by monitoring the air leaving the mine for the presence of carbon monoxide. This is done continuously by passing a sample of the air through an infra red absorption detector. This would not be a suitable way of monitoring the air discharged from the ventilating system of a building because of wide fluctuations in the concentration of carbon monoxide. This can be due for example to cigarette smoke or to the presence of exhuast gases from vehicles.

Small particle detectors

Methods are available for the detection of very small particles, less than 0.1 micron, which could therefore be used for combustion products detection. One is a particle-counting method in which the air sample is ionized by a radioactive source, but the positive ions are then removed. The negative ions accumulate as a static charge, the voltage of which provides a measure of the particle concentration. In some laboratory tests the device was able to detect alcohol fires (which the other particle detectors cannot). It also responded to the burning of a 13 cm^2 piece of paper in a room, and to the heating of a 13 mm length of PVC insulated wire.

A second method of detection depends on the ability of very small particles to act as nuclei for the formation of water droplets in a saturated atmosphere. The drops can then be detected by the normal light scatter method.

Semiconductor gas sensors

Smoke is released from a fire because of incomplete combustion. It contains combustible gases as well as solid particles. These gases can be detected by a suitable device (a Tagachi detector for example). This is a semiconductor composed of a number of metal oxides, and shows a large decrease in electrical resistance in the presence of fuel gases. Unfortunately it will also respond to vapours from things like furniture polishes, solvents, and fly sprays.

Ultrasonic detector

The velocity of sound in air varies with the pressure, temperature and humidity. The velocity through other gases, for example methane, is quite different from that in air. Changes in velocity can be detected by transmitting frequency-modulated ultrasonic beams and then comparing the received (or reflected) frequency relationships with those of the transmission. This provides a sensitive and continuous measure of the time taken for the beam to travel through a particular volume. Changes occurring in this volume due to the presence of smoke, flames, hot air, vapour or even intruders would provide a signal. An entire room could be flooded with direct and reflected ultrasonic transmissions and the received signals could be processed in a small computer. The whole system could then go through a learning process so that the computer could be programmed to provide warnings, and take appropriate action following the detection of overheated plant, smoke, fires, gas leaks or intruders. (See British Patent 1 523 231 'Gas Detection', 31.8.78 and US Patent 4, 119, 950 – R. J. Redding). The method should also be compared with the laser beam gas detector system described below. It is to be hoped that further development will be attempted of this very attractive method.

17

Flammable gas detectors and standards

If a flammable gas is leaking into an enclosure, mixtures with air may be formed which are within the limits of flammability, and which may therefore explode. If action is to be taken in time to avert this, the gas leak must be detected before a lower limit mixture can be formed. Neither of these statements is strictly true because mixtures covering the range of 0 per cent to 100 per cent gas must be formed very quickly in the immediate vicinity of the leak. What we are concerned with is the formation of a sufficiently large volume of an explosive mixture to constitute a hazard, and in practice this means that a gas detector protecting the volume must respond to a concentration which is only a fraction of the lower limit of flammability.

GAS DETECTION

Pellistor detectors

Most of the available gas detectors depend on measuring the heat released from the oxidation reactions between the gas and the air. The reaction can proceed, even though the fuel concentration is below the lower limit, if the mixture is heated to a sufficiently high temperature (although not necessarily to the flame temperature). It can proceed at lower temperatures in the presence of a catalyst (a substance which assists a chemical reaction but is unchanged at the end of the reaction). Early detectors depended on the ability of a heated platinum surface to catalyse combustion reactions. A small spiral of platinum wire was heated in the detector to about 900°C by an electric current. Since the fuel/air reactions took place at the surface of the wire, much of the heat of reaction was transferred to the wire. This increased its temperature and also its electrical resistance. This change in resistance reduced the current flow and this was detected by a Wheatstone bridge. This simple but sensitive device was developed by Sir Charles Wheatstone in the early years of the last century. It was much used for fault finding in the telegraph system.

The platinum catalyst had a short life because metal was lost by evaporation, particularly from areas where reactions were taking place, and the wire soon burned out. Catalysts operating at lower temperatures were developed by encapsulating the heater wire in a ceramic bead (perhaps only 1 mm in diameter) and coating this with a catalytic material. This often now comprises a layer of thorium oxide on to which palladium has been deposited. This is effective at a temperature of 400° to 500°C and is usually called a pellistor. In the detector head a pellistor is connected in series with a ceramic bead assembly which is identical except that it has no catalytic surface. These form two arms of the Wheatstone bridge (which is still used in most detectors), and this arrangement ensures that changes in ambient conditions do not affect the response. In the circuits normally used with pellistor detectors a change of 1 per cent in the concentration of gas results in a signal change of about 25 mV, which is readily detected. A typical design of detector is shown in Figure 17.1. The purpose of the two discs in the figure is explained later.

Although the pellistor has a longer life than the original platinum coils, its sensitivity gradually decreases with time, due mainly to the formation of deposits on the catalytic surface. There can also be an irreversible poisoning of the catalyst, which is discussed below. An improved pellistor has been developed which has about ten times the useful life of the conventional type, and without suffering from a gradual reduction in sensitivity. It has a larger number of catalytic centres and this ensures that a gas/air mixture encountering the device will immediately react, surrounding it with combustion products. The rate at which the rest of the mixture reacts will then depend on the rate of diffusion through this layer and not on the catalytic activity of the pellistor. The sensitivity will therefore remain unchanged until a critical proportion of the catalyst has been lost.

The catalyst can be poisoned by gases which are adsorbed strongly on to its surface. A few parts per million of Halon 1211 or 1301 in the air can reduce the sensitivity of a detector by 50 per cent. However, the catalyst will recover in time when it is exposed to fresh air. Irreversible damage can be caused by substances which are oxidized at the surface of the catalyst and

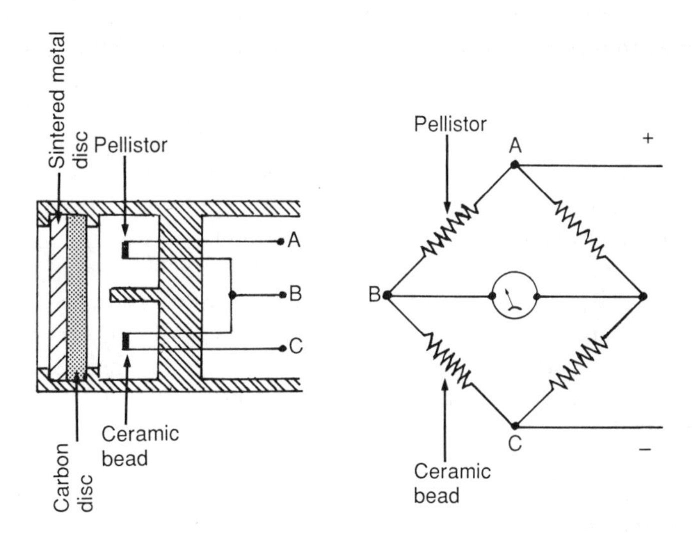

Figure 17.1 Pellistor flammable gas detector and Wheatstone bridge.

leave behind a layer of solid oxide which is not decomposed at the temperature of the pellistor. Silicon compounds are particularly troublesome, and are contained in many aerosol lubricants. These are used particularly to loosen tight threads, and should never be discharged near gas detectors. Other materials which behave in this way include phosphorous compounds, hydrogen sulphide (sometimes present offshore) and metals. The metals may be present in the fumes from welding, or as tetraethyl lead in petrol. The carbon filter shown in the figure is intended to adsorb these poisons. It can be used effectively where leakages of methane or hydrogen are to be detected, but sensitivity to materials like petrol vapour would be lost because some could be adsorbed in the filter. The purpose of the sintered metal disc is to act as a flame arrestor. This is necessary because the hot catalytic surface is a potential ignition source for an explosive gas/air mixture. The carbon disc has insufficient strength on its own to contain an explosion.

The output from this type of detector is often displayed on a meter which is calibrated from 0 to 100 per cent of the lower limit of flammability. Since the amount of heat released from the oxidation reactions is directly proportioned to the concentration of fuel, and since the change in resistance with temperature is linear, the scale is also linear. It will be remembered from Chapter 1 that the lower limit is associated with a critical flame temperature below which the combustion processes cannot be supported. Since the combustion products of many fuels are very similar in composition, it follows that the heat released per unit volume at the lower limit is much the same for most fuels. Indeed this can be demonstrated by multiplying the heat of combustion of a fuel by the value of the lower limit concentration: the result is roughly the same for a wide range of fuels. It would be reasonable to deduce from this that the calibrations on the meter would be correct no matter which fuel was being sampled. This would be sufficiently true for practical purposes were it not for another factor which affects the output from the detector: the diffusivity of the gases. The rate at which the sample can diffuse into the detector through the protective discs, and penetrate the gases surrounding the pellistor, varies from one gas to another. It is inversely proportional to the square root of the density (which itself is dependent on the molecular weight, and on both pressure and temperature). Thus, a detector system which has been calibrated for a particular gas may not read correctly for another gas if the densities are very different. The difference may not be a very great because only the diffusivity and not the lower limit value is important, but if different gases are to be checked with the same meter it may be necessary to apply a correction factor to some of the readings. An indication of the range of factors is given in Table 17.1, which applies to a meter which has been calibrated with pentane.

Table 17.1
Correction factors for gas detector calibrated with pentane

Gas	Correction factor
Methane	0.51
Hydrogen	0.54
Carbon monoxide	0.62
Ethane	0.68
Propane	0.84
Pentane	1.0
Benzene	1.13
Toluene	1.15
Heptane	1.20
Octane	1.24
Hexane	1.25
Nonane	1.47

It should be noted from this table that if a detector is calibrated with methane, as might be done for offshore use, the meter will give a low reading for all other common gases.

The manufacturer of the detector will calibrate the system for the particular gas (or mixture of gases) which is likely to be encountered and will normally supply a list of correction factors for other gases. He can also provide information on the correction to be applied if the detector is to be used at a higher gas temperature than the one at which it was calibrated. A change in temperature affects the rate of diffusion, the lower limit and, to a lesser extent, the heat of combustion of the gas. If a detector has been calibrated at room temperature and is then used to check a lower limit gas mixture which is at 65°C, it will indicate a value of about 85 per cent of the lower limit. Pressure has very little effect on the meter reading because the rate of diffusion is inversely proportional to the pressure and the number of molecules present in the gas is directly proportional to pressure: these effects cancel out.

In some situations, flammable gas detectors can give readings which are grossly inaccurate. Suppose that a detector is exposed to an atmosphere in which the concentration of gas is gradually increased from 0 to 100 per cent. The signal from the pellistor will increase steadily until it reaches a maximum for the stoichiometric mixture. The meter, which shows 0 to 100 per cent of the lower limit, will then be off scale. The pellistor signal will then begin to fall and will drop steadily to zero as the gas concentration reaches 100 per cent.

However, in mixtures containing a few per cent of air, the meter readings will have been on scale again. The reading could for example show 50 per cent of the lower limit when the mixture contains over 90 per cent of gas. If a fixed detector is being used continuously to monitor the atmosphere in an enclosure, a massive gas leak will cause it to alarm correctly as the increasing concentration passes through the present level. However if a small portable device is being used to check the atmosphere in a tank, it may not be able to discriminate between very low and very high concentrations of gas. There are two ways of overcoming this problem. In the first, a special probe is used to draw the sample into the detector. One reading is taken in the normal way and a second reading is obtained during which air is entrained into the probe and mixed with the sample. If the second reading gives a higher value than the first, the mixture in the tank is richer than stoichiometric. An alternative method employs a third sensing device in the detector head, which is also included in the Wheatstone bridge. Heat losses from this device increase with an increase in the thermal conductivity of the sample, which would result from the presence of excessive amounts of gas.

This causes a change in resistance which unbalances the bridge and gives an off-scale reading.

Fixed, continuously operated detectors should be located near to the ceiling if the gas to be detected is lighter than air, and near to the floor if it is heavier. In siting the detectors, allowance must be made for movements of air within the enclosure. It is usual to arrange for the detectors to alarm at two different levels, perhaps 10 per cent and 20 per cent of the lower limit of flammability. At the first alarm a search will be made for the leak but at the second the plant will be shut down and depressurized. It is possible to reduce the number of detectors needed to protect an enclosure by drawing samples from a number of positions and passing them sequentially into a detector. It should be arranged that an interruption of flow in the sample line causes the system to alarm. Small portable devices are also available, sometimes called explosimeters. A sample can be drawn into them by a manual pump, or in some designs the sensing head can be lowered on a wire into the vessel to be tested. Very small devices are used for personal protection and usually give both a visual and an audible warning.

These detectors will fail if the sintered (or carbon) disc becomes blocked with dirt, moisture or paint. They may false alarm in an air current if one pellistor is cooled more than the other. They can be tested quite readily (and safely) by filling a compressed gas cylinder with a gas/air mixture which is say 10 per cent of the lower limit, and releasing a small amount near to the device.

Laser detection of gases

The pellistor devices which have just been described are clearly point detectors. There may also be some delay before an escaping gas cloud encounters a detector: the energy released by a fire is not available to cause mass movements in the atmosphere of an enclosure. A large number of units may therefore be necessary to ensure the early detection of a leak, especially outdoors. There is however available a surveillance detector which can scan a large volume and detect the presence of as little as one part per million (ppm) of gas. This is the EAGLE system (Elliot absorption gas laser equipment).

Most gases will absorb infra red radiation at a characteristic wavelength, and indeed this is a well used method of gas analysis. A laser beam is a powerful coherent beam of radiation at, or near to, a single wavelength. If the wavelength is that at which a particular gas absorbs, the presence of the gas can be detected by projecting the beam in the appropriate direction and monitoring the strength of the radiation reflected from solid objects or from the ground. The presence of gas anywhere along the transmitted or reflected paths can be detected as a reduction in intensity at the receiver. Changes of as little as 1 ppm in gas concentration can be detected, with a response time of 10 m sec.

A prototype system based on this effect was tested for a year on a large outdoor ethylene plant. This gas has a strong infra red absorption at 10.5 microns, and a laser was constructed which radiated at this wavelength. A second laser was made, radiating at 10.1 microns, a wavelength which is not absorbed by ethylene. The two beams were arranged to be coaxial so that they both illuminated the same area: this was circular and with an area of about 1 m^2. Because the wavelengths of the two beams were close together, the reflected radiation from any object was about the same. Also the two beams would be equally affected by the presence of dust or mist. However in the presence of an ethylene leak some of the first was absorbed but none of the second, and the difference in the intensity of the reflected radiation could be detected. This was done by using a single receiver and switching rapidly from one wavelength to the other.

The response time of only 10 m sec means that it is possible to scan the beam rapidly over the protected area, and it permits a velocity of 360 km/hr. By scanning the beam both horizontally and vertically over the area and then processing the signals in a suitable computer, it is possible to provide a picture of the area on a visual display unit. The picture is built up in exactly the same way as that on a television screen. Some detail is lost but the ethylene plant was perfectly recognizable on the VDU used in the early trials. A cloud of gas appears on this image as an opaque area. The normal picture can be built up in the memory of the computer, which can then be programmed to alarm at any selected concentration of gas. Also a transparent container of an ethylene/air mixture of known composition can be included in each complete scan. The reading obtained from this can be used as a continuous means of calibration.

The range of measurement demonstrated in the trials was 1 to 30 000 ppm/m (but could be different for other gases). This rather unusual measurement means for example that 10 ppm/m can be provided by a cloud which is 10 m thick and contains 1 ppm of gas, or a 1 m thick cloud which contains 10 ppm of gas. In some experiments in which the circuit was used to measure the gas concentration, an accuracy of 1 per cent of the reading was obtained.

The single detector which was used to protect this large ethylene plant performed satisfactorily for a considerable time, and indeed paid for itself by giving an immediate warning of even small leaks of this expensive gas. It did have the disadvantages that it was necessary to cool the detector with liquid nitrogen, and also that the complete unit was rather large. Later, much smaller versions of the device were developed and these were fitted with detectors of a different design which had their own built-in refrigerators. Detectors are now available which respond to a wide range of gases. VDU displays can be provided which indicate the concentration of the gas in the cloud by different colours. A three-dimensional representation of the cloud is also possible. The equipment should work effectively using laser beams up to 1 km long. The detector could therefore be used from a helicopter to check for leaks from a pipeline.

There is very little lateral spread of radiation from a laser beam and the received intensity at the target is much the same at any normal distances. This intensity is around 10 W/m^2. This would have no effect on the skin of anyone receiving it, nor on the eyes of a person looking at the laser. Indeed the human body radiates infra red at about 50 W/m^2.

Detectors of this kind have not been specifically designed for fire detection. However both carbon dioxide and water vapour are included in the list of gases which can be detected, and units which are sensitive to these gases must inevitably be able to detect products of combustion.

INTRUDER ALARMS

Malicious ignition has become the largest single cause of fires in industry. Fires may be started in an attempt to claim insurance; or by disgruntled employees, vandals and terrorists, or psychopaths. Clearly security is an important part of fire protection and indeed on many large industrial sites fire, safety and security are combined in one department. Some sites are patrolled by men covering all three aspects of the work, using small vehicles which have been converted for use as rapid-intervention fire appliances.

The obvious precautions should be taken to prevent unauthorized entry through doors, windows and roofs. Various methods are available to control the entry of people into the premises. Unauthorized entry can be detected by alarms on the fences, on doors and windows, and within the buildings. Infra red, ultrasonic and radio devices are all used as surveillance detectors. It is beyond the scope of this book to provide details of this equipment or of its installation and maintenance. However both the ultrasonic and the laser beam devices described above would be able to detect the presence of intruders. It may be that in the future plants and buildings will be protected by systems which provide a warning of a range of unwanted happenings: fires, intruders, leaks of gases and liquids, and plant malfunctions.

STANDARDS

BS 5445, Specification for components of automatic fire detection systems, covers heat detectors in Part 5, smoke detectors in Part 7, high temperature heat detectors in Part 8, and methods of test in Part 9. This is the English language version of the European Standard EN 54, (developed by CEN: Comité Européen de Normalisation). However the European Standard contains an additional Part 6, which covers rate of rise detectors which do not have a fixed temperature response. It was mentioned above that these detectors are used in some countries, particularly France. If rate of rise detectors which comply with EN 54 are offered for use in the UK it is important to check that they conform to the requirements of Part 5 and not Part 6.

The sensitivity test for heat detectors consists of exposing the detectors in a wind tunnel. The air flow is at 0.8 m/s and the temperature of the air is increased at a number of constant rates up to a maximum of 30°C/min. Three grades of detectors are recognized, each with a different sensitivity. Maximum times are given for the response of each grade at different rates of rise of temperature, and there is a minimum time below which no detector should alarm. For example, at the highest rate of rise, 30°C/min, the minimum response time is 15 sec, and the maxima for the three grades are 1 min 34 sec (at 47°C), 2 min 3 sec (at 64°C) and 2 min 42 sec (at 81°C). At the lowest rate of rise used, 1°C/min, the maximum times are surprisingly long: 37 min 20 sec, 45 min 40 sec, and 54 min.

Tests for smoke detectors are also included in EN 54 but these differ from tests which had been developed earlier in the UK and are described in BS 5446 Part 1 (Smoke detectors for use in residential premises). The EN tests are not regarded as being entirely satisfactory and for this reason the BS has not been withdrawn. Optical beam detectors are covered in BS 5839 Part 5 and there is no equivalent European standard. Any future additions to EN 54 will be included as parts of BS 5445, unless the UK records a NO vote, as it did for Part 6. Additional work originating in the UK and not adopted by CEN will be included in BS 5446. The installation of detectors, as opposed to their design and testing, is covered by the Code of Practice which is discussed in the next chapter.

The five parts of BS 6020 cover different groups of instruments for the detection of combustible gases, and BS 4166, 4737 and 5979 are all concerned with intruder alarms.

NFPA standards 72A to 72E contain brief specifications for detectors and are concerned mainly with the complete systems. The testing and approval of detectors in the USA is undertaken by the Underwriters' Laboratories and the Factory Mutual System.

No detailed standards have been prepared covering the use of sampling detector systems, probably because the experience gained with them so far is small compared with that obtained from the use of the more conventional devices. There is however a brief mention in the UK Code of Practice (outlined in the next chapter) where they are described as aspirating systems.

Many of the CEN standards are likely to be adopted by ISO. There are somc diplomatic problems here because the USA in particular (which has some excellent national standards) will not want to be out-voted by a strong European bloc. Under pressure from the USA, ISO is currently working independently of CEN on a standard for self-contained smoke detectors. The Americans are of course well ahead of any other country in the widespread use of these devices in homes.

DETECTORS AND THE ARCHITECT

The positioning of detectors is discussed in the next chapter. They must necessarily be located at a distance beneath the ceiling and away from walls and beams. There is usually no way in which they can be hidden, although flame detectors can be installed in a dark corner of, for example, the roof of a church. Architects may well object to the appearance of the normally available heat and smoke detectors, but manufacturers are usually able to supply them in a range of colours, if a sufficient number is ordered. The ones shown in Figures 16.8 and 16.9 have even been supplied with a gold-plated finish.

18

Detector systems

The very simplest detector system comprises one or more normally open detectors (or manual call points) connected in parallel, and forming part of a circuit which connects one or more sounders to a battery. When one of the detectors actuates the circuit is completed and the alarm is given. A system like this could be unreliable for a number of reasons. First, the detector and its connecting wires are necessarily exposed to a fire and may soon be damaged. If the circuit is broken because of this, the alarm will stop. Second, complete failure could also result from an open circuit fault anywhere in the system, and a short in the detector wiring would give a false alarm. Finally, if the battery is exhausted at the time of a detection, the system will again have failed. Many early systems were designed like this and unfortunately their reliability could be improved only by regular testing. However there are a number of simple modifications which will considerably reduce the probability of failure.

A latching arrangement will ensure that once a signal has been received by the control box, the alarm will sound continuously until some action is taken to stop it. In some early systems it was arranged that a small shutter was held up by a solenoid. The receipt of a signal stopped the flow of current through the solenoid and the shutter fell, closing the alarm circuit and exposing the word 'FIRE'. The system could be restored to its original state only by energizing the solenoid again and then lifting the shutter. With normally open detectors the supply to the solenoid was switched by a relay. A continuous circuit including the solenoid could be used with normally closed detectors and in this arrangement the results of fault conditions would be reversed: an open circuit would give a false alarm and a short would render the system inoperative. The problem of the power supply could be overcome by arranging that normal operation would be from the mains (through suitable rectification and voltage reduction) and that adequate stand-by capacity was available from an accumulator which was normally charged from the mains (or more simply, the battery could be floating on thc DC supply).

With these improvements, a system still remained vulnerable to open and short circuit faults. These could result from mechanical damage, corrosion, or poor contacts where connections were made to the various components. A simple method of checking for these faults was developed using the end-of-line device shown in Figure 18.1. It will be seen that each detector, which is normally open, has a 400 ohm resistance in series with it and that the distant ends of the detector wires are closed by a 4000 ohm resistance (there is no significance in these values: it is only necessary that the first should be appreciably smaller than the second). Suitable circuitry in the control box can continuously monitor the resistance between the two wires, and provide an appropriate response to each of four conditions. In addition, suitably coloured lights can give an indication of the status of the circuit:

4000 ohms	Normal condition	Green light: SYSTEM HEALTHY
400 ohms	Detector actuated	Red light: FIRE
4000+ ohms	Open circuit	Amber light: FAULT
0–400 ohms	Short circuit	Amber light: FAULT

The number of detectors on one circuit is usually limited and a large building or plant is split up into a number of zones, each on a separate

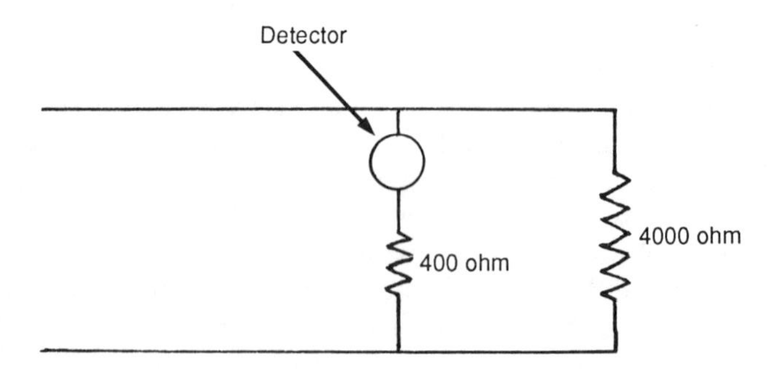

Figure 18.1 End-of-line device.

circuit. An indication is given on the control panel of the zone in which a detector has actuated, and the panel or a replica of it is located near to the entry which is used by the fire brigade. It is usual to arrange that the sounder at the control panel can be silenced when the alarm is accepted, but there should be an indication on the panel that this has been done. Also it is important that the sounder can still be actuated if a detector in another zone alarms.

The great advances which have been made in recent years in electronic equipment, and particularly in the use of microprocessors, have facilitated the development of detectors and systems which are more versatile and more reliable than those discussed so far. The rapidity of this technological advance is reflected by the frequency at which the BS Codes of Practice have been issued. The first, which was in the building series, appeared in 1951. It was replaced by an electrical engineering code in 1972. This was taken over by the fire standards committee who produced a new version in 1980, and another new code in 1988. Before the most recent publication they had issued five substantial amendments to the 1980 version.

The newly developed equipment comprises what are sometimes described as intelligent systems. This is not a strictly accurate name because the systems do no think for themselvs: they are better called programmable systems. This description could also be applied to the techniques discussed earlier by which a system could go through a learning process. The most important advance has been in the use of addressable detectors, which can be individually interrogated by the control panel.

The addressable versions of the three detectors shown in Figures 16.5, 8 and 9 can each be given an address during installation by setting a seven-segment switch. These codes are programmed into the memory of the control panel. The panel can then interrogate the detectors in any order and as frequently as necessary. When a detector is contacted it sends back its own address, the value of the smoke density or air temperature which it is sensing, its output condition (for example, whether or not its indicator LED is on), and type of detector. With this particular system up to 126 detectors can be on a single circuit and they can be scanned at a rate of 32 per second (or 56/sec if a limited amount of information is returned). Despite the large amount of information being handled the system will respond to the actuation of a manual call point in 0.1 sec.

In a system like this the detector circuit can be arranged in a loop. This ensures that if there is a break in the circuit the detectors can still operate. If there are two breaks in the loop, the detectors between the breaks will not be able to signal back to the control panel. This will be indicated as a fault and the lost detectors will be identified. A short anywhere in the loop could make all the detectors inoperative, and this could involve up to 126 detectors. The loss can be greatly reduced by means of isolators. These act as automatic opening switches as soon as the line voltage drops to 12 V or less from the normal 14 to 28 V. If a short occurs between any two isolators, both will switch to the open condition and the length of wiring between them will be isolated. The detectors in this length will not be able to respond, and this will be indicated at the control panel. As soon as the fault is cleared the isolators return automatically to their normal condition (which is why only the two on either side of a fault remain open). The number of detectors between two isolators can conveniently be the number required to cover a zone (see below). Any spur which is taken from the loop should also be protected.

The advantages of a system of this kind are the almost immediate indication of a fault, freedom from interference by transient surges and ripple, fast scan speeds, the facility for remote testing, and the ability to arrange a voting procedure (by which for example two detectors in a zone must respond before certain actions are taken, like discharging a Halon system). Since analogue detectors are used (giving a continuous signal rather than simply an on/off condition) the system can be arranged to give a pre-alarm warning. If for example a number of detectors have sensed an increase in temperature which has not reached the alarm level, it may be that something has happened for which corrective action is needed. The system can draw attention to the condition, which can conveniently be displayed as a bar chart on a VDU. In this display the height of a number of vertical lines would indicate the temperature at various detectors. The system's memory could also be used to check the recent history of temperature changes in the area. The system also interfaces well with the microprocessors which are used to control processes and firefighting systems, and a number of different actions can be programmed.

Where a larger number of detectors are employed, on an offshore production platform for example, the provision of separate wires to each detector results in an inconveniently large number of connections into the control room. The use of loops of addressable detectors would greatly alleviate the problem. However the wiring could be eliminated almost completely by the use of a radio link between the detectors and their control. Local power supplies would still be necessary and in the offshore situation the steel walls of the modules would present a problem: an aerial would be needed outside each of them. The method becomes more attractive in a large factory building, or on a large site with a number of scattered buildings. There is really nothing new in the technique: computers around the world are linked by radio.

Again, the rapid handling of sufficient information becomes possible by the use of microprocessors. Using the RAFT system it is possible to link several thousand detectors in 2000 zones into a control panel. The system can handle up to 130 million different pieces of information, although in practice only a small fraction of the capacity is utilized. The system is so arranged that it is almost impossible for it to be jammed by, for example, a CB radio. Even so, the frequency used is continuously monitored for other signals. In the UK there is the disadvantage that the transmitter power must be strictly limited, but tests with the RAFT equipment gave a range of 1200 m in the open, with the transmitter aerial 6 m high and the receiver aerial 1 m high. It is difficult to predict the range which could be obtained in a building because this would be greatly influenced by the construction materials. Repeater units can be used in areas where the signal strength would otherwise be too low. As will be seen, the UK code of practice does include information on systems using radio links.

Detector systems, and particularly those which include gas detectors, may need to be installed in areas where an explosive atmosphere could result from a gas leak. The electrical equipment in these areas will be flameproof (see Chapter 13) and it is important that no part of the detector system should be capable or producing an incendive spark. The normal functioning of a fire detector in the presence of a fire clearly cannot be a problem, but some detectors could spark as a result of a fault, or during testing. Sparks could also result from a short circuit of the wiring, or a poor contact. If it is necessary to have capacitive or inductive components in the detectors used in a hazardous area, these can be encapsulated, and connections to them can be made intrinsically safe by means of a Zener barrier at the interface between the control circuitry and the hazardous area. This device ensures that both the voltage and the current in the detector circuit are limited to values at which hazardous sparks cannot be produced. It will be remembered that the pellistor gas detectors must inevitably contain an ignition source but that they are safe because of their substantial construction and the presence of a sintered metal flame arrestor.

The versatility of the modern control panels for detector systems permits the use of programmable variations in the sensitivity of the system, referred to in the code of practice as time-related systems. In a busy workshop for example it can be arranged that reliance is placed on the human detector (and call points) during working hours, and that the automatic detectors are used only after hours. Alternatively, a delay can be provided in the alarm circuit so that emergency procedures are not initiated until someone has had time to check that they are necessary. In some manufacturing areas there may be an increase in temperature or a release of combustion products during working hours, and it may be necessary to increase the alarm thresholds to avoid false alarms. Alternatively it could be arranged for example that only heat detectors are used during the day, to be supplemented with smoke detectors at night. The purpose of these programmes is to reduce the number of false alarms, which are still a major problem with detector systems. The individual monitoring of detectors in an addressable system can also help because faulty detectors can be identified and ignored (the system's memory may assist in this identification).

The very simplest detector system provides an audible warning of the presence of a fire so that people can escape (although some of them may first take appropriate action to mitigate the effects of a fire). Most modern systems are used to initiate additional actions. These may include the shut-down of certain equipment, or of an entire process; the actuation of extinguishing systems; the closure of dampers and shutters; the snuffing of flares; the release of magnetically held fire doors; and the turn-out of the fire brigade. This latter function is of considerable importance and may be accomplished in a number of ways.

If the alarm system includes a continuously manned control room, the fire brigade can be alerted by a 999 call (or the equivalent in other countries). This is generally a fast and reliable route, although in some rural areas 9 is the first digit of some local numbers and could sometimes be unavailable. A private line to the brigade control room would be the safest route but this can very rarely be arranged. At the time when most fire brigades were small, and each had a continuously manned control room, it was usual to find that alarms from local factories and other establishments terminated in the control room. When these brigades were amalgamated into large regional organizations, most of the new fire authorities insisted on the removal of these facilities. They are very reluctant to consider the provision of any direct contact with the regional control rooms. Contact on the brigade's radio frequency is also possible but again it is very unusual for such arrangements to be made (facilities of this kind are at the discretion of the chief fire officer).

During the time when a private control room is unmanned a 999 auto-dialler can be used to contact the brigade. Recorded messages are used to instruct both the telephone exchange and the brigade control room. These devices have been unreliable in the past, due mainly to the fact that they cannot be tested (this is not permitted on the 999 system). Also, mechanical devices are inherently less reliable than electronic ones. Their use is discouraged by many fire brigades, but this may be partly because they cannot charge for false alarms received through the 999 system. However there is no doublt that new technology has provided better means for the automatic transfer of information which may eventually replace these devices.

The most frequently used route from a detector system to the fire brigade is through a central alarm station. This name is given to commercially operated control rooms which have been established throughout the country by a number of fire engineering and security firms. The most secure connection to a central station is by a private wire, but the rental charges can be very high, particularly in country areas where the station, or its satellite station, is some distance from the customer's premises. It is, however, possible for a number of customers to share a private line, through which each of them is monitored in turn. These 'omnibus' circuits have much lower rentals. A digital communicator provides a less expensive (and less reliable) route to the central station. It is like a 999 caller in that it uses the public telephone service to make contract. It then transmits an encoded message which provides the central station with information on the location of the incident.

British Telecom offers two methods of connection to control stations. The Red Care system is currently available only in London and the ABC system (Alarms by Carrier), which has been installed in a few areas can currently be used only for intruder alarm systems.

The connection from the central station to the fire brigade control room can be made in various ways, again dependent to some extent on the decisions of the chief fire officer. A private wire is clearly the most reliable route, but this may not be acceptable. The 999 system can obviously be used, but a brigade will sometimes provide an ex-directory number which gives direct access to the control room. Some brigades insist on a much less satisfactory route through their normal adminitrative telephone number. In some locations it may even be necessary for one central station to route its calls through another station.

The reliability of an automatic turn-out system varies greatly from one method to another, and from one part of the country to another. The best route is probably provided by a private line to the fire brigade control (where this is permitted), and the worst by the use of a digital communicator and ordinary telephone lines to an inefficient central station which has a poor connection to the brigade. It is unfortunate that human failure has a significant effect on reliability. At some central stations the majority of incoming signals are from security systems and there have been instances of fire alarms being incorrectly processed (or being ignored, on the assumption that they are false alarms).

The basic reason for this very unsatisfactory situation in the UK may well be the lack of standardization. There are excellent standards covering detectors and detector systems but there is nothing covering the vital link between the system and the fire brigade. The 1980 code did have some appendix material discussing the problem, but this has been omitted from the latest version (although there is some discussion within the Standard). There is however in preparation a Loss Prevention Certificate Board Standard for central stations. This will permit the evaluation of the performance of a station and will to some extent be in parallel with the British Security Industry Association's scheme for the certification of central stations. It is to be hoped that a British Standard covering the whole area of the handling and transmission of fire alarms will eventually be written. By that time the fire brigades may have agreed to a standard method for alerting their control rooms, perhaps by the use of one of the more recent techniques like data networks (used for the exchange of data between computers), or radio.

STANDARDS

The range of NFPA standards covering detector systems has already been mentioned. Central Station Systems are dealt with in NFPA 71, Local Systems in 72A and B, Remote Station Systems (which include connections to a fire brigade control room) in 72C, and Proprietary Systems in 72D (these include for example the control room on a large factory site). Detectors themselves are specified in 72E and fire brigade communications are covered in NFPA 73. Many other countries have comparable standards. In the UK, the FOC provides detailed and quite strict Rules for Automatic Fire Alarm Installations. Some insurers may insist on compliance with the Rules in order to obtain a premium reduction: a careful study of the Rules would then clearly be necessary. However in this section we will be dealing with the most recently published standard, BS 5839 Fire detection and alarm systems for buildings Part 1 1988: Code of Practice for systems design, installation and servicing. It will be remembered that this is a complete revision of the 1980 standard and that it became necessary because of the considerable advances in the use of microprocessors, stored programmes, analogue and addressable detectors, radio linkages and similar developments.

Like the 1980 Code, the new one is divided into sections each of which wll be of interest to different users. Section 1 contains matters of general interest, but Section 2 is aimed particularly at the systems engineer responsible for the design and to some extent the installation of the system. It is the largest section but not quite complete in itself: some of the explanatory material is in other parts. Section 3 deals with the provisions which need to be made for the system in the design of new buildings or in the modification of existing buildings.

Section 4 provides the user of the system with vital information on its proper maintenance. This is followed by seven appendices, of which A has a rather general discussion on connections to the fire brigade; F provides guidance on the integration of detector systems with other systems (like process control and security); and H comprises the text of a Home Office booklet on 'Smoke alarms in the home'. Some of the more significant topics covered by the code are discussed below, in the order in which they appear, but without the numbering. It should be remembered that the code has to deal with all types of systems, from the very simplest to the most complex of the recent developments, covering the range which has already been defined above.

Types of system

An interesting innovation in the code is a classification of systems depending on the type and usage. The most important consideration is whether the system is intended to protect life or property (designated L and P). Either of these types can provide complete or partial protection of the premises, and this is indicated by a number. Thus, a P1 system provides total protection of the property and a P2 system covers only a selected area (but see below for the L classification). A type M system is operated manually and provides only an alarm. The important feature of an L system is that adequate warning should be given to everyone in the building (possibly including people who are asleep) so that they have plenty of time to use the escape routes before they are penetrated by smoke or fire. A P system on the other hand operates in unoccupied areas and does not have to alert people who need to escape. It must however provide prompt information to people who are going to fight the fire, and may additionally actuate systems which extinguish, isolate, and shut down. The main function of a P system is to limit the fire damage (which ideally will be confined to the room of origin).

Exchange of information

This section indicates the questions which need to be asked before a specification for a system can be drafted. Some of these concern the 'action to be taken in the event of an alarm'. Here, the action which needs to be taken by a member of a works fire brigade may be contrasted for example with that of a customer in a shop. Also included is useful list of interested parties who will need to be consulted, like perhaps the local authority, the insurers, British Telecom, and the other people in a multi-occupancy building.

System planning

An examination of the actions to be taken following an alarm is again required during the planning stages, when detailed consideration of the design may have indicated problem areas or possible improvements. At this stage too it is necessary to consider servicing arrangements for the system because they may be costly or may cause interruption to manufacturing processes.

Extent of cover

The decision between P and L cover is usually obvious, but it may be difficult to decide whether full or partial cover will be appropriate. There is little doubt that the insurers will offer premium reductions only for property which is totally protected. They may however be prepared to consider strong evidence that complete coverage is unnecessary. Alternatively it may be cost-effective to pay higher premiums to avoid the expense of both the installation and the maintenance of an unwanted system. If for example an area is continuously manned it would be reasonable to place total reliance on the superior efficiency of the human detector. Other factors which need to be considered are:

- The probability of ignition
- The rate of fire spread within the room of origin
- The value of the contents
- The probability of fire spread to other areas
- The time taken for this to happen.

All these must be weighed against the value of the time which can be gained by the installation of a system.

The most limited life-protecting system is type L3, which covers only the escape routes. To comply with earlier editions of the standard these would have been protected only by detectors in the routes themselves, on the assumption that if there was a fire in a room adjacent to the route, any smoke which escaped would rise to the ceiling and be detected. Recent research has shown that the smoke which emerges through the cracks around a door is relatively cool and therefore collects near the floor. The corridor can become smokelogged before the ceiling detectors alarm. The new code therefore calls for the provision of detectors in any rooms which open on to an escape route, even for this very limited system.

If day rooms or lounges are occupied it might be argued that people there will become aware of a life-threatening fire, and will make their way to an escape route. However if the occupants are elderly (and possibly asleep) or handicapped, they may be at risk. The risk could be increased by the presence of materials likely to cause a rapid spread of fire, or by the absence of good fire and smoke separation from the rest of the building. The extention of an L3 system to include such rooms would change it to an L2 system.

An L1 system would include every room and this would greatly enhance the probability of a safe escape, even for people who were asleep in the room of origin. However it is important not to consider the detection system in isolation. Fire safety is dependent on many other factors: on simple precautions like the use of self-closing doors, for example.

Zones

In earlier editions of the standard there were requirements for a building to be divided into a number of zones, each supplied by a separate circuit. The size of a zone, and hence the maximum number of detectors on a circuit, was limited. Also it was necessary that an indication was given of the zone in which a detector had alarmed.

The 1988 edition is the first standard to recognize the new systems based on addressable detectors. As we have seen, these can include a large number of detectors in a single circuit and can readily identify the particular detector which has responded. However the standard still requires that buildings should be divided into zones and that the main indication should still be the location of the affected zone. It is considered that a display identifying individual detectors could be confusing, but such a display is permitted provided that it is subsidiary to the main one. There is no limit to the number of zones which can be included, and the display need not necessarily be at the control panel.

Since each zone of the older systems had its own circuit, a fault in one zone could not affect another. The addressable detector systems may cover several zones with one looped circuit, which may contain a number of isolators. To ensure that the effects of faults are minimized the code has three requirements: that a fault in one zone does not affect others, that a single fault should not affect an area greater than that allowed for a zone, and that two faults should not affect an area greater than 10 000 m^2.

The code defines a maximum area for a zone: 2000 m^2. In addition it requires that the 'search distance' in a zone should not exceed 30 m. This is the distance which a person has to walk within the zone in order to 'determine the position' of a fire. In an open plan office it may only be necessary to open the door, when the search distance will be no more than 1 m. In a well stocked warehouse it may be necessary to walk several hundred metres to find a fire. The warehouse might have to be split up into a number of zones, or alternatively some remote indicators for the detectors could be positioned by the door (it would then be possible to 'determine the position' of the fire).

If a multi-storey building has a total floor area of less than 300 m^2, the entire building can be one zone. If the total area is greater than 300 m^2 each floor must be a separate zone (but there are two exceptions to this requirement).

In a multi-occupancy building the zones should not include more than one occupancy. However if it is an M type system, with only manual call points and sounders, there may well be an advantage in treating the whole building as a single zone, but arrangements should then be made for the person using a call point to be able to tell the fire's location to whomever is responsible for the building.

The code quite rightly points out that there are many situations in which a zone area less than the permitted maximum should be used. This could apply for example if flammable liquids are being handled, if valuable material is at risk, if there is a threat to life, or if searching could be difficult because of clutter.

Action following an alarm

When an alarm sounds, most people will follow an escape route, usually to a place outside the building. Others may have a duty to perform: they may be members of a fire team, first aiders, fire marshalls, control room staff, or they may be needed to shut down machinery or remove valuable property. Some people may need to be alerted to the existence of a fire even though they are in no immediate danger. Some who are at risk may be unable to take any action because of illness or disablement, and must wait to be rescued. The alarm must also be used to summon assistance, usually that of the fire brigade.

Fire brigade communications

There is a discussion in the code of the various methods of communication which were mentioned above. It is pointed out that a 999 call is the simplest and usually the fastest method (with the exception of a private line). However it is essential that someone is present to make the call at all times. It is important that alarms which sound in the immediate vicinity are not too loud to prevent the use of the telephone. The private line is considered to be the most reliable method but clearly it is necessary to obtain the permission and the cooperation of the brigade.

Sounders

The code covers the use of bells, horns, sirens and similar devices. The information is similar to that given in Chapter 7. Comprehensive advice is given on the use of public address systems, which received only a brief mention in the earlier editions of the standard.

Manual call points

In some of the systems using addressable detectors the scanning time is several seconds. If a manual call point is included in the circuit, there can be an appreciable delay between the time when the glass is broken and the sounding of the alarm. The person using the call point may therefore think that it is faulty and will start looking for another one, and this could be dangerous. The standard therefore calls for a delay of no more than three seconds, but allows systems with delays of no more than eight seconds to continue in use until the end of 1989. There is an additional problem in that even with a short delay people may think that the alarm has not sounded because they are supposed to press the button in the call point. If the design of the system is such that this action will prevent the sounding of the alarm, then the person may again be endangered by standing there and holding in the button. For systems of this kind the maximum delay allowed in the standard is one second.

The maximum travel distance (which may not be the same as the straight-line distance) to a call point should be no more than 30 m.

Detector types

The code includes a description of the different types of detector. The uses of heat, smoke and flame detectors are discussed, and aspirating systems are included.

Siting of detectors

This important part of the code includes advice on the positioning of detectors in many different situations and should be studied in detail. It covers both P and L systems, but there are only a few additional requirements for L systems. These are mentioned at the end of this section.

It will be remembered that the smoke and hot gases produced by a fire rise to the ceiling as a plume and then spread out as a layer beneath the ceiling. The velocity of the gases in this layer is less than that in the plume, and much of the time taken in reaching a distant point on the ceiling is due to the travel time in the layer. The code is concerned with limiting the delay before the gases encounter a detector, and it is for this reason that the maximum travel distance beneath the ceiling is limited. The wording used in the code is 'the horizontal distance from any point in the area to the nearest detector' should not be more than 5.3 m for heat detectors, or 7.5 m for smoke detectors. In practice this means that if you draw circles with these radii around each detector, no part of the ceiling should be outside a circle. Some overlapping of the circles is therefore necessary, as can be seen from Figure 18.2(a). This shows the conventional square pattern for siting the detectors, but it is interesting to note that the triangular pattern in (b) gives better coverage (because the area of overlap is less). In a suitable situation the use of the latter arrangement can result in a reduction of about 10 per cent in the number of detectors needed. The other important dimensions, which appear later in the standard, are these: detectors should not be closer than 25 mm from the ceiling nor further away than 150 mm (heat detectors) or 600 mm (smoke detectors). The minimum distance from a wall should be 500 mm. All these distances are based on the known behaviour of the hot gases near walls and ceilings, and particularly to the formation of 'dead' zones where there is little movement and the temperature is relatively low.

A sloping ceiling will reduce the time taken by the gases in reaching the apex, and the radius covered by detectors at the apex can be increased by 1 per cent for each degree of slope, up to 25 per cent. The gases will also move more rapidly down a corridor than when they are spreading over an open ceiling. In corridors less than 5 m wide, the radius can be increased by half the difference between the actual width and 5 m. Thus, in a corridor 2 m wide the radius can be increased by 1.5 m (half of (5–2)). The gas movement will be slowed by the presence of for example a beam, and the radius should therefore be reduced by twice the depth of the beam (and the detector should be at least 500 mm from the beam).

The height of the ceiling has a considerable effect on the size of fire needed to actuate a detector, and on the delay before it actuates. The maximum heights allowed for grades 1, 2 and 3 point heat detectors are 9 m, 7.5 and 6 m respectively. The lowest height also applies to high temperature heat detectors. Point type smoke detectors may be used at heights up to 10.5 m. An increase in height has less effect on an optical beam detector because although the plume is diluted and cooled it also increases in volume, and the beam therefore traverses a greater length. For this reason these detectors can be used up to 25 m.

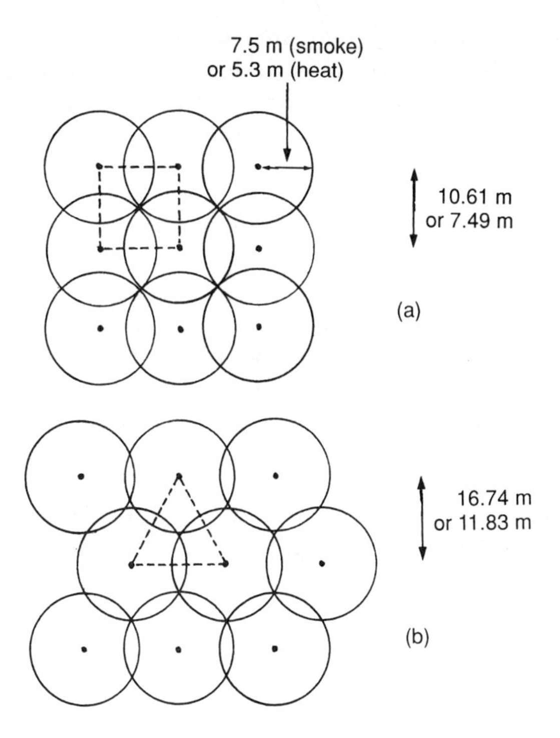

Figure 18.2 Siting of detectors (a) conventional square pattern, (b) triangular pattern.

If there is a direct connection to the fire brigade and the attendance time is no more than five minutes, longer detection times can be accepted. Maximum ceiling heights can then be raised to 15 m for smoke detectors and 13.5 m, 12 m and 10.5 m for the three grades of heat detector. Increased heights are also allowed if a raised part of a ceiling forms a smoke reservoir.

Special consideration must be given to the use of detectors in confined spaces, like ceiling voids. Stratification also presents problems: the presence of a layer of hot air beneath a ceiling can increase the time taken for the plume to reach the first detector. The code also gives advice on the siting of optical beam smoke detectors; on the effects of ventilation; and on the use of smoke detectors in ventilation ducts.

In general the siting of detectors in life safety (L) systems follows the same rules as those for property protection. There is an exception for L3 systems (covering escape routes). The increased spacing of detectors permitted in corridors less than 5 m wide does not apply unless all of the rooms opening into the corridor are fitted with smoke detectors. In addition, where the floor has been penetrated by an opening, like that for a hoist, the ceiling above should have a detector. Detectors are also needed on the ceiling above a stairway, and on the landing ceilings, but the vertical distance between them can be up to 10.5 m.

Control equipment, wiring, power supplies, radio links

When the standard was being written a decision was taken not to include the requirements for the functioning of the control equipment, which had been covered in previous editions. This is now in BS 5839 Part 4, with which the control equipment should comply. However systems completed before the end of 1989 are acceptable if they comply with BS 3116 Part 4. There is also a relaxation for small manual systems: the details are in Appendix G of the code.

Reliable power supplies and adequate wiring are covered by the code and there is a brief description of the advantages and disadvantages of radio links. Methods of monitoring these links are also covered.

Installation

The rules covering the installation of a system are similar to any others concerned with electrical wiring. There is a reference to BS 6651 1985 for guidance on the separations which are needed to avoid arcing if a building is struck by lightning.

Testing and commissioning

Before a system is handed over, the insulation should be tested, and checks made of all the equipment connected to it. The correct response of the control equipment should be checked, and the linkage, if any, to the fire brigade. Where a radio link is used in the system there is a requirement to check for adequate signal strength. The handover of the system should preferably involve the responsible person mentioned below.

Responsible person

If a system is to respond reliably when there is a fire, it is essential that it is properly maintained for what may be a period of many years before this happens. The expenditure on the system will have been completely wasted if, through neglect, it is faulty on the day that it is needed. The difficulties involved in motivating people to carry out (or supervise) regular testing and maintenance were recognized by the committee which wrote the standard. They very rightly decided that the best way of ensuring that the job is done properly is to appoint a 'responsible person'. The standard asks for the appointment of such a person but in case this is not done it says that 'the person having control of the premises' is the person responsible for the system and for the observance of proper procedures. The duties of the responsible person are mentioned in various parts of the code, and are summarized here:

1 To organize procedures for dealing with alarms and other happenings. These should be discussed with the fire brigade and should be appropriate to the premises.
2 To ensure that everyone who has to use the system is instructed in its use. For many people this might simply mean a knowledge of the escape routes to be used when the alarm sounds. Members of fire teams, and fire marshalls, would need to know how to use the system to locate the fire.
3 To ensure that work being done in the building does not adversely affect the system. Contractors, sometimes through no fault of their own, are notorious for their effect on both active and passive fire protection.
4 To ensure that the efficiency of the system is not reduced by obstructions which affect the movement of fire products to the detectors, or the access of people to manual call points.
5 To maintain and up-date the system drawings and operating instructions.
6 To keep a record of every significant event affecting the system.
7 To prevent or reduce the frequency of false alarms.
8 To ensure that the system is properly reinstated after any work on it has been completed.
9 To ensure that the correct maintenance procedures have been followed at the correct intervals.
10 To ensure that the system is serviced after it has alarmed or has been repaired.
11 To carry a stock of suitable spares, so as to ensure the rapid replacement of any faulty equipment.

Although the appointed person is responsible for the work he can of course delegate, and indeed will probably need to because he is unlikely to have all the necessary skills. The code gives details of the tests and maintenance which are needed at various intervals (from daily to five-yearly). The responsible person should be able to do the daily, weekly and (where appropriate) monthly checks but it will probably be necessary for other work to be done by a competent engineer under a service contract.

FURTHER READING, PART FIVE

Bryan, J. L., *Fire suppression and detection systems*. Macmillan, New York.

Cullis, C. F. and Firth, G., *Detection and measurement of hazardous gases*. Heinemann, London.

Todd, C. S., *Guide to UK central stations*. Paramount Publishing Ltd., Borehamwood.

Home Office, *Manual of firemanship, Book 9: fire protection of buildings*. HMSO, London.

Home Office, *Manual of firemanship, Book 10: Fire Brigade communications and mobilizing*. HMSO, London.

PART SIX

AUTOMATIC SYSTEMS

19

Sprinkler systems

One of the earliest fire extinguishing systems was installed in the Theatre Royal in London. It was covered by Patent no. 3606 granted in 1812. The system was designed by William Congreve and comprised a large tank from which water could be pumped into a network of pipes. The water could be run into various parts of the system by opening the appropriate valve and was discharged through a number of half-inch holes. The great value of systems like this for the protection of buildings and ships was quickly recognized. However the system had the disadvantage that it wetted areas which were not involved in the fire and, in the Theatre Royal design, it was necessary to have someone there to open the valves. A number of inventors recognized the advantage of having the discharge holes normally plugged and arranging for them to be opened by the softening of a low melting point material: gutta-percha, solder, and mixtures of waxes and resins were all used for this purpose. The systems were filled with water at a suitable pressure, often being connected to a gravity tank. These systems contained the essential features of a modern sprinkler system.

The actuators of these early systems usually comprised fairly complicated 'valve boxes' which turned on the water to one or more nozzles. The earliest practical sprinkler head, in which the detector and the nozzle were combined, was invented in the UK by Stewart Harrison in 1864. It comprised a hollow brass sphere perforated by a large number of small holes. The orifice connected to the water supply was closed by a soft rubber valve supported by a plunger. The plunger was soldered to the bottom of the sphere, where it could readily be heated in a fire. The heat softened the solder, allowing the water pressure to push down the plunger. Very little interest was taken in this excellent design and much of the development of sprinklers began some time later in the USA.

Because of a number of a disastrous fires in the USA, particularly in Chicago (1871) and Boston (1872), there was a substantial rise in the premiums charged by the insurance companies. A piano manufactuer, Henry Parmelee of Connecticut, was determined to protect his factory with one of the new systems so that he could argue for a reduction in insurance costs. After trying a number of designs, in 1875 he produced a simple downward facing perforated nozzle covered by a brass cap. The cap was attached to the base by a solder with a melting point of 70°C. Because of the mass of metal and the cooling provided by the water, it took several minutes before the solder melted and the cap was pushed off by the water pressure. The earlier Harrison design would have given a faster response. However the Parmelee sprinkler proved to be reliable and effective, and an improved version, with a rotating head, was in production from 1878 to 1882. During this time some 200 000 were sold both in the USA and the UK.

A number of different designs were produced which gave faster response times, most of them being dependent on low melting point solders. The first design to use a liquid-filled bulb was marketed by the Grinnell company in 1922. This overcame a serious problem which had been encountered with the soldered metal designs: that of corrosion. However it was difficult with the Grinnell design to adjust the loading of the valve on to its seating. The problem was solved by the development by Mather and Platt Ltd in 1925 of a bulb shaped like an elongated barrel. Glass is very strong in compression and this design enabled the proper load to be applied to the valve by means of an adjusting screw in the bottom of the sprinkler. An increase in temperature causes the liquid in the bulb to expand and reduces the volume of a small trapped bubble. At a certain temperature the bubble disappears and any further expansion of the liquid shatters the glass. The seal is then released and the emerging jet of water hits an impact (or 'deflector') plate which is supported by two yoke arms. A typical design is illustrated later, in Figure 21.2. This type of glass bulb, usually called a quartzoid bulb, has formed the basis for the design of almost all current sprinklers, with the exception of course of those using soldered struts. Other designs, including a fast-acting glass bulb, are discussed later.

Sprinklers are actuated by being immersed in the layer of hot gases which collects at the ceiling above a fire. The delay before they are triggered depends on the temperature of the gases in contact with the sprinkler, and the sprinkler's sensitivity. The liquids in the glass bulbs are coloured to indicate the temperature at which they will burst under specified test conditions. Soldered sprinklers are also colour coded, but not by the same series of colours. Even with the most sensitive of the standard sprinklers there may be a considerable delay before they respond to the presence of a small fire. In general, a sprinkler will not be actuated unless the flames are between one-third and one-half of the ceiling height (depending on the position of the fire beneath the sprinkler). The response time can therefore be reduced by lowering the ceiling height. As a rough guide, if the ceiling height is halved, the size of fire needed to trigger a sprinkler is reduced to between one-half and one-quarter (again depending on the exact location of the fire beneath the sprinkler).

The impact of the jet of water on the deflector plate causes it to break up into a wide range or drop sizes, with a mean diamter around 1.5 mm. These drops function in three different ways, depending on their size. The smallest droplets are vaporized by the hot gases. This limits the lateral spread of fire because the temperature of the gases is reduced. The medium sized drops are able to fall in the area surrounding the fire so that they wet the surface of the uninvolved fuel, again limiting the lateral spread. Only the larger drops (up to 6 mm in diameter) have sufficient momentum to penetrate the plume of hot gases, so that they can cool the surface of the burning fuel. Some of these will have dropped down from the wetted area of the ceiling.

People concerned with sprinkler research in the USA did not accept this concept of sprinkler function. They considered that a more uniform drop size around 1 mm in diameter would be preferable, and developed the range of 'spray sprinklers' in the 1950s. These can usually be recognized by a large deflector plate with comparatively small teeth around the edge. There is little difference in the effectiveness of the two types against normal fires, but some recent work in the USA has indicated a need to revert to the 'conventional' design, which is still used in the UK. This resulted from tests with fires involving aerosol cans with flammable fills. This work is again discussed later, in the section on Recent Developments

From this discussion it will be clear that a sprinkler system may not necessarily be able to extinguish a fire. It can still perform a vital role, first by detecting the fire and raising the alarm, and second by holding the fire in check. Where a considerable amount of damage could be done before the arrival of the fire brigade (possibly involving the loss of an entire building), a sprinkler system will ensure that only a fraction of this loss occurs. Sprinkler systems can therefore be very successful in preventing major losses even in high risk situations (like the Parmelee piano factory). Insurance companies have therefore always encouraged the installation of systems (by substantial premium reductions) and many have contributed to research and testing facilities. A group of insurance companies in America established what is now the Factory Mutual System soon after the earliest sprinklers were made, and built the first laboratories in the world to be devoted solely to fire research. Research still continues in many countries, in centres which also provide a rigorous testing facility for system components.

One of the first requirements of a research programme was to establish the application rate of water which would control a fire. This was found to vary with the type of fuel and the way in which it was distributed. A vast amount of information on the performance of sprinklers has been collected and analysed over the years, and this has enabled risks to be placed into a number of broad categories, each needing a particular rate and method of application. The rate for most normal fire situations was established many years ago as 0.1 gal/ft^2/min (roughly 5 l/m^2/min). This is about fifty times the theoretical (and practical) minimum discussed in Chapter 6, but as we have seen, much of the water applied from a sprinkler fails to reach the burning surface. This is due to the updraft from the flames, the obstruction caused by some of the fuel, and the evaporation of some of the water by the hot gases. The application rate of 5 l/m^2/min is often expressed in the UK as 5 mm/min: the rate at which the depth of water would increase over the area covered by the sprinkler.

If a system is to be both successful and reliable, a great deal more than simply the application rate needs to be controlled. For their own protection the insurers found it necessary to draw up rules covering every detail of the design and construction of sprinkler systems. Later, standards and codes of practice were established in many countries. In the UK the first set of rules was written by Sir John Wormald in 1885. The third revision of these rules was accepted by the Fire Offices' Committee and became the first edition of their 'Rules for Automatic Sprinkler Installations'. The issue of the FOC rules did much to encourage the wide use of sprinklers throughout the country. In America a group of insurers met in New York in 1895 to start work on their own sprinkler rules. This was the foundation of the National Fire Protection Association and their first standard on sprinklers was published in 1896. The use of the FOC rules and the NFPA standard spread to the many countries in which the UK and the USA had interests, and they are still applied extensively around the world (Australia for example accepts either).

Both standards have been revised and extended many times as a consequence of fire loss experience and the results of research. We are concerned here only with the current version of the FOC rules: the 29th edition, which was published in 1968. This was a major revision which brought the rules closely in line with the rules drafted by the CEA (the European Insurers' Committee). The FOC rules are comprehensive in that they cover the classification of occupancies, water supplies, the spacing and location of sprinklers, and the design of pumps, pipework and valves. Minimum requirements for four different types of system are included:

- wet pipe systems, which are always filled with water;
- dry pipe systems, which contain air under pressure until the system is actuated, and are used at low temperatures;
- alternative wet and dry systems, which can be used at both summer and winter ambient temperatures;
- pre-action systems, which are dry pipe systems into which water is admitted by the actuation of a separate detector system, before the opening of the sprinkler heads.

The rules also cover deluge systems, which have open sprinklers or spray heads to which water is admitted following the detection of a fire.

It should be remembered that the FOC rules are exactly what the name indicates: a set of rules. They are not intended to be a complete guide to the design of a sprinkler system. Indeed on first reading they may appear to be confusing and lacking in explanation. It is necessary to read right through the book to appreciate the areas which are covered and the order in which the information is presented. The following chapters will provide a guide to this reading, and a help in understanding both the basis of design and the functioning of sprinkler systems. The rules are frequently amended and there is no doubt that a 30th edition will eventually be published. No attempt should be made to design or check a system unless the very latest information has been obtained. Sprinkler systems are designed to last and there will still be some installations which were completed before the 29th edition of the FOC rules was published, in 1968. If it is necessary to make simple additions or modifications to these systems, this should be done in comformity with the rules of the 28th edition, copies of which are still available. The main difference between the two editions is the introduction of a detailed hazard classification. There was for example no recognition in the 28th edition of the increased hazard resulting from high-piled storage. More stringent conditions were also imposed for water distribution.

20

Sprinklers: water supplies

Automatic sprinkler installations are designed to hold a fire in check and to give warning of danger. They do not necessarily extinguish a fire, but very often do so. Looked at from the point of view of insurance interests, an automatic sprinkler installation provides immediate protection once a given temperature is reached and it has a positive capability of limiting the maximum potential loss to an acceptable level. They are not subject to any of the limitations of the firefighting services such as smoke or a delayed call.

Objections are often raised that automatic sprinklers cause a great amount of water damage. This is very often an excuse rather than a fact. A sprinkler installation is no more likely to leak than a domestic water supply; probably less – because the installation will have been made by specialists and in any case will do less damage, as a serious leak will actuate the alarm signal thus drawing attention to the leakage. If the damage referred to is meant as damage in extinguishing a fire, that damage is surely justified rather than have a total loss by fire. In any case, without a sprinkler installation the fire must ultimately be extinguished by the heavy jets of the fire service resulting in all probability in far more water damage than that of the limited use of the sprinkler head.

REQUIREMENTS OF A WATER SUPPLY

The requirements for water supplies for automatic sprinklers are very stringent and though the following notes are given for guidance, the latest edition of the FOC rules must always be consulted. The installing engineers and the insurance company will, of course, always give expert advice and guidance.

The water supply must be of sufficient quantity, pressure and flow to be capable of meeting the requirements for the different classes of hazard as laid down in the rules. The supply must not be subject to freezing, drought conditions or anything that might serously deplete the supply. It should be under the direct control of the occupier or the rights of supply must be suitably guaranteed.

The water must be free from fibrous or other material in suspension that might cause corrosion by accumulation in the pipework. Under some conditions, water that might not otherwise be acceptable may be taken through a 'jackwell'. A jackwell is a gridded enclosure forming a pumping pit to enable water supplies to be taken from a river, canal, lake or similar supply in which there may be floating debris. There is an illustration of a jackwell in Figure 28.1.

The use of brackish water is not normally allowed, but under some conditions it may be approved provided that after having had brackish water in the installation, the distributing pipes are flushed out and the whole system charged with fresh, clean water.

Generally speaking, no other pipes may be connected to the supply main but there are certain exceptions laid down, particularly for hose reels. Wherever such connections are permitted, a stop valve must be fixed close to the connection to the supply pipe and this valve must be suitably labelled. A test of any such stop valve should be maintained and the label regularly inspected. No such connections are permitted on the distribution lines beyond the control valves.

Hazards by occupancies

The supplies of water for sprinkler requirements depend on the fire hazards of the occupancy of the premises. There are three main classes of occupancy laid down by the FOC for this purpose. Detailed lists into which the occupancy of buildings are divided for fire hazard classification can be obtained from the FOC, but it is the insurance company which makes the final decision in to which class an occupancy falls. Early consultation with the insurance company is therefore necessary. If insurance brokers are employed, the consultation will, of course, be conducted through them.

The three main classifications of hazard occupancies are as follows:

1 Extra light hazard applies to non-industrial occupancy
2 Ordinary hazard for commercial and industrial occupancy

3 Extra high hazard for commercial and industrial occupancy having abnormally high risks.

Extra light hazard class

This is for the lightest risks such as hospitals, hotels, office buildings, etc. In parts of the premises such as attics, basements, boiler rooms, kitchens, laundries, storage areas, workrooms, etc., it is necessary that the density of water discharged should be increased. This is achieved by spacing the sprinkler heads closer to each other.

Ordinary hazard class

There are four subdivisions in this hazard class:

Group I Light ordinary hazards such as breweries, cement works and restaurants.
Group II Medium ordinary hazards such as bakeries, confectionery manufacturers, light engineering works, medium sized retail shops (under 50 assistants).
Group III High ordinary hazards such as boot and shoe manufacturers, corn mills, large department stores.
Group IV Special. These are group III hazards where there is a danger of flash fires and consequently a greater area has to be covered by water spray. These would include such risks as cotton mills, film studios, flax, jute and hemp mills, match factories, oil mills. In some areas, special types of protection will be required.

The rate of water discharge that is required for the ordinary hazard class is greater than that required for extra light hazards and is the same for all groups. Each group, however, assumes a greater area of discharge and this in turn affects the quantity of water that will have to be supplied to the installation.

Extra high hazard class

These are divided into two types of risks:

1 Process risks such as aircraft hangers, firelighter and firework manufacturers, foam plastic and foam rubber manufacturers, paint and varnish manufacturers and wood wool manufacturers.
2 High-piled storage risks. These are divided into four categories according to the nature of the stock and the height to which it is stacked.

The importance that high stacking plays in the destruction of a building by fire is very great and any raising of the height of stored materials, even materials waiting for and undergoing process, is taken into account.

It should be noticed that high stockpiling will automatically turn any class of hazard into an extra high hazard class although, if the overall storage height does not exceed certain specified limits in each category, the occupancy may be considered as an ordinary hazard class.

Some relief can be obtained from the severe imposition of water supplies in the extra high hazard class if sprinklers are installed at intermediate levels. Beyond a certain height of stockpiling, the installation of sprinkler heads at intermediate level is compulsory. In some cases part of an installation may be designated as a different hazard class. This means that distribution arrangements have to be increased in these parts and calculations as to the quantity and pressure of water supply are affected. But the same set of control valves may be used for the complete installation provided that the total number of permissible heads is not exceeded.

When setting out the data for sprinkler systems it is important to make due allowance for future expansion and for any possibility of a change of occupancy hazard, particularly if the change is liable to involve storage or increased storage, because by increasing storage, pipe sizes may have to be increased. If extra and larger fittings (elbows, tees and so on) are put in when the installation is made, very little extra cost will be incurred when compared with the cost of alterations or increased insurance rates if the premises are converted to some other use or storage at a later date.

Classification of existing installations

In an existing installation, a regular watch should be kept for any changes in the use of a building or the increase of storage that could affect the classification of occupancy hazard, particularly raising the height of stockpiles. Any such change should be notified to the insurance company and their advice or instructions obtained.

Having obtained the hazard occupancy classification, the next step is to ascertain the pressure required under given conditions of water flow. These are laid down in the rules, but to interpret these the technical services of installing engineers should be obtained.

SOURCES OF WATER SUPPLIES

The following are the acceptable sources of water supply:

1 Town mains

2 Elevated private reservoir
3 Gravity tanks
4 Automatic pumps
   (a) Drawing from purpose-built reservoirs or almost unlimited natural sources
   (b) Boosting from town mains or elevated private reservoirs
5 Pressure tanks

Fire brigade pumping inlets are now compulsory under the rules for gravity tanks and pressure tanks. It is advisable to fit them to all systems irrespective of the primary source of supply, but if town mains are used as a source of supply it is unlikely that any water authority would permit such an inlet if the pumped supply was other than from mains containing potable water.

Town mains

The main must be capable of supplying the required flow and pressure at all times. It is particularly important to take the lowest record when there is a heavy demand on the main, such as during the day in an industrial district.

The town mains must be fed from a source of at least 1000 m^3 and if they are required to supply an extra high hazard system the capacity of the source must be increased by an amount laid down according to the type of high hazard class.

Terminal mains or branch 'dead end' mains of less than 150 mm diameter are not accepted for ordinary hazard, group III and group IV special or extra high hazard class.

Where ring mains exist on the premises, it is advisable to include valves so that sections can be closed down for repairs or alterations without closing down the supply to the sprinkler installation.

It is recommended that isolating stop valves on ring mains should be of the interlocking type of control valve. With this type of valve the key cannot be removed unless the valve is open and the removable key is the only means by which the valve can be turned. Existing valves can sometimes be fitted with the device and it may not even be necessary to disturb the valve in the pipe line in order to do so. Some form of reminder that the key is out of place should be adopted. Figure 20.1 consists of diagrams of an interlocking key type valve in the open and closed positions.

It is good policy to examine any existing water main system on the premises to see whether a short extra piece of main could be inserted to form a ring main, or whether an extra valve or more could be added to protect the

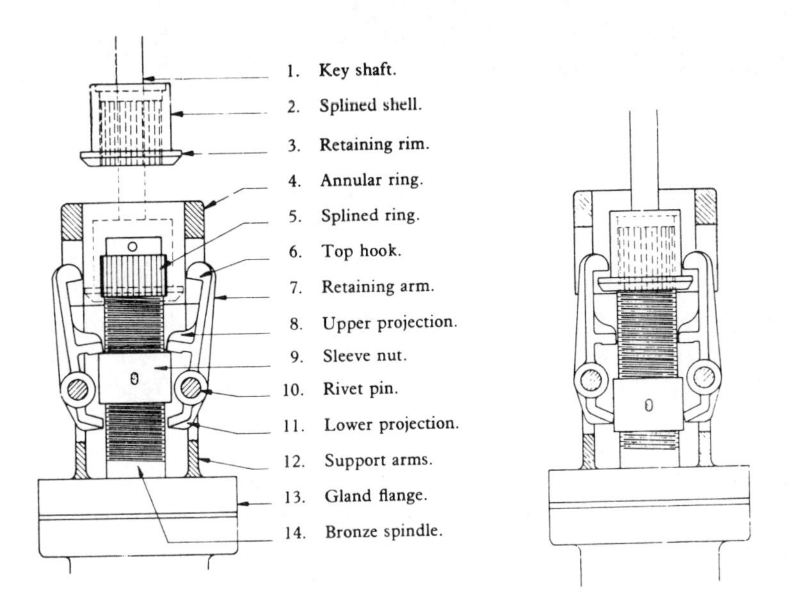

Figure 20.1 Isolating stop valve with interlocking key (a) Valve open, (b) Valve closed. (Mather & Platt Ltd.).

supply to the sprinkler installations in this manner. Plans of all such valves and mains should be posted up where they are readily available to the engineer and to the fire brigade, and in large installations it is also recommended that the plans should include instructions on which valves to operate to close or open the various sections.

Any stop valves, other than those under the control of the water authority, must be under the direct control of the occupier and they must be strapped and padlocked open.

Elevated private reservoirs

An elevated reservoir is a ground reservoir but situated at a higher level than the protected premises and which is under the sole control of the owner of the installation.

If the reservoir is used to supply water for purposes other than a sprinkler installation, it must have a constant capacity exceeding 500 m^3 if used for an extra light hazard class; 1000 m^3 if used for an ordinary hazard class and if required for an extra high hazard class then it must be at least 1000 m^3 plus the extra capacity required for the capacity design density of that particular class. Pressure flow tests under these circumstances must be carried out when the demand is at its greatest.

Gravity tanks

A gravity tank is one that is specially built and erected on the premises and at such a height as to provide the required pressure and flow at the installation control valves. It must be adequately protected against freezing and if not totally enclosed in a tower it must be covered over to exclude daylight and solid matter. A substantial permanent ladder or stairway must be fitted to provide easy access. The water must be kept clean and free from sediment. The tank must be cleaned out every three years and if necessary repainted.

It must have a capacity of at least that required for minimum quantities of stored water as laid down for the particular class occupancy hazard. If it exceeds the minimum quantity, it is permissible to use the surplus for other purposes. But this surplus may only be drawn off by a pipe entering the side of the tank above the level of the maximum sprinkler requirements. An indicator must be fitted permanently so that it will show the depth of water.

The level of the water must be automatically maintained and if it is the only supply for the installation the tank must be capable of being automatically refilled in six hours. If the rate of input is too low to achieve this a larger tank must be used. The tank must be provided with an adequate overflow pipe.

The use of one tank to supply installations in two or more buildings under separate ownership is not allowed.

Automatic pumps

When an automatic pump is to be used, there must be a sufficient quantity of water available to keep the installation supplied with water at the required design pressure.

Minimum capacity of stored water supplies

The rules contain tables of the quantities of water that must be stored for the various occupancy hazard classes. In the case of extra light and ordinary hazard classes the minimum storage capacity has to be greater if the height of the sprinkler heads above the control valves is more than 15 m and less than 30 m, and still further increased if the height is between 30 and 45 m.

For extra high hazards the minimum capacity rises to 875 m^3 when the design density reaches 30 mm/min. (Water application densities, which in imperial units are expressed as gallons per square foot per minute, are usually expressed in metric units as cubic decimetres per square metre per minute or millimetres as per minute.)

Boosting from town mains

Where permitted by the water authority, a pump may draw directly from the town mains provided that the latter is capable of providing water at all times at the maximum rated output of the pump. In such cases, there should be a bypass round the pump with a back-pressure valve on the bypass. The bypass should be at least the same diameter as the water supply connection to the pump.

Pump suction tanks

The minimum quantity of water to be stored exclusively for the use of automatic sprinkler installations is laid down in the rules and varies according to the hazard classification and the height of the sprinkler heads above the control valves.

If an inflow of water can be relied upon at all times to start refilling the tank automatically, considerably smaller storage capacity is required. The inflow plus the capacity of the storage must be sufficient to supply the pump when running at its full load for a specified time. This time limit is also laid down in the rules and is related to the minimum storage capacity referred to above.

Where the capacity of the storage tank is less than 500 m^3, it is laid down that the supply to the tank must be capable of refilling it in less than six hours or, if the rate of input of the supply to the tank is less than that required to refill within six hours, the capacity of the tank should be increased by the amount of the shortfall.

Effective capacity of pump suction tanks

In calculating the effective capacity of a pump suction tank whether depending on inflow or not, the depth is taken as the measurement between the normal water level in the tank or reservoir and the low water level 'X' shown in Figure 20.2 which is the level above the inlet to the suction pipe determined by dimension 'A' in column 2 of Table 20.1. Low water level 'X' is calculated to be the lowest level before a vortex is created and the pump begins to draw air.

Figure 20.2 is an illustration of the effective capacity of pump suction tanks. Table 20.1 is a table of suction pipe sizes and other dimensions for use with Figure 20.2.

When the suction pipe is taken from the side of the tank as in examples (*a*)

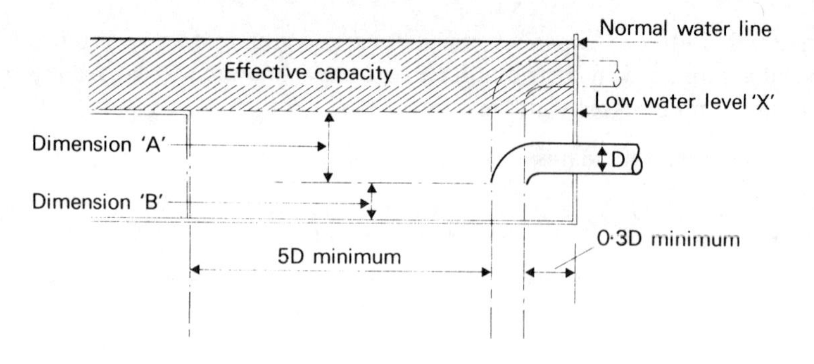

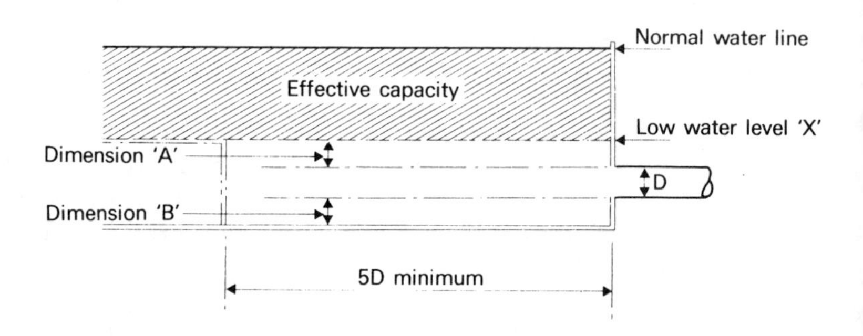

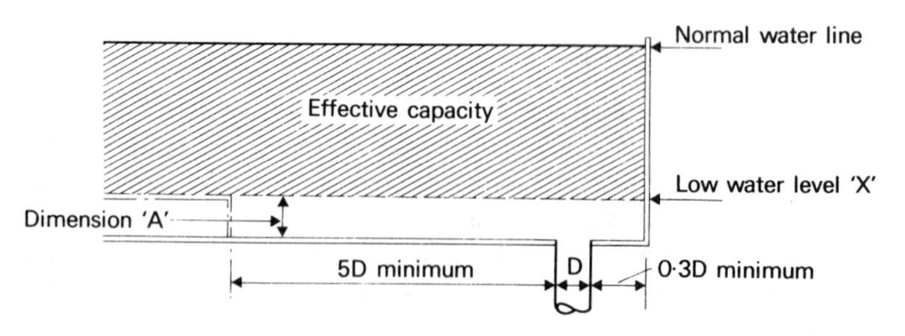

Figure 20.2 Effective capacity of a pump suction tank, (a) with a curved suction pipe taken from the side of the tank, (b) with a straight suction pipe taken from the side of the tank, (c) with the suction pipe taken from the bottom of the tank (FOC rules).

Table 20.1
Suction pipe sizes and other dimensions for use with Figure 20.2 (referred to in the Rules as table 2440)

1 Nominal diameter of suction pipe D (mm)	2 Dimension A (mm)	3 Dimension B (mm)
65	250	80
80	310	80
100	370	100
150	500	100
200	620	150
250	750	150

and (*b*) in Figure 20.2, there must be a clearance of not less than that shown as dimension 'B' in col. 3 of Table 20.1, between the base of the tank and the lowest level of the pump suction pipe.

Where the suction pipe draws water from a sump at the base of a suction tank, dimensions 'A' and 'B' specified in Table 20.1 and the minimum length dimensions specified in Figure 20.2 apply; the position of the sump is indicated by broken lines in each example of Figure 20.2. In addition, the sump width must be not less than $3.6D$, where D is the nominal diameter of the suction pipe. The point of entry of water to the suction pipe must be located centrally across the width of the sump.

The pump characteristics to give the required pressure and supply are laid down in the rules for the different classes of occupancy hazards. To achieve the correct pump characteristics, it is sometimes necessary to restrict the outlet by means of an orifice plate which is not integral with the pump. Rules for the size of the orifice and plate are specified and, in particular, part of the plate must extend outside in the form of a tag giving hydraulic information necessary when calculating the delivery characteristics. These calculations are made by the installing engineers prior to their quotation.

In the actual design of the pump, means must be provided in the upper part of the pump casing for the relief of air which might become trapped there and, if the pump operates under suction lift conditions, the release must be automatic. Means must also be provided to allow a continuous flow of water through the pump at a sufficient rate to prevent overheating of the pump when it is running light.

The pump must have a detail plate giving its nominal rating as specified by the rules and giving the hydraulic details if an orifice plate has to be used. The plate will also give the manufacturer's details.

Pump suction pipe

Where possible, the centre line of the pump should always be 600 mm below the level of the water supply and therefore under positive head conditions: this is shown at 'X' in Figure 20.2. Under these conditions, a table is drawn up showing the sizes of suction pipe for the various hazard classes.

The length of the suction pipe must not exceed 30 m and while it should be as straight as possible each elbow or bend must count as 3 m of length.

The suction pipes of pumps working under positive head conditions may have a fall from the supply to the pump provided that air cannot be trapped at any point.

Where pumps are installed with a centre line above or less than 600 mm below the level of the water supply, that is under suction lift conditions, the vertical height between the low water level and the centre line of the pumps must not exceed 3.7 m.

The suction pump must be fitted with a foot valve at its lowest point, and the size and the point of entry to the foot valve must comply with a table of nominal sizes of suction pipes and effective capacity of suction tanks. The suction pipe must be truly horizontal or with a continuous rise towards the pump to avoid the possibility of air locks. There is a different table of sizes for suction pipes when suction lift conditions prevail.

Pump priming arrangements

The priming of the pump has to be carried out by an elevated cistern automatically maintained with water obtained from an independent source. For extra light hazards the tank has to contain a minimum of 100 dm^3 and the priming pipe has to be a minimum size of 25 mm. The tank for ordinary and extra high hazards must hold a minimum of 500 dm^3 with a pipe of 50 mm for priming.

Arrangements must be made so that, if there is a serious leak from the pump, the suction hose or the foot valve and the priming system is being drained more quickly than automatic refilling can take place, then the pump itself must automatically start.

In the next section, it is recommended that when a pump comes into operation a signal is made. The reason for the signal must be investigated and the possibility of a fault in the priming arrangements or a leakage causing the fault should not be overlooked.

Water for priming purposes may only be taken from a gravity tank if there is an excess of capacity over the minimum capacity requirements and only then in such a manner that the level of the water in the gravity tank docs not fall below the specified minimum requirements.

If the water for priming the pump is provided by a connection from a branch from the town mains forming a supply to the sprinkler system, such connection must be made on the town mains side of the back-pressure valve upstream of any tapping for the pump starting devices.

Except where the abstraction of priming water is permitted from a gravity tank, arrangements must be such that they do not afford any outlet by which water can pass directly or indirectly from a pressure tank or gravity tank into pump suction pipe.

The following paragraphs are an abstract of the rules of the Fire Offices' Committee.

Automatic starting arrangements

The pump should have a direct drive and must start automatically. The automatic starting device for pumps must be of an approved pattern and set to operate when the pressure in the trunk main has fallen to a value not less than 80 per cent of the pressure in the trunk main when the pump is churning. Means should also be provided for manual starting by reproducing the pressure reduction. Once started the pump must run continuously until stopped manually.

Where an automatic pump forms the sole supply a fall in water pressure in the sprinkler system, which is intended to initiate the automatic starting of the pump, shall at the same time provide a visual and audible alarm at some suitable location, for example, in the gatehouse or by the installation control valves.

It is recommended that where the pump is situated remote from the protected premises visual and audible indication of the pump operation should be provided at some similar suitable location.

Power for the above warning systems must be taken from a separately switched sub-circuit to that feeding the pump in the case of an electric pump. In the case of compression-ignition engine driven pumps the power for the warning system may not be taken from the battery which is used for automatic starting. The term 'sub-circuit' means a circuit connected to a switch supplied directly from the low or medium voltage bus-bars on the main distribution board of the premises.

A test for the automatic starting of the pump must be carried out weekly. Means must be provided to reduce the applied water pressure to the starting device to simulate the condition of automatic starting at the requisite pressure.

If the engine is a compression-ignition type, the test must be for a period of at least 10 minutes each week. Where closed circuit cooling systems are used the water level in the primary system must be checked at the time of carrying out each test and if necessary water shall be added during the course of the test procedure.

Power unit

Power sufficient to drive the pump at the required pressure must be available at all times through the year.

Pumps must be driven either by an electric motor or an approved compression-ignition type engine.

Electric motor-driven pumps

The electric supply must be obtained from a reliable source, preferably from a public supply. Where a suitable public supply is not available full particulars of the generating plant must be submitted to the fire insurers for approval.

The electrical connection must be such that a power supply is always available for the motor when the switches for the distribution of other power throughout the premises are open. Any switches on the power feed to the motor must be clearly labelled '*sprinkler pump motor supply – not to be switched off in the event of fire*'.

A tell-tale indicator lamp or lamps must be provided to show that there is a power supply available for the motor. Where the supply is AC the failure of any one phase of the supply must be indicated. The indicator must be in the vicinity of the pump and so placed that it can be readily seen by maintenance personnel. All indicator lamps must be in duplicate (duplicate filament lamps are allowed).

An automatic warning of power failure to the motor starting switch (in the case of AC supply, of any one phase of the supply) shall be given visually and audibly at some suitable location, for example, in the gatehouse or by the installation control valves. Power for this warning system must be taken from a separately switched sub-circuit to that feeding the motor; or alternatively where the power is taken from a battery instead of the mains supply, the battery must be trickle-charged and have a capacity sufficient to provide an alarm for 72 hours' duration. The power may not be provided from any battery used for automatic starting of a compression-ignition engine driven pump.

Fuses

High rupturing capacity fuses must be installed and capable of:

(a) protecting the cable connections to the motor

(b) carrying the stalled current of the motor for a period of not less than 75 per cent of the period after which such a current would cause the motor windings to fail.

No-volt releases

Any no-volt release mechanism must be of the automatic resetting type such that on restoration of the supply the motor can restart automatically.

Magnetic thermal overload trips

These overload trips are not allowd.

All wiring associated with the electrically driven pumps including that for the monitoring circuits must be in accordance with the IEE wiring regulations. Motor supply cables inside buildings must be either PVC or rubber-insulated, and armoured; or mineral-insulated copper-sheathed. External overhead cables are not permitted within 6 m of any window, door or other opening in (a) any of the sprinkler-protected buildings and (b) any other building which is within 15 m of any of the sprinkler protected buildings.

Engine

The engine must be:

(a) of the compression-ignition mechanical direct injection type, capable of being started without the use of wicks, cartridges, heater plugs or ether, at an engine room temperature of 7°C and must accept full load within 15 seconds from the receipt of the signal to start;
(b) naturally aspirated, super-charged or turbo-charged and either air- or water-cooled. Charge air cooling is not allowed;
(c) capable of operating continuously on full load at the site elevation for a period of six hours;
(d) provided with a governor to control the engine speed within 4½ per cent of its rated speed under any condition of load up to the full load rating.

In addition, any manual device fitted to the engine which could prevent the engine starting must return automatically to the normal position.

The coupling between the engine and the pump must allow each unit to be removed without disturbing the other.

Cooling system

The following systems are acceptable:

(a) Cooling by water from the sprinkler pump direct into the engine cylinder jackets via a pressure-reducing device to limit the applied pressure to a safe value as specified by the engine manufacturer. The outlet connection from this system must terminate at least 150 mm above the engine water outlet pipe and be directed into an open tundish so that the discharge water is visible.
(b) A heat exchanger, the raw water being supplied from the sprinkler pump via a pressure-reducing device, if necessary, to limit the applied pressure to a safe value as specified by the engine manufacturer. The raw water outlet connection must be so designed that the discharge water can be readily observed. The water in the closed circuit must be circulated by means of an auxiliary pump driven from the engine and the capacity of the closed circuit must not be less than that recommended by the engine manufacturer. If the auxiliary pump is belt driven, there must be multiple belts such that should half the belts break the remaining belts must be capable of driving the pump.
(c) A frame- or engine-mounted air-cooled radiator with a multiple fan belt driven from the engine. When half the belts are broken the remaining belts must be capable of driving the fan. The water in the closed circuit must be circulated by means of an auxiliary pump driven by the engine and the capacity of the closed circuit must be not less than that recommended by the engine manufacturer.
(d) Direct air cooling of the engine by means of a multiple belt driven fan. When half the belts are broken the remaining belts must be capable of driving the fan.

Air filtration

The air intake must be fitted with a filter of adequate size to prevent foreign matter from entering the engine.

Exhaust system

The exhaust must be fitted with a suitable silencer and the total back pressure must not exceed the engine maker's recommendation. Where the exhaust system rises above the engine, means must be provided to prevent any condensate flowing into the engine.

Engine shut-down mechanism

This must be manually operated and return automatically to the starting position after use.

Fuel system

The engine fuel oil must be Class A to BS 2869: 1957 or equivalent. There must be kept on hand at all times sufficient fuel oil to run the engine on full load for six hours in addition to that in the engine fuel tank.

Fuel tank

The fuel tank must be of welded steel conforming to BS 814: 1961 for mild steel drums. The tank must be mounted above the engine fuel pump to provide a gravity feed. The tank must be fitted with an indicator showing the level of the fuel in the tank. The capacity of the tank must be sufficient to allow the engine to run on full load for

Extra light hazard	3 hours
Ordinary hazard	4 hours
Extra high hazard	6 hours

Fuel feed pipes

Any valve in the fuel feed pipe between the fuel tank and the engine must be placed adjacent to the tank and it must be locked in the open position. Pipe joints must not be soldered and plastic tubing must not be used.

Auxiliary equipment

The following must be provided:

- a sludge and sediment trap,
- a fuel level gauge,
- an inspection and cleaning hole,
- a filter between the fuel tank and fuel pump mounted in an accessible position for cleaning.
- means to enable the entire fuel system to be bled of air. Air relief cocks are not allowed; screwed plugs are permitted.

Starting mechanism

Provision must be made for two separate methods of engine starting, i.e.

(a) *Automatic starting* by means of a battery-powered electric starter motor incorporating the axial displacement type of pinion, having automatic repeat start facilities initiated by a fall in pressure in the water supply pipe to the sprinkler installation. The battery capacity must be adequate for 10 consecutive starts without recharging with a cold engine under full compression.

(b) *Manual starting by*:
   (i) crank handle, if engine size permits, or
   (ii) electric starter motor.

Note: The starter motor used for automatic starting may also be used for manual starting provided there are separate batteries for manual starting.

Battery charging

The means of charging the batteries must be by a two-rate trickle charger with manual selection of boost charge and the batteries must be charged in position. Where separate batteries are provided for automatic and manual starting the charging equipment must be capable of trickle charging both batteries simultaneously. Equipment must be provided to enable the state of charge of the batteries to be determined.

Tools

A standard kit of tools must be provided with the engine and kept on hand at all times.

Spare parts

The following spare parts must be supplied with the engine and kept on hand:

- two sets of fuel filters, elements and seals,
- two sets of lubricating oil filter elements and seals,
- two sets of belts (where used),
- one complete set of engine-joints, gaskets and the hoses,
- two injector nozzles.

Pressure tanks

Pressure tanks are accepted as a sole supply in the case of extra light hazard class and ordinary hazard group I class only; in the case of ordinary hazard, groups II, III and group III special they are only acceptable as one source of a duplicate water supply.

General requirements

Pressure tanks must be housed in a readily accessible position in a sprinkler-protected building or in a separate building of incombustible construction used for no purpose other than for the housing of fire protection water supplies. The tank must be adequately protected against mechanical damage. The temperature of the room should be maintained above 4°C.

Where the building housing the pressure tank is remote from the sprinkler-protected premises, such that it is impracticable to provide sprinkler protection (where it is required) from one or other of the installation control valve sets in the premises, the sprinkler protection in such a building may be supplied from the nearest accessible point on the downstream (installation) side of the back pressure valve (where one is fitted) on the supply pipe from the pressure tank. In such cases there must be a controlling stop valve (secured in the open position) fitted on the supply pipe to the sprinklers together with an approved alarm device with visible and audible indication of the operation of the sprinklers provided at some suitable location, e.g. in the gatehouse or by the installation control valves. A 15 mm drain valve must be provided downstream of the flow alarm to permit a practical test of the alarm system.

When used as a single water supply, the tank must be provided with an approved arrangement for maintaining automatically the required air pressure and water level in the tank under non-fire conditions. The arrangement should include an automatic warning system of approved design to indicate failure of the devices to restore the correct air pressure and water level within a reasonable period, both visually and audibly at some suitable location, e.g. in the gatehouse or by the installation control valves. Power for this warning system must be taken from a separately switched sub-circuit to that feeding the air compressor and water pump supplying the tank.

The arrangements referred to in the previous paragraph are also advocated in cases where the tank provides the duplicate supply; otherwise daily inspections must be carried out to verify the water level and air pressure and such deficiencies as may be found immediately remedied.

The tank must be fitted with air pressure gauges in duplicate and a gauge glass to show the level of the water. The second pressure gauge will not be required where the air pressure is automatically maintained. Stop valves are to be fitted on both connections to the gauge glass and they should normally be kept shut.

Stop valves and back pressure valves must be provided on both the water and air supply connections to the tank and they must be fixed as close to the tank as practicable.

Safety valves fitted to pressure tanks must be of an approved type and fixed in such a position that the valve seating is water-sealed. A connection to the valve from the air space above the water line should be provided to permit the rapid escape of air in the event of the valve opening. The setting of the safety valve for the correct working pressure must be carried out by the installing engineers and the valve must be so constructed that it can be tested without the setting being interfered with. The setting mechanism should be protected against alteration by unauthorized persons. The outlet from the relief valve must be an open end so that any leakage will readily be detected.

Pressure tanks must be examined thoroughly every three years when they must be cleaned and painted both internally end externally if necessary.

The use of one pressure tank to serve installations in two or more buildings under separate ownership is not allowed.

Minimum quantity of water to be maintained in the tank

When sole supply:

Extra light hazard class	7 m^3
Ordinary hazard class – group I	23 m^3

When duplicate supply:

Extra light hazard class	7 m^3
Ordinary hazard class – all groups	15 m^3

Air pressure

The minimum air pressure to be maintained in the tank depends on:

- the proportion of air in the tank;
- the minimum pressure at the highest sprinkler when all water is expelled from the tank; and
- the static pressure loss corresponding to the static heat when the base of the tank is below the level of the highest sprinklers.

GRADING SPRINKLER INSTALLATIONS

Sprinkler installations are graded according to the number and type of water supplies available and on the grading depends the premium discount that can be obtained.

Having examined the various systems of water supply and decided which are available, a decision can be made as to the possibilities of obtaining a good or better grading, but it must be emphasized that the insurance surveyor and the installing engineers should be called in for discussions as early as possible if any alteration to an existing installation is considered.

There are three grades of sprinkler installation:

Grade I having either 'duplicate' water supplies or one 'superior' water supply provided the total numbers of sprinklers (excluding those in concealed spaces) in the protected building or range of buildings does not exceed 2000, with not more than 200 (excluding those in concealed spaces) in each separate risk. A separate risk is defined as a risk separate from the rest of the building by construction conforming to the requirements of a fireproof building.

Grade II having one 'superior' water supply without the limitations to the number of sprinkler heads mentioned aboved.

Grade III having a single water supply as undernoted:

- town main
- elevated private reservoir
- gravity tank
- automatic pump

Grade III water supplies are not normally accepted for extra high hazard systems.

Duplicate supplies

Where a duplicate water supply is required, it must be capable of providing the same pressure and rate of flow and have the same capacity as is required for the primary supply. In the case of extra light and ordinary hazard classes only, a pressure tank will be accepted as one source of a duplicate water supply.

The following combinations of water supplies are regarded as affording a duplicate water supply service:

1 Two town mains:
   (a) The mains must be either independent or form part of an interconnected system having stop valves so arranged that in the event of a breakdown anywhere in the system, at least one of the mains to the installation can remain operative.
   (b) The system must be fed from more than one reservoir.
   (c) There must be a branch connection from each main carried separately up to the premises containing the installation. Two or more installations on any premises in one ownership may have the

second and subsequent installation supplied by a single pipe taken from a point of interconnection of the installation.

2 Town main and pressure tank (extra light and ordinary hazard classes only).
3 Town main and elevated private reservoir or gravity tank.
4 Town main and automatic pump. Where the automatic pump draws from a suction tank that requires an automatic inflow of water to maintain sufficient quantity to run the pump in accordance with the requirements in the rules, the town main which forms the duplicate supply must not be used to supply the balance.
5 Automatic pump and pressure tank (extra light and ordinary hazard classes only).
6 Automatic pump and elevated private reservoir or gravity tank, provided that the latter does not form the source of supply for the automatic pump.
7 Two elevated private reservoirs or gravity tanks. One double capacity reservoir or tank will be acceptable if it is suitably subdivided with separate down pipes from each division. The point of connection of each down pipe to the sprinkler main must be as close as possible to the protected premises and the common main must not traverse ground not under the control of the owner of the installation nor under the public roadway.
8 Automatic pump supply drawing from:
   (a) A virtually inexhaustible source, such as a river, canal, lake.
   (b) Two limited capacity reservoirs. The primary reservoir must have a holding capacity equal to that required for the particular hazard class. The secondary reservoir may be of small capacity with automatic inflow provided that it meets the requirements of the rules in this respect. In the case of extra high hazard systems, the primary reservoir may also be of smaller capacity provided it meets the requirements. It will not be necessary in such cases to provide separate automatic inflow facilities for each reservoir if the rate of inflow is available for either reservoir and meets the requirements. Consideration will be given to the acceptance of a single limited capacity reservoir where the circumstances are considered favourable. Such circumstances will include:
   (i) an effective stored capacity not dependent upon automatic inflow, but less than 450 m^3;
   (ii) the use of potable water;
   (iii) the reservoir being lined with a material which will not be conducive to corrosion, flaking or similar deterioration;

(iv) the reservoir being completely enclosed to exclude daylight and to ensure that the water does not become contaminated with extraneous matter, having either two automatic pumps, one of which at least must have a compression ignition engine drive and any two must be capable of providing in the aggregate the necessary pressures and flows for the respective hazard class; or three automatic pumps, of which at least two must have a compression ignition engine drive and any two must be capable of providing in the aggregate the necessary pressures and flows for the respective hazard class. In both arrangements the pumps must be capable of operating in parallel, i.e. they must have similar flow characteristics.

Where pumps are under positive head conditions a common suction pipe is permitted; otherwise there must be independent suction pipes and priming arrangements. In the case of the second arrangement provision must be made for the pumps to draw from either tank so that when one tank is rendered inoperative for cleaning, etc., the other tank is available for all three pumps. It is also recommended that in the case of the first arrangement both pumps should be able to draw from either tank.

9 Elevated private reservoir and pressure tank (extra light and ordinary hazard classes only).

'Superior' water supplies

The following are approved as superior supplies.

Town main

The town main must fulfil the following requirements:

1 It must be fed from two ways by mains each of which must be capable of furnishing the pressure and flows required.
2 There must be duplicate connections from the main carried separately up to the premises containing the sprinkler installation with a stop valve (open or closed) on the main between the two branches so that, in the event of a fracture in the town main or if it is necessary to close down part of the main, the stop valve could be operated to ensure that the installation could be supplied from the unaffected part of the main.
3 The mains referred to in 1 above must not be directly dependent on a common trunk main anywhere in the town main system.

4 The town main system must be fed from more than one source. If it is not possible to provide the duplicate connections referred to in 2 above, special consideration will be given to the waiving of the requirements if there is a stop valve (secured open) on the town main immediately on each side of a single branch connection.

Automatic pump supply

This must consist of either:

1 two automatic pumps, one of which at least must have a compression ignition drive and each capable of providing independently the necessary pressures and flows for the respective hazard class; or
2 three automatic pumps, of which at least two must have compression ignition engine drive and any two must be capable of providing in the aggregate, the necessary pressures and flows for the respective hazard class.

Where pumps draw directly from the town mains or from a suction tank which requires the inflow from a town main to provide the requisite capacity, the town main must be fed from both ends and conform to the requirements specified in 2, 3 and 4 and the relevant note in the previous paragraph. Where the pumps are under positive head conditions, a common suction pipe is permitted. Otherwise, there must be independent suction pipes and priming arrangements.

Elevated private reservoir or gravity tank

This is approved if adequately protected against freezing.

Pressure tank

This is approved for extra light hazard class and ordinary hazard group I class only provided that:

1 the water capacity is not less than:
7 m^3 for extra light hazard class
23 m^3 for ordinary hazard group I class
2 there is an approved arrangement for maintaining automatically the required air pressure and water level in the tank under non-fire conditions.

21

Sprinklers: water distribution

SPRINKLER HEADS

Water is distributed by sprinkler discharge heads, consisting essentially of a water jet impinging on a deflector which breaks the jet up into a drenching spray. The water jet is held in check by a valve, the valve being released either by the melting of a solder joint which breaks up a supporting strut or breaking of a glass bulb containing a heat expansible fluid.

Figure 21.1 is an exploded view of a soldered, fusible type of sprinkler discharge head and Figure 21.2 is a similar view of a quartzoid bulb type.

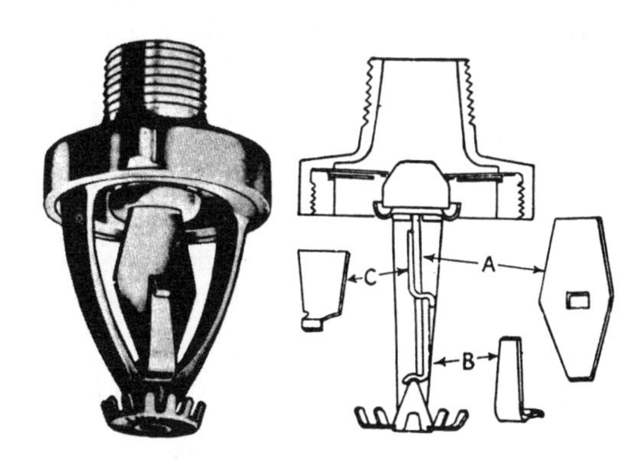

Figure 21.1 Exploded view of a soldered fusible type of sprinkler discharge head. (Mather & Platt Ltd.)

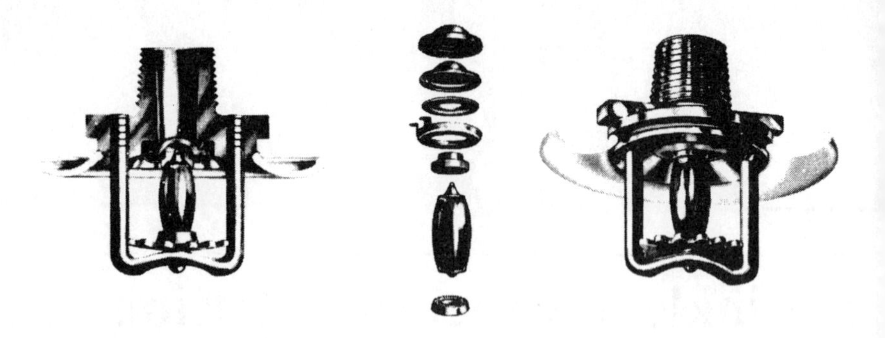

Figure 21.2 Exploded view of a quartzoid bulb type of sprinkler discharge head. (Mather & Platt Ltd.)

The normal sprinkler head releases at 68°C (155°F) but there is a range of heads which will release between 57 and 260°C (135 and 500°F). Generally speaking, the temperature rating of the heads should be as low as possible, but not less than 30°C (86°F) above the highest anticipated temperature conditions. Where there is any doubt, a maximum type recording thermometer should hung in a position to coincide with that of the sprinkler head. If the installation is already in use, it may be necessary to put in a high temperature head while the tests are being carried out.

The temperature rating is marked on the deflector of every head and, in addition, quartzoid bulbs are coloured differently for the different ratings. The standard rating is red. Solder types have different coloured yoke arms, the standard being uncoloured. The following is a list of the colour codes for sprinklers of different temperature ratings. Note especially that the colours are not the same for solder type sprinkler heads as for quartzoid bulbs.

Soldered and fusible type °C	*Colour of yoke arm*
68–74	Uncoloured
93–100	White
141	Blue
182	Yellow
227	Red

Quartzoid type

°C	
57	Orange
68	Red
79	Yellow
93	Green
141	Blue
182	Mauve
204–260	Black

Multiple jet control

These devices are virtually stop valves held closed by a fusible link or quartzoid bulb device. They are to be found in places where the spread of fire across a floor or area can be very rapid.

Where a multiple jet control is to be used, the sprinkler heads will be open, that is uncapped, and the result of the operation of a multiple jet control is to release water over a very much greater area than that covered by a single sprinkler head.

They are also used to protect areas where there is heavy electric switchgear or electronic apparatus. In this case, it is usual to enclose the switchgear or apparatus in a fire-resistant compartment so that sprinkler discharge heads are not required. To maintain the warning given by the sprinkler system, use is made of a multiple jet control installed in the compartment but this is so arranged that instead of the sprinkler heads being in the compartment, the multiple jet control releases water into a pipe where it operates an alarm and it is drained away safely.

If the walls of the compartment are of combustible material (but the ceiling must always be of non-combustible material) the discharged water is taken to a series of open drencher heads along the combustible walls on the outside of the compartment. Figure 21.3 is an illustration of a multiple jet control.

Protection of sprinkler heads

Sprinkler heads must not be painted and must not be obstructed in any way. Sprinkler heads that are painted will not operate at their correct rating and may not operate at all. Once a sprinkler head or fusible link has been painted, it must be replaced as there is no satisfactory method of removing paint without the possibility of affecting the operation of the sprinkler head on some future occasion. If necessary, sprinkler heads can be protected

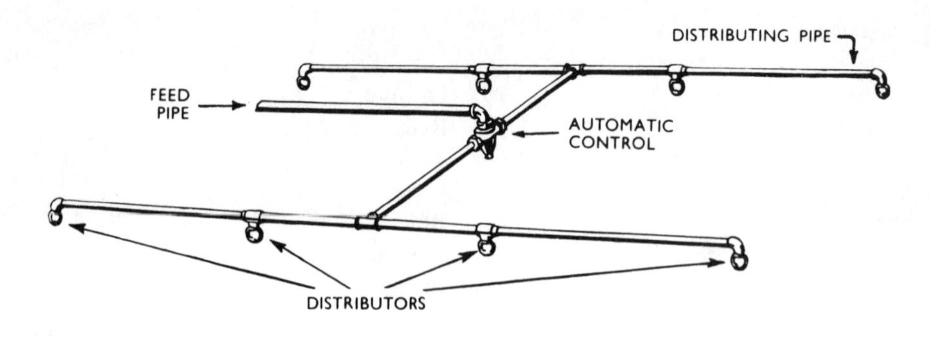

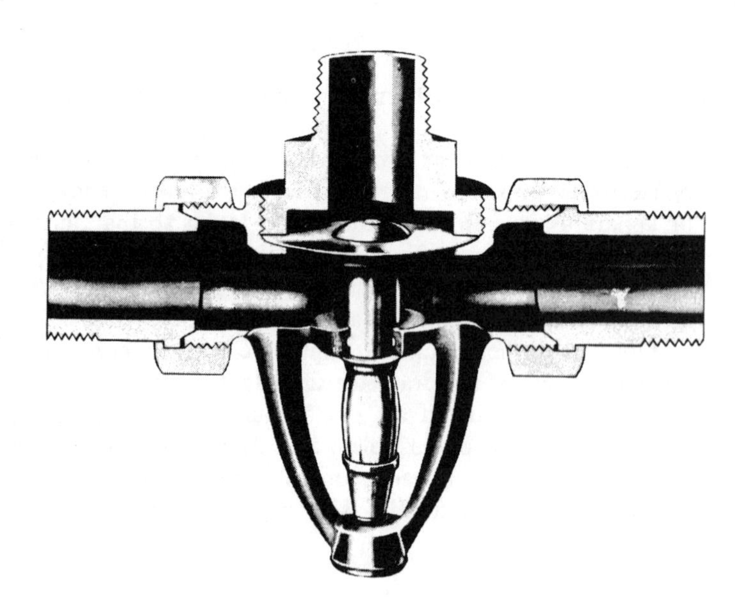

Figure 21.3 Multiple jet control. (Mather & Platt Ltd.)

while painting is in progress in a nearby area by means of a paper bag. The bag must be removed immediately the surrounding area has been painted. The bags should not be in place for more than a few minutes.

Sprinkler heads that are liable to mechanical damage should be protected by special guards supplied by installing engineers.

No goods may be stored within 0.5 m of a sprinkler head and for high-piled combustible stock the clearance must be 1 m or more.

Spare sprinkler heads

A stock of spare sprinkler heads should be kept, and it is suggested that six spare heads are required for an extra light hazard installation, 24 for ordinary hazard installations and 36 for extra high hazard installations. Double this quantity is called for if more than one installation is maintained. If there are any special temperature or other special types or multiple jet controls, spares of these should also be maintained. Suitable spanners for removing and replacing sprinkler heads should also be kept with the spare heads, and it is good pactice also to keep suitable spanners for use when draining the installation.

When replacing heads after a fire, it may be necessary to replace heads in the surrounding area that have been affected by heat even though they might not have actually operated.

Where corrosive vapours are present, the manufacturers will supply specially treated sprinkler heads. Alternatively, the heads may be coated with petroleum jelly which must be entirely cleaned off periodically before being renewed.

Normally sprinkler heads are installed in a pendant position but, if pipes are to be drained against frost, they are fixed upright above the supply pipe so that no water is left in the joints. Most deflectors are universal, but occasionally special deflectors to operate in a particular way are used.

DESIGN OF A SPRINKLER INSTALLATION

A sprinkler installation has to be hydraulically designed to provide an appropriate density of water discharge by an assumed maximum number of sprinkler heads that may all be open at the same time. The number of heads required is expressed as an area in square metres of water coverage. The density of discharge and the maximum area to be covered affect the pressure required, the flow and the capacity of the water supplies that will be required.

Spacing of sprinkler heads

The rules for the number and spacing of sprinkler heads are complicated. The spacing is dependent on the covering and water discharge rate, which in turn depend on the classification hazard of the occupation of the premises.

Pipework

The layout of the pipework and the sizes necessary to feed the specified number of heads must also be calculated. Guidance is given in the rules for a number of different arrangements, in which the water may be fed to the end or to the centre of an array of heads. For large installations it is usually possible to use smaller, and therefore less costly, pipes if they are arranged in a number of loops or networks. The calculation of pipe sizes is then very complicated but computer programs are available which will develop the most economic design.

The program will also provide a list of the pipe lengths and other components (bands, tees and so on) which will be required. It may be necessary to have calculations of this kind approved by the insurers before a system is installed, to ensure compliance with the overall requirements of the rules.

If pipework is being installed for an alternate installation, it must be sloped at a sufficient angle to allow for drainage. As far as possible drainage must be back to the main drain at the valve controls. Where this is not possible, subsidiary drain points will be required. To assist in drainage operations, subsidiary drain points should be numbered and listed.

Pipes must not be embedded in concrete; neither should they be in a position where it is difficult for any work in connection with extensions or alterations to be undertaken.

Where corrosive conditions exist, pipework must be thoroughly cleaned and protected by suitable means. It is suggested that two coats of bitumastic paint should be applied and renewed at intervals of one to five years, according to the severity of the conditions.

Where pipework is liable to mechanical damage it must be protected by adequate guards, and vertical risers and valves must be protected by guard rails where necessary.

If any water supply is not of the highest purity, flushing points must be installed at the end of the distribution pipes. Flushing points are usually closed like drain plugs with a valve and a plug. Water pipe plugs have square ends and special spanners should be kept for them.

Hangers for sprinkler pipes

There must be ample hangers to support sprinkler piping and it should be unnecessary to add that nothing should be hung from sprinkler pipes. As a general guide, there should be one hanger between each sprinkler head. There should be a hanger not more than 900 mm from the end of each distribution pipe.

When inspecting hangers, make sure that there is no undue strain on the pipework, that pipes are not pulled so that they do not drain correctly and make sure that down pipes for drainage purposes are properly secured.

Certificate of compliance with the FOC rules

It is laid down in the rules of the FOC that automatic sprinkler installations will not be approved unless the erection of them has been carried out by an approved manufacturer, using only approved materials, or by contractors specially approved by the manufacturers. Even then, the manufacturers must be satisfied that the installation complies with the rules and that the standard of design is correct and the workmanship is satisfactory. Lists of approved manufacturers can be obtained from the Fire Offices' Committee. Even if the system is not being installed to meet insurance requirements, it is advisable to check the technical credentials of any firm offering a sprinkler protection installation. If the work is being carried out in accordance with FOC rules, the manufacturer has to supply the insured with a completion certificate in a special form stating that the installation, extension or alteration of the installation has been carried out in accordance with the current printed rules of the FOC for automatic sprinkler installations.

Although many of the salient points of a sprinkler installation will be described in this chapter, the subject is so complex that only specialists in sprinklers can fully interpret the rules and the advice of the insurance company should be sought at the earliest stage of consideration.

BUILDINGS TO BE SPRINKLER-PROTECTED

The building itself

Under the FOC rules, every part of the building must be sprinkler-protected There are a few exceptions such as a fireproof storey, a fireproof room or enclosed fireproof stairways, provided that these are enclosed and constructed in accordance with the rules of the FOC relating to their standards of construction. The term 'fireproof' is used here in its meaning within the FOC rules.

Unless the FOC rules are carried out in full, no approval will be given and no insurance discounts allowed.

Under certain building regulations, local councils can demand the provision of sprinkler protection in basements and ground floors, and the size of a permitted building compartment is increased if sprinklers are installed. These installations do not, however, rank for FOC approval unless the entire building is fitted with them in accordance with the rules.

Adjoining buildings

To obtain FOC approval, buildings which communicate with the protected building must also be sprinkler-protected. In this respect a building which communicates is one that is not a separate fire compartment as laid down in the FOC rules of standards of construction. There are certain exceptions.

Exposing buildings

Exposing buildings must also be sprinkler-protected. An exposing building is one of greater capacity than 150 m^3 within 10 m of the protected building and which, by reason of its occupation and construction, constitutes an exposure hazard to the protected building.

There are again exceptions to this rule but, where compliance is difficult, it may be possible to obtain protection by installing drenchers over the openings or combustible sections facing the exposure hazard or by other means. For details of the exceptions, the rules should be consulted and where there is any real difficulty the insurance company may be able to help.

OTHER TYPES OF INSTALLATIONS COVERED BY THE FOC RULES

Drencher installations

Drencher installations are used to protect openings in walls and combustible parts of walls and roofs from an exposing hazard. They are also used for example to keep the safety curtain in a theatre cool. When used for this purpose in a theatre, the control will be found on the off-prompt side of the stage near the safety curtain release gear.

The installation of external drenchers is governed by special rules of the Fire Offices' Committee. Briefly, these rules require all sprinkler heads to be of an approved open pattern under a master control. This master control must be so situated that it can be operated in safety.

The rules cover the protection of all window openings on the top two storeys and of alternate storeys below these. Roofs have to be protected by a line of drenchers on the apex of the roof. Drenchers may not be more than 2.5 m apart nor more than 1.2 m from the window jamb. The piping must be arranged so that it will drain naturally when the water is shut off.

As the drenchers are all open, it is easily understood that water supplies must be exceptionally good. The installation may only be fed from a practically unlimited and approved source that will give an efficient operating pressure at the level of the highest head. The installation must be provided with a fire brigade pumping inlet. The installation must be regularly tested and inspected.

High velocity spray installations

These consist of specially designed drencher heads which produce a fine spray of water travelling at high speed. High velocity spray installations are used to extinguish fires involving burning liquids.

Systems like this are effective only against fires involving liquids with high flash points, preferably above about 60°C. They function by cooling the surface layers of the liquid to a temperature at which insufficient vapour is released to form a flammable mixture. The high velocity is needed to ensure that the drops can penetrate the plume of hot gases. Very large drops would be able to do this even if they were simply falling under gravity, but they would then penetrate rapidly through the liquid and would achieve very little cooling. The small drops do not have sufficient momentum to penetrate the liquid in this way but they do manage to churn up the surface layers, and this ensures that cooler fuel is mixed into this layer together with the water. Cooling also results from the evaporation of some of the finer drops, which reduces the flame temperature and hence the radiation to the surface. The combined effect of the reduced radiation and the direct surface cooling by both fuel and water reduce the surface temperature to a value below the fire point. Burning can then no longer continue. It is sometimes suggested that the formation of an oil/water emulsion at the surface is important in ensuring extinction of the fire. It seems unlikely that this does happen, but there is certainly intimate mixing of oil and water, which ensures rapid heat transfer. Burning high flash point fuels can be extinguished by the application of a fine spray from a branchpipe, and here there is very little possibility that an emulsion is formed.

Medium velocity spray installations

These sprinklers are very similar in action to high velocity sprinklers and are controlled in much the same way. The purpose of these sprinklers is to keep tanks and pipes containing flammable liquids or explosive gases cool during fire conditions. To achieve this the droplets of water are of such a size that they will be able to penetrate the plume more efficiently and impinge directly on to the surface to be protected. Where a larger number of projectors is required, the water supply is controlled by means of a deluge valve (see Figure 21.5). Figure 21.4 is a diagram of a deluge installation.

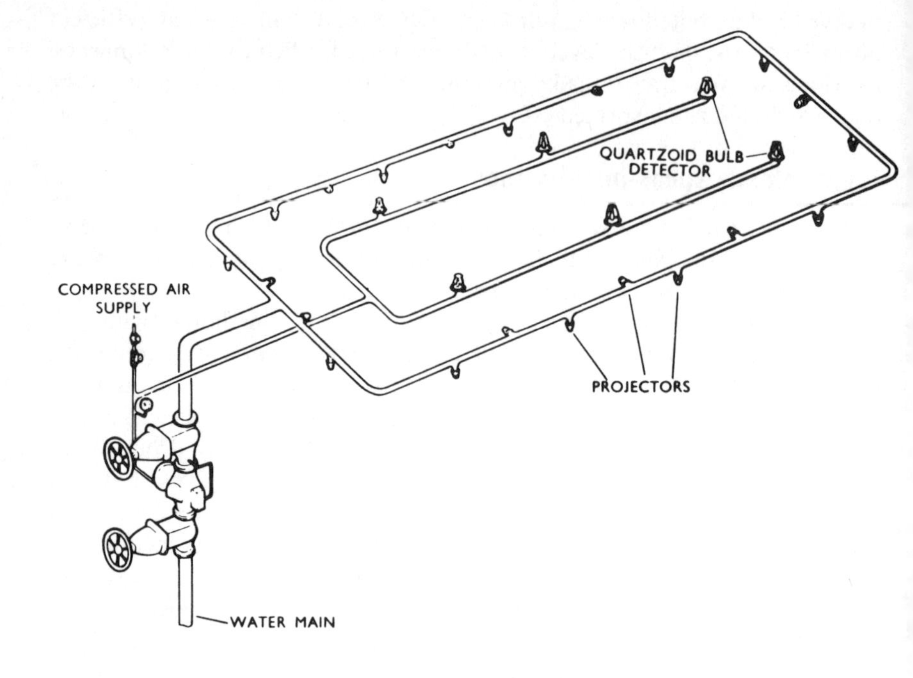

Figure 21.4 A deluge valve installation.
(Mather & Platt Ltd.)

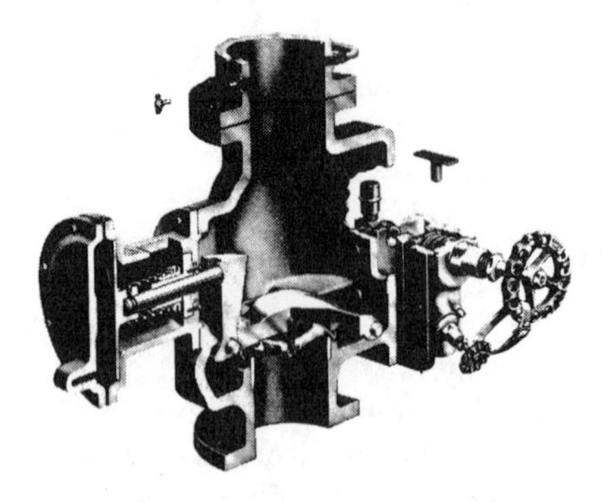

Figure 21.5 Cut-away deluge valve.
(Mather & Platt Ltd.)

Because the use of medium velocity sprinklers involves large quantities of water, very often for a long time, the design of bunds and drains should be given careful consideration. In some cases, it may be possible to build dwarf walls with sufficient capacity to hold all the liquid in the tanks or pipes. But more often it is necessary to make arrangements to drain away flammable liquids to a safe area together with the water from the sprinklers.

The pre-action device

Heat detectors and other automatic fire detectors can be set to detect fire conditions at a much earlier time than that possible by sprinkler heads. By using detectors of an FOC approved type in conjunction with a dry installation, the installation can change over to water in readiness for a sprinkler head to open, thus saving a few valuable seconds in getting water to an incipient fire.

The automatic detector may be connected to the fire station or to a local control point for a private firefighting team. An alarm can then give given directly to the fire brigade as well as the alarm that is given by the sprinkler gong. It is possible that an early alarm may permit first aid firefighting to take place before the sprinkler installation operates to release water.

The deluge valve

Figure 21.5 shows a cut-away deluge valve. The hinged clack valve is held down against the pressure of water by a latch. The latch can be tripped by a bolt connected to a diaphragm. The installation is filled with compressed air and is so arranged that if the pressure drops below a certain level, the diaphragm moves over, thrusting the bolt forward to trip the latch and release the water.

There is an ordinary sluice valve above and below the deluge valve and both these valves must be strapped and padlocked open when the deluge valve has been set and is ready for operation.

PRECAUTIONS TO BE TAKEN WHEN AN INSTALLATION IS SHUT DOWN

The following are the rules for the precautions to be taken when an installation is to be rendered inoperative as laid down by the Fire Offices' Committee.

Permission must be obtained from the fire insurers before rendering a sprinkler installation inoperative.

Alterations and repairs to the installation or its water supplies should be carried out during normal working hours as far as practicable, and all expedition must be used so that the sprinklers may remain inoperative for as short a time as possible. As much of the installation as practicable must be kept operative during the progress of the work. If the work cannot be completed in one day particular attention must be paid to this point when the premises are left each day.

Before the water is turned off a thorough examination of every part of the premises must be made to ascertain that there is no indication of fire. Smoking should be prohibited during the progress of the work.

When an installation is rendered inoperative during working hours, foremen or heads of departments must be notified so that, in case of fire, the best possible use may be made of the hand-extinguishing appliances.

When an installation is rendered inoperative and is likely to remain so outside working hours, all the fire-extinguishing appliances must be held in special readiness for immediate use with a sufficient number of trained personnel available to handle them. If possible such arrangements should be made before the water is turned off and also as much as possible of the installation should be kept operative outside working hours by blanking off the inoperative section or sections.

In the case of manufacturing premises, when the alterations or repairs are extensive or it is necessary to 'break' a pipe exceeding 40 mm (nominal) diameter or to overhaul, repair or remove a main stop valve, alarm valve or back pressure valve, every effort must be made for the work to be carried out when the machinery is stopped.

In the event of an installation having to be rendered inoperative as a matter of urgency or becoming inoperative through accident, the above precautions, so far as the same are applicable, must be observed with the least possible delay and the fire insurers should be notified as soon as possible.

Drought conditions

Where town main supplies are curtailed through drought special attention should be given to the maintenance, in an efficient condition, of any other supplies. All fire-extinguishing appliances must be held in special readiness for immediate use. In the case of large premises it is recommended that a watchman should be on duty throughout the night.

A very serious position arises when a sprinkler installation has to be closed down, because the insurance company will be making an allowance for protection that does not exist when the installation is out of commission.

The first consideration must therefore be to obtain the consent of the insurance company or to notify them if the shut-down is caused by an emergency.

Copies of the precautions to be taken can be obtained as leaflets on written application to the Fire Offices' Committee. They should be readily available and the conditions rigidly observed. They are automatically a condition of the insurance policy and failure to comply with them might invalidate the policy.

So important is it to maintain sprinklers operative for as long a time as possible, to leave only the least possible area unprotected at any one time and to put them back into commission as quickly as possible, that it is suggested that a notice should be placed on the desk of the director in charge of fire protection and also on the desk of the fire precautions manager, until the installation is again completely operative.

One of these cards might also be prepared for exhibition while waiting for minor repairs, alterations and extensions to be carried out.

PROVISION OF PORTABLE FIRE EXTINGUISHERS

One of the rules of the FOC for automatic sprinklers is to the effect that fire extinguishers must be provided and kept in working order in all sprinkler-protected buildings. Specific numbers or types of extinguishers are not stated in the sprinkler rules, but it is laid down that they are to be in accordance with the requirements of the insurance company concerned.

SPRINKLER FIRE ORDERS

The fire orders for sprinkler installations must be drawn up to meet local conditions and individuals. Where patrolmen and watchmen are employed, they must be carefully coached in what to do when they hear a sprinkler alarm, or when an electric signal is received from the installation.

The most important point of all is to call the fire brigade or, if an electric signal exists with the fire brigade, to confirm with the brigade that they have received the call. The next point is a negative one: on no account must the main stop valve be shut down until the whole premises has been fully searched for fire. It is not sufficient to discover that a leak has developed or that a sprinkler head has operated for a fire. The whole building must be searched before the main valve is shut down.

The most effective method of carrying out this instruction is for a man to be posted immediately at the main valve with instructions that the valve must not be turned off until he has been informed by the chief officer of the fire brigade that the building has been searched and that the fire is out or that there is no fire.

After a fire, a man should remain posted at the main valve until any damaged sprinklers can be replaced or a damaged section of pipe plugged off, and the main valve is turned on again. In the event of a fire while a sprinkler installation is shut down, even if there is still an open pipe, it is possible to open up the main valve and obtain some value from the sprinkler protection. Additional water damage is covered by insurance and is preferable to a total loss by fire. In an emergency such as this, regular sharp taps on a sprinkler pipe with a metallic bar will be plainly heard at the valves as a signal to turn the water on again.

In the rare cases when there is a failure of a sprinkler installation, the most frequent reason is that the water is shut off when the fire occurs, and included under this condition are the times when the sprinkler installation has been shut down too early. It is good practice not to shut down the main valve until a charged line of hose is available at the seat of the outbreak, even though it may be considered that the fire has been extinguished.

At every fire, however small, a man should remain on the site of the outbreak until the sprinkler heads have been replaced and the installation has again become fully operative. This man should have with him suitable fire extinguishers and preferably a charged line of hose or a hose reel under water.

22

Sprinklers: control valves

There are three types of automatic sprinkler installations. They are:

1 Wet installations, in which water is kept in the pipes at all times. A wet installation may only be used in a building which can, and must, be kept above freezing point at all times. Figure 22.1 shows typical control valves in a wet sprinkler installation.
2 Alternate, or wet and dry installations, in which the water in the pipes is replaced by air during the winter months. Figure 22.2 shows typical control valves in an alternate sprinkler installation.
3 Dry installations in which water does not enter the pipes except under alarm condition. These installations are only found in countries that have extremely low temperatures, or in cold rooms and ovens. They are controlled in a similar manner to alternate installations and no further description of them will be given.

WET INSTALLATIONS

All the control valves found in a wet installation will also be found in an alternate installation.

The main valve

The main valve controls the water entering the installation. It is a sluice valve (wedge valve) and is illustrated in Figure 22.3.

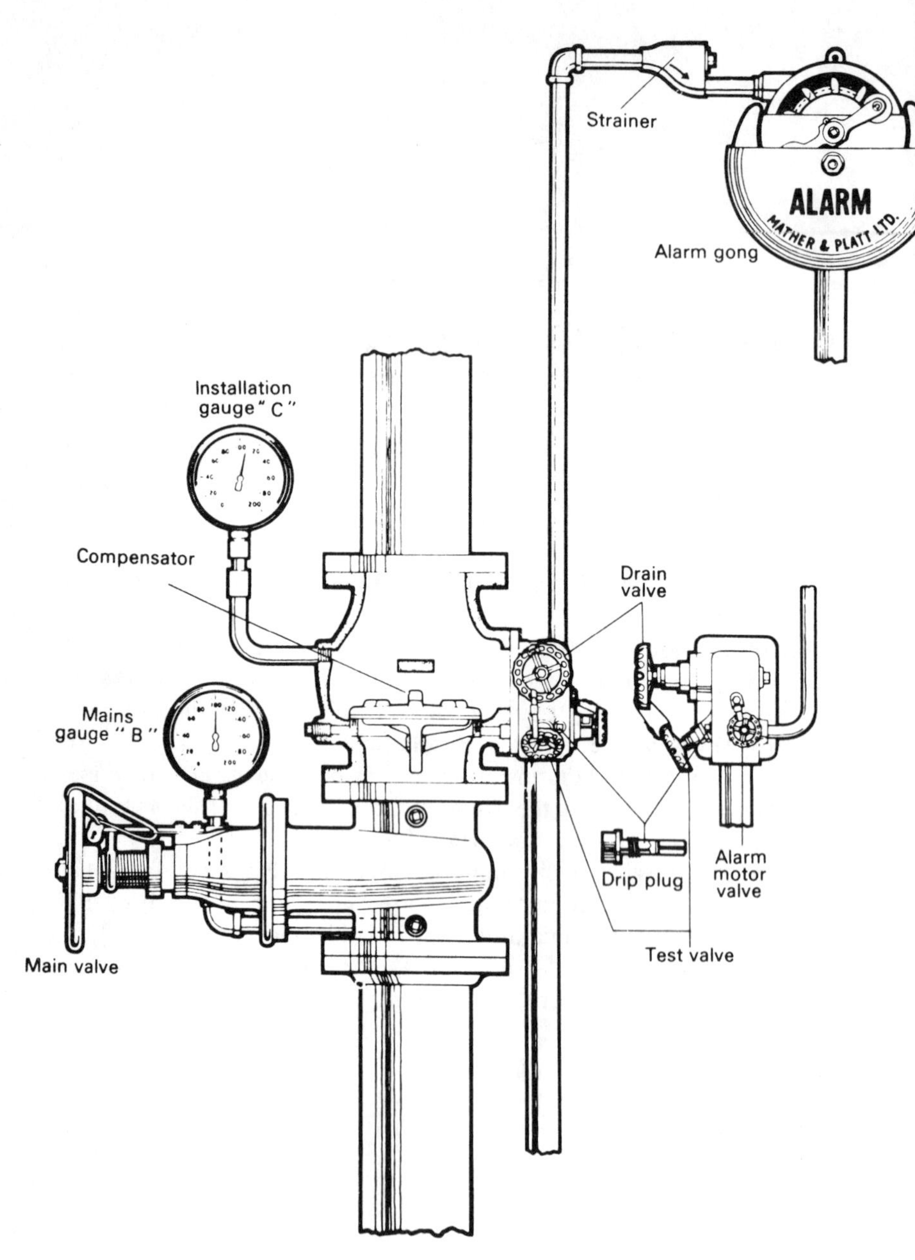

Figure 22.1 Control valves of a typical wet sprinkler installation. (Mather & Platt Ltd)

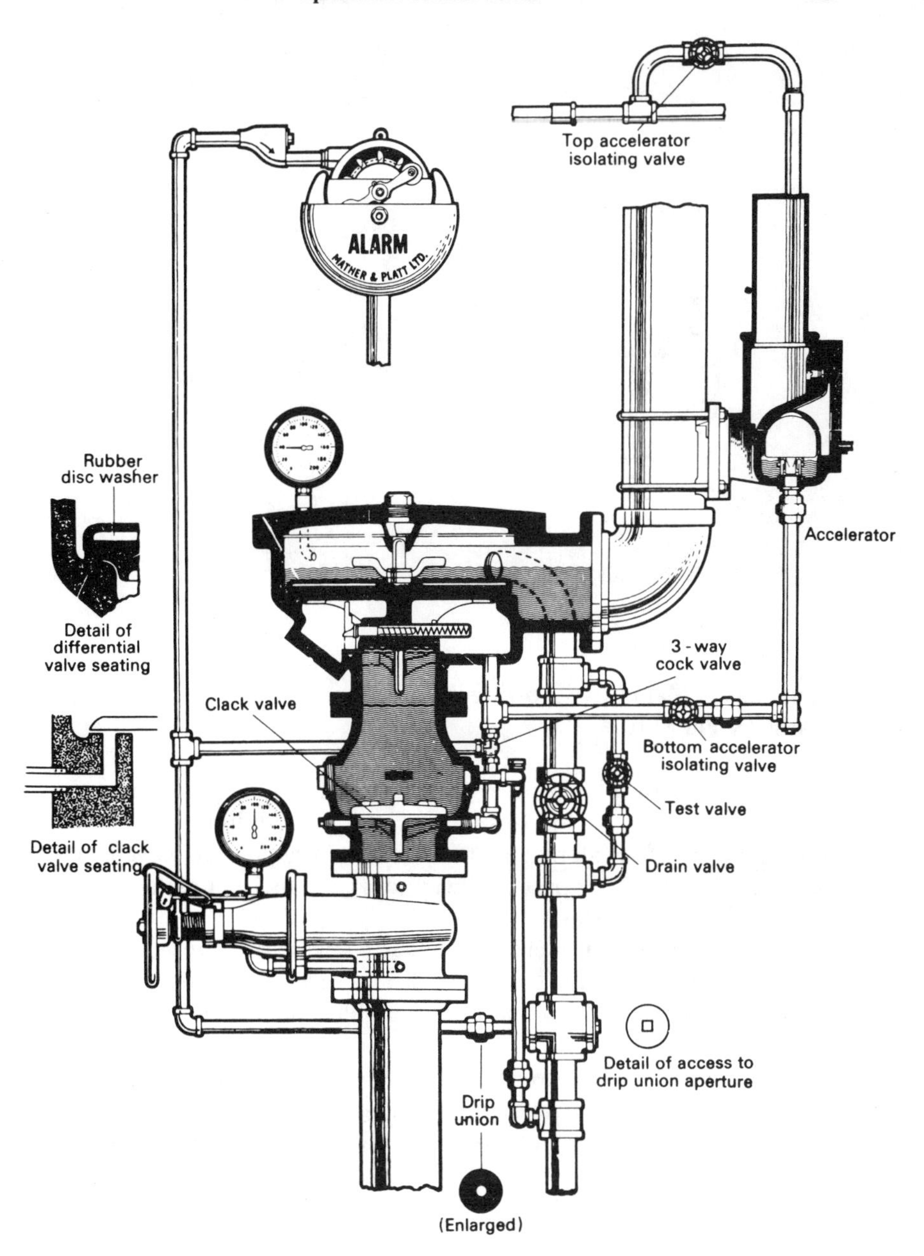

Figure 22.2 Control valves of a typical alternate sprinkler installation. (Mather & Platt Ltd)

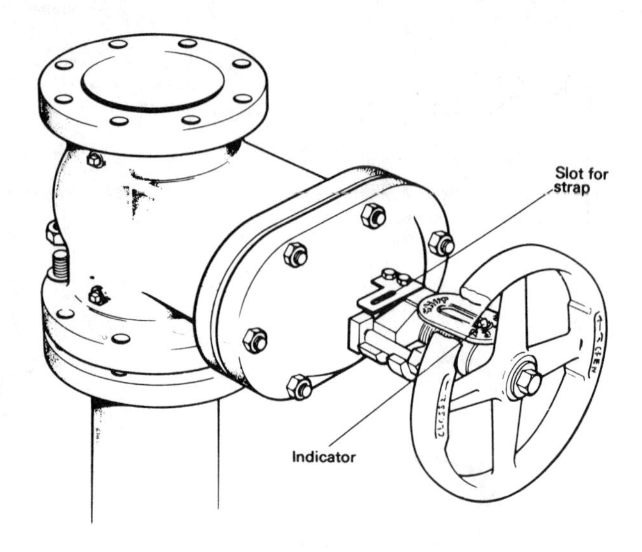

Figure 22.3 Main valve showing indicator
(Mather & Platt Ltd.)

It is operated by a hand wheel, and the direction of turning it to close or open it is shown by an arrow on the wheel. In addition to this, there is a pointer which moves along a plate to indicate whether the valve is open or closed. In Figure 22.3 a part of the hand wheel has been cut away to show the position of the indicator.

It is important that everyone connected with fire precautions should form a habit of looking at this indicator every time they approach a sprinkler installation. If the main valve is shut, the whole installation is out of order.

The main valve should be strapped and padlocked in the open position. It should not be chained because, in an emergency, the strap could be cut and a knife is hung up near the main valve for this purpose. The key to the padlock should be kept under strict control so that no unofficial alterations can be made to the installation.

The main valve must not be closed until it is certain that there is no fire risk. It must not be closed during a fire except under the orders of a senior fire officer, and it must not be closed for any other purpose unless the rules given in Chapter 21 have been carried out.

Alarm gong and motor

It is laid down as one of the principles of sprinkler installations that warning is to be given of danger. This warning is given by means of a loud sounding gong which is operated by the flow of water into the installation. This is an important feature and, although electrical warning systems may be added, the positive signal is that of the alarm gong.

The alarm gong is rung by means of a loose hammer driven by a water wheel of a type known as a 'pelton wheel' – that is, it consists of a series of cups on the rim of a wheel on to which a jet of water impinges.

Before entering the alarm motor, the water passes through a strainer. The strainer and pelton wheel can be seen clearly in Figure 22.1. This strainer should be examined at regular intervals, and cleaned as necessary. The length of time between examinations will depend on the state of the water system and the amount of loose material deposited from it.

The water to drive the pelton wheel is obtained from the alarm clack valve and can be controlled by the alarm gong shut-off valve.

Alarm gong shut-off valve

A valve is included in the conrol system to allow the alarm gong to be shut off. This valve has to be kept open, and should be strapped in the open position. On no account should it ever be closed except during a fire. If it is closed, one of the important functions of the installation – that of giving a warning of danger – is lost.

Alarm clack valve

The alarm clack valve immediately follows the main valve. It consists of a flat metal circular disc having a valve stem which fits loosely into a guide, allowing vertical movement only. The disc has a suitable washer.

The clack valve fits on to a special seating, consisting of two parts, an outer and an inner seating, forming a groove between them. When the valve is pushed upwards off its seating, water can flow into the installation and also into the groove. The groove leads to the pipe which supplies the alarm motor. The alarm clack valve and the detail of the annular groove leading to the alarm motor can be seen in Figure 22.2.

When the main valve is turned on, water passes up through the alarm clack valve (a small part going into the groove, and then on to the gong motor) until the installation is fully charged. The pressure of the water will then be the same above the alarm clack valve as it is below and the clack valve will drop on its seating, stopping any further flow to the gong motor.

Compensator

In the centre of the alarm clack valve there is a small automatic valve known as a 'compensator'. This valve permits a small quantity of water to pass into the installation without raising the alarm clack valve. The object of the compensator is to make allowance for small fluctuations of pressure such as might be due to temperature changes, and even for small fluctuations in the mains. Where main surges are greater than the compensator can pass, special arrangements have to be made. These are described later.

Drip plug (or drip plate)

If water were allowed to remain in the alarm motor and the pipe leading to it, it is possible that it might freeze and put the alarm gong out of action. To prevent this, a special plug is inserted into the controls. This plug has a small hole in it to allow the water in the pipework to the gong motor to drain away. The plug is generally associated with the drain valves and can be seen as a square-headed bolt. A plumber's spanner with a square end should be used to remove the plug. Any other type of spanner will eventually ruin the head of the bolt. The same type of spanner will be required for other purposes.

The bolt should be withdrawn every six months for inspection and cleaning, if necessary.

In older systems, the drip plug is in the form of a hole in a plate embodied in a union leading into the 50 mm drain. To examine the hole, it is necessary to remove the cover in the drain pipe. This cover has a square lug by which it can be unscrewed. Again a plumber's spanner should be used to prevent damage to the plug.

When looking for leaks in the valve systems, it is well to remember that water from the alarm gong system will be draining down for some time after the gong has ceased to operate. Only by inspection of the drip plug or drip plate can it be determined whether the water is coming from the installation or not.

Installation drain valve

A drain valve is fitted to the control valves to enable the system to be drained for repairs and for re-fixing spinkler heads after a fire. In the case of dry and alternate installations, it is used when changing from water to air.

The installation drain valve is also used by the installing engineers when proving the capacity of the water supplies. The capacity of the installation has to be checked periodically to ascertain whether water supplies have deteriorated. The calculations are very complicated and should be left entirely to the installing engineers.

The size of the valve is 50 mm for ordinary and extra high hazard systems, but may be 40 mm for extra light hazard systems.

Test valve

A 15 mm test valve is also fitted to the control valves. The opening of this test valve allows water to escape from the installation, and so rings the alarm gong. It is the equivalent of opening one sprinkler head.

The alarm gong must be tested every week, and the results of the test recorded on a chart supplied by the insurance company.

Gauges

Associated with each set of control valves are two or more gauges. One gauge records the pressure in the installation, that is the pressure above the alarm clack valve. This gauge should be lettered 'C'. Another gauge shows the pressure below the main stop valve and this should be lettered 'B'. If there are any back pressure valves on the town mains, there will be gauges to show the pressure on the town side of the back pressure valve. These gauges will be lettered 'A' together with a number if there is more than one town main with a back-pressure valve.

It is important that all patrolmen should be able to understand the significance of any variation in the indication given by the pressure gauges. This is discussed under inspection and testing procedure.

ALTERNATE AND DRY INSTALLATIONS

The essential difference between an alternate installation and a wet installation is the addition of an extra valve known as a 'differential' valve, and certain changes in the control valves that are necessary for testing purposes when the installation is on air.

Differential air valve

This valve can be seen in Figure 22.2. From the illustration it will be seen that the air pressure acts on the top section of the valve which has a water seal. This section is eight times larger than the bottom section of the valve which holds the water back. Theoretically, the air pressure can hold back a water pressure eight times greater, but in practice it is a little less. As a safety precaution the air pressure is usually kept at 40 lbf/in^2 (2.76 bar).

As soon as a sprinkler head opens, the air pressure commences to fall until it can no longer hold the differential valve shut against the pressure of the water. Water then enters the installation pipework. Once the differential valve has been raised by the flow of water, it is held open by a spring catch.

As it is difficult to keep an installation absolutely sealed against leakage of air, it is necessary to watch the air pressure gauge,and occasionally boost the air pressure. In most modern systems this is carried out automatically by means of a small air pump. While this system is very good, it tends to make the patrolman careless so that he does not watch for the approach of trouble.

When the installation is on air, a different method of obtaining water to operate the alarm motor has to be used. This is by taking the pipe for the alarm motor from the atmospheric chamber between the upper and lower sections of the differential valve.

Three-way cock valve

In an alternate installation, the different method of obtaining water for the alarm motor when the installation is on air requires a special three-way cock valve to effect the changeover between wet and dry working.

The setting of this cock valve is very important as it is possible to leave it in a position that will shut off the alarm completely. As it has to be operated each week under test conditions when the installation is on air, there is quite a possibility of this happening.

Figure 22.4 shows the various settings of the three-way cock valve. As the warning bell can be put out of action if the three-way valve is incorrectly set, it is most important that the valve should be regularly inspected to see that it is correctly set, and to see that it is kept strapped in the correct position.

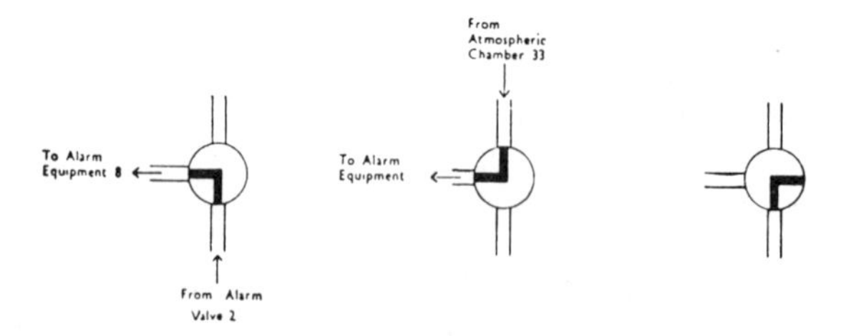

Figure 22.4 Three-way cock valve (*a*) on water (*b*) on air (*c*) in the off position
(*Mather & Platt Ltd*)

Test valve (on air)

If the test valve used when the installation is on water is opened when the installation is on air, air pressure will be lost, and water will enter the installation pipework. This is very undesirable because it is a lengthy operation to drain the installation of water and re-set the differential valve. During this time the installation will be inoperative.

A special test valve is provided for use when the installation is on air. But before the gong motor can be supplied with water, the three-way cock has to be turned to wet working conditions. After the test, this three-way cock valve must be restored to the air working position.

Occasionally, difficulty is experienced in getting the alarm motor to stop after a test, but if the three-way cock valve is momentarily turned to the off position, the valves throughout the installation will usually settle down.

Test valve (on water)

On the alternate installation, the test valve, when the installation is on water, will be found as a direct link between the installation pipework and the drainpipe.

Air accelerator

To reduce the time that the water takes to overcome the air pressure when a dry installation is operated, a device known as an 'air accelerator' is fitted (see Figure 22.5). This device consists of an air chamber which is separated from the installation pipework by a diaphragm. Connected with the diaphragm is a plunger. When the air leaks from the installation, the diaphragm is deflected to one side, moving the plunger which, in turn, trips a heavy bob weight and lifts a valve. This valve is connected between the installation pipework and the atmospheric chamber in the differential valve. The air pressure is then balanced on each side of the top section of the valve and the pressure of the water can immediately lift the differential valve, and flow into the installation.

There is a small orifice which allows a slight leakage between the installation and the air chamber to allow for normal changes in temperature which might otherwise cause a pressure difference and give rise to false alarms. This does occur when the orifice is blocked.

Tail-end valves

Where a small number of sprinkler heads, not exceeding 100 in the group, are required in a section of the premises subject to freezing, but where the

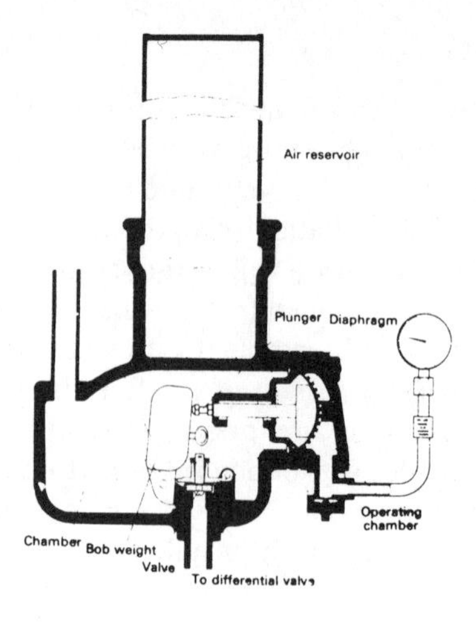

Figure 22.5 Air accelerator (*Mather & Platt*)

rest of the installation can be maintained at a satisfactory temperature, the use of a tail-end valve on a wet system is permitted.

A tail-end valve is, in effect, an alternate system, enabling a small number of sprinkler heads to be under air pressure. The advantage of the system is that it maintains water on most of the heads and brings water much closer to any heads that might operate on the air section. Tail-end valves can often be used in refrigerated rooms or low temerature ovens.

Subsidiary stop valves

Valves are not allowed in the distribution pipework as a general rule, but the following exceptions are permitted:

1 A valve to assist the regular testing of a permanent dry pipe installation.
2 A valve to control a small group of sprinklers in an area not exceeding 100 m^2, which are exposed to frost conditions, such as small loading bays, outside staircases or gangways. This exception is not intended to allow goods to be stored against outside walls of the building. This is particularly dangerous, as a fire in such stored goods could attack the building at unprotected windows and eaves, and so many sprinkler heads would open that the supply would be weakened and the building destroyed. It is particularly important to drain those sprinkler heads that are shut down not only at the beginning of the

cold season, but regularly during the winter months, to discharge any accumulated water. The heads must, of course, be installed in the upright position, and the pipes sloped for drainage.

At one time it was common practice to tie paper bags over exposed heads, but this is not now permitted as it could seriously interfere with the operation of the sprinkler head.

3 A valve is permitted in connection with hoods over the drying ends of papermaking machinery and in similar situations where work has to be carried out under conditions which might damage the sprinkler head. In such cases, the valve must be strapped and padlocked in the open position when the hood is in place. Special arrangements should be made to include an inspection of this valve in the daily inspection of the premises.

Two other valves are permitted, but these do not control any sprinkler heads. They are:

4 Valves at the end of pipes used for draining or flushing out the pipework.
5 Valves in connection with the outlets at the highest level of sprinkler heads. These valves are used when testing the flow of water at these points.

PRESSURE SURGES IN PUBLIC WATER MAINS

The pressure in street mains varies considerably between day and night, and in particular it tends to build up in a series of surges during the early evening. Although the alarm clack valve is fitted with a small compensator valve to allow for slight rises in pressure, the surges are often too great for this to prevent the alarm gong ringing for a few seconds at every surge. This can be very annoying to local inhabitants and precludes the use of an electrical warning system. To overcome the problem, a retarding device, consisting of a larger tube, is sometimes installed in the alarm gong pipework. But the most satisfactory method is to install a pump to bridge the alarm clack valve. By this means, the pressure in the installation can be kept above that of the highest surge. Both systems delay the operation of the alarm gong for a short period, but do not otherwise interfere with the working of the system. Because of this delay in the ringing of the gong in the event of fire, the permission of the insurance company must be obtained before it is installed.

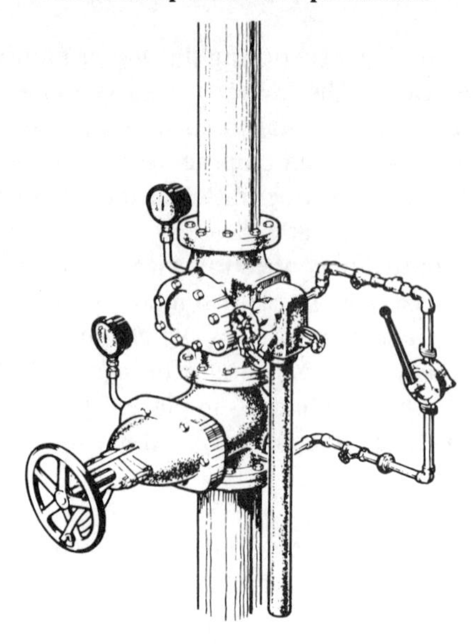

Figure 22.6 Booster pump connections (Mather & Platt Ltd)

Figure 22.6 shows the way a booster pump is connected. In modern installations, the booster pump is automatic and comes into operation when the pressure drops below a predetermined value, this setting being above the normal highest surge. The pump control is so designed that it does not supply water at so great a rate that the normal operation of the clack valve is affected.

IDENTIFICATION OF AN INSTALLATION

Rules for automatic sprinklers specify that there is to be a plan showing the area protected by sprinkler installations, with the position of the control valves. This plan has to be exhibited where it can readily be seen by the fire brigade on arriving at the premises. The plan is of particular importance when there is more than one installation. The position of the plan should be permanent, and the attention of the local fire brigade should be drawn to it when they visit the premises.

Where there is more than one installation on the premises, each installation must be numbered, the number to be shown on the control valves, the associated gong and on the plan. In addition to showing the position of the control valves on the plan, it is useful to have a different colour for the area protected by each installation. If possible all the pipes on that installation should be painted a coinciding colour, but where for the sake of appearance this is not desirable, possibly small bands of colour could be put on the pipes at suitable intervals to assist in tracing out an installation when necessary.

In addition to the plan kept on the premises, the rules call for a large metal plate to be fixed outside the premises as near to the main stop valve as possible displaying the words ‘sprinkler stop valve inside’.

It is permissible to connect an approved pressure operated switch at the point where the water supply is taken off for the alarm motor. No shut-off valve is permitted between the pressure switch and the installation. The connection from this switch can be taken directly to the alarm receiving station, and extra insurance premium discount may be allowed if this is done.

Where there is more than one installation, an indicator board will be required. The electrical circuits should be similar to those described later under automatic fire alarm systems.

Instead of the pressure switch, a water flow alarm valve can be placed in the water main. This, as its name implies, signals whenever water is flowing in the main. In very large installations, it is sometimes of considerable help to be able to ascertain quickly which part of the system is operating. By the insertion of water flow detectors, signals can be brought back to the area of the control valves, and even to a watch room.

PATROLMAN'S SPRINKLER INSPECTION

The training of patrolmen should include instruction with regard to the inspection of the automatic sprinkler installation, and also how to read pressure gauges.

Immediately on taking over duty, the patrolman should visit the sprinkler control valves. He will observe that the main valve is strapped and padlocked in the open position. He will check that the three-way cock valve is in its correct position for the state of the installation. If the installation is on air, he will also check that the accelerator-isolating valves are open. He will next make his first check of the gauges, and on every subsequent round he will again observe the gauges, and if there is any variation he will take the action laid down in his instructions.

It is suggested that a small diagram whould be hung near the control valves, showing the position of the three-way cock. It would be very helpful if the acceptable range of the pressure readings on the gauges could also be added to the diagram, together with local instructions as to the procedure if there is any variation. The diagram would be changed when the installation was switched between wet and dry working. A warning should be given to ignore the position of the handle on the three-way cock valve as this is often a loose fit and may have been changed round. The position of the cock valve is marked on the valve stem (see Figure 22.4).

If a notice is not used, a small red mark can be painted on the installation gauge (gauge C) for use when the installation is on air. This will show the lowest point to which the pressure may fall before definite action has to be taken. It is preferable that the patrolmen should understand the gauges and interpret what is happening.

All patrolmen should be taught the significance of sprinkler installation gauges and it must be impressed on them that they must report immediately if the gauges show any unusual variation or if there is any setting of the valves with which they are not familiar. For instance, if the gauge C was lower than B or A, it would indicate that the main valve was shut off. If A was lower than usual (or B if there was no gauge A), this would indicate that the main supply was shut off from outside – a very serious state of affairs. Apart from the action to be taken the local fire brigade should be notified because it might mean that the street fire hydrants would also be out of action. If warned, the fire brigade might be able to take preliminary precautions in case they had to set up a water relay in the event of an outbreak of fire.

Arrangements must also be made to ensure that an official of the company is informed, so that steps can be taken to inform the insurance company as stated in the rules for the action to be taken when the water supplies to a spinkler installation are out of action. Where possible emergency measures should be taken such as bringing up a motor trailer pump and laying out hose to the nearest hydrant or hydrants.

Gauge C also has an important function when the installation is on air, because if the pressure in the installation falls more quickly than the relief aperture in the accelerator can give compensation, the valve will trip. This will turn the installation back to water working and the gong will ring continuously, giving warning of a fault condition.

It is most important that the gauge C should be regularly surpervised when the installation is on air, particularly after it has been turned over from water. A leakage of air more often occurs at that time because air will find its way out even when water may not do so. The interference with the installation by changing over will also tend to find any weakness in the distributing pipes.

During the first hour after changeover, the gauge should be read at least twice and if there is a drop in pressure of more than 2 lbf/in^2 (0.14 bar) the installing engineers must be called back immediately. If the gauge remains fairly steady after the first hour, the next visit should take place in about two hours. The rate of fall can then be estimated, and the engineers called back to good time before air pressure has dropped to such an extent that water can enter the installation.

It is usual in modern installations to include an automatic air pressure booster pump which will pump the installation back to 40 lbf/in^2 (2.8 bar) but so slowly that its action does not interfere with the sudden drop in pressure that occurs when a sprinkler head is opened. In older installations, the air pump has to be started as required.

If a slow leak is allowed to continue, water will eventually enter the installation and in doing so will operate the alarm valve. When the gong of an alternate system starts to ring, a quick test to determine whether it is a major leak, such as a sprinkler head opened by fire, or neglect of a slow leak can be made by turning the three-way cock to wet operation. If the gong stops it is a sure sign of a slow leak. This is because the flow of water to the gong is self-sustaining, and the clack valve cannot re-seat itself while water is flowing to the gong. Care must be taken to ensure that the three-way cock valve has not been left in the off position.

Immediately after the installation has turned over of its own accord to wet working, the installation must be opened up, and either the composite valve pushed up to its full travel and held there by the ratchet device within the chamber or the installation must be re-set to working on air.

WEEKLY TEST

A test of the alarm motor and gong has to be made every week, a different procedure being adopted for an alternate installation on air.

Although the drip plug should remove the water from the motor and feedpipe, it is advisable to postpone the tests if very severe freezing conditions exist. It is good practice to prepare a checklist of the various items comprising the control valves, ticking them off while checking that they are correct.

PERIODICAL CHECKS

1. The strainer on the alarm motor supply should be opened up at about six-monthly intervals according to the amount of silt that collects in it.
2. In older installations the hammer of the gong and the motor bearings require oiling about every six months. Modern gongs have nylon bearings which do not require oiling.
3. The drip plug or the hole in the orifice plate should be inspected

occasionally. If there is a leakage of water not coming from the alarm motor supply pipe, the installation being on water, the clack valve is not seating correctly and the system must be closed down, drained and the clack valve and its seating cleaned and re-set.

If the installation is on air setting, water is leaking past either the top or bottom of the differential valve. The installation must be shut down and the differential clack valve removed and cleaned. When checking the differential valve, give particular attention to the seating of the top valve, including the rubber washer.

4 Every six months remove the lock and straps from the main stop valve, and run the valve up and down, grease the threads if necessary. Check that the gland is watertight, but do not overtighten the gland nuts. If leakage is persistent, it is better to renew the gland packing.

5 When an alternate installation is on air, all drain points should be opened for a short while and any accumulated water blown out. Be careful not to let out so much air that the installation changes over to water working, nor at such a rate that the accelerator valve is tripped.

Drain points will be found at the low level parts of the sprinkler pipework and should be fitted with a valve and a plug. The plug will have a square end, and only a square ended (plumber's) spanner should be used on it.

6 By putting a special orifice in the drain pipe, the installing engineers can provide a chart to show the pressure that is to be expected when the drain valve is fully opened. If any variation is noted, the full calculated tests are required to determine whether the water supply and pressures are still ample. These tests are now very technical and can only be carried out satisfactorily by installing engineers. No definite times are demanded, but once a year is suggested.

ELECTRICAL SUPERVISION

Control valves

It is possible to maintain a sprinkler installation under electrical supervision. This is carried out by installing switches at various points on the control valves.

Mention has already been made of switches to indicate water flow and these constitute a fire alarm signal. Further switches can be introduced and the signals brought back to an indicator board associated with the installation or installations. From this indicator board, a fault signal can be sent back to an alarm receiving station, or the home of a duty engineer. The main concern is that if a fault signal is given there must be means for obtaining attention to rectify the fault.

The following points, which are not given in the FOC rules, could be specified:

1 The main valve. Fault signal to be given when the valve is not fully open.
2 Mains pressure gauge. Fault signal to be given when the mains pressure falls below a predetermined level.
3 The installation pressure gauge. Two switches are required, one for when the installation is on water, and one when it is on air. Fault signal to be given for a predetermined low reading on the gauge in use.
4 The three-way cock valve. Two switches are required, one for use when the installation is on water, and one when it is on air. Fault signal to be given when the setting is incorrect.
5 Accelerator isolation valves. There are two valves, one on each side of the accelerator, which, when closed, enable the air accelerator to be disconnected from the installation so that tests and maintenance can be carried out without releasing the air from the installation. Fault signal to be given when either valve is not fully open, because if either valve is shut, the accelerator is out of commission.

Note that items 3, 4 and 5 will be pre-set for air or water when the installation is changed over at the beginning of each season.

Pumps

The FOC rules call for visual and audible warning of a fall in mains pressure when the fall is intended to start a pump, if the pump is the sole supply.

The rules also call for a tell-tale lamp to show that there is electric power available if the pump is driven by an electric motor.

If the pump is driven by a compression-ignition engine, it is also advisable to know that the pump has started. If it has not started automatically, it means that manual starting must be immediately attempted. A signal is therefore required to show that the pump is delivering water into the installation.

The sequence of signals should therefore be: first a signal that water is flowing from the mains into the installation, followed by a signal to show that pressure has dropped to 80 per cent in the mains, and finally a signal to show that there is pressure in the delivery from the pump.

To assist the starting of older types of compression-ignition engines, a decompression device was fitted, enabling the engine to be turned more easily until it fired. The device is thrown out of gear automatically after a number of revolutions of the engine have taken place. If this device is not re-set after each start, starting is more difficult. An indicator for giving warning when it has not been re-set will prevent this from being forgotten.

Where more than one supply of water is used, valves and back-pressure valves are required for certain aspects of control. If desired, these controls can be fitted with indicator switches to show when they are on their normal setting.

23

New sprinkler developments

FLUSH AND RECESSED SPRINKLERS

Many architects object to the appearance of ordinary sprinklers and a number of manufacturers have developed heads which are less obtrusive. Some are recessed into the ceiling and are designed to drop into a lower position when they are actuated. With this type the temperature-sensitive element, sometimes a horizontally mounted quartzoid bulb, is exposed just beneath the ceiling. Other designs are completely hidden by a metal disc.

It will be remembered that sprinklers are actuated by being immersed in the layer of hot gases which forms beneath the ceiling. The temperature immediately adjacent to the ceiling is usually less than that a few centimetres away, mainly because of heat losses to the ceiling. For this reason, sprinklers located at or above the ceiling will take longer to respond than those in the more usual location. A possible compromise is to use some of the more recent versions of the conventional design, which are small and either chromium-plated or fabricated in plastic materials (white plastic being quite inconspicuous).

ON-OFF SPRINKLERS

Figure 23.1 shows a sprinkler which opens at a pre-set temperature (like all other sprinklers) but then closes when the temperature has fallen below a second predetermined value. The temperature-sensitive element is a dished bimetallic disc shown as a snap disc on the figure. In the off position the

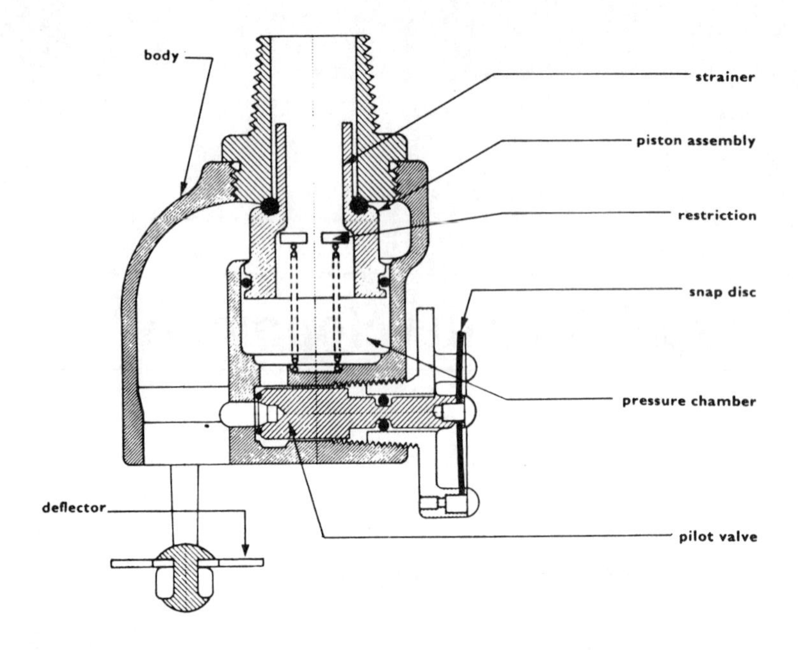

Figure 23.1 On-off sprinkler head

water is held back by the piston assembly, which pushes the upper O-ring against its seating. The piston is pushed up by the spring, and it is also arranged that the area of the lower part of the piston is greater than that at the top. When the critical temperature is reached the disc snaps over so that its outer surface is convex instead of concave. This movement opens the pilot valve, and water runs out of the pressure chamber beneath the piston. Water cannot flow through the restriction in the piston assembly fast enough to keep pace with this loss, and the pressure acting on the upper surface of the pison forces it open. When the temperature drops and the disc snaps back to its original position, the sequence of events is reversed.

The head can open and close several times if necessary, to control a fire with the minimum application of water. The same effect can be obtained where the water supply to a group of open nozzles is controlled by a suitable detector system.

FAST-ACTING SPRINKLERS

The response time of sprinklers has gradually been reduced with the introduction of new designs over the years. A quite dramatic reduction to a

Figure 23.2 Fast-acting sprinkler heads
(*Viking Microfast, Total Fire Protection Co. Ltd*)

response time about one-sixth that of conventional designs has been achieved by the Viking Corporation. One of their 'Microfast' sprinklers is illustrated in Figure 23.2. It will be seen that the glass bulb is very thin, only 2.8 mm in diameter, so that its thermal mass is very low. Because of the high compressive strength of glass, and particularly as in this design when it is in the form of a column, an adequate loading can still be applied to the valve seat. Indeed these sprinklers are tested hydraulically to 34 bar.

HIGH RACK STORAGE

In modern warehouses, particularly automated ones, goods are stacked to quite considerable heights. They then automatically require an extra high hazard classification under the FOC rules no matter what is being stored. A great deal of research has resulted in the concept of separate groups of open sprinkler heads located within the racking and actuated by a suitable detector. A line detector (discussed in Chapter 16) is particularly useful: it can be threaded around each bay of the racking.

AEROSOL STORAGE

Because of the possible deleterious effects of CFC (chlorofluorocarbon) propellants on the atmospheric ozone layer, many manufacturers have changed to hydrocarbon propellants. The one most frequently used is butane. Some of the contents of aerosols are already flammable (hair lacquers and furniture polishes for example) and the addition of butane has obviously increased the hazard. This is particularly true when one considers that the CFCs are chemical extinguishing agents (although not particularly efficient ones – see Chapter 6). Aerosol cans are usually packed in tens or dozens in cardboard boxes or on carboard trays, and then formed into pallet loads by shrink wrapping. A fire involving these packing materials can cause the rupture of a can after only a minute or so, and the fire will then grow rapidly in intensity. The burst cans may be ejected quite violently, but surprisingly the fire is not spread by them: the contents burn as a small BLEVE at the moment of rupture. Fire spread is by radiation from an intense fire (50 MW or more from a single pallet load), and the ignition of other materials in excess of 6 m away froam the fire occurs very rapidly. Clearly the introduction of flammable aerosol propellants can considerably enhance the fire risk in a warehouse, and a great deal of testing, both full scale and in the laboratory, has been undertaken. Work at the Factory Mutual Test Center (USA) sponsored by the Chemical Specialties Manufacturers Association has demonstrated the need for rapid detection and also for the penetration provided by large droplets (which are not produced by spray spinklers). The Viking fast-acting sprinklers were included in the tests and also an ESFR (early suppression fast response) system.

ESFR SPRINKLERS

It should perhaps be emphasized that the Factory Mutual Test Center (FM) did not develop the ESFR sprinklers solely for the protection of stocks of flammable aerosols (see above). They had started a programme of research on the optimization of sprinklers in 1968. The main objective of the ESFR work was to develop a system which would deal with a rapidly developing fire by rapid detection followed by complete suppression (rather than simply control). Ideally this should be achieved using no more than four sprinklers. The success of the system depends on a combination of a fast response with a high water application rate and with a high proportion of large droplets. A standard was issued in April 1987 (FM Loss Prevention Data 2–2) which was based on the work which had been completed at that time. Testing has continued since that date.

COMPUTER PROTECTION

Surprisingly, sprinklers have been used to protect some computer rooms, particularly where a great deal of paper is used. The computer units themselves can be dried out successfully after a discharge.

HIGH RISE BUILDINGS

Sprinklers are of particular value in preventing the spread of fire in high rise buildings, where the potential for multiple fatalities is high. Data collected during a 3½-year period in New York showed a 99 per cent success rate for sprinklers installed to the NY Building Code. The buildings were used as offices, hotels, shops, showrooms, flats, warehouses and factories. Of 661 fires, 654 were controlled. In the seven fires which were not controlled, the system had been shut off completely in five cases, was partly inoperative in one case, and the final fire involved a very hazardous situation. It is interesting to note that in 70 per cent of these fires only one sprinkler operated, and 10 or more sprinklers operated in only 2 per cent of the fires.

If these buildings contained offices, flats or hotels, the NFPA13 standard would require a maximum coverage per sprinkler of 21 m^2 and a minimum density of 3.78 mm/min. The FOC and CEA rules would require the same coverage but a density of only 2.25 mm/min. This is just one of the many differences between the standards adopted in different countries.

ELECTRICALLY ACTUATED SPRINKLERS

A 'Metron' actuator comprises a small metal cylinder with a tight-fitting piston. The cylinder contains a small quantity of a gas-generating pyrotechnic composition which can be electrically fired. If one of these is screwed into a suitably modified yoke of a sprinkler head, the firing of the actuator will cause the emerging piston to shatter the quartzoid bulb. This arrangement allows a sprinkler to operate in the normal way, but also enables a number of heads to be opened simultaneously on receipt of a signal from a detector or by the manual operation of a switch.

24

Other water systems

Mention has already been made of systems other than sprinklers which are covered by FOC rules. Many systems, particularly those intended for cooling, are not intended to comply with the rules: in particular they will frequently rely on water being pumped into the system by the fire brigade. It is still necessary of course to design systems with standards of reliability similar to those of a sprinkler system.

The main purpose of water cooling is to buy time: to prevent the failure of steelwork until a fire has been extinguished (or has burned out). Steel loses completely its mechanical strength at a temperature of about 500°C. At this temperature structured steel will collapse and a pressurized vessel will rupture. The application of only a very small flow of water in most fire conditions will considerably increase the time taken to reach this critical temperature. If we assume a particular rate of heat transfer to the steel (usually by radiation from flames), there is a minimum water application rate which will ensure that the temperature never does reach 500°C. At some lower rate than this the temperature may eventually reach 500°C, but after a considerable delay. Preferably sufficient water should be applied to keep the temperature below 100°C otherwise the bulk of the water may fail to wet the surface. Water which is applied to storage tanks is also intended to ensure that the contents are not overheated by radiation from a fire. For example, the walls of a tank which have been wetted internally by fuel oil should not be heated to a temperature above the flash point, or an explosive mixture will be formed in the vapour space. In this situation it may be necessary to keep the temperature to below say 40°C.

In some early fire trials which were done by the NFPA in America it was found that when a steel vessel was immersed in the flames of a hydrocarbon test fire the heat transfer rate was around 20 000 BTU/ft^2/hr (about 63 kW/m^2). From the experimental results it was decided that the input should be limited to 6000 BTU/ft^2/hr (about 19 kW/m^2). They found that this could be achieved by the application of water at the rate of 0.25 US gal/ft^2/min, which converts to 12.2 l/m^2/min. NFPA/ANSI standard 15 was written as a result of this work, and contains detailed specifications covering water supplies, pipes, nozzles, methods of testing and so on. For vessels the standard requires this application rate for all exposed surfaces, including the bottoms of spherical or cylindrical vessels (which should be wetted by sprays and not simply by run-down). The same rate is needed for vertical structural steel members, but horizontal members and metal pipes require only 0.1 US gal/ft^2/min (4.9 l/m^2/min). The standard also provides details of systems to protect electrical transformers, and conveyor belts.

These values are considerably in excess of the theoretical requirements, and are about 25 per cent more than the experimental rates. However, a considerable excess is needed because of uneven distribution and also because the water cannot penetrate the surface, as it can for example when burning wood is extinguished. Rates very similar to these have been adapted for exposure cooling in many countries; the legislation covering the protection of UK offshore installations requires a minimum application of 12.2 l/m^2/min. It should be remembered however that this rate is based on work with hydrocarbon pool fires. Heat transfer rates could be greater in confined conditions, and particularly if a high pressure jet flame is impinging on to steelwork. Some practical tests have been done (again in the USA) which simulated the conditions likely to exist where a jet flame fed by a leak from an offshore wellhead is involving a neighbouring wellhead. It was found that much higher water application rates were needed to prevent the failure of the second wellhead, and the recommended value was 100 US gal/min (380 l/min) to each wellhead likely to be involved in such a fire.

Systems to protect fixed roof storage tanks frequently comprise a simple pourer above the centre of the roof, and some means of ensuring an even distribution over the surface and down the vertical walls. The discharge should be from an open-ended pipe close to the roof, otherwise the water may be deflected in a high wind. This would cause serious problems if the next tank up-wind was on fire. The dry riser is sometimes brought through the bund wall to a point where a fire brigade connection can be made, or it may be connected by a valve to the fire main. A system comprising a number of fixed spray nozzles will ensure a more even distribution of water. It has the additional advantage that, if it is divided up into two or more sections, only the side which is exposed to the fire needs to be cooled, and this will save water. Some systems are arranged to be turned on automatically by the actuation of a detector: usually by the method shown if Figure 21.4. However, in a typical tank farm situation it is not usually necessary to arrange for such prompt action. Cooling water can of course be applied manually, or from a portable monitor, but the advantage of a fixed system is that once it is in operation the firemen are released to get on with the firefighting.

Systems are also used to deluge process plant, particularly those containing flammable liquids. It is sometimes inconvenient to have an array of nozzles over the plant and use is made instead of oscillating monitors. These are arranged around the edge of the plant, and sprays or jets are swept over the area. These, and other systems, can be made more effective where spills of flammable liquids are possible, if arrangements are made for the injection of an AFFF concentrate into the water supply.

25

Total flooding and local application systems

TOTAL FLOODING

Automatic systems using agents other than water have been in use for many years. A CTC system for the protection of aircraft engines was successfully demonstrated in 1926 (at the Royal Aircraft Establishment). Work on an NFPA standard on carbon dioxide systems began in 1928. These were both predated by inert gas systems on ships (which are discussed later). The ships' system comprises a generator from which inert gas is ducted to a hold, but in the other systems discussed here the agent is stored in one or more vessels under pressure, from which it is discharged through pipelines and nozzles.

Standards are available in many countries for carbon dioxide and, more recently, Halon systlems. The only Halons sufficiently volatile for total flooding applications are 1301 and 1211. The NFPA/ANSI standards for the three agents are 12 (first issued in 1929), 12A (1965) and 12B (1971). These standards (but not the UK standards) cover also local application systems, which are discussed below. BS 5306, the code of practice for fire-extinguishing installations, covers sprinklers in Part 2, carbon dioxide systems in Part 4, Halon 1301 systems in Part 5 Section 5.1, and Halon 1211 systems in Part 5 Section 5.2. These follow the NFPA standards quite closely, as do the corresponding ISO standards.

Extinguishing and inerting

It is the purpose of a total flooding system to convert the whole of the atmosphere in an enclosure into a critical concentration of an extinguishing agent. The enclosure may be anything from an individual computer cabinet to a complete offshore module. Two different critical concentrations may be used depending on whether the hazard is a fire or a potential explosion. Any fire which is burning there is dependent on the enclosed air, and if this brings with it an extinguishing concentration of the agent, the fire will go out. On the other hand, if a sufficient volume of flammable gas, or a liquid at a temperature above its flash point, is released into the enclosure, an explosion is possible. If the total flooding system then discharges an inerting concentration into the enclosure, the vapour/air/agent mixture will be incapable of ignition and an explosion will be averted. If the leaking gas or liquid has ignited as it escapes, a fire will be established, and the possibility of an immediate explosion will again be averted. However, if this fire is extinguished by total flooding and the leak continues, explosive conditions will eventually be established unless the agent concentration is sufficiently high to prevent this, and continues to be above the critical value until the leak has been stopped.

Clearly we are talking about the diffusion flames of a fire and the premixed flames of explosions which were discussed in Chapter 1. Methods of extinguishing these flames were discussed in Chapter 6, where it was shown that higher concentrations of agent are needed to extinguish premixed flames than are needed for diffusion flames. The two concentrations are usually called (as above) the inerting concentration, and the flame-extinguishing concentration respectively. In designing a total flooding system it is safe to use the flame-extinguishing concentration only when it can be shown that the maximum release of fuel into the enclosure will be insufficient to create an explosive atmosphere. Surprisingly small quantities of flammable liquids can be hazardous. Table 25.1 shows the approximate weights of various fuels which, when completely vaporized, will produce a

Table 25.1
Approximate weight of fuel to make one cubic metre of a lower limit mixture

Fuel	Quantity g/m^3
n-Butane	45
Carbon disulphide	32
Ethyl alcohol	58
n-Heptane	31
Propane	42
White spirit (naphtha)	51

cubic metre of a lower limit mixture with air (it will be remembered that for most fuels twice this amount will create the stoichiometric mixture, which is the most hazardous).

Clearly the safest way is to design the system for an inerting concentration, which will protect against both fires and explosions. With a relatively cheap agent like carbon dioxide this is not a problem. Halons are more expensive and if a very large system is being planned, a saving can be made by designing for the flame-extinguishing concentration, if it is safe to do so.

Leakage

Once a flame-extinguishing concentration has been established there will be an immediate and complete extinction of the flames. It will be necessary to maintain this concentration until there is no possibility of re-ignition of the fire, which usually means waiting until the fuel has cooled to a temperature below the auto-ignition temperature or, if a source of ignition is present, to a temperature below the fire point. An inerting concentration may need to be maintained until a leak has been isolated, or until some other action can be taken. If it is possible for air to enter the compartment after a discharge, then clearly there can be a steady reduction in agent concentration as the air diffuses into the mixture. As it happens, all the agents used for total flooding have a density higher than that of air, and so the mixture which is created by a discharge will also have a higher density than air. This means that if there are holes at the top and bottom of the protected room, the agent/air mixture will flow out at the bottom and will be replaced by air entering at the top. If there is a hole at the bottom only, the mixture will flow out through the lower half of the opening and air will enter through the upper half. In this case there will be considerable mixing of the air into the inhibited atmosphere. There will also be good mixing and rapid dilution if an air-conditioning system or a simple ventilating system is operating in the room.

Before a discharge occurs any air-conditioning or ventilating systems must be shut down and the dampers closed. All openings into the room should also be closed. This should be done between the time of detection of a fire (or of the presence of flammable gas) and the beginning of the discharge. There is however a possible problem here: if the enclosure is completely sealed and a quantity of agent is injected, the pressure will be increased accordingly. If for example sufficient agent is used to provide a 10 per cent concentration in the volume, then when it has all vaporized and the temperature has had time to return to its ambient value, the pressure will be 100 m bar higher than it was originally. This is more than enough to demolish a brick wall. A normal building, with walls designed to withstand winds up to 140 mph, should not be subjected to a higher pressure than about 25 m bar. It has been found that in practice the gaps around doors, windows and other openings can provide sufficient venting, even for structures like refrigerators and ducts. Where a compartment has been tightly sealed because it contains hazardous materials, it will probably also have been provided with explosion venting, and this will relieve any overpressure from a discharge. However, if venting has not been provided, it must be done, and the area of the vent which will be needed to prevent an increase in pressure above a safe value must be calculated. The various standards on total flooding systems provide suitable formulae, which are based on the discharge rate of the agent and the maximum safe pressure within the enclosure.

When venting is provided, either naturally or by means of a vent of calculated area, some of the atmosphere in the room must inevitably be lost during a discharge. This is allowed for in the calculations of agent requirements, and it is assumed that what is lost will be an air/agent mixture at the full design concentration, thus ensuring that any errors are on the safe side.

Compensating for leakage

If the leakage of inhibited air from an enclosure cannot be prevented, because there are openings which cannot be closed or because of forced ventilation, there are three ways of ensuring that the design concentration is maintained for the required time. Methods are provided in the standards of calculating the rate of loss through an opening (the rate increases both with the size of the opening and the concentration of the agent). Methods are also given of calculating for the corrective measures detailed below. Calculations are often simplified by the provision of nomograms and graphs.

One method depends on discharging an overdose of agent initially to ensure that the concentration has not fallen below the design concentration before the end of the protection period. It is necessary to stir the atmosphere in the enclosure during this time so that the ventilation cannot form pockets of partially inhibited air which can support combustion.

Alternatively, a slow flow of agent can be maintained after the end of the main discharge to maintain the concentration against the leak. Again it is necessary to stir the atmosphere, or to discharge the additional agent in such a way that is provides sufficient mixing.

It is possible also to make use of the fact that if air is allowed to enter the top of the enclosure at the same rate at which the agent/air mixture is being lost from the bottom, very little of the fresh air will mix with the inhibited air. There will be an interface between the two layers which will descend at a velocity which depends on the discharge rate through the bottom holes (again this can be calculated). If the risk which is being protected occupies less than half the height of the enclosure, and if the interface will have descended no more than half of the height during the protection period, this method can be used. With this method it is important that there should be no stirring of the atmosphere, so that any mixing across the interface will be by diffusion only (which in this case will be quite slow).

Design concentration

Methods of measuring flame-extinguishing and inerting concentrations for different fuels are given in Chapter 6, and the laboratory apparatus which is used is shown in Figures 6.3 and 6.4. The design concentration values for the Halons in both the UK and US standards are obtained by increasing these values by 20 per cent and then setting in addition a minimum value for any system. The laboratory measurements were obtained in conditions which were ideal for combustion and which therefore required the maximum agent concentrations. However, a practical engineer finds it difficult to accept that a small flame burning in a laboratory can be more difficult to extinguish than a full-scale fire: hence the generous safety factor. The safety factor will of course be useful in compensating for the engineer's errors, and also for possible leakages of agent from the system. What is surprising is that the standards provide no warning about the one situation in which the laboratory values may genuinely be too low: flames burning hehind an obstruction in an airflow. This is also discussed in Chapter 6 where it is shown that there is an increase in agent requirements at quite low air flows, which might well be encountered in some industrial situations.

When a system is discharged the agents are delivered to the nozzles as a liquid. Carbon dioxide cannot exist as a liquid at atmospheric pressure (see Chapter 6) and so as it passes through the nozzle it flashes off into a mixture of vapour and solid (at -78°C, called carbon dioxide snow). At room temperature, the Halons are well above their boiling points (–4.0°C for 1211, –57.6°C for 1301). The liquid discharge cools down to these temperatures because some of the agent flashes off as vapour, leaving the remainder as a coarse spray. As the discharge progresses into the protected volume, the heat needed to sublime the CO_2 snow or to vaporize the Halon droplets must be obtained from the air. A calculation will soon show that as the temperature of the air falls to that of the solid or liquid agent, it releases only just enough heat to convert the whole of the agent into vapour. For this reason the nozzles must be so arranged that the agent is discharged into, and mixes with, the entire volume of air in the enclosure. It may be tempting to direct the discharge towards the risk, but if this is done and solid or liquid agent impinges on say a horizontal metal surface, it will soon remove the available heat from the metal, and may then remain as snow or as a liquid puddle on the surface. Vaporization will then be very slow because it will depend mainly on convection currents in the atmosphere. Admittedly this is less of a problem in the presence of a large fire, but if there is an incipient fire, or if the discharge is intended to inert an explosive atmosphere, the critical concentration may not be achieved for some time, and may never be achieved throughout the whole of the enclosure. It should be remembered that the purpose of a total flooding system is to convert the atmosphere into an extinguishing or inhibiting concentration of the agent. Other systems are designed to apply the agent directly on to the risk: these local application systems are discussed below.

Agent quantity

The quantity of agent which will be needed to produce the design concentration
in a particular volume will depend on the temperature of the air into which the agent is discharged. This is because the specific vapour volume (the volume of vapour released by a given weight of agent) varies with temperature. For Halon 1301 for example the specific vapour volume at −10°C is 0.142 m^3/kg and at 40°C it is 0.170 m^3/kg. 1 kg of this agent discharged into a 3 m^3 enclosure at −10°C would be sufficient for a concentration of 4.5 per cent, but at 40°C it would produce 5.4 per cent. Looking at this effect from the system engineer's point of view, if he wants a 5 per cent concentration in this 3 m^3 enclosure, he will need 1.11 kg of Halon 1301 if the ambient temperature is −10°C, and only 0.93 kg if the temperature is 40°C. In practice he will design for the lowest temperature which will be encountered in the protected volume.

At first sight it may be puzzling that different weights of agent may be needed to protect the same volume. It is important to remember that our real concern is neither weights nor volumes, but numbers of molecules. As it happens, 5 per cent is not far from the peak concentration value for ethane (see Chapter 6 and Figure 6.1 for an explanation of peak concentration). At the peak, the fuel/air ratio will be nearly stoichiometric, and the composition of the mixture by volume will be roughly 5 ethane + 90 air + 5 Halon. Equal volumes of gases at the same temperature and pressure contain equal numbers of moleculess, so we can say that a stoichiometric mixture of ethane and air will be inerted if there is one molecule of Halon for every molecule of ethane. We can also say from the same argument that ethane cannot form explosive mixtures with air if there is at least one molecule of Halon with every 18 molecules of air (again this is clearly not a completely accurate statement, but is near enough to illustrate the point). It is this molecule ratio which must be maintained. When the temperature of a gas is reduced the density increases, so there is a greater weight of gas, and hence a greater number of molecules in the same volume. A greater weight of Halon will be needed to maintain the correct balance of molecules.

The actual weight of a particular agent which will be needed to produce a given concentration in a given volume and at a known minimum temperature can be calculated very readily from tables, graphs and nomograms in the standards. In these, allowance is made for the air which is displaced from the enclosure by the agent. As explained above, it is assumed that the air takes with it the full design concentration of the agent. (The figures given above were based on data of this kind.)

Discharge time

It will be remembered from the discussion on the Halons in Chapter 6 that they must necessarily decompose as they enter a flame, so that they can interfere chemically with the combustion reactions. The end products of the inhibition reactions contain halogen compounds which are more toxic than the original agent, and are also potentially corrosive. To minimize the production of these breakdown products it is important that the time during which the Halon vapour is entering the flame should be as short as possible. This means that the discharge time should be as short as is practicable so that the critical concentration of the agent is achieved in the minimum time. For this reason the NFPA standard calls for a discharge time of 10 seconds or less, and the British standard says that the rate of discharge should be so selected as to minimize the exposure time of the agent to the flames.

Carbon dioxide does not decompose in a flame but clearly it is advantageous to achieve the design concentration as quickly as is practicable. Because of the higher concentrations which are needed, the recommended discharge times are between 30 seconds and one minute.

Having calculated the agent quantity and knowing the required discharge time, the total flow rate can be found. The flow through each nozzle can then be determined. Hydraulic calculations can then be made to determine the dimensions of the nozzles and of the pipelines which are needed to achieve these design requirements. One of the problems encountered in making these calculations is discussed next.

Agent flow

Whenever a liquid is flowing through a pipe the pressure at the discharge end will be lower than that of the input. This drop in pressure is due partly to the work done in moving the liquid, but mainly to the work done in overcoming the friction between the liquid and the wall of the pipe. The loss of pressure can be calculated using equations which take into account the length and diameter of the pipe, the velocity of the liquid, and a friction factor, which varies with the roughness of the wall.

The discharge rate of a liquid through a simple nozzle can also be calculated from a knowledge of the diameter of the orifice and the pressure at the nozzle. Thus, the discharge rate of water from a branchpipe or from a sprinkler can readily be calculated if we know the dimensions of the hose (or pipes), the friction factor and the input pressure. Unfortunately a straightforward calculation like this cannot be used in designing a total flooding system. This is because normal ambient temperatures are well above the boiling points of the agents, which must therefore be kept under pressure as liquefied gases. If we take carbon dioxide, for example, when the liquid is being stored in a closed vessel at a temperature of say 15°C, the pressure in the vapour space above the liquid will be about 51 bar. If now the liquid is allowed to discharge so that a flow is established in a pipeline, a pressure drop will occur along the length of the pipeline. Vapour will be released from the liquid in an attempt to restore the original vapour pressure, and two-phase flow (liquid and vapour) will be established in the pipeline. If the pressure falls to 5.1 bar, the triple point pressure, three-phase flow will occur, and the pipeline, and particularly nozzles, may become blocked with solid carbon dioxide. Clearly this must be avoided, but two-phase flow is inevitable and this makes it very difficult to calculate nozzle discharge rates and total discharge times. There is no danger of three-phase flow in a Halon system, but the problems of two-phase flow are made more complicated because the containers are pressurized with nitrogen (see below). Nitrogen is quite soluble in Halon, and the solubility increases with increasing pressure. This means that the reduction in pipeline pressure during a discharge causes a release of dissolved nitrogen as well as vapour.

Normal hydraulic calculations cannot therefore be used in the design of total flooding systems but again data are presented in the standards which make calculation possible, although complicated. Computer programs simplify the task, and these are available through manufacturers of both the Halons and the systems.

Agent supply

The agent supply to a system is usually stored in containers which are of a suitable size to be manhandled. In a large system several containers will be manifolded together. It is sometimes arranged that a single bank of containers provides protection for two more enclosures, suitable valves being used to direct the agent into the appropriate system. As a more attractive alternative, if each hazardous area has its own system, cross-over arrangements are sometimes provided so that a second shot can be used in a stricken area by sacrificing the protection of an unaffected area. This is done for example with the systems which protect aircraft engines.

The coefficients of expansion of the three agents are quite high. It is possible therefore that if the temperature increases after a container has been filled, the liquid may expand sufficiently to burst the container. Guidance on the maximum safe filling density, in kg of agent per litre of volume, for different ambient temperatures, is given in the standards. An increase in temperature will also result in an increase of pressure in the container. For carbon dioxide this change will simply follow the vapour pressure curve up to the critical temperature (31°C). With the Halons the situation is more complicated because of the presence of nitrogen. As the volume of the vapour space is reduced by the expanding liquid, the pressure of the nitrogen will increase. The solubility of the nitrogen in the Halon increases with increasing pressure, but it also decreases with increasing temperature. Fortunately the Halon manufacturers have made measurements of the pressures generated in containers over a range of temperatures, for different fill ratios and different initial nitrogen pressures. Again, the data are available in the standards.

Pressurization of the Halons with nitrogen is used to achieve the 10 seconds discharge time without having to resort to inconveniently large pipes and nozzles. Also, the variation of pressure with temperature is less than it would be if the agent vapour pressure only were present (due to the compensating changes in the solubility of nitrogen in the Halon). NFPA 12A, the standard for Halon 1301 systems, allows two filling pressures (measured at 21°C): 25 bar and 42 bar. The latter pressure simply ensured that components which had been developed for carbon dioxide systems would also be suitable for the Halon systems. The pressures prescribed for Halon 1211 systems are 10 and 25 bar, the lower one being admitted because systems using that pressure were available in the UK when the standard was written. All these pressures have been reproduced in the BS standards.

Agent containers are nearly always provided with dip pipes, which are connected at the top to an actuator. Many of the carbon dioxide systems are completely mechanical in their operation, a cutter being driven through a metal diaphragm (which seals the container) by a falling weight. The weight itself is held up by a wire connected to one or more fusible links which are located above the risk. The mechanism becomes less complicated if only one of a bank of containers is opened in this way. The remainder are opened by a cutter which is attached to a piston in a cylinder, and this is pressurized by gas released from the first container.

The actuators of Halon systems are usually electrically operated. This can be done by again having cutters attached to pistons, but by pressurizing the cylinder with gas generated by a pyrotechnic material (gunpowder for example). Ignition can be by the same kind of electric fuse that is used in detonators. Systems will occasionally be found in which containers originally designed for carbon dioxide have been filled with Halon. These are opened by the gas released from a small carbon dioxide container which is itself opened by an electrically operated device. This arrangement has the disadvantage that the failure of the electrical actuator or the loss of the contents of the small container will result in the failure of the entire system. If an actuator had been fitted to every container, the failure of one actuator or the loss of the contents of one container would result only in a reduction in the amount of Halon discharged. This could well be compensated for by the safety factors which were applied when the system was designed.

Pipes connected to the containers for the distribution of the agent must be able to withstand the full pressure in the containers. In addition, the three agents all have liquid densities greater than that of water and they will acquire considerable momentum as they are driven into the pipes, particularly in systems at the higher pressures. When there is a change of direction at a bend, the impact of the liquid will apply a high loading to the pipes. For both these reasons the pipework must be substantial. The standards provide guidance on this, usually by reference to standards covering the mechanical properties of pipes.

Banks of agent containers are often connected to a manifold by lengths of flexible high-pressure hose. When a container is being connected to a manifold it is important that the hose should be joined to the manifold first and then to the container. The actuators of many containers are fitted with a lever which clan be used for a manual discharge, and if this is accidentally operated an unattached hose can whip violently. Men have been killed by such accidents.

Deep-seated fires

Combustion of this kind is discussed in Chapter 1. Early editions of the

NFPA standards for carbon dioxide and Halon 1301 contained soaking time curves for dealing with these fires. These showed the times during which it was necessary to expose a deep-seated fire to a particular concentration of the agent to ensure complete extinction. For example the standard for Halon 1301 showed that it would take 35 minutes to extinguish a deep-seated fire at a concentration of 5 per cent, 15 minutes at 20 per cent, and 7 minutes at 60 per cent. However, when some additional tests were made, difficulty was experienced in obtaining reproducible results. A laboratory test was devised which enabled a large number of experimental results to be obtained. These showed that the time to extinguish was dependent not only on the agent concentration but on the type of fuel, its degree of comminution, its distribution within the enclosure, the time during which it had been burning, the ratio of the area of the burning surface to the volume of the enclosure, and the amount of ventilation. The standards now contain a warning that the required time is very difficult to assess, and that it may not be possible to maintain the concentration for a sufficient time to ensure that a deep-seated fire is extinguished. If a deep-seated fire is established before the flame-extinguishing concentration has been achieved, other methods of extinction will have to be used. This will probably mean raking over the fuel and applying water to any smouldering surface which is exposed.

In some of the later full-scale tests, measurements were made of the time taken for fires involving a number of different fuels to become deep-seated. A deep-seated fire was defined very simply as combustion which had not been extinguished by a 30-minute exposure to the flame-extinguishing concentration of the agent. In each test 3 kg of fuel was ignited in an open-basket garden incinerator, ignition being provided by burning a small quantity of petrol beneath the basket. The fuel was either computer paper or computer cards, and these were either stacked or crumpled or shredded. Wood wool was also used. Table 25.2 shows some of the surprisingly short times which were recorded. The first of the two numbers is the longest burning time which did not result in a deep-seated fire; the second is the minimum time which did.

These results show that if shredded paper or a similar fuel is ignited in a protected enclosure, the fire will almost certainly be deep-seated before the system can create a flame-extinguishing concentration. This could also happen with crumpled paper if there were any delay in detecting the fire. There are two possible consequences of this. First, when the agent concentration has decayed, the smouldering could revert to flaming

Table 25.2
Time for the development of deep-seated fires

Fuel	Time to become deep-seated
Stacked paper	5–10 min
Crumpled paper	2–5 min
Shredded paper	17–20 sec
Stacked cards	5–10 min
Crumpled cards	1–2 min
Shredded cards	15–20 sec
Wood wool	15–20 sec

combustion. Second, if a Halon is being used, some of it will be pyrolysed by the smouldering fuel and this will increase the concentration of breakdown products. It will be remembered that these are potentially corrosive. If parts of the enclosure away from the smouldering fuel are cold, water vapour produced by the fire could condense out on to metal surfaces. The products could then dissolve in the film of water, making it acidic. Clearly, where Halon systems are used to protect computer and similar equipment it is essential to ensure that nothing is kept in the room which could burn as a deep-seated fire. The prompt ventilation of the room after a fire is equally important.

One interesting result of the laboratory work was that the Halons were more effective against deep-seated fires that was nitrogen. The specific heats and the diffusivities of the vapours of both agents are lower than those of nitrogen, and it seems possible therefore that they did have some chemical effect on the combustion processes. It is suggested in Chapter 1 that smouldering may be a two-stage process:

$$2C + 0_2 = 2CO$$
$$2CO + 0_2 = 2CO_2$$

The second reaction, in the vapour phase, proceeds very slowly unless it is catalysed by the presence of water molecules. This action of the water could presumably be inhibited by a Halon.

Safety

Anyone who is in a room during a total flooding discharge, or who enters the room soon after the discharge, will be breathing the achieved concentration of the agent. The toxicity of the agents has already been discussed in Chapter 6 and it will be clear from the data that an inhibiting concentration of carbon dioxide would be rapidly fatal. The concentrations needed for total flooding are more than twice the concentration (12 per cent) which causes death in a few minutes. A person exposed to a Halon 1211 discharge should be able to walk out at once quite safely, and it should be possible to remain in some Halon 1301 concentrations for several minutes. The precautions which are required by the standards, and by the HSE in the UK, are discussed below. It is however most unlikely that any advantage could be obtained by discharging a Halon system into an occupied area.

People are much better at detecting fires and much more efficient at extinguishing incipient fires than is the most sophisticated automatic system. The great majority of fires are small when they start and it is difficult to imagine how a large fire could suddenly develop in say a computer room. Suppose for example that a fire is started in a waste paper container in a computer room, perhaps by a cigarette. This could be extinguished rapidly and efficiently by a hand extinguisher or even a glass of water. The discharge of perhaps 200 kg of Halon would be extremely wasteful; it would leave the room unprotected until the empty containers were replaced, and it might not even put the fire out (it could be deep-seated). In addition it would be unwise to assume that exposing people to a Halon discharge will always be safe. There could be high concentrations near to a discharge nozzle, people could be injured in the rush to escape (the noise from a discharge can be alarming), and there is the possibility of cardiac sensitization (discussed in Chapter 6).

If there is a real possibility of an instantaneously large and life-threatening fire in an occupied area, the use of a fast-acting Halon flooding system could be justified. It would be better of course to eliminate the hazard. Total flooding with Halon might also be acceptable in for example the missile control room of warships in action. These are exceptional situations and it is difficult to visualize any other occupied areas in which any advantage could be gained by having a Halon system in the automatic mode. A changeover from automatic to manual operation should therefore be made whenever a protected area is occupied. This can often be incorporated with the security locking arrangements.

If a fire occurs within a computer cabinet, due to an electrical fault, it can be extinguished quite readily by the discharge of a Halon 1211 hand extinguisher into the cabinet. This can be done through a suitable injection point: sometimes a bayonet coupling is used to which the end of the extinguisher's hose can be attached. This simple arrangement enables the total flooding of the cabinet to be achieved quite safely. Alternatively a

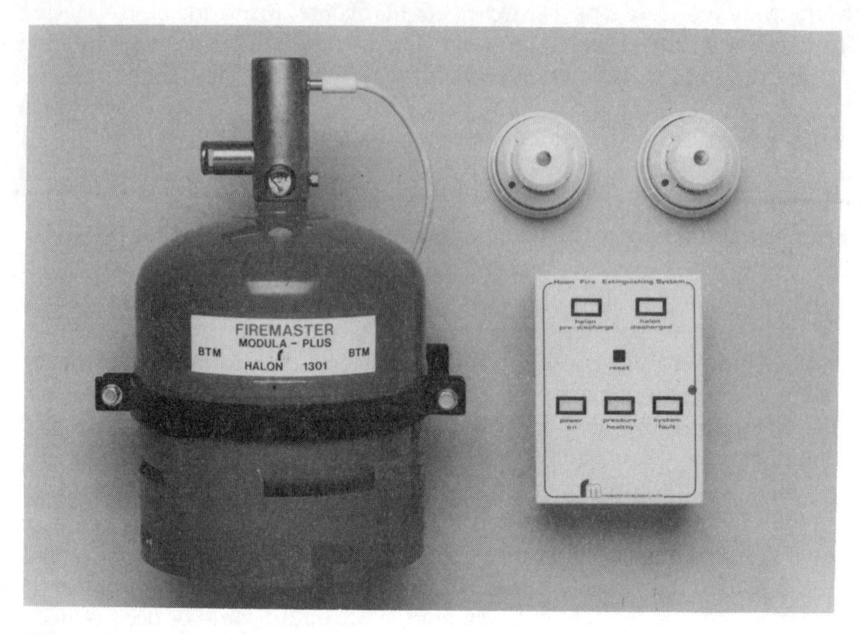

Figure 25.1 Small automatic Halon system
(*Firemaster Extinguisher Ltd*)

small automatic systam can be installed within the cabinet, actuated by its own detectors. Such a system is illustrated in Figure 25.1 (the indicator lights on the small control panel are discussed below). It is unfortunate that many computer manufacturers are reluctant to use these systems.

Because of the hazards associated with total flooding systems (particularly those using carbon dioxide), it is essential that the status of the system should be immediately obvious to anyone approaching the protected area. Coloured lights are frequently used, together with suitable information and instructions. These could include: system armed; system locked off; automatic operation: do not enter; manual operation: safe to enter; and system discharged: do not enter. For carbon dioxide systems it is usual to indicate only three situations: armed, locked off, or discharged.

LOCAL APPLICATION SYSTEMS

If a relatively small fire risk exists in a large enclosure it may not be practicable to consider using a total flooding system, and particularly if there is a number of unclosable openings. In this situation a local application system can be used: an arrangment of fixed nozzles around the risk, each of which will function in much the same way as a hand extinguisher. An oil quench tank in a large engineering workshop would be a typical example.

A large number of these systems has been installed over the years but their design has been based mainly on the manufacturers' know-how. Very little basic research has been done and no-one has developed simple and reliable performance measurements like those used to measure flame-extinguishing and inerting concentrations. However, the results of a few simple tests which were made with the Halons are summarized in the appendices to NFPA 12A and 12B.

There must be a minimum application rate ($kg/m^2/sec$) which is needed to extinguish a particular fire, corresponding to the vertical part of the quantity/rate curve in Figure 6.6. There must also be a maximum area which can be covered by a particular nozzle. If the nozzle is arranged vertically above a liquid pool fire there will be an optimum height at which this coverage can be achieved. If the nozzle is too close its spray pattern will cover only a small area, and some of the fuel may be splashed out of its container. At too great a height, some of the discharge may fail to penetrate the plume of hot gases (and in this respect, the distribution of drop sizes produced by the nozzle will be important).

It could be argued that the application rate itself need only be just sufficient to convert the flow of air which is feeding the flames into a flame-extinguishing concentration. If this is so then at the moment at which this is achieved, the flames will go out. When the tests mentioned above were planned this was the original assumption. Pool fires were burned in a tray with a single nozzle located above it at a height such that the spray pattern was considerably wider than the tray. A number of fires were burned and the discharge rate (kg/sec) was increased after each successive test in anticipation that at a certain critical flow rate there would be an immediate extinction. This did not happen and it was found instead that the time needed to extinguish the fire decreased as the discharge rate was increased. It was thought at first that this was a random effect due to pulsations and other variations in the flame size, and to variations in the agent spray pattern. However, it was found that these results were quite reproducible and that a smooth curve could be drawn showing the variation of time to extinguish against agent flow rate. For a particular nozzle located at its optimum height above a 1.39 m^2 tray of n-heptane, an application rate of 1.074 $kg/m^2/sec$ extinguished the fire in 1.5 seconds, but at 0.292 $kg/m^2/sec$ the time was 7 seconds. Clearly something was happening which took longer to be effective at low application rates. A possible explanation could be that

Halon drops were penetrating the surface of, and mixing with, the fuel. A critical condition was reached when there was a sufficient concentration of Halon vapour in the fuel vapour which, combined with the concentration brought in by the air, was capable of extinghishing the flame. A time like seven seconds might be needed to build up a sufficiently high concentration in this partially back-to-front extinguishing situation (back-to-front extinguishing is described below).

It is unfortunate that this experimental work was discontinued, because there is still a need for a fundamental approach to the design of local application systems. All that can be done at the moment is to characterize the performance of each nozzle for different fires. This is done by plotting the minimum flow rate required for different nozzle-to-fire distances, and indicating the maximum rates which can be used without causing the splashing of liquid fuels. The area which can be covered at various distances then needs to be measured, and finally the minimum application rate needed for these areas (again indicating splash conditions). The selected application rate should not be less than the optimum rate obtained from a quantity rate curve, to which value a suitable safety factor should be applied: 1.5 in the NFPA standards. Having in this way selected the position and the number of nozzles, and the flow of agent through each, the pipework can be designed based on two-phase flow calculations.

Because of the higher pressures in the systems, discharges of carbon dioxide and Halon-1301 are more likely to cause splashing of liquid fuels than is a discharge of Halon 1211. All the systems need to be carefully designed if this problem is to be avoided. For the particular risk mentioned above, an oil quench tank, it is usual to have tankside nozzles rather than overhead ones, so that the discharge is directed over the surface of the oil.

It is of course possible to use a powder in a local application system and this is sometimes done in outdoor situations, where less of the agent will be lost in a high wind. Powder can be discharged successfully through quite complicated pipe and nozzle arrays. The use of powder does have the advantage that most of the discharge will penetrate to the fire. Some of the vapour of a Halon, for example, may be lost, particularly in a discharge from an overhead nozzle.

Back-to-front extinguishing

On an offshore production platform most of the gas which has been separated from the crude oil is either pumped ashore, used to pressurize the formation, or burned in the gas turbines. Because of the nature of the processes involved, a small quantity of gas has to be released. This is burned in the familiar flares. These are constructed so that the flame is well away from the platform, but if there is a very large leak of gas the cloud could drift into the flare and be ignited. A flame could then flash back to the source, possibly resulting in an explosion. It is necessary therefore to extinguish the flare rapidly as soon as a gas leak is detected. This could be done by a Halon 1211 local application system and it will be instructive to see how this could be designed.

The gas is methane and the flare is a diffusion flame, so the mixture in the reaction zone will be stoichiometric: 10 per cent fuel. The flame-extinguishing concentration is quite low, only 2.8 per cent, but if we follow the requirements of the standards we will use the minimum design concentration, which is 5 per cent. So at the moment that the flame goes out there will be in the reaction zone 10 parts of methane to 90 parts of an air/agent mixture. Since there is 5 per cent of Halon in the mixture, there will be 85.5 parts of air and 4.5 of Halon. The obvious way of getting the Halon to the flame is to inject it at the bottom of the flare stack, so that it travels up with the methane. However, if we are to preserve the fuel/air/agent ratios of 10/85.5/4.5 we cannot do this by putting 5 per cent of Halon into the fuel. We must have 4.5 parts of Halon mixing with 10 parts of fuel, so that the concentration of Halon in the methane is 31 per cent. So we have the choice of putting an array of nozzles around the flame which will produce a 5 per cent concentration in the air, or to create a 31 per cent concentration in the methane. In theory we could use the same amount of agent for both methods but in practice we would have to allow for losses from the air system: it is after all a local application system. What can sometimes be done in situations like this is to discharge the agent very quickly into the stack so that a slug of pure agent travels up to the flame. Care must then be taken to ensure that the temporary interruption of the methane flow will not cause problems upstream of the stack.

Another situation in which back-to-front extinguishing might appear to be attractive is a storage tank fire. With a floating roof tank the agent could be injected into the small volume beneath the seal. A fire in a fixed roof tank can happen only after the roof has been removed, or opened up, by an explosion. The agent could then be injected into the space above the fuel and beneath the flames. These methods would have the advantage that no agent would be carried away by the wind. However the required concentration of agent would again be very high. The stoichiometric concentration of many hydrocarbon vapours is only about 2.5 per cent, so to produce the design concentration of Halon 1211 in the reaction zone of the flame it would have to be present as a 64 per cent concentration in the vapour. There would also be the additional problem that some of the liquid Halon from a nozzle would mix with the liquid fuel, thus reducing the amount of vapour. Halon systems are not used to protect fixed roof storage tanks. It is usual to find fixed foam systems (or monitors) for firefighting and water deluge systems for cooling. Local application Halon systems are however used to protect the seals of floating roof tanks, the agent being discharged from a number of nozzles located above the seal.

Back-to-front extinguishing is not usually a practicable method of extinguishing diffusion flames. Where its use is considered, careful calculations should be made based on the stoichiometry of the flame reactions.

CARBON DIOXIDE SYSTEMS

Figure 25.2 shows the actuating and lock-off mechanisms for a typical carbon dioxide system. A fire in the protected area will cause the solder in the frangible link to melt. The main operating weight, which has been suspended by a wire which includes the link, will then drop. A catch on the wire will then pull down the lever marked 'A' in the figure and this will cause a cutter to open the container. It will be noticed that the catch is slightly above the lever. This ensures that the weight has gained some momentum before the lever is struck. The system can also be actuated manually from a release box, usually simply by detaching the end of the wire from a shackle. The system can also be discharged by pulling lever A. The locking-off device is simply a bolt which slides over to prevent the main operating weight from dropping should the system be actuated either manually or automatically. It is of course possible to use any type of detector system but it is necessary then to rely on actuation by electrical devices. Additional actuation by a manually operated mechanical method may also be provided in case there is a fault in the electrical circuits.

Large quantities of carbon dioxide may be needed for some systems and this may require the installation of an inconveniently large number of cylinders. These could be replaced by a single large container, but large pressure vessels are expensive to make and would be awkward to move. An attractive alternative is the use of refrigerated storage. The liquid carbon dioxide is stored at a temperature of -17°C with a corresponding pressure of 20 bar. The vessels are large but relatively cheap, and the carbon dioxide can be delivered from road tankers.

Systems must be so designed that the lowest pressure is always well above the triple point pressure of 5.1 bar. Because of the high pressure, the discharge velocity through the nozzles is high. This results in the entrainment of a large volume of air into the jet, which can be a disadvantage particularly with local application systems. For this reason the nozzles may be fitted

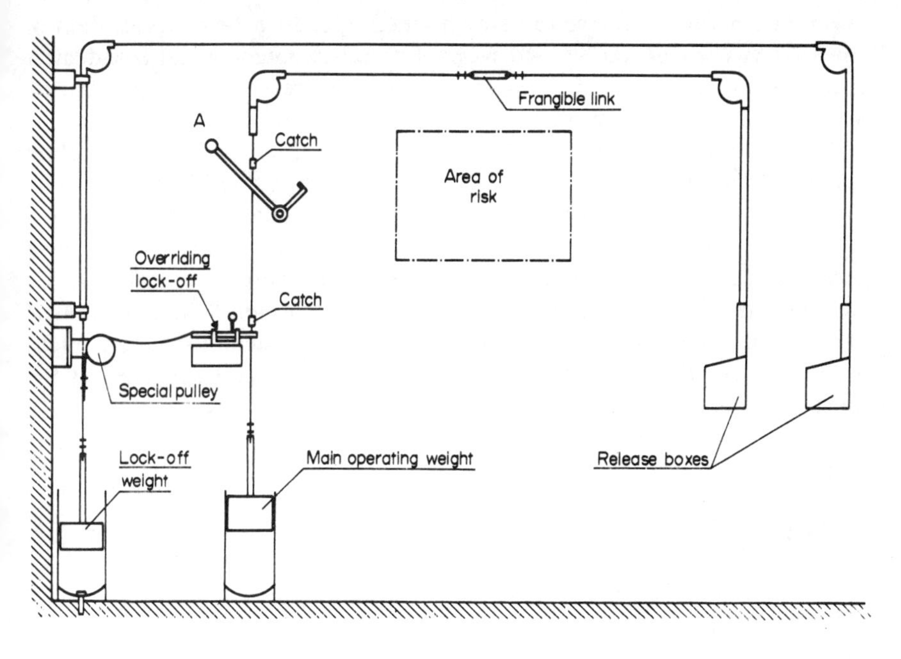

Figure 25.2 Schematic layout of overriding lock-off (*Angus Fire Armour*)

with horns so that the jet has expanded and its velocity has fallen before it encounters the air. The entrainment of some air is necessary so that sufficient heat can be provided to sublime (evaporate) the snow. Unfortunately the snow acquires a high static charge which cannot be relaxed safely. For this reason it is usually recommended that carbon dioxide systems are never discharged in an attempt to inhibit an explosive mixture: the mixture may be ignited by a static discharge before it has been inerted. However, some research on the generation of static charges by carbon dioxide systems has indicated that a hazardous discharge should not result from the release of less than 90 kg of agent (to produce a 40 per cent concentration) or if the smallest dimension of the protected enclosure is less than 5 m. This information has not yet appeared in any of the standards.

As we have seen in Chapter 6, the mechanism of extinction by carbon dioxide is the application of a thermal load within the flame, which reduces the temperature to a critical value. It may therefore be puzzling to find that some of the design concentrations in NFPA 12 are calculated from the 'residual oxygen values' (the limiting oxygen concentrations). This is really a misnomer because the results were obtained by noting the concentration of an inert gas which was needed to inhibit the combustion. Different inert gases would have provided different values (in relation to their specific heats) but the 20 per cent safety factor would take care of this.

It should be remembered that the design concentrations for carbon dioxide total flooding are based on the inerting rather than the flame-extinguishing concentrations. This is because of the relatively low cost of the agent.

Perhaps the most important point to remember about a carbon dioxide system is that the inhibiting concentration would be rapidly lethal. Anyone who was at a place remote from the exit when the system discharged might well be killed. The chances of escaping safely would also be reduced by the thick fog which is formed, and by the fact that carbon dioxide increases both the rate and depth of breathing (thus increasing the rate at which it is being ingested).

HALON 1301 SYSTEMS

It will be remembered from Chapter 6 that testing with human beings had indicated that the concentrations of Halon 1301 likely to be used for many total flooding systems would not be incapacitating to inhale for exposures of several minutes. Because of this NFPA 12A allows total flooding to be used in occupied areas at concentrations not exceeding 10 per cent, provided that the area is evacuated immediately the discharge begins. If evacuation cannot be completed in less than one minute, the concentration must not be greater than 7 per cent.

In areas which are not normally occupied, concentrations between 10 and 15 per cent may be used provided that people can get out in less than 30 seconds. For concentrations greater than 15 per cent, 'provision shall be made to prevent inhalation by personnel'. The guidance in the British standard follows advice given by the HSE and is more restrictive. The distinction is made between the design concentration and the achieved concentration. The latter is calculated from the anticipated net volume of the enclosure when allowance has been made for the volume of the contents. It should also be remembered that the manufacturer of the system will be reluctant to supply Halon containers with non-standard fills. He will prefer to use the next larger container from his range. The standard says that where the achieved concentration is not more than 6 per cent the system can be in the automatic mode at all times. To allow people to evacuate the area before a discharge, a delay (usually not exceeding 30 seconds) can be incorporated, with a suitable alarm. When this is done, a switch should be provided in the area so that the discharge can be prevented if it is not needed (because the fire has been extinguished manually, or because there has been a false alarm).

If concentrations between 6 and 10 per cent are needed in a normally occupied area, the system must be on manual control while the area is occupied and can be switched to automatic control when it is not. If concentrations in excess of 10 per cent are needed, the system should not be an automatic operation at any time. There are no restrictions in unoccupied areas.

Thus, the American standard allows concentrations of up to 10 per cent by automatic discharge in occupied areas provided that people leave the area at once. The British standard permits only 6 per cent in these conditions, and in addition advises the use of a pre-discharge alarm and pre-discharge evacuation.

The discharge of a Halon 1301 system inevitably generates static charges but the risk of producing an incendive discharge may be acceptably low. The research mentioned above indicated that the risk is negligible if less than 50 kg is discharged or if the smallest dimension of the enclosure is less than 5 m. Again this is not included in the standards. Whatever the sizes the whole of the system should be adequately bonded and earthed.

Both the UK and US standards provide detailed guidance on the design and strength of the components to be used in assembling a system.

HALON 1211 SYSTEMS

In the human testing with Halon 1211 the earliest detectable symptoms were experienced after a one-minute exposure to concentrations of between 4 and 5 per cent. This was therefore accepted as an indication of the maximum safe exposure. Total flooding systems are designed for at least a 5 per cent concentration so clearly their use in occupied areas needs to be controlled. In this, the NFPA standard is more stringent than the BS standard, since it does not permit the use of Halon 1211 in occupied areas at all. Its use is allowed in normally unoccupied areas only when people can escape in less than 30 seconds (this was the time noted in the human tests with both agents as the minimum delay before the appearance of any symptoms).

The UK standard quite reasonably permits the use of Halon 1211 in normally occupied areas provided that the system is on manual control whenever people are present.

Halon 1211 was found to generate rather greater static charges than Halon 1301. Discharges were considered to be safe if the quantity was less than 15 kg or if the smallest dimension of the enclosure was less than 3 m. Again, the system should be adequately bonded and earthed.

The UK standard deals only with total flooding systems. The US standard covers local application as well, although much of the information is in the appendix. Halon 1211 is particularly well suited to this use because lower pressures can be used and because the relatively high boiling point of -4°C permits the agent to be thrown for a considerable distance as a liquid. It is less likely to cause splashing of liquid fuels, and has good penetration into a fire, where the liquid may be able to accomplish some surface cooling. If a system is being considered for use in a small enclosure, calculations should be made of the final concentration of agent vapour in the air. If this is found to be greater than 4 per cent, the system should be on manual operation only when the area is occupied.

Again, the standards provide detailed guidance on mechanical design.

SUSPENDED UNITS

When Halon systems are installed problems are sometimes encountered in finding suitable locations for the containers and for the pipework connecting the containers to the nozzles. It is possible to eliminate the pipework completely and to connect an actuator and a small container to each nozzle. This may conveniently be done in for example a computer room by locating the containers in the ceiling void. The nozzles are arranged to protrude through the ceiling. The electrically fired actuators are all connected to a suitable control box to which a detector system is also connected. For this application an adequate supply of agent for each nozzle can be contained in a spherical vessel which can be readily accommodated in a ceiling void. Installation work is simplified by this arrangement, but the replacement of containers after a discharge can be difficult.

A simpler arrangement comprises a small container of Halon to the bottom of which is attached a normal sprinkler head. These can be suspended in for example a fume cupboard or a flammable liquid store. The use of the 'Metron' actuator mentioned in Chapter 23 (Sprinklers) enables a number of these units to be linked together. It can be arranged either so that all are released by the actuation of one unit in a group, or by the actuation of a separate detector. The suspended unit illustrated in Figure 25.3 has its own pressure gauge. The 'Metron' actuator can be seen mounted in a specially designed sprinkler head.

An even simpler device was developed for use in television sets, after a number of fires had been started in hotels by component failures. A length of plastic tube about 10 mm in diameter was sealed at one end and filled with Halon 1211. The other end was sealed to a normally closed pressure switch. The tube was then threaded around the more vulnerable parts of the

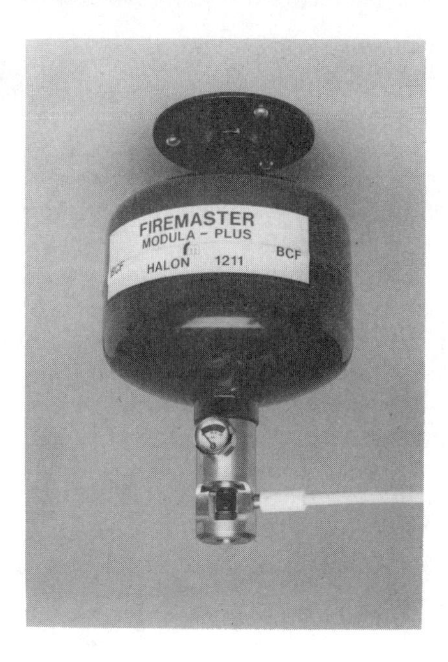

Figure 25.3 Suspended Halon unit with 'Metron' actuator
(*Firemaster Extinguishers Ltd*)

set and the positive side of the mains connection was brought through the switch. A fire in the set would melt through the nearest part of the tube and release the Halon, causing the total flooding of the cabinet. At the same time the mains would be isolated by the opening of the pressure switch. Tests showed the device to be reliable in action and to have an acceptably long operating life (it would of course fail safe). Unfortunately the television manufacturers did not think that the relatively small additional cost of fitting this protection could be justified.

PORTABLE SYSTEMS

It is sometimes necessary to arrange for the temporary protection of a vulnerable piece of equipment. For example, a laboratory experiment involving flammable liquids may have to be left running overnight, or the fire hazard on a ship or an aircraft may have been increased during repair work. One of the suspended units described above might be used, for preferably a 'suit case' system. One of these is illustrated in Figure 25.4. It comprises two detectors, a combined power supply and control unit, and a Halon container with an attached nozzle. A number of additional items is

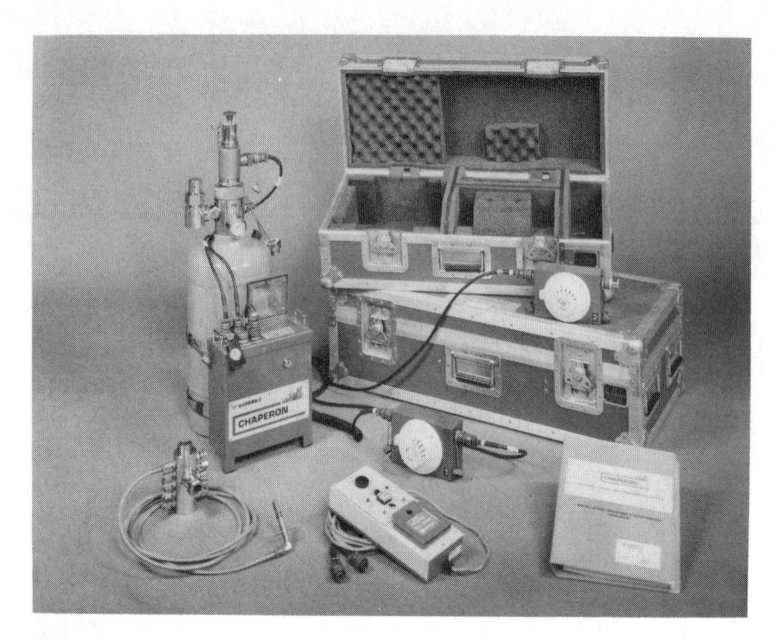

Figure 25.4 'Suit case' portable Halon system
(*Wormald Ltd*)

available, for example a manifold and hoses for direct injection into electrical cabinets. This can be seen at the bottom of the figure, on the left, with next to it a remote indicator and control panel.

TESTING, INSPECTION AND MAINTENANCE

When a total flooding Halon system has been installed it is necessary to demonstrate not only that it complies with the relevant standard, but also that is functions correctly. The correct operation, both automatic and manual, of the actuators can be checked without discharging the system. The pipes can be blown through to show that they are free from obstructions and debris, and can also be pressure-tested for leaks. The achievement of a uniform distribution of agent at the design concentration, and within the 10-second discharge time, can be demonstrated only by a trial discharge. It is necessary then to record the agent concentration at a number of points in the protected enclosure, and to monitor its rate of decay. The correct functioning of dampers and shut-down devices can also be checked, and the pipelines can be examined for damage or displacement.

A test like this like be quite expensive if a large volume of Halon is involved, and there is also the problem of atmospheric pollution (like the halogenated aerosol propellants, the halogenated extinguishants are suspected of depleting the ozone layer in the atmosphere). Halon 122 (CCl_2F_2, boiling point -29.9°C) can be used as a substitute for Halon 1301. It is considerably less expensive but may still deplete the ozone. Halon 121 ($CHClF_2$) could also be used, and is not suspected of affecting the ozone layer. Carbon dioxide is even cheaper, but it too causes atmospheric problems (the 'greenhouse' effect). Unfortunately none of them has physical properties which are sufficiently close to those of Halon 1301 to produce either the same concentration, the same throw, or the same discharge time. If there is any doubt about the achievement of the correct distribution or concentration, a test with the agent itself will be necessary.

When the system has been accepted, regular maintenance is necessary to ensure that it will function correctly when needed. Each week all pressure gauges should be checked and a visual inspection made of all operating controls and indicators. The piping and nozzles should be checked for damage or displacement. Every six months the contents of all of the Halon containers should be checked either by weighing or by the use of a liguid level detector. A container showing a loss of more than 5 per cent of agent or more than 10 per cent of pressure (having allowed for temperature differences) should be replaced. Functional checks should be made of all components but without discharging the system, unless the inspection has indicated that a discharge test is necessary. Some systems are so arranged that the containers are continuously weighed and an alarm is given when the 5 per cent permitted loss has been exceeded. For this purpose, a strain gauge is incorporated in the attachment of some suspended units.

Carbon dioxide systems should also be regularly checked. A weekly inspection should be made for damage and displacement, and the liquid level of low-pressure storage tanks should be checked. When the total loss exceeds 10 per cent they should be refilled. The liquid contents of high-pressure cylinders should be measured every six months either by weighing or by the use of a liquid level indicator. Any which have lost more than 10 per cent of their contents should be replaced, as should any which are due for a hydraulic pressure test. Pressure gauges are not usually fitted to these cylinders. They could not provide any warning of a leak until all the liquid had been lost, simply because the full vapour pressure would be maintained so long as a liquid surface, however small, was present in the cylinder.

FOAM SYSTEMS

Low-expansion foam is used against two-dimensional (pool) fires of flammable liquids. Medium and high-expansion foams can in addition be used against three-dimensional fires (but these fires can also be tackled with AFFF even at very low expansions: see below). Foam is produced by mixing a suitable concentrate with water, and then mixing sufficient air with the solution to produce the desired expansion. The properties of different types of foam are described in Chapter 6. For outdoor risks foam systems are often manually controlled. Automatic systems are available, actuated by different detector types, including sprinkler bulbs, and find particular application indoors.

Small systems, perhaps with a single nozzle, are frequently supplied by a pressurized tank of solution. Many concentrates in 4 to 6 per cent dilution have a storage life of several years, and since the replacement cost is low this is a convenient way of avoiding the need for a more complicated system. The detector may well be a sprinkler bulb, with a suitable head which will entrain and mix the air. A system based on a standard 9-litre extinguisher body, filled with fluoroprotein foam, would cope with a 6 m^2 fire and would therefore be adequate to protect a small oil quench tank.

Larger systems usually depend on separate water and concentrate supplies. The solution can be formed by means of an inductor or a metering pump and good mixing can be ensured by passage through the main pump. The concentrate may be stored locally in suitable tanks or it may be convenient in a large installation to have a mobile tanker (often a converted petrol tanker). For very large installations separate water and foam concentrate mains may be provided throughout the site. The made foam may be discharged from fixed or oscillating monitors, or through a number of fixed pipelines. These may deliver the foam for example to the rim seal of a floating roof tank; through foam pourers into a bund; or through special low-velocity pourers on the inside of a fixed roof tank. With the latter systems it may be necessary to have a seal to prevent the loss of vapour into the foam lines. This is usually a thin sheet of glass which can be broken by the pressure of the foam.

Foam may also be injected into a storage tank through the normal product lines. Special high back-pressure generators are needed for these base injection systems. Ordinary protein foam cannot be used because it can pick up sufficient fuel on its passage through the tank contents to be flammable when it reaches the surface. It is necessary to use fluoroprotein or similar concentrates.

Foam systems are used to protect many large aircraft hangars. AFFF is often used, with floor coverage of between about 5 and 8 $1/m^2$/min of solution. It is usual to cover the under-wing areas either by means of pop-up nozzles in the floor, or oscillating monitors. In addition there are other nozzles (possibly special sprinklers) at truss level. AFFF systems can also be used for three-dimensional fires in process plant, for example in offshore production platforms. This can be applied through the nozzles of a normal water deluge system: it is not necessary to expand the foam completely to obtain the AFFF effect. The system can then be used both for cooling and for firefighting, depending on the need.

High-expansion systems can be used for 'total flooding' in areas where manual firefighting would be difficult or dangerous. They are used for example in underground storage areas, particularly where the layout is labyrinthine. They have the disadvantage that the whole enclosure must be filled for even a small fire, and a considerable time may then be needed to disperse the foam.

HiEx generators are available with outputs up to about 2500 m^3/min of foam, which is passed through ducts up to 3 m in diameter. At one time these generators were commonly located on the roof, simply because this was a convenient place for them. Unfortunately this can result in the entrainment of smoke into the generator, and if this contains the acidic breakdown products from platic materials (particularly PVC), the stability of the foam will be affected. The generators are better located at floor level, although there are limitations to the vertical and horizontal flow of the foam. Beyond a certain travel distance the foam will break down as fast as it is being generated.

If a large building is to be filled with HiEx, it is important that the filling rate can overtake the vertical spread of fire. If this is not achieved, flames will be able to spread horizontally at the ceiling, and the radiation from these will break down the foam. It is likely then that the fire would not be controlled. In general, if goods are stacked higher than about 3 m, the foam system should be backed up by a sprinkler system. This will ensure that fires at the top of the stacks are controlled while the room is being filled with foam. As with all other flooding systems it is necessary to control leakage through openings in the bottom of the enclosure. For HiEx this can usually be done by covering the opening with wire mesh. An opening at the top is necessary to vent the displaced air, and this will be useful in providing an indication that the volume has been filled.

It must be remembered that it is dangerous for people to be trapped in HiEx. Initially this is because of the disorientation caused by the complete loss of vision and hearing. Prolonged exposure will result in the accumulation in the lungs of a lethal amount of water. The enclosure must therefore be evacuated before the discharge, if sufficient foam is available to produce a layer greater than 1 m in depth.

Many of the water deluge systems offshore can be converted to AFFF if there is a need for firefighting in addition to cooling. This is done simply by injecting the concentration into the water feeding the system: as we have seen, the ordinary spray nozzles will produce sufficient aspiration for the AFFF effect to occur. The water for the system will be obtained from the fire main, and the pressure in the main is likely to fluctuate because of other firefighting activities, or as additional fire pumps are started up. The correct concentration of agent in water can be maintained despite these fluctuations by using a balanced pressure tank. The AFFF concentrate is stored in a plastic bag inside a metal tank, and the volume outside the bag is connected directly to the fire main. This ensures that the concentrate is always at the mains pressure. The correct concentrate flow can then be induced through a suitable orifice plate by an in-line venturi.

INERT GAS GENERATORS

Fires in the accommodation areas of ships are tackled by conventional firefighting methods: hand extinguishers, hosereels, hoses and branchpipes, and sprinklers. Additional protection is needed in engine rooms, cargo holds, and the tanks of tanker vessels. These are provided with fixed systems. Carbon dioxide and Halons can be used but many vessels are equipped instead with inert gas systems. These comprise separate generators or units processing exhaust gases from the funnels. These systems are a requirement on tankers and are used to inert the vapour space when the tanks have been filled, and also the empty tanks while they are being cleaned. Cleaning is done by means of high-pressure water jets and these are capable of generating static charges. A number of explosions have occurred in crude oil tankers when unprotected tanks were being cleaned.

In some older ships provision is made for the injection of steam into cargo spaces, from the main or auxiliary boilers. The minimum rate is usually 1.3 kg of steam/m^3 of the largest compartment/hour. This is not a very effective method of fire control because unless the compartment is very hot much of the steam will condense out. The atmosphere will not then be inerted, but some of the uninvolved fuel may have been slightly dampened. However, the use of steam is no longer accepted by IMCO (the Inter-Governmental Maritime Consultative Organization) and such systems have not been installed on ships built since the 1960s. They have been replaced by inert gas systems, which are required to be provided on most ships of over about 1000 tonnes. Legislation differs from one country to another, but mainly follows the IMCO recommendations. In the UK control is by statutory instruments issued under the Merchant Shipping Acts of 1894, 1949 and 1964. Regulations in the USA are drawn up by the Coast Guard, and fixed firefighting equipment is covered by Navigation and Vessel Inspection Circular 6–72.

The exhaust gas from diesel engines cannot be used for inerting because the oxygen content is too high, particularly when the engine is running on a light load. Combustion is more nearly stoichiometric in a boiler, and a typical composition of funnel gas is:

Carbon Dioxide (CO_2)	12 to 14 per cent
Oxygen (O_2)	2 to 4 per cent
Sulphur dioxide (SO_2)	0.3 to 0.5 per cent
Nitrogen (N_2)	remainder

This gas is taken from the boiler uptake through isolating valves to a scrubbing tower, which it enters through a water seal at the base. Water sprays scrub out the SO_2 and any solids, and also cool the gas. The gas then passes through a water separator and then to the fans which feed to the inert gas main through a water seal. The two water seals ensure that there can be no flow of flammable gases from the tanks back to the funnel. On a tanker, the flow rate of the gas must be at least equal to 125 per cent of the combined capacities of all the cargo pumps, so that tanks can be inerted as fast as they are being emptied. The UK regulations require that the maximum oxygen concentration is no more then 6 per cent, and the corresponding US Coast Guard requirement is 5 per cent. Both are well below the limiting oxygen concentration, which for most fuels is about 11 per cent. They are also above the concentration which is normally present in the funnel gases (see above). On ships where the funnel gas cannot be used for inerting, separate oil-fired inert gas generators are used.

Fires in ships' holds are very likely to become deep-seated and it is necessary then to maintain an inerting atmosphere for a considerable time. This is usually done at least until the vessel is in port, the effect being monitored by temperature measurements in the stricken hold. Then as soon as it appears safe to do so, the cargo is removed in the presence of the fire brigade, who will deal with any smouldering combustions which is exposed.

During the 1960s a great deal of work, including full-scale testing, was undertaken at the UK Fire Research Station on a mobile inert gas generator. This comprised an aircraft gas turbine engine fitted with an afterburner and a cooler. The gas leaving the engine exhaust contained 17 per cent oxygen and was at a temperature of 400°C. Combustion in the afterburner reduced the oxygen to 14 per cent. Water was then sprayed into the gases in the humidification section so that they were cooled and diluted with the water vapour. The gas emerging from this section was at a temperature of 100°C and contained only 7 per cent of oxygen. This gas was used successfully to extinguish a number of large building fires. However, it was necessary that the temperature of the gas did not fall much below 100°C, or some of the water vapour would have condensed out. Once the oxygen concentration reached 11 per cent the gas would no longer be an effective extinguishant. Because of this the fire brigade representatives at the trials thought that the method would be unsuitable for their use. Men could be protected from the lack of oxygen by breathing apparatus, but would be unable to work for more than a few minutes in air with 100 per cent relative humidity and at a temperature of 100°C. Clearly the method would be attractive where men did not have to enter the building, and where other methods of firefighting were difficult or inappropriate.

26

Explosion suppression systems

During the Second World War the majority of the aircraft which were lost, on both sides, came down in flames. Many of the fires were started by an incendiary round passing through a fuel tank. Because of the variations in both pressure and temperature at different altitudes there was an explosive mixture in the vapour space during most flights, no matter what fuel was used. The problem of preventing fuel tank explosions was studied at the Royal Aircraft Establishment by William Glendinning. He developed the first practical explosion suppression system.

The technique is possible because of the relatively slow rate of rise of pressure in the early stages of an explosion. This allows the presence of the explosion to be detected, and the premixed flame extinguished, before the pressure has increased sufficiently to rupture the tank. This can be understood if we imagine a spherical flame propagating away from an ignition source. The radius of the sphere will increase linearly with time (at the burning velocity of the mixture), but the volume of the sphere, and hence the pressure, will increase with the cube of the time (because the volume is equal to $4/3\ \pi r^3$). This is not completely true, as was explained in Chapter 1, but it does account for the shape of the curves in Figure 26.1. These show the observed pressure/time relationships following the ignition of stoichiometric hexane/air mixtures in vessels with capacities of 4.5, 45 and 450 litres. In each case the maximum pressure would have been about 8 bar, but we are not concerned with that end of the curves. The important thing is the initially slow pressure rise.

The method devised by Glendinning (British Patent 643188, 1948), and

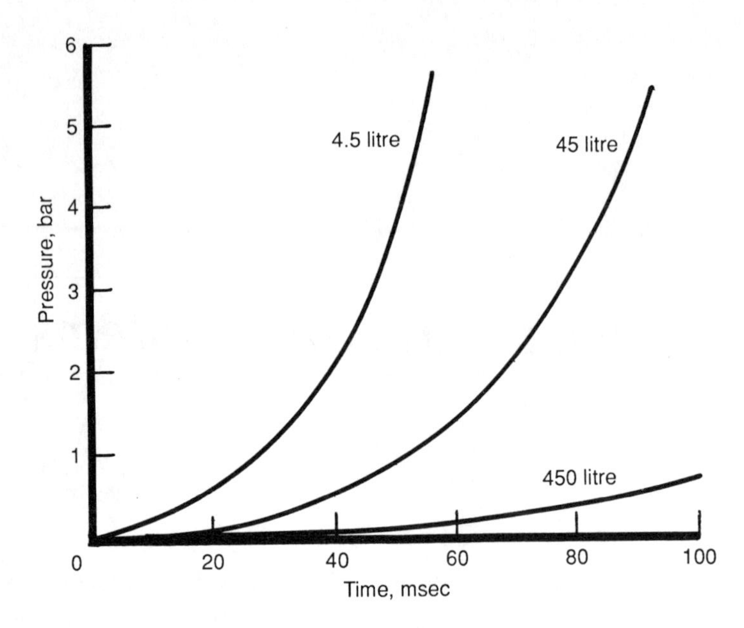

Figure 26.1 Pressure/time curves for hexane/air explosions in closed vessels

now used extensively in industry, is illustrated in Figure 26.2. This shows part of the expanding flame front in three successive positions. The increase in pressure resulting from this combustion is picked up by the detector (the use of the term 'pressure waves' in the figure is intended to indicate that the

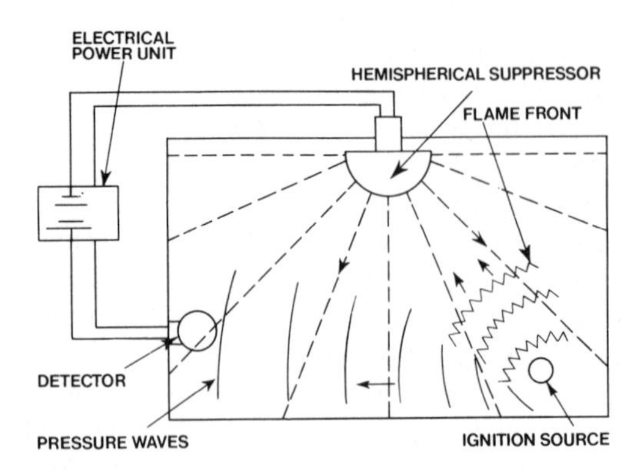

Figure 26.2 An explosion suppression system
(*Graviner Ltd*)

sound waves resulting from the ignition processes, or from irregularities in burning, can also trigger the device). The completion of an electric circuit allows current to flow through a detonator located inside a hemispherical container of Halon. The firing of the detonator bursts the container and discharges the Halon at a velocity considerably in excess of the burning velocity of the flame. The flame is extinguished before the pressure in the tank has exceeded the bursting pressure.

The detector comprises a flexible metal diaphragm mounted in a metal box, which is bolted into the side of the protected vessel. The detector is shown in Figure 26.3. It will be seen that there is a small bleed hole in the diaphragm which ensures that a slow rate of rise of pressure (due to atmospheric fluctuations for example) will not operate the detector. A rapid rise, resulting from an explosion, will move the diaphragm so that it snaps over to a new position where the electrical contacts are closed. If the explosion itself produces only a slow pressure rise, the contacts will close at a preset static setting because of the rubber diaphragm, which does not have a bleed hole (this diaphragm also protects the metal one from possible corrosive attack). This device is extremely sensitive and will respond to very small changes in pressure, but for practical purposes it is usually set to respond to a rate of rise of about 0.7 bar/sec or a static pressure of about 35 m bar. These settings can be adjusted according to the volume of the vessel.

The detonator fires about 1 m sec after the contacts in the detector have closed. The Halon is discharged at a velocity of about 60 m/sec and the time taken to reach the flame will obviously depend on the size of the vessel.

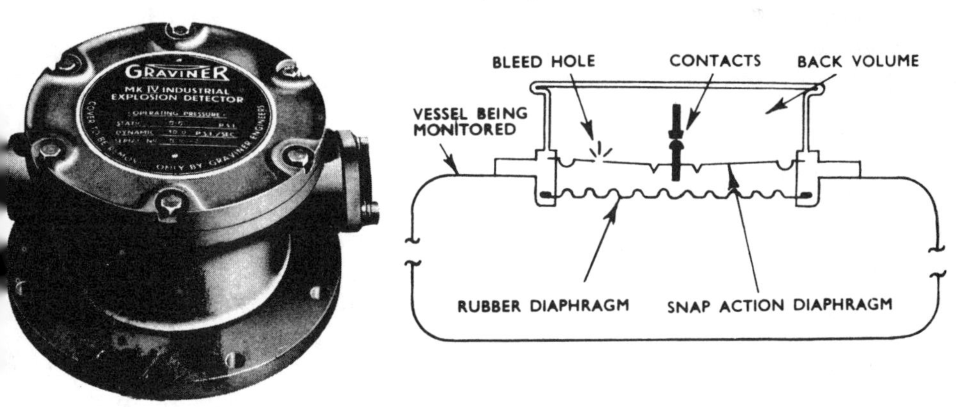

Figure 26.3 Detector of explosion suppresion system (*Graviner Ltd*)

Table 26.1
Explosion suppression in different vessels

Event	4.5 l vessel		4500 l vessel	
	Time m sec	Pressure m bar	Time m sec	Pressure m bar
Ignition	0	0	0	0
Detector operates	7	17	60	14
Detonators fire	8	34	61	15
Flame extinguished	11	136	85	69
Maximum pressure	—	136	—	69

Some typical times for the various events in this sequence are shown in Table 26.1. The values on the left-hand side were observed during the suppression of a stoichiometric hexane/air explosion in a 4.5 l vessel (the pressure/time curve for an unsuppressed explosion is in Figure 26.1). It will be seen that in this experiment the detector contacts closed 7 m sec after ignition, when the pressure was 17 m bar. 1 m sec later the detonator fired and the flame was extinguished 3 m sec after that. The maximum pressure was 136 m bar. In this very small vessel, some of this final pressure can be attributed to the combustion of the detonator explosive, and to the vapour pressure of the Halon. In a larger vessel these effects would be negligible, and the slower rate of rise or pressure (as in Figure 26.1) would ensure that the maximum pressure was lower. This is shown by the results on the right of the table. These show that in a 4500 l vessel it took 60 m sec before the rate of rise was sufficient to trigger the detector. The detonator delay was again 1 m sec, and it took another 24 m sec to suppress the flame (by which time the Halon could have travelled about 1.4 m). Despite taking eight times as long to suppress the explosion, the maximum pressure was half that in the smaller vessel.

Halons with high boiling points are usually used in these systems because of the need to project droplets at high velocity into the flame. Halon 1011 (CH_2BrC1) and Halon 1202 (CBr_2F_2) are frequently chosen. However it seems likely that the mechanism of extinction is not simply chemical inhibition but is due partly to flame stretch and partly also to the creation, by the droplets, of an omnidirectional flame arrestor (both effects are described in Chapter 6). It is possible to suppress an explosion by dispersing the liquid fuel instead of a Halon, but this would leave behind an explosive atmosphere instead of an inhibited one. Powder is also used as a suppressant, but again it is incapable of inhibiting the atmosphere.

Two types of suppressor are used in industrial systems: the hemispherical unit already described, and a high rate discharge bottle in which the agent is stored under pressure. In this latter suppressor a metal sealing disc is opened by the shock wave from a detonator. These bottles are often used for advanced inerting, to ensure that a flame cannot travel from the vessel in which the explosion occurs into other interconnected parts of a plant. The system can also be used to open very fast-acting explosion vents, which comprise a sheet of 'Armourplate' glass which is shattered by a detonator. High-speed isolation valves are also available. These are also explosively operated and a valve 0.2 m in diameter for example closes in about 80 m sec.

The very sensitve detectors can be triggered by mechanical vibration. Where this could cause false alarms, two detectors can be mounted on the plant at a right angle to each other and wired in series. The system will then be actuated only when both detectors respond. In a completely enclosed vessel, with no obstructions, a light-sensitive detector can be used. This was done in the systems developed for military aircraft. It had the advantage of detecting the flash of the projectile passing through the tank wall, before the explosion had even started.

Explosion suppression systems are extremely reliable and their use is being extended to considerably larger vessels than the ones used above as examples. Research and development has been undertaken in a number of countries, particularly the UK, the USA, Switzerland and Germany. Work has started on international standards concerned with the testing of explosive dusts and the design of systems.

Small battery-operated explosion suppression systems are also available. One or more of these can be used to protect plant in which dust or gaseous explosions are possible. It will be seen in Figure 26.4 that the suppressant (in this case 6 kg of Halon 1301) is stored in a spherical high rate discharge container. The detector is the diaphragm type described above. The bursting disc of the container is ruptured by an electrically fired explosive charge. The clean-out plunger shown in the drawing is used to ensure that the pipe connecting the detector to the protected vessel does not become blocked with product. The device is equipped with a test circuit and a pressure gauge. It is easily attached to the wall of the vessel by four bolts and requires a single hole of only 9 cm × 32 cm for both pressure detection and Halon discharge.

FURTHER READING, PART SIX

Nash, P. and Young, R.A. *Automatic sprinkler systems for fire protection.* Victor Green Publications, London

Marryatt, H.W. *Automatic sprinkler performance in Australia and New Zealand.* Australian FPA, Melbourne

Bryan, J.L. *Fire suppression and detection systems.* Macmillan, New York

Bahme, C.W. *Fire officer's guide to extinguishing systems.* NFPA, Boston

Bartknecht, W. *Explosions, course, prevention, protection.* Springer-Verlag, New York

Home Office *Manual of firemanship, Book 9: fire protection of buildings.* HMSO, London

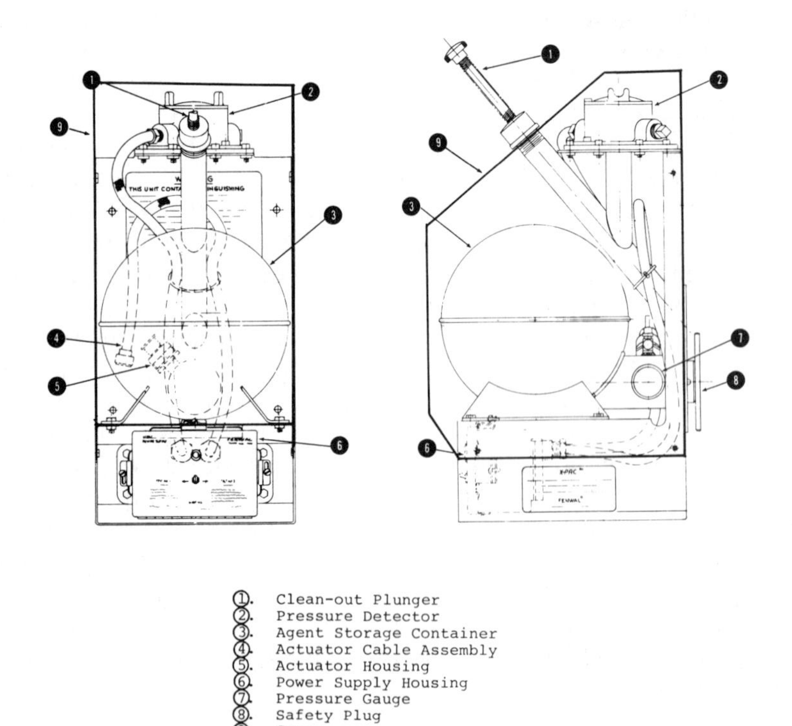

①. Clean-out Plunger
②. Pressure Detector
③. Agent Storage Container
④. Actuator Cable Assembly
⑤. Actuator Housing
⑥. Power Supply Housing
⑦. Pressure Gauge
⑧. Safety Plug
⑨. Cover

Figure 26.4 Small battery-operated explosion suppression system (*Fenwal International*)

PART SEVEN

FIREFIGHTING

27

Portable extinguishers

Very occasionally a fire will be large from the moment of ignition. This could result for example from a massive release of flammable liquid, or could be caused by an explosion. We are however concerned in this chapter with the vast majority of fires, those which are small when they start.

Incipient fires can be dealt with promptly and effectively by the use of first-aid appliances: hand extinguishers, wheeled units, hosereels (and even wet sacks and buckets of water). On a well run factory site as many as 95 per cent of all fires are put out by the people who discover them. The 5 per cent failure rate happens because people are not there, or the wrong attack is used, or the fire is discovered too late (or is too big initially). Reliance must then be placed on automatic systems (if any) and professional firefighting. During the inevitable delay before the arrival of the fire brigade a great deal of damage can be done (although it can be mitigated to some extent by passive fire protection). Loss and damage may continue to happen for some time before the fire is eventually extinguished. Most industrial fires are dealt with by the fire brigade both promptly and efficiently with the minimum of loss, but a large number of the firms go out of business as a direct result of a fire.

Hand extinguishers, and people who know how to use them, are the first line of defence against this disaster. It is difficult to overstate the importance of this simple and inexpensive protection. Appropriate and well maintained extinguishers should be immediately obvious and readily accessible throughout any working area. People should be regularly trained so that they know how to actuate an extinguisher, they know its capabilities against a fire, and have the knowledge and confidence to use it effectively.

Regular training can sometimes be incorporated with the extinguisher maintenance programme. For example, powder extinguishers have a limited useful life because the powder becomes compacted, particularly in the presence of vibration. These extinguishers can be withdrawn from service say every six months and replaced by units which have recently been serviced and refilled. The withdrawn units can then be discharged against practice fires. This arrangement provides a regular supply of extinguishers for training and also enables the safe life of an extinguisher on a particular plant to be checked.

Extinguisher design

The method of actuation of an extinguisher should be immediately obvious and preferably all the extinguishers on a site should be operated in the same way. With many of the currently available designs it is necessary only to remove a safety clip and then to push down a plunger. The nozzle may be mounted on the body of the extinguisher or may be at the end of a short length of hose. With the latter arrangement it is usual to have a control valve at the nozzle so that the discharge can be stopped and restarted as necessary.

All the extinguishers which comply with modern standards are used in the upright position and are therefore provided internally with dip pipes. Some people, remembering an older generation of extinguishers, may attempt to use them inverted. This will allow the pressurizing gas to escape through the dip pipe, but the agent will remain in the body. However this is a useful way of stopping a discharge from an extinguisher which does not have a control valve (thus avoiding unnecessary mess, particularly when powder is used). The instructions on modern extinguishers will frequently say: 'Hold upright. Strike knob. Aim at base of fire.'

Some models, particularly the smaller sizes, may be actuated by a trigger mechanism in the handle. Others have a suitcase handle, again incorporating a trigger. Their method of operation appears to be obvious, but training is still needed to ensure that everyone on site can use an extinguisher without having to read the instructions.

The contents of an extinguisher are expelled by means of a gas which is under pressure in the body of the extinguisher. Pressurization can be achieved in three different ways. In one of the earlier designs the gas (carbon dioxide) was generated, when it was needed, by the reaction between sulphuric acid and a solution of sodium carbonate in water (which comprised the main contents of the extinguisher). The acid was contained either in a glass bottle, which was broken by the plunger, or in an open receptacle. With the latter arrangement the whole unit had to be inverted to release and mix the acid: this was a turn-over soda-acid extinguisher. The disadvantages of the soda-acid design are: the problems of handling fragile acid bottles, the fact that the solution is corrosive both before and after discharge, and the high conductivity of the jet if it encounters live electrical equipment. An early design of foam extinguisher contained two solutions, one acidic and one alkaline, which again reacted to produce carbon dioxide when they were mixed by inverting the extinguisher. A suitable foaming agent was included in one of the solutions. Modern extinguishers produce mechanical (rather than chemical) foam in an aspirating nozzle. Any remaining water or foam extinguishers of this kind should be scrapped and replaced by more modern designs. There are two alternative designs: stored pressure and cartridge-operated units.

Stored pressure extinguishers are suitable for water, foam, powder and Halon. The first three are usually pressurized with air, but nitrogen is always used for Halons. Some designs of water and powder units can be refilled by putting in the right amount of agent and then connecting the extinguisher to a compressed air line. Some powder and Halon extinguishers can be refilled only by the manufacturers, but suitable filling equipment is available for most designs. A stored pressure extinguisher does have the disadvantage that a number of seals must remain gas-tight for a considerable time, otherwise the unit will fail to discharge. However, the pressure within the body can be indicated by a gauge or other suitable device.

In the alternative design the pressure is supplied by a small cartridge of liquid carbon dioxide. This is opened, usually by a cutter, to pressurize the extinguisher and discharge the agent. The pressure generated in the body is sufficient to open any seals which have been used to prevent the ingress of contaminants. Leakage is possible only through the cartridge seal, and the contents of the cartridge can be checked by weighing (the empty, or 'tare' weight is stamped on the cartridge). The cartridge can be returned to the manufacturer for refilling, although it is possible to fill them with the equipment used for carbon dioxide extinguishers. Careful control is essential to ensure that the cartridges are not overfilled, otherwise the expansion of the liquid could rupture the sealing discs.

Stored pressure powder extinguishers may be more reliable than cartridge-operated ones, particularly if they are exposed to vibration (on a vehicle, for example). This is because compaction of any powder is inevitable, given sufficient time. In cartridge operation the gas is released into, or finds its way into, the free space above the powder bed. As the gas penetrates into the bed, entering the small voids between the particles, the difference in pressure through the bed can cause further compaction. On the other hand, if the voids already contain gas at the working pressure and the discharge valve is opened, the flow of powder and gas out of the extinguisher will cause a pressure differential in the opposite direction. This will move the particles apart instead of pushing them together, thus helping to overcome the effects of compaction.

Choice of extinguisher

The extinguishing agents were discussed in detail in Chapter 6. The information can be briefly summarized as follows:

- Water cools the surface of a fuel to a temperature below the fire point. The surface is then unable to release a sufficient flow of flammable volatiles to support a flame. Water even as a spray has little direct effect on a flame.
- Foam floats on top of the surface of a flammable liquid forming a barrier which prevents radiation from reaching the surface and prevents volatiles from feeding the flame.
- Carbon dioxide is brought into the reaction zone of the flame with the air. It acts as a thermal load in the flame, reducing the temperature to a critical value at which the flame can no longer exist. At least 30 per cent in air is needed to achieve this.
- Halons also enter the flame with the air, where they interfere directly with the chemical processes, acting as negative catalysts. As little as 3 per cent in air will extinguish many flames. The physical properties of Halon 1211 make it the best suited for use in extinguishers.
- Powders are also chemically active and present a large surface area within the reaction zone of the flame, again interfering with the chemical processes.

These last three agents act mainly within the flame. Carbon dioxide and ordinary powders can do little to prevent the re-ignition of solid fuels from the smouldering combustion which remains after the flame has gone. Halon 1211, which is a low boiling point liquid, can penetrate to the fuel surface where it does have some inhibiting effect. Certain powders can act in this way as well: they are called ABC or all purpose powders.

Although water is ideal against many fires, it is unlikely to have any effect on burning liquids or gases (apart possibly from spreading a liquid fire, because most liquid fuels will float on water). The flames of these fires can be knocked down very well by Halon 1211 or powder, and less effectively by carbon dioxide. They can be removed completely, but after a longer application time, by foam.

Water should not be used against fires involving live electrical equipment because the firefighter may receive an electrical shock.

Extinguishers are chosen in accordance with the type of fire against which they are to be used. Two different fire classifications are in use, one originating in the USA and the other in Europe. The European version is described in the European standard EN2:

Class A Fires involving solid materials, usually of an organic nature (coal, paper, cardboard), in which combustion normally takes place with the formation of glowing embers.
Class B Fires involving liquids or liquefiable solids.
Class C Fires involving gases.
Class D Fires involving metals.

The American system described in NFPA/ANSI/10 differs from the European standard in that class C fires are those which involve energized electrical equipment. The American system can sometimes be recognized because on an extinguisher the letter A is shown in a green triangle, B in a red square, C in a blue circle, and D in a yellow five-pointed star.

In the European standard the B and C classes are really the same thing: liquids burn only when they have been vaporized and have therefore become gaseous. It would have been more useful to have had:

Class B Fires involving liquids, liquefiable solids and gases.
Class C Fires which can be extinguished by a surface-sealing agent.

This would have enabled the distinction to be made between agents which can tackle three-dimensional fires (powder, Halons and carbon dioxide) and those for two-dimensional fires (foam and AFFF). The first three agents would then have a BC classification, as would AFFF in certain circumstances.

The European classification has been used in Table 27.1, which is intended to help in making the best choice of extinguishers. The performance of the agents has been assessed as poor, good or very good. 'No' simply means that the agent is unsuitable. 'People' fires are those in which a flammable liquid has been spilled on someone's clothing and has ignited. The traditional fire blanket is very difficult to use effectively and it is much better to attack the flames with powder or Halon 1211. Powder will not injure the person but he should not be allowed to breathe the vapours of Halon 1211. Clothing wetted by the agent should be removed. The colour code at the bottom of the table is that adopted by manufacturers in the UK. The body of the extinguisher is painted in one of these colour.

Table 27.1
Extinguisher performance against different fires

Fire classification	Agent					
	Water	Foam	CO_2	Halon 1211	Powder	Special powders
European classes						
A	VG	G	P	G	P	G
B	NO	VG	G	VG	VG	–
C	NO	NO	G	VG	VG	–
D	NO	NO	NO	NO	NO	G
Electrical	NO	NO	G	VG	VG	–
People	P	NO	P	VG	VG	–
UK colour code	Red	Cream	Black	Green	Blue	–

It was mentioned in Chapter 6 that none of the usual agents can be used against class D (metal) fires. They will all react with the hot metal and increase the heat output of the fire. The reaction can be violent, but sometimes it is possible to extinguish a small fire by the massive application of water. More usually special powders are used which flux or sinter on the surface of the metal so as to exclude the air.

AEROSOL EXTINGUISHERS

When aerosol extinguishers were first introduced, some of the early designs quite rightly gained a very poor reputation. This was mainly because they contained a mixture of aerosol propellants, none of which was a brominated extinguishant. The significance of the presence of the various halogen atoms in the Halon molecule is discussed in Chapter 6, where it is shown that Halons containing only chlorine and flourine, like the propellants, have only a very limited performance. It is possible to demonstrate that they do have a firefighting capability by using very small trays, but any attempt to establish a rating against a standard fire will soon prove their limitations. These aerosol devices were demonstrated indoors against petrol fires only a few centimetres in diameter, and the implication of the sales talk was that a similar performance could be obtained on larger fires. The sales literature did reveal the composition of the Halon propellant mixture but claimed also the presence of a magic ingredient to which a superior extinguishing capability was attributed. It has been suggested that this was an aromatic oil (with no firefighting capability at all).

It was recognized by extinguisher manufacturcrs that an acrosol containing Halon 1211 would have an acceptable performance, but at a higher price. It was necessary first for the aerosol manufacturers to develop a high rate discharge (HRD) valve which could provide the same flow of agent as that from a small conventional extinguisher. It was then possible to make aerosols which could obtain the same fire ratings as these extinguishers, up to the maximum weight which could be contained in a standard can. The only disadvantages then were the less robust construction, and the limited shelf life (compared with that of conventional extinguishers), which is common to all aerosols.

A number of excellent Halon 1211 aerosols was developed, and also a few filled with powder, and these were placed on the market. Unfortunately a range of poor performance units was also available, and to enable the public to discriminate between the two it was decided to prepare a British standard: BS 6165: Specification for small disposable fire extinguishers of the aerosol type. There are no other national or international standards covering these extinguishers.

The standard allows a maximum capacity of 820 ml, which is the largest standard can, and required a minimum fill of 80 per cent. The agent can be Halon 1011 (CH_2BrCl) or 1211 ($CBrClF_2$), or a powder. The propellant should be carbon dioxide, nitrogen, Halon 122 (CCl_2F_2), or Halon 1301 ($CBrF_3$). Most of the available extinguishers contain an 80/20 mix of Halons 1211 and 122. A storage test is specified in which the Halon units are kept for eight months at a temperature of 20°C. The weight loss should be no greater than a rate of 5 per cent a year, and they should all operate correctly after storage. There is also an impact test, a drop test, and a pressure test.

The important requirement of the standard is in the performance against standard CEN fires. A minimum 8B rating is required, and if a class A performance is claimed, then the minimum should be 3A. If the largest permitted aerosol container has the minimum fill of Halon 1211, this will be about 1.2 kg of the agent. This quantity is more than enough to obtain a 21B rating. A 1.5 kg extinguisher has obtained a 3A rating, which might just be within the capability of the large aerosol. A 3A crib contains 56 wooden sticks each 38 mm square and making a stack 300 × 500 × 560 mm high.

The attraction of aerosol extinguishers, apart from their low cost, is that their method of operation is obvious even to untrained peopel, and they do have a very creditable performance. They are ideal for use in an office, where a typist would not hesitate to use one against a fire in a photocopier, or involving spilled solvent, although she might be reluctant to operate a full-size extinguisher. The aerosols should of course be used in addition to the normal extinguishers and not as a replacement. They may conventiently be placed on shelves and desks throughout an office or workshop area.

Obsolete extinguishers

The soda-acid and chemical foam extinghishers which had to be inverted have already been mentioned. These turn-over extinguishers should be regarded as obsolete. Any which are still in use are probably very old and unreliable, and their presence can cause confusion about the correct method of operation of other extinguishers. These extinguishers should all be withdrawn and scrapped (or given to a museum).

Small methyl bromide extinguishers are occasionally seen. This is a Halon with quite a good performance, but it is extremely toxic. These extinguishers should also be withdrawn, discharged safely in the open air, and replaced by Halon 1211 units.

HOSEREELS

In areas where class A risks predominate, hosereels provide excellent first aid protection. They have the advantage of providing a continuous supply of water instead of the 10 litres or so from a hand appliance. They should be so sited and of sufficient length that a jet from at least one of them can be directed on to combustible materials in any part of the protected area. The requirements of BS 5306:l Part 1 are typical of other standards and rules. The hose should be not more than 45 m in length, but less if the likely routes for the hose are tortuous. There should be at least one reel for every 800 m^2 of floor, and it should be possible to take the nozzle to within 6 m of any risk. The water supply should be such that if the two hoses which are hydraulically the most remote are turned on, they should both provide a jet about 6 m in length, with a flow of at least 30 1/min. For a typical 30 m hose fitted with a 6.35 mm nozzle, a minimum pressure of 1.25 bar would be needed. With a 4.8 mm nozzle the required pressure would be 3.0 bar. If sufficient pressure is not reliably available from the mains, a booster pump should be provided.

The reels may be conveniently provided with a swivel so that the hose can be run off in any direction. It is possible to have reels of this kind inside a cupboard. The water can be controlled in one of three ways. An ordinary control cock can be turned on before the hose is run out, and a second control used at the nozzle. It is sometimes arranged that the action of removing the nozzle from its housing turns on the supply. The water can also be turned on automatically by the rotation of the reel: as soon as a few metres of hose have been pulled out, the valve will be open. Some early designs were unreliable because it was necessary to wind off the full length of hose, and unfortunately it was also possible to throw loops of hose off the reel. No matter what type of hosereel is provided, regular practice in their use is necessary.

Hosereels are sometimes connected to a sprinkler system water supply. This has the advantage that an alarm will sound when one is turned on (and this will also help to discourage their use for washing down). It is more usual to connect hosereels to the domestic mains, but the water authority should be consulted before this is done.

Hosereels, like extinguishers and manual call points, should be located in conspicuously marked positions. In a warehouse or a factory, at least one should be immediately visible and readily accessible to a person working anywhere in the area. It can often be arranged that all the equipment is available at a fire point, and that a notice indicating its position is clearly visible. (The location of extinguishers is discussed below.)

OTHER FIRST AID EQUIPMENT

A steam lance is a traditional piece of firefighting equipment on a chemical works. It comprises a length of steel pipe with a thermally insulated handle, connected by a flexible hose to a steam main. The steam pressure should not be more than about 6 bar or too much air will be entrained into the jet. An experienced operator can make very effective use of a lance. If the flames from a gas leak have been extinguished, the lance can be left in position to prevent re-ignition and to help in dispersing the gas.

Another traditional method is the use of wet sacks. These are kept in a drum of water (to which a fungicide is usually added). They are useful for small fires, again typically resulting from gas leaks, where the discharge of an extinguisher could cause the interruption of a manufacturing process.

Buckets of water may still sometimes be seen. They are ineffective because of the limited throw and the difficulty in hitting the target. An extinguisher enables much more effective use to be made of the same volume of water, and from a greater range. A requirement for buckets does still appear in the rules of some insurers, but they should always be replaced by the equivalent number of extinguishers, or hosereels. Buckets of sand are even more difficult to use effectively. The sand can however be poured over small spills of flammable liquid, or small metal fires (if the sand is dry). Again, they should be replaced by appropriate extinghishers.

It should not be forgotten that one way of fighting a fire is to starve it of fuel. An emergency plan for dealing with a fire in a warehouse for example might well include the provision of fork lift trucks, which can be used to move stored goods away from the fire.

TRAINING

The vital importance of the first line of defence has already been stressed. Most fires are small when they start, and people are potentially both good detectors and good firefighters. All they need is the right equipment, and adequate training.

Class A fires, against which water extinguishers can be used, can conveniently comprise a simple heap of lengths of wood or cardboard boxes. The fire can be ignited by a small tray of flammable liquid and allowed to burn for several minutes until it is well established. It should be tackled first from the up-wind side, the jet being directed from side to side over the whole surface of the fire. The operator should then walk slowly around the fire until all the burning surfaces have been wetted. Some of the water should be retained in the extinguisher and the embers watched to see whether any smouldering develops. This kind of fire can conveniently be used to demostrate the class A performance of a Halon extinguisher, or an all purpose (ABC) powder.

Class B fires can be burned in round or square trays. The fuel should be floated on a layer of water to minimize the distortion of the trays. Almost any fuel can be used, but commercial n-heptane, or a fuel with a similarly narrow boiling range, has two advantages: the fires will correspond to standard test fires and, because there is little fuel used in each test, the fire can be relighted repeatedly, and will still be the same size. The class B fire should be attacked from the up-wind side only. Using the Halon or powder extinguisher, the technique is to detach the flame from the nearest edge of the tray by sweeping the jet of agent along this edge. The agent should not be applied at too high a velocity or it will penetrate too far into the fire, or may splash fuel out of the tray. The attack should be started sufficiently far back from the fire that, when powder is used, the jet has broken up into a cloud before it enters the flame. The flames should then be pushed back over the fuel surface, again by a sweeping action, and finally detached from the back edge. If foam is used, it is best to arrange a vertical target, conveniently a sheet of metal, at the back of the tray. The jet of foam is directed through the flames to hit the target, from which it will spread over the fuel surface. Alternatively, the jet can be aimed at the ground just in front of the tray from where it can bounce over the edge and fall on to the fuel. When Halon and powder are used, fires on square trays will be slightly easier to extinguish than fires of the same area on round trays. This is because the flame cannot be chased into a corner in a round tray.

All employees should be trained in the use of extinguishers, although office workers may not need to attempt anything other than quite small fires. If fire teams have been established they will need extra training, preferably with hoses and other equipment. Theoretical training should also be given to ensure that all the employees understand simple fire precautions, what action to take if they discover a fire, how to respond to a fire alarm, what escape route to take, which extinguisher to use against a particular fire, and the fire hazards they are likely to encounter on the site.

NUMBER AND LOCATION OF EXTINGUISHERS

A broad guide is that extinguishers and manual fire call points should be at an identified fire point, one of which should be visible and accessible from any part of for example a warehouse with a maximum travel distance of 30 m. At least one of these points should be immediately visible and readily accessible to a person working anywhere within the warehouse area. In a small room the fire points may be conveniently sited at the fire exits. This ensures that the person using an extinguisher can retreat quickly to the exit if he is threatened by the fire.

The extinguishers should not stand on the floor but should be on a wall bracket. The extinguisher handle should be about 1 m from the floor. The extinguishers may therefore be obscured when viewed from some positions in the room and it is important that the words FIRE POINT (or a similar description) are clearly visible above their position.

The number, type and locationof extinguishers may be decided by local management, and may follow advice given by the fire brigade. Insurance companies will frequently have minimum requirements for the provision of extingishers. Guidance can also be obtained from codes of practice, and that given in BS 5306 Part 3 is discussed below.

The number of extinguishers which will be needed in an area can be related to their performance. The performance of an extinguisher can be measured by using it against a standard range of fires and noting the size of the largest fire which can reliably be put out. In the USA the Underwriters' Laboratories (UL) have a range of cubical wooden crib fires for class A extinguishers, and they also use a vertical wooden panel and some excelsior (wood wool). Their class B fires are rectangular metal trays ranging in size from 2.5, 5, 12.5, 25 up to 1600 ft^2. In Europe a CEN standard defines a similar range of test fires, except that the larger crib fires are rectangular and the class B fires are in circular trays. These trays have diameters ranging from 0.56, 0.72, 0.91, 1.2 up to 3.9 m and greater.

The rating number which appears on the extinguisher is derived from the size of the largest tray with which it can cope. In the UL system the area of the tray in square feet is divided by 2.5. This is intended to represent the difference in performance between the unskilled person who has to use it and the fully trained man who obtained the rating. Thus, the first four UL fires mentioned above correspond to these extinguisher ratings: 1B, 2B, 5B and 10B. The unskilled person should be able to use them successfully against flammable liquid fires of 1, 2, 5 and 10 ft^2 respectively.

The first four CEN ratings are 8B, 13B, 21B, and 34B. This series of numbers is obtained by adding together the previous two numbers: thus 34 is the sum of 21 and 13. The number itself indicates the number of litres of fuel which are burned in the fire. The surface area of the fire in square metres can be found by dividing the rating by 31.8. The CEN class A fires also follow the same numbering sequence but in this case the number indicates the length of the wooden crib. A 3A fire is 0.3 m long, a 5A is 0.5 m long but of the same cross-section, and an 8A is 0.8 m long. The largest in the series is a 55A. The UL fires are simply numbered 1A, 2A, 4A up to 40A. The 2A fire corresponds very roughly to the 13A in the CEN series.

We can use these ratings to calculate the number of extinguishers which will be needed in a given location. For class A fires the number is simply related to the total floor area and for class B fires it is related to surface area of flammable liquid. The figure which is obtained is the total rating. Suppose for example this is 40A in the UL series. We can obtain this coverage by using 40 extinguishers each with a 1A rating, or one large wheeled unit with a 40A rating. It is very probable that neither choice would be correct and that the best protection would be provided by fewer than 40 units, each with a rating higher than 1A. It should also be remembered that these assessments of extinguisher requirements will indicate the minimum provision, which will frequently need to be exceeded.

For class A risks the UK code of practice (BS 5306 Part 3) requires a total rating of 0.065 times the floor area in square metres on each storey of a building. In no case should the total rating be less than 26A, which corresponds to two 9-litre (2 UK gallons) extinguishers (each with a 13A rating). These extinguishers have nearly the same capacity as 2½ US gallon extinguishers, which can obtain a 2A UL rating.

The requirements for class B fires (again in the UK code of practice) are obtained by applying a factor to the B ratings to allow for the difference between the unskilled user and the skilled operator who obtained the ratings. The required rating is obtained by multiplying the anticipated surface area of the fire (in square metres) by 80. As we have seen above, the rating number of a CEN fire is 31.8 times its area in square metres, so the increase from 31.8 to 80 represents a factor of just over 2.5, which is very nearly the same as the UL factor.

Thus, the minimum rating to protect a class B risk with powder or Halon extinguishers is obtained by multiplying the fire area in square metres by 80 for CEN rated extinguishers, or by using the actual area in square feet for UL rated extinguishers. Foam extinguishers are less dependent on the skill of the operator and the required rating when these are used can be found by multiplying the area in square metres by 50 (CEN), or (using the same ratio) by multiplying the area in square feet by 0.625 (UL). However, these calculations give the minimum rating of the extinguishers which should be use against the fire, and not the actual number of extinguishers. This number can be obtained for CEN rated extinguishers by multiplying the fire area by 150 and dividing by the extinguisher rating. For example, suppose we want to protect a single open tank with a surface area of one square metre. The minimum extinguisher rating is given by:

$$80 \times 1 = 80$$

The next highest rating on the CEN scale is 89B, so this is the minimum acceptable rating for the extinguishers. The number needed can be obtained from:

$$150 \times 1 = 150$$

This can be achieved by using two 89B extinguishers, or alternatively by using a single 183B extinguisher. If foam is used, 55B extinguishers can be used, and three of these will be needed.

When making these calculations, each room should be considered separately as should risks separated by more than 20 m. If there is a number of separate risks which are closer than 20 m apart, guidance on the calculation of ratings is given in a table in the code of practice. For spillages up to 8 mm deep the rating is found by multiplying the anticipated volume of the spillage, in litres, by 10. This calculation can be used in providing protection for drums of flammable liquids: it is assumed that the entire contents of the largest drum can be lost (unless it is possible for a larger number to be damaged, by a fork lift truck for example). If the drums are in a bund, the calculation can be based on the area of the bund, if this gives a lower rating.

It should be emphasized again that the numbers obtained by these calculations should be regarded as minimum values. It should be remembered that the extinguisher ratings are obtained by using simple hydrocarbon fires (n-heptane). Lower ratings would have been obtained with fuels like ehtyl ether, carbon disulphide and methanol. If these, or similar materials, are being used, test fires should be burned to determine extinguisher requirements and to provide training for personnel.

MAINTENANCE OF EXTINGUISHERS

In many small factories and offices extinguishers are the only means of fighting a fire. They will be very rarely used but in an emergency it is essential that all of the equipment functions properly. There will not be a second chance of stopping the fire while it is still small. Reliability can be ensured only by regular inspection and maintenance. Sometimes this can be done under contract, either by the manufactorer or by a local firm. Whatever arrangements are made, someone at management level should be responsible for ensuring that the job is done properly and that adequate records are kept. The date of inspection should be noted on a label attached to each extinguisher (not stamped on the extinguisher itself). The extinguisher should be withdrawn from service at regular intervals for a full examination and for refilling if necessary. (It has already been noted that this can conveniently be combined with practical fire training.) Any local requirements covering the testing of pressure vessels should also be observed, and this applies particularly to carbon dioxide extinguishers.

In drawing up an inspection and maintenance schedule the manufacturer's instructions should be considered as well as any requirements of the insurers. The appropriate standards and codes of practice should also be studied. These documents will be considerably more detailed than the brief notes given below. Reference is made here to soda-acid extinguishers because many of these are still in use and indeed they are still being manufactured.

Extinguishers should be inspected at least once a month to check that they are in their correct location, have not been discharged, are at the correct pressure (if a gauge if fitted) and have not been damaged. Every extinguisher should have a more thorough check once a year. Powder extinguishers, especially if they have been subjected to vibration (on a vehicle for example) may need to be checked every few months.

If extinguishers are not to be used for training then it is not essential (although still desirable in some cases) to discharge them every year. They should, however, be withdrawn from use and examined every year. The contents of Halon and carbon dioxide extinguishers can be checked by weighing. The condition of a powder can be checked by pouring it out of the extinguisher and returning it. If these and other checks are satisfactory, the units can be returned to use. All extinguishers should be completely discharged at regular intervals. The longest intervals which should be allowed for extinguishers which are exposed to ideal conditions are:

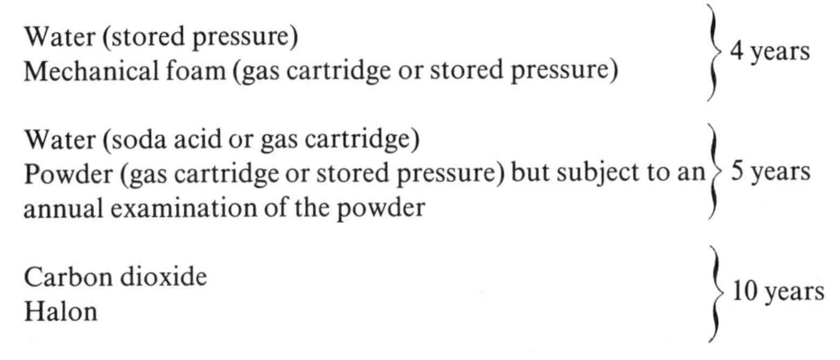

Water (stored pressure) Mechanical foam (gas cartridge or stored pressure)	4 years
Water (soda acid or gas cartridge) Powder (gas cartridge or stored pressure) but subject to an annual examination of the powder	5 years
Carbon dioxide Halon	10 years

The carbon dioxide extinguisher should be subjected to a hydraulic pressure test after the discharges. This should be done again at the end of 20 years and then every 5 years.

ROUTINE CHECKS

The notes below indicate some of the checks which should be made at an annual (or more frequent) inspection, but a relevant standard and the manufacturer's instructions should be consulted for full details. Where necessary, suitable tools should be obtained from the manufacturers. If the body of an extinguisher shows signs of corrosion or damage it should be replaced. Similarly, damaged, faulty or corroded components should be replaced by components supplied or approved by the manufacturers. Illegible labels should also be replaced.

Soda-acid water extinguishers

When the actuator on this type of extinguisher is pushed down it breaks a glass bottle of acid. The acid reacts with an alkaline solution which fills most of the body. The released gas pressurizes the extinguisher and forces out the liquid. The handling of acid bottles can be hazardous and the alkaline solution is corrosive. It is for these reasons that soda-acid units are often replaced by gas cartridge or stored pressure designs. The main points to be covered are:

- Check the acid bottle and replace if cracked
- Check the liquid level and pour out the liquid
- Examine inside and outside surfaces for corrosion and damage
- Check the venting device
- Check the dip tube, strainer, hose and nozzle
- Check the actuator for free movement. Rectify if necessary
- Check all sealing washers
- Refill, reassemble and replace the safety pin or seal.

Cartridge-operated water extinguishers

These contain a cartridge of compressed carbon dioxide. The gas is released to pressurize the extinguisher when a seal on the cartridge is ruptured by the actuator. The filled weight and tare weight of the cartridge are stamped on the body. Point to be covered are:

- Weigh the cartridge and replace it if more than 10 per cent of the contents have been lost (or as recommended by the manufacturer)
- Check the cartridge for corrosion and damage
- Proceed as for the soda-acid extinguisher
- Check manufacturer's instructions about anti-freeze additives and corrosion inhibitors before topping up any lost water.

Stored pressure water extinguishers

These extinguishers are pressurized from a compressed air-line, or a similar source, after they have been filled. They are fitted with a pressure gauge or indicator. Points to be covered are:

- Weigh the extinguisher to check for loss of contents
- Discharge completely and check that the gauge or indicator is functioning correctly
- Follow the same procedure as for the other water extinguishers.

Cartridge operated foam extinguishers and stored pressure foam extinguishers

These are checked in the same way as the corresponding water extinguishers.

Cartridge-operated powder extinguishers and stored pressure powder extinguishers

The contents of a powder extinguisher should be examined only in a dry room and the powder should be exposed to the air for as short a time as possible. Care should be taken to ensure that the powder is not contaminated by a different type of powder. Oil and grease should not be used on the actuating mechanism. The contents of a stored pressure extinguisher can be examined if it is inverted and the discharge valve is opened to release the compressed gas (this does not apply to factory sealed units). The main points are:

- Empty out the powder and examine it for caking, lumps or contamination. If any of these are found the powder should be discarded
- All relevant checks previously mentioned should be applied.

Carbon dioxide extinguishers

The annual inspection comprises weighing and external examinations. At the periods indicated, the extinguisher should be discharged and pressure tested hydraulically. The main points are:

- Weigh the extinguisher. If more than 10 per cent of the contents have been lost (or as recommended by the manufacturer) the extinguisher should be discharged and refilled
- Examine the body for damage and corrosion
- Examine all other components and clean or replace where necessary
- If the actuating mechanism can be removed (usually by unscrewing from the body) it should be removed and checked.

Halon 1211 extinguishers

Athough the Halon 1211 has an appreciable vapour pressure the larger extinguishers are always pressurized with nitrogen. This is to reduce the variation in discharge rate over the operating temperature range. Any loss of nitrogen would reduced the efficiency of the extinguisher and for this reason any unit which shows only a slight loss of weight should be rejected (unlike a carbon dioxide extinguisher for which up to a 10 per cent loss may be acceptable). At an annual inspection:

- Weight the extinguisher. If there has been any loss in weight it should be rejected
- Proceed as for carbon dioxide extinguishers.

28

Private water supplies, hydrants, hoses and pumps

When all else has failed, a fire will finally be extinguished by the massive supplies of water poured on to it by the fire brigade. The quicker and earlier that the brigade can obtain the water they require, the sooner they will get it under control and the less loss there will be from the effects of fire and of the water.

Bulk supplies of water for sprinkler systems are described in Chapter 20. These supplies may not be used for any other purposes but the same requirements should be applied to supplies of water for other fire-extinguishing purposes. It must be emphasized that the two systems are to be kept absolutely separate although, subject to the rules for automatic sprinklers, fire hydrants will be found on the public mains that also supply sprinkler systems. At this point, the interests of the insurance company and those of the fire brigade appear to cross and, although it is essential to keep the water supply feeding the sprinkler system, there may come a stage in the fire when so many sprinkler heads have opened that they are no longer effective. They then constitute a severe drain on the water supply that could be better used by the fire brigade under manual control. The decision when to shut down a sprinkler installation during a fire must rest with a senior fire officer only and cannot be lightly taken.

PUBLIC WATER SUPPLIES: THE LEGAL POSITION

A water authority has a statutory duty to lay water mains for the supply of water for domestic purposes only. They have no duty to lay mains for firefighting purposes. However, having decided to lay a main, the water company is bound to inform the fire service and to provide them with a plan showing the proposed position of the main. The fire brigade will indicate what hydrants they require and the water authority will install them and charge the brigade accordingly. They are not bound to install fire hydrants for private consumers unless the brigade requires them to do so.

A water authority, however, is bound to supply water free of charge for firefighting purposes. This gives rise to a complication when fire hydrants are connected to mains on private premises because the water authority may want to charge for water that is used for trade and purposes other than firefighting. To make their charge, they will install water meters in the mains feeding into the premises. Modern meters offer little obstruction to the flow of water but some of the older and smaller meters can form a serious obstruction.

There are two methods of overcoming this difficulty. One method is to insert a valve system known as a 'bypass' valve in the main supply and to lead a small pipe round it, the smaller pipe being metered; when the bypass valve is opened the meter is short-circuited and no water flows through it. The other method is to seal the hydrants.

When a bypass valve is fitted, the turncock's tools that are required to open the valve must be kept close to it in plain view. A cupboard with a glass front is very suitable and an indicator plate must be fitted giving the position of the valve in a similar manner to a hydrant plate but clearly marked as a bypass valve.

Sealed hydrants

With modern large meters the obstruction to the flow of water is so slight that it is often not worthwhile fitting a bypass valve and so the sealing of each hydrant is carried out. Quite frequently, a separate main is laid for fire hydrants and this main has no other connections on it. The big disadvantage of this system from the fire service point of view is that the main can be shut down and no one will know about it until an emergency arises. Then it could become one of the factors of a major fire loss, whereas if the hydrants are on the domestic supply there will be a quick call if the supply is cut off.

RING MAINS

Wherever possible, dead end mains should be avoided and the mains should be joined to each other and brought together in the form of a ring. In large installations, these rings will form a matrix and an ample number of valves should be fitted so that sections may be isolated so that as little as possible of an installation is put out of commission for any purpose at any one time.

Depth of mains

Where possible, mains should be laid at a depth of not less than 1 m to avoid frost damage. Due consideration should be given to the weight of traffic likely to go over the main and the ground strengthened accordingly. If it is possible to lay the main where heavy traffic will not pass over it, this should be done. The main must not be embedded in concrete and when it passes through a building it must not be subject to freezing. Plans of all mains, valves and hydrants should be readily available, preferably exhibited in such a place that they can be consulted freely.

Contamination of public water supplies

There must never be any possibility of contaminated supplies being able to mix with potable supplies. This means that mains that can be supplied from the public supply may not have any arrangement of inlet connections that would enable a supply of water from an outside source to be used. Neither non-return valves nor any arrangement of closed valves can be accepted by a water authority.

SOURCES OF WATER SUPPLIES

There are two sources of water supply other than those described under water supplies for sprinkler systems in Chapter 20. These are:

1 Rivers and canals
2 Wells and boreholes.

Before taking any practical steps to obtain a water supply from other sources, a legal investigation of the position must be made. Very often the local fire brigade will have knowledge of any restrictions and in any case it is advisable to discuss with them the best methods of obtaining auxiliary water supplies. These supplies may be required for the supply of special fire mains or for the direct use of fire brigade pumps.

Rivers and canals

Where direct access to a river or canal can be made for fire service vehicles, all that is required is a hard standing for the vehicles and suitable drainage for the cooling water discharged from the engine. (The engines of fire service pumps are cooled by taking water from the main pump, passing it through the cooling system and allowing it to run to waste. Unless suitable drainage is provided this water can bog the vehicle down.)

Where access is difficult because of buildings on the banks, it may be possible to lay a suction main through the building ending in a suction hydrant with a suitable outlet for the brigade suction hose. In all cases, the fire brigade will best be able to describe their requirements and should be consulted before making definite plans.

If the source of supply contains considerable quantities of fibrous or other objectionable material in suspension which could obstruct the strainer of the suction pipe, the use of a 'jackwell' is recommended. Figure 28.1 shows the construction of a jackwell.

Wells and boreholes

The strata below the surface of the earth can roughly be divided into two classes: those that permit the passage of water and those that do not. Rain falling on the ground sinks through until it eventually comes to an impermeable stratum. It is then soaked up in the permeable stratum above, forming a reservoir of water.

By sinking a well into the permeable stratum water can be brought to the surface. If the permeable stratum lies in a basin, the water may come to the

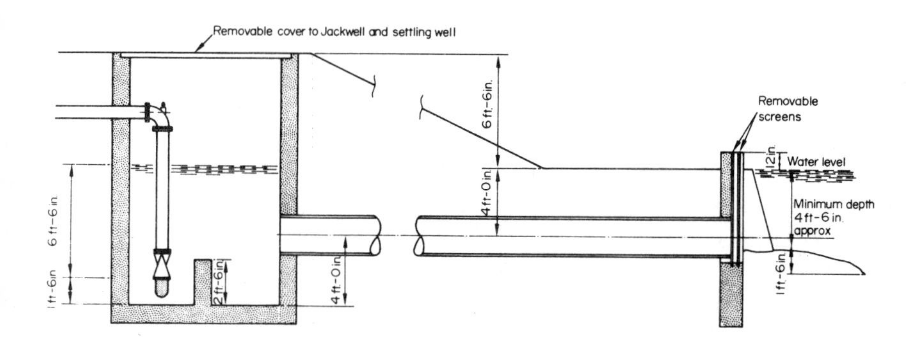

Figure 28.1 A jackwell

surface of its own accord in the form of a spring, but if a pipe is sunk and the water reaches the surface it is known as an 'artesian well.'

Where the water is at a considerable depth, only a comparatively small pipe is driven down instead of digging a well; this constitutes a 'borehole'. The borehole needs only to be of sufficient size to enable a submersible pump and electric motor to be lowered to the bottom of the shaft.

There are many factors, including legal permission, to be taken into account before sinking a well or borehole and experts on the subject should be consulted.

STATIONARY PUMPS FOR PRIVATE WATER SUPPLIES

The FOC rules for pumping sets, that is the pump and power unit for use with automatic sprinkler installations, have been described in Chapter 20. Although no rules apply to pumping equipment used for private hydrants and water supplies, it is good policy to adhere to the same rules relevant to pumps for supplying sprinkler installations under FOC rules.

Hydraulic ram pumps

These pumps are not suitable for use as fire pumps, but are very useful for replenishing or topping up reservoirs from streams and rivers. They are very simple in construction, cheap to install, cost nothing to run and need practically no maintenance. Figure 28.2 shows a hydraulic ram pump. The principle is very simple. As the water flows in from the source it will have acquired some momentum dependent on the rate of flow of the stream or

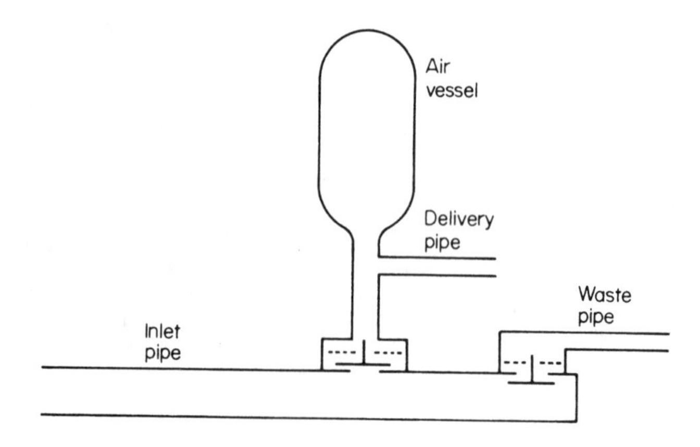

Figure 28.2 Hydraulic ram pump

river; it therefore closes the waste valve. The closure of this valve sets up a shock wave which can produce a very high pressure. The momentum carries the water forward opening the valve into the air chamber. Having now lost its momentum the valve to the air chamber will close, the waste valve will open allowing a drop in pressure until the water again acquires momentum. This action raises the water in a series of pulses sometimes as frequent as 60 to 80 times per minute.

A ram pump such as this may drive water vertically to a height of up to 75 m.

HYDRANTS

Rising mains in high buildings

Rising mains or 'risers' are either dry or wet. A dry riser is one that is normally empty but ascends throughout the building, being fitted with outlets at every storey. At ground level, it has a fire brigade pumping inlet. A wet riser is one that is normally charged under pressure, the pressure being supplied and maintained by its own pumps.

Dry risers

A dry riser should be installed in any building over 18.3 m in height up to 60 m; above that height wet risers should be installed. Horizontal mains can be laid from the riser up to a distance of 12 m to the outlet. If any part of the floor is more than 60 m from the rising main, a second dry riser should be installed. The riser should be a 100 mm pipe.

The outlets known as 'landing valves' (see Figure 28.3) should be matched to the outlets of the local fire brigade and the riser should be topped with an

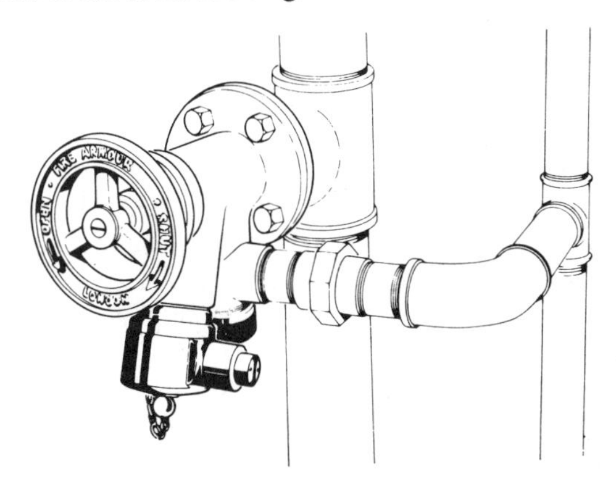

Figure 28.3 High pressure landing valve

automatic air vent valve. Outlets should be so arranged that they do not cause an obstruction and are not liable to mechanical damage. Where possible, they should be in the room to be protected and not on the stairway. If they are on the stairway the hose, if abandoned, might obstruct the closing the door. The outlets should be controlled by a wheeled valve and the direction of turning to open the valve should be plainly marked on it.

The inlets to the risers should be standard with those of the local fire brigade and should have a drain valve at the lowest point. The inlets should be contained in a glass-fronted box, locked if necessary and plainly marked 'Dry riser fire main'. These inlets should be so placed that they are not subject to mechanical damage. They should be situated horizontally within 12 m of the actual dry riser. If possible, the inlets should be situated fairly close to a street fire hydrant having a suitable hard standing for a fire brigade pump.

Wet risers

Wet rising mains should be fitted in all buildings over 60 m in height. The same rules as to the distances should be observed as for dry risers. A suitable drain cock should be situated at the base of the riser. Each rising main should be fed by two automatic pumps, one of which should have a compression ignition engine drive, each pump being capable of providing indefinitely 1365 l /min at sufficient pressure to provide 4.1 bar with two 20mm nozzles at work at the highest outlet.

The pumps should generally comply with similar conditions to those which supply sprinkler installations under the FOC rules, that is housed in a fire-resistant compartment under positive head suction conditons with automatic starting by pressure drop and once started, to run continually until manually stopped. Automatic audible and visual warning signals should be given at a suitable point on the starting of the pumps. Power supplies should be laid down under similar conditions to the FOC rules both as regards electric supply and compression-ignition engines.

Outlets (landing valves) should be fitted with automatic pressure reducing valves so that the pressure on the hose cannot exceed 4.1–4.8 bar. The device should also give relief so that dangerous pressures cannot be built up when a shut-off branch is closed suddenly. The water from the relief device can be taken to waste or returned to the supply side of the pump. Figure 28.3 shows a landing valve.

Landing valves on wet and dry risers should always be closed with blank caps chained to the valve; otherwise all sorts of rubbish may be pushed into the outlet. Hoses should not be left connected because on a wet riser the valve may leak and damage the hose. On a dry riser, a valve may have been turned on unofficially with the result that, if water is let into the riser, pressure may be lost and the supply reduced below requirements in an emergency.

Water supplies for wet risers

Both pumps should be able to draw from two separate sources of supply. The two sources of supply should be a tank of not less than 11 400 l into which a towns main can supply 1600 l/min through a float valve. A second supply should be a storage tank of not less than 45 500 l which can be supplied from a towns main at a rate not less than 455 l/min. Each water supply should be connected by a 150 mm main to a fire brigade inlet collecting box.

To prevent the hose on the lower storeys from bursting, due to the high pressure needed to supply water to the top of the building, 'break' tanks used to be installed at intermediate levels. However with modern landing valves, incorporating a pressure relief device, break tanks are no longer required.

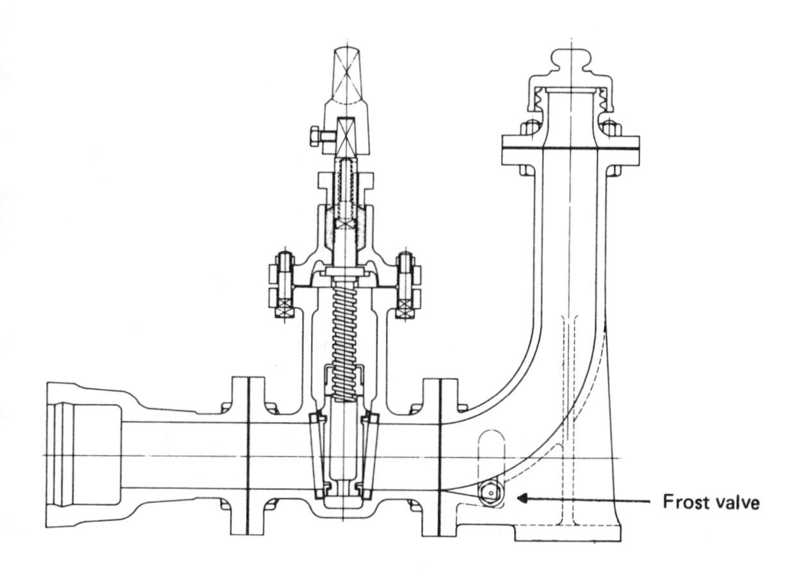

Figure 28.4a Sluice valve hydrant (*Angus Fire Armour*)

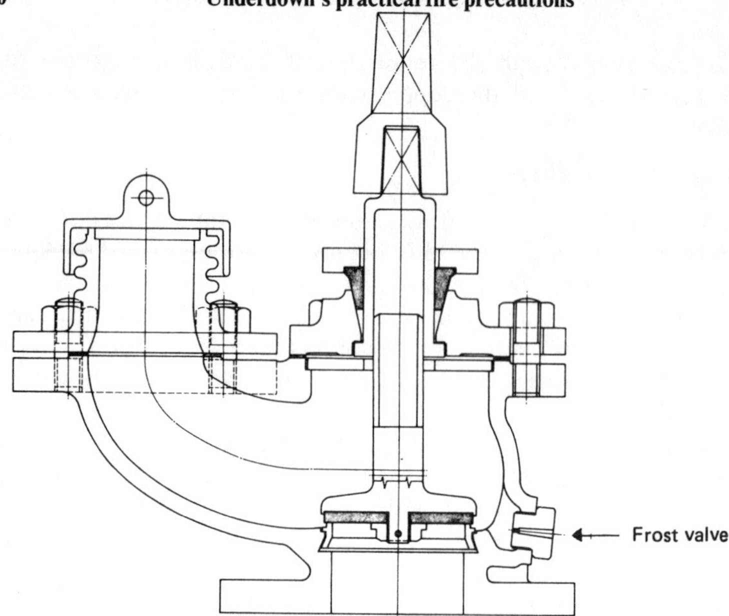

Figure 28.4b Screw-down hydrant (*Angus Fire Armour*)

GROUND HYDRANTS

There are two main types of ground hydrants now in general use, sluice valve hydrants and screw-down hydrants. Older types such as the ball valve hydrant should be condemned and replaced with modern types because, if the pressure in the main falls or the water is shut down, the ball is liable to drop and permit contamination of the mains water. Even material large enough to block the main or prevent the ball from reseating can enter the main. It is also difficult to ensure that they do not leak when shut down. Figures 28.4a and b show the two main types.

Before installing hydrants, the local fire brigade and water authority should be consulted so that similar types to those in use in the district can be installed. Where there is a choice, preference should lie with the screw-down type which is cheaper to purchase and to install. Theoretically, they do not give such a full flow but in practice it is impossible to tell the difference. The only factor against the screw-down type is that there is a tendency to screw them down too tightly and so damage the washer. Washers are quite easy to replace and provided they are changed immediately they have any tendency to let water leak past, no serious damage will occur. But, if this tendency is neglected, the hydrant will be screwed down with ever-increasing force until the spindle is broken.

Where a hydrant is in constant use, it is better to use a sluice valve type. British Standard 750 1984 sets out the standards for both types of hydrants and for the dimensions of the surface boxes.

Hydrant outlets

There are three main types of hydrant outlets: bayonet, round thread and fire brigade V-thread. Fire brigade V-thread is an odd size having $5^1/_5$ threads to the inch. The coupling at the bottom of the stand pipe must match the type of hydrant outlet, which should be the same as those in use by the local fire brigade. All outlets on the same premises should be identical with each other. The three types of hydrant outlets, together with matching bottoms of the stand pipes, are shown in Figures 28.5a, b and c.

Where a difference of outlets is found, it is quite easy to change them without even shutting down the main, because the hydrant outlet is on a flange which is bolted to the body of the hydrant. The new hydrant outlet should be obtained with a blank flange and then, if the old outlet is unbolted it can be used to obtain a template for drilling the flange of the new outlet.

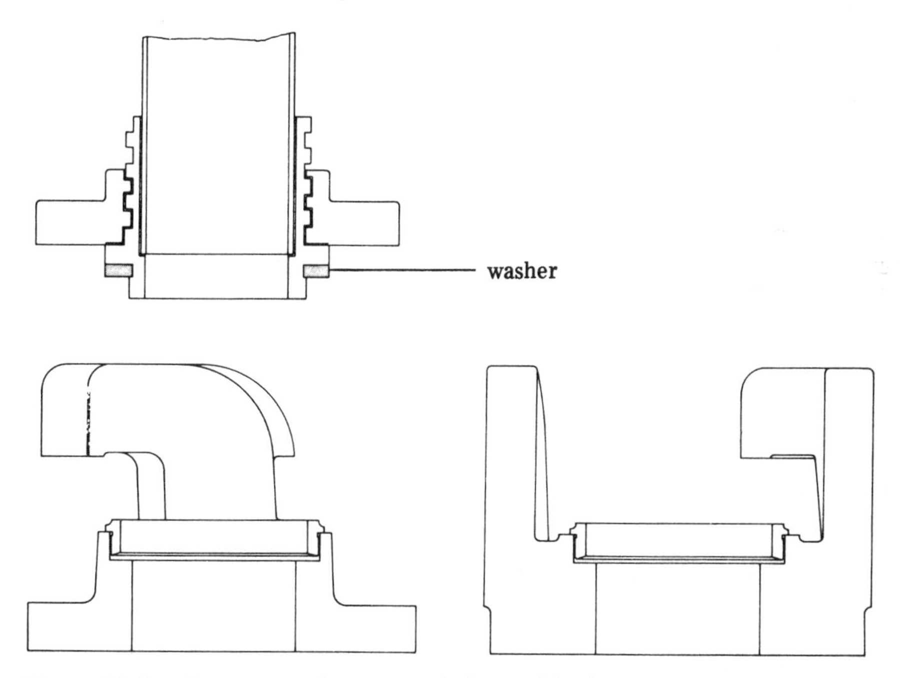

Figure 28.5a Bayonet outlet on stand pipe and hydrant
(*Angus Fire Armour*)

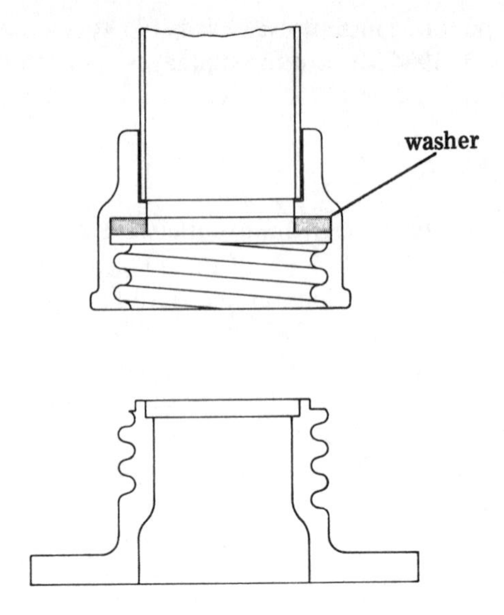

Figure 28.5b Fire brigade round thread on stand pipe and hydrant (*Angus Fire Armour*)

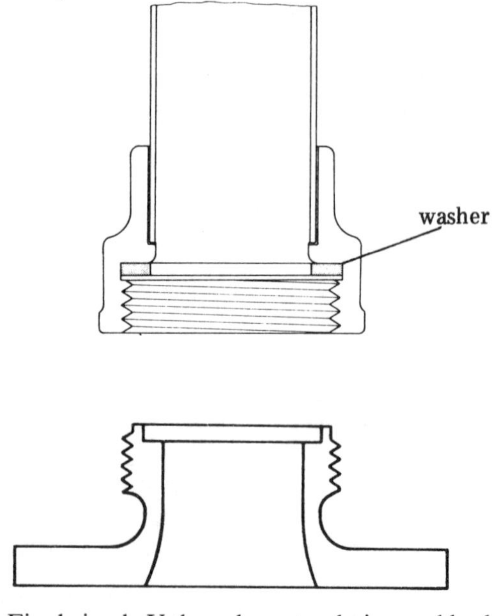

Figure 28.5c Fire brigade V thread on stand pipe and hydrant (*Angus Fire Armour*)

The hydrant outlet is often protected with a loose-fitting blank cap to prevent mud and rubbish from dropping into it. It is necessary to remove this cap before the stand pipe can be fixed.

Hydrant stand pipes

Although stand pipes are part of the equipment of a fire service, it is useful to consider them with hydrants. The bottom of the stand pipes corresponding with the types of hydrant outlets are shown in Figure 28.5. Particular attention is drawn to the washer in each case. If this washer is not in place, there will be severe leakage and the stand pipe if loose may easily damage the hydrant. The washer on the bayonet stand pipe is more vulnerable than the others.

It is very important to see that the lugs on the bayonet stand pipe are screwed down to their lowest point before any attempt is made to fix the stand pipe on the hydrant. It is a fault that occurs easily because when the stand pipe is removed from the hydrant outlet it will lift out after a fraction of a turn. If left in this condition, the lugs will not engage under the outlet lugs and, if it is not noticed in time, the lugs on the stand pipe will be further raised before the fault is noticed. It should be a matter of drill to see that the lugs are immediately re-set after use.

STAND PIPE HEAD

Figure 28.6 shows the top of the stand pipe and attention is drawn to the grips which are rigidly welded to the body of the stand pipe. The purpose of these grips is to enable the stand pipe to be screwed into the hydrant outlet. They are necessary because the head of the stand pipe is made to turn within the body to enable the hose to be taken straight away from it without any sharp bends. After fixing the stand pipe by means of the grips, the head should be rotated in the same direction as is used to tighten the stand pipe, otherwise the stand pipe might be loosened.

The head of the stand pipe rotates on a special washer and if the head becomes too tight (or too loose) it can be removed by undoing the holding down bolts and replacing the washer. Only the special washers obtainable from the manufacturers should be used.

Hydrant valve spindles

To prevent corrosion, the valve and valve spindle of a hydrant are made of

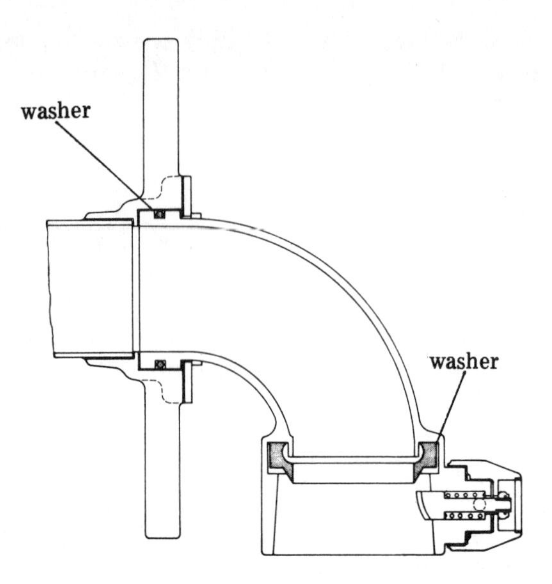

Figure 28.6 Stand pipe outlet (*Angus Fire Armour*)

brass or gun metal, but these metals are soft and the constant use of a steel key, especially if used carelessly, will quickly damage the square head of the spindle. To avoid this damage, a steel thimble of 'false spindle' is placed over the head of the spindle and held in place with a set screw. It must, of course, be possible to remove the false spindle; otherwise the gland could not be re-packed. In some older types of hydrants, the spindle was deliberately made smaller to ensure that steel keys would not be used without the false spindle. If, therefore, the false spindle was missing the hydrant could not be used. A note should be kept of all hydrants that have a false spindle so that a special check on them can be made when hydrants are being inpected.

Spacing out hydrants

The spacing of ground hydrants will depend partly on the premises at risk and partly on the supply of water. The nominal spacing of hydrants in streets in a housing district is about 130 m but this distance should be much less for industrial and commercial premises. Obviously, where possible, hydrants should be situated near points of entry into buildings, but due consideration must be given to the possibility of structural collapse of the building. This consideration should overrule the disadvantage of hose having to cross a road because under fire conditions the road will probably be closed to general traffic and in any case the hose could be ramped where it crosses the road.

Hydrant pits

The hydrant pit supports the weight of the hydrant and the frame for the cover. It usually consists of a slab of concrete as a base with brickwork to form the walls. The hydrant frame should be slightly higher than the surface of the ground so that water tends to drain away from it rather than into it. Unless there is danger of water seeping into basements, the pit should have a drainage hole so that water cannot accumulate in the pit.

There is a modern tendency to use pre-cast concrete pits. These are much cheaper than brick-built pits and can be installed by unskilled labour. Where there is any danger of drainage from a hydrant pit the walls can be rendered with an impervious lining or the hydrant sealed into a cast iron tank. When constructing a pit in brickwork it is usual to allow extra room in the pit so that spanners can be used on the hydrant without having to break up the walls.

Hydrant indicator plates

Hydrants should be marked by using the same standard marking plate as is used by the local fire brigade. The plate should be fixed about 1 m above ground level, but where snow could drift and obscure a plate it should be slightly higher.

Turncock's tools

These consist of a hydrant key and bar. They can be separate items or combined in one tool, the separate items being easier to use. Although lightweight tools are available and are useful when they have to be carried around for test purposes, forged steel keys are much to be preferred for firefighting use. The reason is that lightweight keys usually whip considerably and are likely to spring if a hydrant is hard to turn, much of the turning torque being absorbed by the key and bar instead of the hydrant spindle.

Use of hydrants

A hydrant should always be opened slowly; otherwise the sudden rush of water may jerk the branchman off his feet. It is also liable to cause damage to the hose, especially if it is kinked at any point. Similarly, a hydrant should always be closed slowly, particularly the last few turns. If a hydrant is shut down rapidly the sudden stoppage of the flow of water sets shock waves known as 'water hammer' which can reach pressures considerably greater than that of the main and which may then burst the main. The damage will occur at the weakest point of the main, not necessarily at the hydrant but up to a mile or more away. The same danger occurs with the use of a shut-off cock on a control branch which, if operated too quickly, could burst the hose.

Before removing the hydrant key, see that the hydrant is completely shut down, but do not use undue force if there is a slight leakage. In this case, report the hydrant as defective. If the hydrant pit is flooded, scoop some water out of the pit to see if it is refilling or if there is any movement in the water after it has had time to settle down. Remember to place the protective cap over the outlet and finally check that the false spindle is in place. If it is missing look on the end of the hydrant key!

Maintenance of hydrants

Ground hydrants should be examined every three months. If the hydrants are sealed and the seal is unbroken, testing need only be carried out at six-montly intervals althouth inspection should still be carried out quarterly. The water authority must be notified before the seal is broken and, needless to say, water should not be released during frosty weather.

When inspecting a hydrant, the first point is to bale or pump out the pit if it has water in it and then to remove any mud, stones or weeds. Check the spindle gland to see that it is not leaking. If a false spindle is necessary, check that it is correctly in place and of the correct size to fit the hydrant key. Ship the stand pipe and see that it fits correctly and easily. Put a blank cap fitted with a pet cock on to the stand pipe outlet, turn on the pet cock and then turn on the hydrant. When the air has been released, turn off the pet cock. Check that there are no leaks from the hydrant or foot of the stand pipe. If there is any leakage from the foot of the stand pipe, check the washer and its seating. If there is any leakage round the blank cap, check the washer on the stand pipe outlet. If there is any damage to the stand pipe outlet, it is probable that the blank cap would not go into the outlet or at least it probably would not stay in.

If the valve is at all stiff, it should be exercised by running it up and down. (The blank cap sould still be in position, otherwise water hammer might be set up). If the valve is still stiff the gland may need repacking. It is not necessary that the gland should be absolutely watertight when the valve is open and over-tightening a dry gland can cause damage to the spindle.

When the inspector is satisfied, the hydrant should be shut down and the water pressure released by means of the pet cock. The blank cap should now be removed and the hydrant flushed out and then shut down. Check that it is not leaking. Some hydrants are fitted with a frost valve or plug so that water cannot be left in the outlet bend. This can be seen in Figure 28.4. It is difficult to tell if these hydrants are leaking until all the water has drained out of the bend, and sometimes it is necessary to leave them and re-visit later to check for leakage. During frosty weather or where it is difficult to allow quantities of water to flow freely, the stand pipe and blank cap (with pet cock) should be fitted and the valve exercised only to ensure that it does not become set. Such hydrants should be flushed during summer months adding a length of hose to the stand pipe if it is necessary to take the water clear of surrounding property. At the beginning of winter, the edges of the hydrant pit frame should be well coated with grease or the special mixture that can be obtained to prevent the lid from becoming frozen in.

Testing water mains

To determine the flow of water from the mains, special testing equipment is required. This is attached to a stand pipe on a hydrant, and when the hydrant is turned on it gives the actual flow of water from the hydrant. To be of true value, the test should be made when other hydrants in the immediate neighbourhood are fully opened as they would be under fire conditions. This may only be done with the consent of the water authority if the hydrants are attached to public mains and it is always advisable to enlist the help of the local fire brigade. The local fire brigade may already have carried out tests and be able to supply the information required.

Although these tests disclose the possibilities of the supply of water from the mains for the use of fire brigade pumps, a more practical test for use of hydrants without a pump is by means of a flow-pressure test taken in the actual flow of the water from the nozzle. This is carried out by means of a 'pitot tube'. This consists of a small tube of about 1.5 mm diameter which is held by a clip directly in the stream of water from the nozzle. The other end of the pitot tube is connected to a pressure gauge. The branch pipe with the pitot tube should be fitted at the end of a line of hose of the probable length that might be required for use.

Hose boxes

Near to each hydrant but protected from the weather there should be a well constructed hose box made of seasoned timber and having small ventilation holes at the top and bottom. If necessary, hose boxes can be kept locked with a key in a glass safe but maintenance is less likely to be skimped if the box can remain unlocked. Where hose boxes are to be used inside a building, a glass front is advisable so that if any of the contents are misplaced the fault can be seen immediately. But, where the box has to be in the open air, it should have a solid front because all hose is seriously affected by long exposure to the action of light, particularly strong sunlight.

The hose should be supported on wooden pegs rather than metal struts which tend to corrode and damage the hose. The hydrant key and bar should be hung up and although metal supports can be used it is better to use wood. The metal of the tools should not be allowed to touch the hose and it should be unnecessary to add that tools should not be put away dirty or wet. Similarly with the branch pipe and spare nozzles and nozzle spanners, these should be properly supported on wooden pegs in such a manner that they are unlikely to fall or be knocked down. If the hydrant has a false spindle, a spare one should be kept in the hose box.

HOSE

Hose is a very expensive item in fire precautions equipment and more hose is ruined by lack of care and bad maintenance than is ever worn out in use.

Selection of hose

When new hose is being bought, selection should be made of one of the double-covered types in which the reinforcement is synthetic fibre rather than cotton. The reinforcement is totally embedded in the synthetic rubber which is used to protect the outside as well as the inside of the hose. This type of hose is not only resistant to wear but is also resistant to petrol, oil, grease, alkalis, acids, most chemicals and sea water. It is completely immune to mildew. Where cost is a critical factor and thc hosc is unlikely to be used for firfighting, cheaper hose can be installed but it is essential that rubber-lined or impervious hose is used in buildings. Where hose is only likely to be used outside, unlined hose could be installed. However, with hose, the best is cheapest in the long run and a good quality hose properly maintained has an almost unlimited life.

Care of hose

The common causes of damage to hose are from mechanical injury, shock, heat, mildew, petrol, oil, chemicals and exposure to daylight, particularly strong sunlight.

Mechanical injury, consisting of tears and abrasions, is caused by dragging the hose over rough ground, by traffic passing over the hose when it is charged or uncharged with water, and by jumping on coiled hose to get the coils even. Mechanical damage also occurs when hose is laid out too quickly as from a hose carrier or when hose is run out from flaked storage so that the couplings crash down on to the hose. It is inadvisable to keep hose in a flaked condition because of the sharp bends in it as well as the damage when it is paid out. Mechanical damage can occur when hose is taken round or over sharp corners such as round a wall, over a curb or over a coping stone. Some form of padded protection should always be given against the abrasive action of the hose at all such points.

Damage by shock occurs when control branches are shut down too quickly, setting up a shock wave in the trapped water in the hose. It could be severe enough to burst the hose but it will certainly weaken it by overstraining. Damage by heat mainly occurs at fires when the hose is passing over hot or burning material. The danger is worse to lined hose because there is no percolation of water through the walls of the hose.

Mildew occurs in cotton-covered hose when the textile is not properly dried and aired or the hose is stored under damp conditions. If rubber-lined hose is not properly drained and dried, an acid is formed inside which can be very destructive. Petrol, oil and grease damage all hose very rapidly except that which has an outside protection of synthetic rubber. If possible, hose should not be laid where it might come into contact with petrol, oil and grease but where this is unavoidable the hose should be well scrubbed with soap or detergeant and water as soon as possible.

Damage from chemicals usually occurs at fires and generally such hose has to be condemned and replaced. Hose coated with synthetic rubber is proof against most chemicals and a quick wash down is all that is usually required. Hose can be damaged in the hose store by acid from chemical extinguishers and acid batteries where these are stored or used in the same room. Where there is sufficient hose to warrant such use, a separate store room is advisable. When hose is taken into a garage or motor workshops (to use the vulcanizing apparatus for repair work) particular care needs to be taken to avoid damage by acid, petrol and oil.

MOBILE POWER PUMPS

Ejector pumps

These are not firefighting pumps but are very useful for raising water from depths beyond normal suction lift, for acting as low-level suction strainers – they will raise water from the last 15 mm of depth – and they are particularly useful for pumping out cellars and basements as they are extremely small, easily portable and do not create any poisonous atmosphere as there are no exhaust fumes.

They are operated by forcing water through a jet in the main body of the pump. The body is open at the base through a strainer, and the rush of water through the jet draws water in through the strainer. There are no moving parts which can fail and they can be left to work without attention and no damage is done to the ejector pump if the pressure falls, if the strainer fouls up or if the suction comes out of water. Their fundamental action is to convert a small quantity of water at high pressure to a larger volume of water at low pressure.

Floating pumps

These pumps are designed to float on the surface of the water; the impeller then becomes submerged so that there is positive priming giving immediate delivery of water. They will pump to 25 mm depth of water. They are usually fitted with a two-stroke engine and recoil starter but other types of power drive are available. Figure 28.7 is a section elevation of a small floating pump showing the simplicity of its construction. The weight of this pump is only 10.8 kg and at 1.65 bar it will deliver 86.37 l/min sufficient to give a good fire stream with a 10 mm nozzle.

There is a larger version of the pump weighing only 27.22 kg which will give 236 l/mm at a pressure of 2.1 bar and which will give a good fire stream with a 12.5 mm nozzle and a special firefighting version to supply 20 mm nozzle at 2.4 bar.

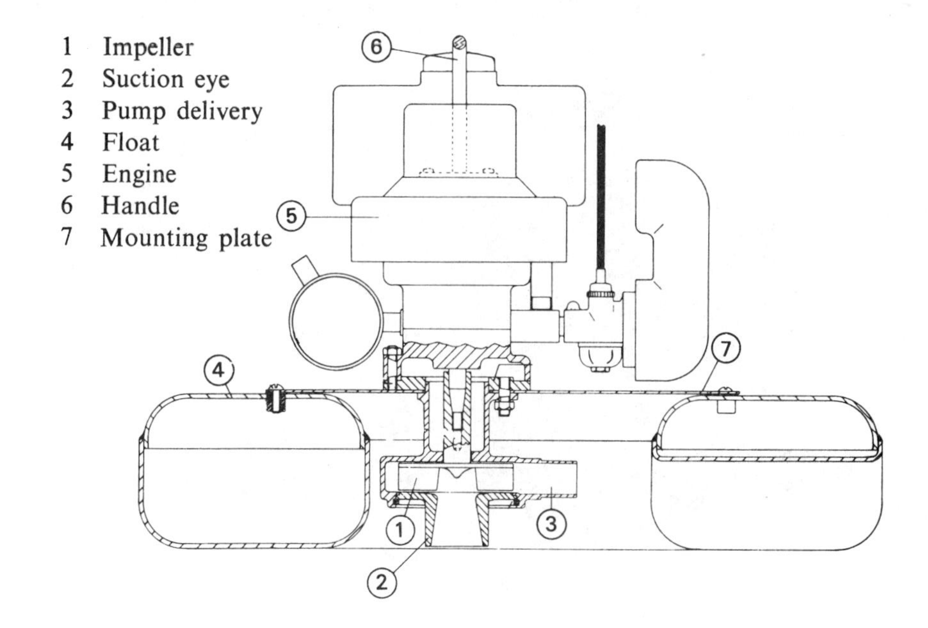

Figure 28.7 Sectional elevation of a floating pump (*L Hathaway Ltd.*)

Portable power pumps

The smallest useful portable power pump is one which will deliver a minimum of 455 l/min at a pressure of 6.9 bar.

Figure 28.8 shows a small pump which will give this performance. It is a single-stage centrifugal pump manufactured in non-corrodable light metal alloy. The shaft is sealed with spring-loaded carbon-faced glands which require no attention under normal working conditions for long periods of time. It has only one outlet, a 65 mm instantaneous coupling which is controlled by a screw-down valve. The priming on this particular pump is by an efficient automatic vane type air pump in conjunction with a water trap. The power unit is a single cylinder, air-cooled, two-stroke engine, fitted with recoil starter and hand throttle. The dry weight is 70.3 kg. Some of these pumps have a closed circuit inter-cooler and some have direct cooling from the main pump. Priming is usually by diverting the engine exhaust gases through an ejector device used to extract air from the main body of the

Figure 28.8 Small portable pump (*L Hathaway Ltd*)

pump. Priming should be at a rate of about 0.3 m per second if the suction side is in good condition.

Pumps which have a larger capacity are generally powered by four-stroke petrol engines, some of which have recoil starters. The one illustrated in Figure 28.9 has a capacity of 1140 l/min at 6.9 bar. Other pumps may have diesel power units, or air-cooled engines.

Small pump of this size are mounted in a tubular chassis frame that can be carried by two men, and for transportation they can be carried in the boot of a car or mounted on a small trailer chassis. A very useful means of transporting them is on a lorry which can also carry a tank of water. Perhaps a warning should be given here that any old lorry will not do for this purpose. It must be a reliable machine and not one that has been taken out of service because of increasing obsolescence. It should preferably be a new machine doing a year of service before it is taken into the industrial fleet.

Figure 28.9 Portable power pump (*Coventry Climax Engines Ltd*)

Trailer pumps

Trailer pumps are designed on their own chassis which, in the version illustrated in Figure 28.10, has a weight of 1524 kg. These heavy duty pumps are sometimes powered by diesel engines and are complete with self-starter, accumulator and generator. The pump performance is 4546 l/min at 6.9 bar working from a lift of 3.1 m. This particular pump can also be fitted with a foam manifold as illustrated. The four foam inductors can each be used with a foam-making branch pipe or they can be used independently to provide part cooling water and part for producing foam. Each inductor can supply a foam-making branch pipe capable of producing 7000 l/min of foam, that is a total of 28 000 l/min. A pump such as this can be used as an emergency pump to stand by when a sprinkler installation or the water supply is shut down.

Figure 28.10 Trailer-mounted pump with foam manifold (*Angus Fire Armour*)

Water/foam firefighting trailers

Firefighting trailers are very suitable for industrial use. They comprise basically a tank holding some 910 l of liquid. The liquid can be either foam concentrate or water or any ratio of foam compound to water provided that it is predetermined and arrangements made accordingly. Also carried on the trailer chassis there can be a portable pump with suction hose, hydrant tools, suction and delivery hose and a foam inductor and foam-making branch pipe. The pump can be removed from the trailer chassis.

Self-propelled pumps

These are the main pumps of the firefighting service. They are only suitable for use where there is an extensive road system and an ample supply of water.

Maintenance of portable and trailer fire pumps

On all occasions when a pump has been used other than from a fire hydrant on potable mains, the pump should be washed out from such a hydrant as soon as possible. The pump should not be allowed to run dry unless the pump bearings are of a type that can withstand such treatment. The pump gland should not be tightened so hard that damage is done to the shaft. A slight water leak from the pump gland is not only permissible but necessary to maintain the water seal in many pumps.

Mobile pumps must be housed in a room in which the temperature is not allowed to fall below 4.4°C (40°F) but as an additional precaution during the winter anti-freeze should be added to the engine-cooling system if permitted by the manufacturer, who will of course advise against this if the engine is direct-cooling. It is also advisable to drain the water from the pump casing where this is possible.

The engine and grease nipples should be regularly greased and oiled in accordance with the manufacturer's instructions and an occasional spot of oil given to the controls such as the throttle control. It is sound practice to grease the threads on the suction inlet and the suction couplings at the same time.

A suitable record book should be prepared showing the history of the machine and all the maintenance and repair work done on it. The record should also contain the number of hours that the pump has been at work both at drill and at fires.

Weekly proving test

The engines of pumps should run every week. The practice of starting and running internal combustion engines for only a few minutes is very damaging to the engine because it will be continually running on the choke and this procedure will cause sooted plugs, dilution of the engine oil with petrol resulting in dry cylinder walls and moisture condensation throughout the exhaust system leading to general corrosion and a rapid deterioration of the machine.

All pumps should be tested every week by pumping from clean water. The best method of doing this is by means of a tank let into the ground. The capacity need not be very high as the water can be returned to the tank after pumping, but on the other hand a large tank will also provide additional storage of water for firefighting purposes. Water should flow through the pump for at least 15 minutes, while maintaining pressure at the rated delivery against a suction head of not more than 3 m. A suitable size of nozzle should be selected to achieve this purpose. The manufacturer will suggest a suitable size if requested, taking into account the type and length of the delivery hose.

Monthly test

After the weekly test on water, all lengths of suction hose should be coupled up, the strainer being removed and replaced with a blank cap on the last length. The hose should then be laid out and inspected for any signs of damage. The pump should next be started and the priming device brought into action. Time should be taken by means of a stopwatch and priming should be stopped when the gauge reads 60 mm of mercury on the compound gauge or at 45 seconds whichever is the earlier. If the gauge does not reach this standard, there is a defect in the suction arrangements.

After stopping the priming device, the engine should also be stopped and the gauge watched for one minute; during this time the gauge should not fall back by more than 30 mm mercury gauge. If it falls back more quickly than this, there is a leak somewhere in the suction side. Defects are most likely to be found in the suction couplings or washers, but leakage may also occur at the pump gland, at the pressure gauge connections or even the delivery valves. Occasionally, leakage will occur in the suction hose itself, usually where the tails of the couplings are bound in. The fault is easy to prove by diconnecting and replacing each length in turn. In severe leakage, faults may occur at more than one point, but if the pump is tested properly each month small leakages should be found before more serious faults develop.

When a fault is difficult to locate it is permissible to replace the suction cap with a suction collecting head and put all the suction hose under positive pressure from the hydrant. Several precautions must be observed when doing this. First, when all the suction hose is connected up, one delivery of the pump must remain open while the hydrant is being opened up slowly. Under no circumstances may the pressure from the hydrant exceed 3.5 bar. If the hydrant pressure is known to be greater than this, a dividing breeching can be put into the delivery of the hydrant and a length of delivery hose with a hand-controlled branch having a very large nozzle connected up. The control can then be used to reduce the pressure on the line going to the pump. The hydrant must be turned on very slowly and the gauges watched carefully. When the pressure is stabilised, the pump delivery can be closed down very gently until the gauge reads 3.5 bar. Any leakage of water will show where air could be leaking in under suction test. If faulty delivery valves are suspected, the test pressure must be very low because the pressure will tend to close the valves. In all cases, allow sufficient time for the water to show itself. A fairly prolonged test may be necessary.

29

Salvage (damage control)

The Salvage Corps which had been established for many years in London, Liverpool and Glasgow were recently disbanded. Thier uniforms and appliances looked very much like those of the fire brigade, but although they attended all large fires they did no actual firefighting. Their function was to minimize the damage caused both by the fire and by the water used for firefighting. Goods which were damaged were removed to their headquarters where they had specialist equipment for drying out and for removing smells. At Liverpool for example there were special ovens for drying out bales of cotton (most of the imported raw cotton was landed at Liverpool). They were concerned mainly with goods in warehouses, which they inspected frequently, but would also attend any fire in their area. They were supported entirely by the insurance companies and had no legal status at all. They were however always welcomed by both the fire brigade and the property owners. When the decision was taken to disband all three corps, their equipment and their duties were taken over by the local brigade. Salvage has always been a part of the fire brigade's duties, but naturally it takes second place to rescue and firefighting. Also, salvage operations cannot be expected to continue for long after the fire has been extinguished, and certainly the brigade cannot be involved in the recovery of damaged goods and equipment. A number of specialist firms is available which will undertake this work, particularly when electronic equipment is involved (a process similar to dry cleaning is used). There is also an organization which specializes in the recovery of fire-damaged documents and records. A surprising amount of information can be obtained from papers which appear to be completely charred. However, if the paper is wet, as it almost certainly will be, it can be irrecoverably damaged by fungal attack very quickly. It is therefore essential to contact the organization immediately and to follow their instructions. If they are unable to remove the documents at once they will probably ask for them to be kept in a deep freezer. (Addresses are in Chapter 33.)

It has already been pointed out that salvage work is an important function of a works fire brigade. They will be engaged in firefighting until the local authority brigade is ready to take over, and then their role will change. Clearly this must be pre-planned and agreed with the brigade, and it is important that at least one of the industrial brigade officers (or a manager) is always available as the incident controller to give information on hazardous processes, access to buildings, water supplies, and so on.

Fire losses are either direct or indirect. Direct losses are those that can be measured, calculated and insured, but indirect losses can be infinitely greater and more disastrous. It is impossible to calculate the loss that occurs when production ceases even for a short while, for the loss of skilled manpower who cannot wait until work is again available to them, or for the loss of purpose-built machines. Nor it is possible to put a value on the loss of a market or of customer relations, the breakdown of delivery dates or the work that is necessary in trying to restore lost records – orders received, goods dispatched, invoices sent out as well as other administrative records and work. Until these troubles have been experienced, it is impossible to realize fully and appreciate their import. It would be a very useful exercise if top management had to visit a colleague the day after his business had been struck by fire.

Salvage considerations should start when the plans and designs of premises are being prepared, for then certain features can be built into the structure. In older buildings, many improvements can be made that will reduce the losses which follow a fire. Though the cost may seem prohibitive, if properly budgeted and spread over a period of time, results may one day prove to be well worth the effort. Some improvements cost nothing more than a little rearrangement of the stock in a store or warehouse.

Fire damage has four distinct sources: the actual destruction of material by fire, the effects of smoke, heat, and the losses which occur from the large volume of water thrown into the building to extinguish the fire, including the operation of automatic sprinkler installations. To minimize those losses is the aim of all salvage work. Salvage work can be divided into three phases:

1 Before a fire
2 During a fire
3 After a fire.

BEFORE A FIRE

Salvage work before a fire can be subdivided into work concerning the building and that dealing with the contents.

Inbuilt salvage protection : drainage

One of the first steps to be taken when planning a building is to ensure that the sewers are large enough to carry away the water from inside and outside the building that might be used in firefighting operations. This is especially necessary when automatic sprinklers are installed. For this reason, the size of the sewers and the pipes feeding into the sewers may have to be considerably in excess of what might otherwise be the normal requirements of the building.

The gullies (drain points) which lead to the main drains should be indicated much in the same way as street hydrants because floating debris tends to be drawn over the gully grid by the flow of water. This stops the flow and at the same time makes it difficult to locate and clear the grid. Gullies inside a factory or warehouse might be indicated by painting a sign on the ceiling above the drain. Internal gullies should be sunk slightly into

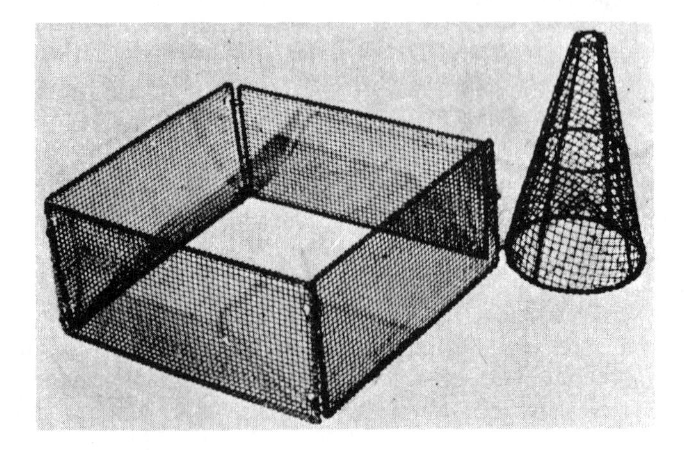

Figure 29.1 Drain guards (*Liverpool Salvage Corps*)

the floor and covered with a loose-fitting wooden plate that will float off leaving the gully clear. Immediately there is any heavy flow of water, drain guards should be put in place to prevent gullies from being blocked. Figure 29.1 shows two types of drain guards. It is very necessary to ensure that nothing large enough to block the drain or the sewer should be washed in during the course of a fire.

Inbuilt salvage protection : floors

The floors of a new building should be made impervious to water and gently sloped so that water will flow to one side where it can be taken to an external door or through wall scuppers to cascade harmlessly outside the building. Sills and ramps should be provided at communicating doors to prevent water from flowing from one compartment to another.

In older buildings, where it is not possible to slope the floors to scuppers in an outside wall, it may be possible to use internal drains. This is especially the case in mill buildings and stores and warehouses where there are wooden floors. Figure 29.2 is a plan of a floor drain suitable for this purpose. In practice, a row of these would be situated in a line down the centre of the floor. In normal use, they lie flat on the floor and do not form any sort of obstruction but, should drainage be required, a water chute can be hung on the four hooks below the floor and taken to a window or door leading to the outside. The plug can then be pushed up from below.

It is unlikely that sills and ramps can be provided in old buildings and where it is not possible the use of weir plates should be considered. Figure 29.3 shows a weir plate plainly showing the method of use.

Where water is liable to accumulate in basements, the slope of the floor should take the water to the lowest point where, if main drainage is not available, a sump should be provided. The sump could be covered over as suggested for gullies and it should also be well indicated above any possible flood level. If the sump is likely to be difficult to access, automatic pumping equipment should be provided. In this case the pump should be of a submersible type and should start automatically as soon as there is water in the sump. This would be done by means of a float switch. The wiring and switchgear to the pump should be well protected against water, properly installed MIC cable (mineral-insulated cable) should be used, and the switchgear should stand out from the walls so that water draining down the wall cannot affect the operation of the control. The feed to the control should be protected throughout its length and it should not be shut down when the main switches are withdrawn.

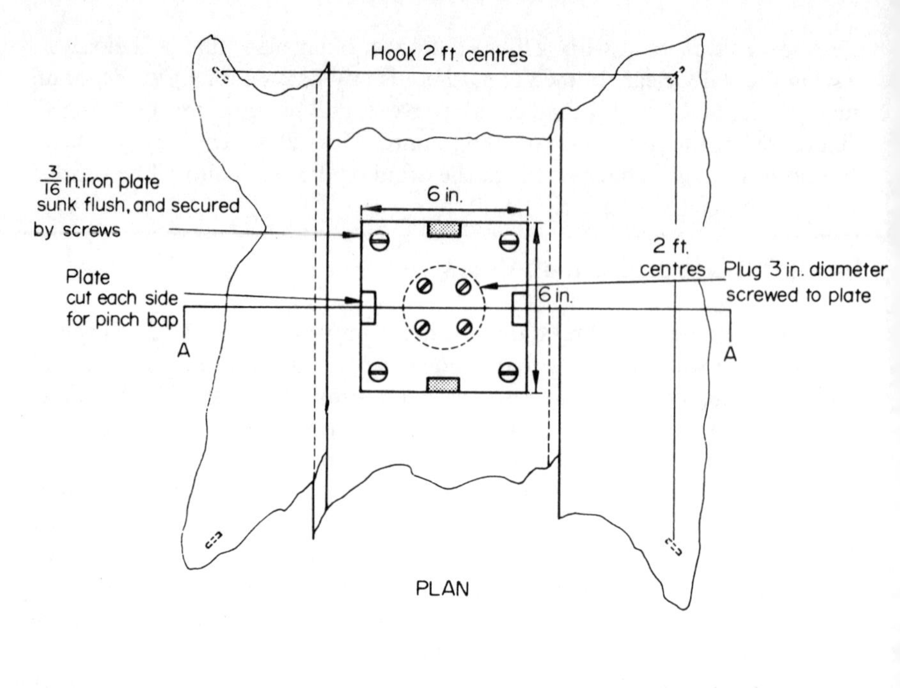

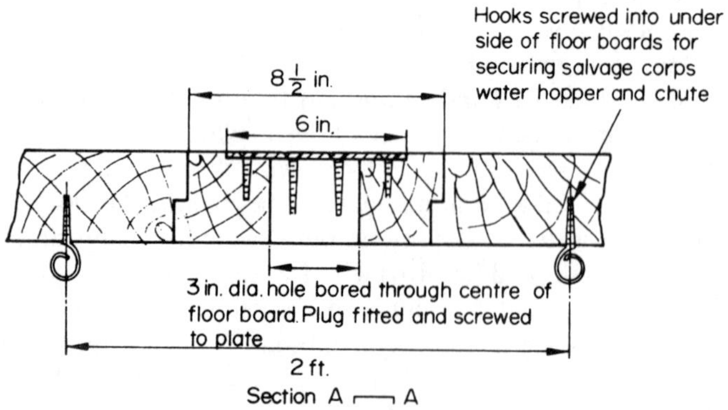

Figure 29.2 Plan of a floor drain (*Liverpool Salvage Corps*)

Inbuilt salvage protection against smoke

As well as water, smoke does an immense amount of damage. This occurs in two ways: first, by obstructing the firefighting services by making it more difficult to find the seat of the fire and secondly by the actual damage it does

Figure 29.3 Weir plate (*Liverpool Salvage Corps*)

to the buildings and even more particularly to the contents of the building, including stock and office records. The smoke ventilation of single-storey buildings has already been discussed and suitable ventilation should be installed to protect the top floor of multi-storey premises.

The installation of smoke stop doors has been detailed under the evacuation of premises, but consideration should also be given to the installation of additional doors and the conversion of existing doors to keep back smoke from other sections of the premises. Where large office areas or large storage sections are involved, it is a big advantage if smoke can be prevented from spreading throughout the whole area.

With modern air conditioning some thought should be given to the possibilities of smoke extraction and investigations are being made at the moment into the creation of a small air pressure in stairways and corridors to avoide smoke logging. The main difficulty is to decide how this should be done without increasing the supply of air to feed a fire.

Safeguarding the electricity supply

One of the first considerations after a fire is to restore the electricity supply which is required for many salvage operations as well as for the provision of light. Very often there is considerable difficulty in restoring the supply because fire and heat have damaged the insulation of the cables, and water and dampness will have rendered the apparatus dangerous. In new buildings, wiring should be carried in sealed ducting or MIC and all switchgear and similar apparatus should be mounted on blocks to hold them well away from the walls to avoid damage from water running down the walls.

In older buildings, a careful inspection of the whole electrical system should be made detailing improvements that could be effected. From a fire prevention point of view,older wiring should be replaced, and for salvage considerations this replacement policy should be accelerated. Minor switchgear should be enclosed in a watertight cupboard, but major switchgear should be housed in a fire-resistant compartment. When selecting the priority of work following the inspection, preference should be given to main cables. It is of little use to protect subsidiary circuits if the mains are unusable.

Safeguarding machinery

It is suggested that every machine should have an individual cover. This applies not only to large machines in a factory but to small tools and office equipment such as typewriters, calculators and so on. The cover should be kept with or near to the machine during working hours and placed over it at all other times. The cover need only be a polythene bag. Small portable machines and equipment of high value such as highly sensitive testing machines should be kept in fire- and water-resistant cupboards or rooms when not in use.

Storage of goods

By giving careful thought to the disposition of goods in stores and warehouses, losses in the event of fire can be considerably reduced. Incompatible chemicals should be kept in separate stores and, as there are a great number of these, the fire precautions manager should always be informed of any new materials being brought into the premises and should be consulted as to the position and method of storage. As well as receiving this information he should, during his daily inspection of the premises, be watching for any new materials or change of storage.

As he may not be a chemist, he should consult the company chemist or the suppliers of the materials. All suppliers of chemicals that are potentially dangerous will give advice on the use and storage of their materials together with suggestions for dealing with them in an emergency.

Where there is no incompatibility of materials, the arrangement of mixed storage can be made to reduce losses by using incombustible materials to make 'fire breaks' between combustible materials. In this way, there would be say a rack of household hardwares between racks of cartons of food products. This system of storage should be adopted in small stores as well as large warehouses.

The method of storage should be carefully considered. Steel racking is to be preferred to wooden racking because it adds no fuel to the fire. Where no racking is used and goods are piled one on top of each other there is a great danger from the bottom layer becoming weakened by the action of water and then of the whole stack collapsing. Where pallets are used, these too should be of steel.

The bottom pile of goods should always be raised slightly above floor level in case of flooding whether by firefighting or from other causes such as roof leaks or burst water pipes. Ample room should be left between all stored goods to assist in firefighting operations as well as salvage work. Racking and storage of goods should never be completely to the ceiling nor be tight up against walls.

The private fire brigade as a salvage unit

The first few minutes at a fire are of vital importance and a period of intense activity for the firfighting services while they are preparing their equipment and getting into position. During this rush of activity the emphasis is on extiguishing the fire and firemen cannot always be taken off this work to start any sort of salvage precautions. Yet this is the period in which it is of paramount importance to start salvage work.

Clearly it is a matter of judgement to weigh up the relative importance of firefighting and salvage activities. The decision can only be taken at the fire. If it is a relatively small fire and is being contained, then clearly some men may be diverted immediately to salvage operations. As soon as the local authority brigade arrives and is able to take over the firefighting, the whole of the private brigade can concentrate on salvage. It is a very useful arrangement if people from an engineering services department are members of a part-time brigade, or can be called in to help a full-time brigade.

Consider for a moment the advantages they have. First, a thorough knowledge of the layout of the premises including the points of fire weakness in the building, the run of the drains, the accessibility to the roofs. Then again they have knowledge of the most valuable machinery, the valuable stocks including those difficult to replace and those most liable to damage by water, by smoke or by the effects of fumes. They would also have a knowledge of production methods and the work necessary to get into production again. But they must be trained and they must be properly equipped.

DURING A FIRE

The incident controller

The incident controller must be at his station when the local authority brigade arrives, and in possession of an up-to-date knowledge of the fire situation. It is essential that he should be able to tell the brigade not only the state of the evacuation but also what firefighting activity is going on. This means that the officer or officers of the private fire brigade at firefighting and salvage positions must speed this information back to the incident officer. It is to be hoped that the confidence of the local fire officer will be such that he will also feed information back through his own control to the incident controller. In this way, requests for information that the fire brigade may need (for example, the state of stocks of flammable liquids or the services of such specialists as chemists or electricians) can be speedily met. Similarly, the local authority brigade will provide information about, for example, damaging accumulations of water.

Start of salvage work

Salvage work should start as soon as possible. The quicker plant and stocks can be protected with waterproof sheets the less damage will occur. At the same time, any doors, particularly smoke stop doors, should be closed to prevent smoke damage to other parts of the premises not affected by fire. The quick covering up of plant and stock is a matter of organization and training in suitable methods. Waterproof sheets should be available in different sizes and easily identifiable as to size so that effort is not wasted in trying to cover a small machine or pile of goods with a large sheet or having to search for a large sheet if not clearly marked.

The next step is to start making dams to prevent water from flowing laterally from room to room if other precautions such as raised sills or weirs have not been constructed, and to direct the water harmlessly to the outside. Here the knowledge and use of salvage dollies will be required. Before attempting to sweep water out it may be necessary to rig up tarpaulins to direct water flowing from upper floors out through windows or doors. To create this diversion it may be necessary to make an opening in a floor or to pierce a ceiling. Neither job should be undertaken until protection has been provided below. Very quickly after the start of firefighting operations, water will be flowing heavily into the drains. This fast flow will carry with it all sorts of loose material which will then clog them unless suitable precautions are taken. Men must be detailed to watch and clear the drains. It is very bad practice to remove the grids over drains as material may be washed into them and stop them up in a position out of reach. It is better to have a man standing by constantly clearing them. The authorities will appreciate being warned that there is a large fire with considerable quantities of water being used. It can be assumed that the water authority will have been informed that there is a serious fire but a second notification will not come amiss.

It should not be general practice to remove property from burning premises although under exceptional circumstances it might be necessary. In some cases, this may already have been done by well-meaning outsiders. In all such cases, any plant or goods must be protected from the elements as soon as possible and removed to safety if necessary. If the company has a security force it will already be operating to prevent pilferage or looting and to exclude strangers from the premises. If there are no security men available, the private salvage corps should be prepared to undertake this watching service.

AFTER A FIRE

As soon as possible after a fire, attempts should be made to get rid of smoke by opening up all doors and windows to provide a through draught. But under no circumstances may this be done without the permission of the officer in charge of the fire, if the brigade is still in attendance. The next step is to protect any openings in the roof that may have been made and then to start mopping up and drying out the building with heating units. The smell of burnt material can be removed by the use of suitable deodorants. Water that has accumulated in basements and cellars must be pumped out. The electrical installation should be inspected and tested as soon as possible and damaged sections isolated so that power can be restored wherever possible.

Sprinkler stoppers

If only a few sprinkler heads have opened in a fire, one of the first actions to prevent further water damage is to stop the flow of water by the use of sprinkler stoppers. Figure 29.4 shows a sprinkler stopper. The purpose of the stopper is to prevent the flow of water without shutting down the main stop valve. This serves a double purpose as it not only stops the flow of water immediately without waiting while the whole installation is drained out but also it enables the sprinkler installation to remain operative until all possibility of re-ignition ceases. The closing down of a sprinkler installation too early after a small fire is a frequent cause of alleged sprinkler failure.

Figure 29.4 Sprinkler stopper (*Liverpool Salvage Corps*)

The method of using the stopper is clearly seen in the illustration. The top plug is placed in the sprinkler aperture and the bottom pad rests on the deflector; then by sliding the bar along the plug is forced into the aperture and the flow is stopped. The operator gets very wet!

There is a grave danger of sprinkler stoppers being left in place after being used at a fire. Some form of check system should be used to account for any stoppers that are used and to see that they are promptly replaced with sprinkler heads.

Mopping up

By this time, members of the management should be in attendance and will be able to indicate any items which may require special treatment. The work of wiping down machinery should be followed by a thin coating of water-repellent oil and the use of light machine oil. Similarly, the wiping down of office equipment including furniture, woodwork and furnishings can save considerable reinstatement. In warehouses, it may be necessary to open up packed goods and dry them out, taking care to see that stocks of the same materials are kept together and identification of the contents is preserved. Extra staff may be required for the work and the authority of the insurance company for this purpose should be obtained as quickly as possible.

Although property damaged by fire automatically belongs to the insurance company, there must be no abandonment of it and there should be no delay in commencing salvage operations nor in notifying the insurance company concerned.

FURTHER READING, Part Seven

Pollak, F. *Pump users' handbook*. Trade and Technical Press, Morden.

Karassik, I. and Carter, R. *Centrifugal pumps*. McGraw-Hill, New York.

NFPA *Fire Protection Handbook*. NFPA, Boston.

Tuve, R. L. *Principles of fire protection chemistry*. NFPA, Boston.

Home Office *Manual of firemanship, Book 3: fire extinguishing equipment*. HMSO, London.

Home Office *Manual of firemanship, Book 7: hydraulics and water supplies*. HMSO, London.

PART EIGHT

FIRE PROTECTION MANAGEMENT

30

Fire teams and fire brigades

Everyone in even the smallest business or industrial establishment should be trained in the use of first aid firefighting equipment. In larger premises, and particularly where hazardous processes are used, fire teams should be formed and provided with suitable equipment and adequate training. Wherever the size of the operation or the severity of the risk can justify the establishment of a private fire brigade, very serious consideration should be given to this. Unfortunately when the profitability of a firm is reviewed there is a temptation to regard the works brigade as an expensive and non-productive luxury. The usual argument is 'We pay our rates, so the local authority can put out our fires. Anyway we're insured, and we haven't had a fire for years.' This ignores the real reasons for having a works brigade: active fire prevention, equipment maintenance, training, liaison, salvage (or damage control), the reduction of fire losses before the arrival of the public brigade, and the provision of expert advice and help to the brigade. The manager who says, 'We don't need a works brigade because we don't have any big fires' has got it back-to-front. He should say, 'We don't have any big fires because we have a works brigade.'

In many industries the need for private brigades has increased. This is because of the change to large single-stream processes which represent a high capital investment. The use of computer control has resulted in a dramatic reduction in the number of people needed to run the plant, and in an emergency no-one can be spared to fight a fire. At the same time, advances in technology have meant that many firms can continue to be competitive only by employing more hazardous processes. The pressure to cut costs does not stop there: it applies also to the fire service. If for example the local brigade finds that its first attendance appliances are more than adequate for 98 per cent of the fires, they may well decide to reduce the number of appliances, and hence achieve a substancial reduction in manning. Our large single-stream plant is likely to have one of the 2 per cent of fires which cannot then be tackled effectively by the first attendance.

The insurance companies too are becoming more competitive. Instead of the old tariff system, by which they all charged the same premium in the same industry, they are now tailoring the premium to fit the risk. This means that they need to have much more detailed information about both the process and the safety procedures. They are likely to base the premium on a scoring system in which the hazardous single-stream plant scores many negative points, which can be offset by the positive points earned by the works brigade. In this way the provision of a brigade can be cost-effective, and with the additional benefit of a reduction in business interruption when they do have a fire, because its initial spread will be limited.

It is however important that the best possible return should be obtained from an investment in a brigade. It is quite wrong to assume that they are only there to put out fires, and must be kept busy with other work for the rest of the time (like painting and gardening). There are some advantages in sharing safety and security functions, but the most important aspects of an industrial fireman's work have already been mentioned.

Some industrial brigades are badly equipped and poorly paid (and in rare cases, badly led). There is no point in making a commitment to a brigade and then reducing its effectiveness by poor funding and inadequate management backing. There is a need to enhance the status of the industrial fireman, and fortunately a great deal of training is now available at all levels. The Home Office has for many years accepted that the graduateship examination of the Institution of Fire Engineers is equivalent to the station officer promotion examination. This provides an excellent way of ensuring that an industrial officer is at the same academic level as his public service colleague. Indeed the possession of the G.I.Fire E. qualification could be made a requirement for promotion in an industrial brigade, with the membership qualification a requirement for a senior position. The badges of rank worn by these officers should preferably be those of station officers and assistant divisional officers in the public service: chief officers' badges can rarely be justified. Practical and theoretical training is usually available in most areas for firemen, and a number of competitions are run at local and national level. The addresses of some of the associations providing these facilities are given in Chapter 33.

FORMATION OF A PRIVATE FIREFIGHTING SERVICE

To assess the size and equipment of a private fire brigade, many factors have to be taken into consideration. These are discussed in the following sections.

No matter how close to the premises a public fire station is, there must always be some lapse of time before the brigade can arrive, not just at the premises but with fire-extinguishing media at the scene of the outbreak. This is the period of delay in which the owner must look after his own interests. To assess this period there are three considerations: the time that it takes the public firefighters to assemble, the time that it takes them to reach the premises and finally the time that it takes to get their firefighting equipment in action at the scene of the outbreak.

Turn-out time

The time it takes the public brigade to assemble depends on the type of the brigade. In a large town the brigade will almost certainly consist of full-fime firemen on instant call, probably working on or around their vehicle and they will leave the station in a few seconds. In a small town or rural area, the brigade will probably consist of part-time men, who go about their business normally and are called to the fire station by the sounding of a siren, house bells or possibly by a pocket radio call system. A delay of several minutes may elapse before the crew is assembled and ready to leave.

Distance to travel

The distance to travel cannot be measures in miles alone, but must be assessed by the possible conditions that will be met and the terrain to be traversed. For example, have the brigade to pass through a busy city centre; is the route blocked by one way streets? In the country, can the route be blocked by holiday traffic? Does the brigade have to travel over railway crossings? Can they be stopped by a road bridge being open over a canal or river? These questions can only be answered by careful consideration of all the local conditions.

In the study of local conditions, consideration should also be given to back-up times, that is the time that additional assistance can be brought in, possibly from some miles away. A few discreet enquiries should be made as to how long it will be before further assistance can reach the premises. Consideration should also be given to how long it would be before the full attendance required could be assembled if the local brigade were already engaged in a large fire. These situations are, of course, allowed for in the predetermined attendance by the mobilization of other public fire brigades under mutual assistance schemes, but delays may occur due to various causes.

The premises

An important feature in the assessment of the requirements of a private firefighting service is the size and type of the premises. Account must be taken of whether there is more than one building to be protected and the nature of the construction. A modern well constructed fire-resistant building will not require so much protection as an older building with a considerable amount of combustible material in the building itself.

Some account must be taken of the situation of the building and whether it be threatened by a neighbouring building on fire. It is possible that the public brigade may be stretched to its limits in the initial stages of a large nearby fire. A delayed attendance is then probable and the function of the private service would be to deal with radiated heat and flying firebrands. Under these conditions, the private fire team or private fire brigade might be on their own for some considerable time.

The effect of the size of a building is very difficult to estimate, but under the worst possible conditions it could be expected that it would be necessary to attack a fire from each external doorway and at strategic points within a building where it can reasonably be expected that a fire can be stopped from spreading.

Hazard classification

The classification of premises by fire hazard has been described in Chapter 20 and although this is specifically for sprinkler protection it is very useful as a guide to fire protection generally. The classification can be roughly determined from the lists of trades available from the Fire Offices' Committee, but is probably more easily determined by asking the company's insurers.

The classification takes two points into consideration: the area the fire is likely to attain in the first few minutes and the minimum quantity of water that will probably be required to control that size of fire, that is to contain it within its place of origin.

The size that a fire can reach in the first few minutes without automatic sprinklers or other automatic protection is dependent on the nature of the burning material and the speed with which a firefighting unit can get to the scene of the outbreak. The speed with which the unit can get there depends in the first place on the method of detecting the outbreak. If there is an efficient automatic fire detector system, and an efficient and effective firefighting unit on the premises, it is possible that the fire can be manually attacked in less time than it would have taken for automatic sprinklers to operate.

Where the hazard is related to the quantity of water likely to be required, it can be assumed that rather less is likely to be required for manual application than for automatic application, as the consideration is not so much holding a fire in check as to extinguish it, after which the supply of water can be stopped. If, however, firefighting cannot be started until after the normal time that a sprinkler system would come into operation, a much larger supply of water will be required.

Fire detection system

No firefighting service can be of any value until it has received notification of the fire. Manual fire alarms must be provided by law but, where there are parts of buildings that are left unattended for lengthy periods of time, whether by night or day, automatic fire detection is essential. The private firefighting organization has a much better chance of success if the automatic fire detection installation is one of the modern systems. Chapter 18 describes the different types that are currently available.

If sprinkler systems are installed, the private firefighting arrangements are still necessary, as instanced by the emphasis which the FOC rules for sprinklers place on the provision of portable fire extinguishers. The role of the private fire service, however, changes in character when sprinklers are installed and under fire conditions it acts as an adjunct to the sprinkler system until the final extinction of the fire and the control of the sprinkler system itself.

Firefighting media required

Water is the main requirement for general firefighting purposes, and bulk supplies have been considered in Chapter 28. The availability of sufficient water in bulk and the rate of its application to a fire are important considerations in the type of equipment that has to be provided and the number of men required to apply that water with sufficient force to the seat of the fire. Some risks, such as chemical works, oil and paint works, especially if the plant is isolated in a rural area, will need large quantities of foam. It is not to be expected, neither is it reasonable to require, that sufficient stocks of foam or other special chemicals will be kept at a small local fire station for individual special hazards. Neither can it be expected that a small part-time station will have specialists or officers with the knowledge necessary for dealing with all types of major hazards. These risks are the special field of the private firefighting service.

Rate of water application required

It is very difficult to assess the rate at which water will be required for firefighting purposes. In the very early stages a 9-litre extinguisher would be ample, but progressively more will be required as the fire grows through lack of detection and delay in getting water to the fire.

Assume first a fire in an office which would normally be classified as an extra light hazard. Detection during the day should be fairly quick and before the fabric of the building is involved. The application of water from a hose reel will probably be sufficient. If, however, the fire started in an unattended stationery store there might be a much larger fire, especially if the person who discovered it left the door open! It now becomes necessary to apply water at a greater rate and a line of hose with a branch pipe fitted with a 15 mm nozzle will be required. If the fire is just beginning to spread from the room of origin, it is probable that an even larger nozzle will be required or preferably two lines of hose and two branch pipes with smaller nozzles.

If automatic fire detection has been installed and there is an effective fire team available on the premises, a fire under these circumstances should not reach the stage that a line of hose is required.

As the hazard class increases, the initial attack will need an ever-increasing rate of water application. In an initial attack on a more serious fire, the aim should be a large number of small jets from different angles and different positions rather than one large jet. Large jets will not be required until the fire has obtained a strong hold, by which time the public fire service should be in attendance. In most extra high hazard classes with a large block of premises completely involved, and once the firefighters have been driven out of the building, water applications may rise as high as 45 000 litres or more per minute.

BS 5908, the code of practice on fire precautions in chemical plants, gives some guidance on possible water requirements. It suggests that a minimum pumping rate of 150 l/s could be assumed, and that sufficient water should be available to maintain this flow of 2½ hours. However considerably more water might be needed, particularly where it was necessary to apply cooling water to surrounding structures. The possible requirements for individual plants can be estimated by working out the surface area of a simple figure like a cylinder or a cube which would enclose the plant, and adding to this the ground area covered by the plant. The water application rate should be $10\ l/min/m^2$ for the total area. The minimum quantity of water required can be based either on the estimated time needed to extinguish the fire, or the burning time of the fuel (with a suitable safety factor).

Quantity of water required for hose reels

Generally speaking, for hose reels, the ordinary water mains will give an ample supply although the pressure may have to be boosted by a small automatic electric pump to give a suitable pressure particularly at the top storey. In estimating quantities of water required for supplying pumps, sufficient water should be available to run the pump for 30 minutes.

FIRE TEAM

Having taken all the previous sections into consideration, it is now possible to estimate the size of the private fire service that will be necessary.

Consider first some small premises used as an office block. Assume it to be not more than two storeys in height, with a full-time public fire station not beyond two or three miles away, and with fire hydrants within 60 metres of the front and rear doors. Under normal conditions the public fire brigade will be in attendance within five minutes so that a fire has to be contained for at least that length of time.

It is unlikely that sprinklers or automatic fire detectors will be installed in this type of building, so that detection will be by employees during the day and the police or passers-by during non-working hours. As there is unlikely to be anyone in the premises at night, there is little that can be done except when the premises are occupied. As fire extinguishers should always be in groups of two or more with a total of not less than four on each floor, there will be at least eight extinguishers available with 16 refills. It is suggested that a fire team only would be required. The team should consist of four to six volunteers from the staff, one of whom will be designated as leader.

It should be borne in mind that if a number of extinguishers can be used simultaneously at different angles on a fire they constitute a far more effective form of attack than a succession of extinguishers used one at a time.

Arrangements should be made that the first two members of the team to arrive should pick up the nearest extinguishers and get to work on the fire immediately. But arrangements should also be made that the next two members of the team bringing up two more extinguishers should wait for the order to get them to work. The leader must then use his judgement whether to use the extra extinguishers immediately or wait for the arrival of still more extinguishers.

A modern water extinguisher lasts for just over a minute and can be refilled in under a minute provided that water is not too far distant. Thus, by working as a team, water could be continuously discharged on the fire until the arrival of the brigade. The leader should preferably not be handling an extinguisher but should stand back and direct operations ensuring the best use of the team and the water supply available. At the same time, the leader should be watching for danger to his team by the spread of fire which might necessitate a change of tactics or possibly calling the team off. In the event of a retreat of this nature doors and windows must be closed as they go.

If hose reels or internal hydrants and hose have been provided, a much larger fire can be fought, but more training is required. Team training should include realistic exercises in maintaining a supply of water for extinguishers as well as for actual firefighting practice.

As premises get larger or other conditions apply such as the distance from the public fire station increasing, more than one team will be required. Fire extinguishers may still be the prime method of attack but more comprehensive training will be required. A problem may now arise that the larger number of personnel involved may have a tendency to get in each other's way and it also becomes more difficult to identify members of fire teams from other people. Suggestions have been made as to the use of armbands or special overalls but neither of these methods is very successful. The best method so far seems to be to instruct every member of the teams to carry a fire extinguisher with him to the scene of the fire. At the fire they rally to their own leader who takes his orders from a nominated senior or possibly the first leader to arrive at the incident.

PRIVATE FIRE BRIGADES

A private fire brigade is formed when the members of the fire teams meet regularly for training with major equipment and join together to obtain instruction in the finer points of fire engineering. It is not possible to lay down exact rules for deciding the strength of a brigade but a careful study of the various points already discussed will have a considerable bearing on the question. Generally speaking, however, an organization of three or four teams of six men, with a leader in charge of each, together with a senior officer and his deputy will be found to be a very useful unit. In a large complex there may be a number of similar units, all of which can be called to a fire, and it will be necessary then to have an officer in overall charge, who could be employed full-time in this capacity. It is essential that a chain of command is established so that there is never any doubt about who is in charge at an incident. Some means of identification should be used, and this can conveniently take the form of distinctively marked overalls.

Full-time, retained or part-time service

The brigade may consist of full-time men who do nothing else but fire brigade duties; it may consist of a nucleus of full-time men backed up by a number of retained men or it may consist entirely of part-time men. In this respect, retained men are considered to be men who agree to be available for specified periods in their own homes and who are living sufficiently close to the premises to be available on call, whereas part-time men in this respect are men who can only be of service during working hours. A brigade formed solely of part-time men cannot be very effective because there is no firefighting unit available for most of the 24 hours.

The most valuable men are retained full-time men, that is men who are engaged during working hours as full-time firemen but are available on call after hours. Whatever form the brigade takes, careful attention must be given to the preparation of duty rosters so that there is always a full crew available.

Senior officer

The term 'senior officer' is used here to identify the officer in charge of the brigade. Titles like company fire officer, or senior fire officer are to be preferred, together with rank badges appropriate to the size of the brigade (usually those of an SO or ADO).

The senior officer should have had considerable firefighting experience, good administrative ability, be a good leader, and be able to communicate well. He is therefore likely to have been a local authority fire brigade officer, and will only be attracted by good salary and conditions. He should be responsible to a senior member of the company, and preferably to a director. If, as sometimes happens, he is supervised by the site services manager or someone in a similar position, his requirements and influence may become subsidiary to those of other departments.

Membership of the private fire brigades

Membership of the brigade should be made as attractive as possible, not necessarily by cash returns but by privilege, to attract the best class of men.

All personnel selected should be physically fit and should pass a medical examination on the basis that the first few minutes at a fire can involve the greatest possible physical strain.

To be accepted as firemen, they must be able to satisfy such conditions as attendance at drills and fires as are required, bearing in mind the necessity of attendance outside normal working hours. The competitions of the various fire services associations can do much to maintain the enthusiasm of the members, but care must be taken to see that the brigade receives training other than purely competition work, which is mainly based on speed. Senior management should regard it as part of their duty to attend such gatherings and give every encouragement to the men by their own personal attitude.

Full-time members of a private fire brigade should be encouraged to study and to obtain the diplomas of the Institution of Fire Engineers.

Equipment of the brigade

Although the equipment of the brigade will vary according to the many factors already discussed, the following is a list of items that will probably be required (pumps and pumping capacity will be discussed in a separate section):

1 Additional hand extinguishers of the various types in use together with spare recharges.
2 Hose including hydrant equipment
3 Shut-off branches and various nozzles and, if the water supply warrants it, dividing breechings of the controlled type.
4 Portable lighting equipment, including electric hand lamps with spare batteries and in larger premises a mobile generator may be justified.
5 Tools for obtaining entry including door openers, crowbars, hand-saw, hammer, axe, ceiling hook, bolt croppers and metal shears. Electric cutting tools might also be justified in very large risks.
6 Ladders. These should be of wood and not metal.
7 Salvage equipment (see Chapter 29 which deals with this subject).
8 Rescue equipment if the risk is of a size that may require it.
9 First aid and resuscitating equipment.
10 If automatic sprinklers are installed, suitable spanners and spare heads, plugs and sockets of varying sizes.
11 Personal protective clothing.

Calling out the private fire brigade

The fire call will be by direct alarm if the brigade is whole-time, but in other cases it will depend possibly on other communication facilities within the premises. The ideal system is by one of the modern internal pocket radio or 'alerter' systems.

Where the initial alarm is given by the sounding of the evacuation bells or sounders, it must be possible for the members of the fire teams to find out where their services are required. If pocket radios can be provided the problem is simple. Otherwise it may be necessary to introduce a special system into the internal telephone system so that a number of simultaneous calls can be made. With modern detector systems mimic displays can be provided at the brigade's assembly points which will indicate the location. If a public address system is available then this can certainly be used to instruct members of the brigade and also to assist in the evacuation of affected buildings and plant.

During non-working hours, the problem may be more difficult. Unless sufficient members of the brigade can be assembled during this time to form an effective unit, much of the value of a private fire brigade is lost.

Where it is possible to offer housing accommodation in very close proximity to the premises schemes can usually be devised for men to be available on rota and a fire warning bell can be installed in the men's homes. The circuit is available through the telephone service and is known as an omnibus circuit. Ideally, sufficient firemen should be employed to provide full cover in three eight-hour watches, additional men being called out only in an emergency.

Transport of the brigade

This must depend on the extent and nature of the premises and the mobilizing point of the brigade. It might vary from a hand-pulled hose cart, to a modified electric truck suitable for negotiating the gangways of a factory or warehouse, to a small motor vehicle. In large complexes, major firefighting appliances, possibly specially designed for the risk, may be necessary.

It is frequently possible to buy second-hand applicances from the local authority brigade. These well maintained low-mileage vehicles are often excellent bargains.

FIRE STATION

This will vary according to the strength and equipment of the brigade but should always include one room where the members of the brigade can meet for training and for social gatherings. If possible, the room should be for the exclusive use of the brigade so that they can display instructional aids and other matters of interest to them.

An office should be available for the use of the senior fire officer personally as well as a small office for use as a watch office, the latter containing a filing cabinet and files for the records and information that should be available to the memberss of the brigade. It should contain a small library of technical books on the subject of fire and fire precautions together with the current and bound issues of the technical fire journals.

In the larger brigades, a control room will be required and in this room the fire alarm equipment will be contained.

TRAINING THE BRIGADE

The training of the brigade must be carried out under strict discipline and at no time should it be allowed to degenerate into anything but the serious undertaking which it is.

A training programme should be prepared in advance covering at least a month. It should consist of instruction and practical drills: half an hour of each will be ample, especially where part-time men are concerned who have probably already done a day's work. Each drill period should keep every man busy; once they are allowed to stand about without activity a drill period can quickly degenerate.

The frequency of the different types of drills will be governed by the equipment available and training conditions. With a part-time brigade having nothing more than hose and hydrants, a drill once a month might be sufficient but where heavier equipment such as pumps are used once a week would be more usual. A full-time brigade would drill four or five times a week. For part-time men, a minimum number of attendances must be laid down to ensure efficiency.

Schedule of instruction

The schedule of instruction should mainly consist of half-hour talks and lectures, although some speakers may require longer. The danger to watch for is that men may become bored and/or tired with inactivity which is understandable at the end of the day. The following is a list of people who could be approached to provide interesting speakers:

- Public fire brigade training officers
- The fire brigade officer from the local fire station, to talk about his local station, its equipment and how it functions
- A fire insurance official
- A fire insurance surveyor
- Speakers from other private fire brigades
- A speaker from the British Insurance Association
- A speaker from the local fire protection association
- Technical representatives from equipment manufacturers
- Chemists and other technicians of the company
- Water board officials
- Sewage board officials
- Someone engaged in combustion research, or education, who can provide an insight into basic combustion technology
- Forensic scientists dealing with fire investigation.

Attendance at meetings of the Institution of Fire Engineers and of other fire brigade associations should be allowed to count as drills, especially if a speaker is present. Variation from speakers can be made by visits to:

- Fire stations
- Other private fire brigades
- Equipment manufacturers.

There are also some excellent films and slides available, some purely training, others of general interest. The Fire Protection Association has a list of films and other material available and is always very helpful. If the company does not own a projector, one can be hired.

Regular instruction should also be given in the topography of the premises and the processes involved, if necessary taking the premises building by building, until the members of the brigade are familiar with every possible entrance and exit, the position of all fire equipment and manual call points, telephones and similar information. They should be shown the special points of danger: where a weak point occurs and fire could spread from one place to another, as well as the places where flammable liquids or compressed gases are stored and used. In some factories, notably paint factories, flammable liquids are piped across floors. These should, of course, have shut-off cocks outside the premises which should be known to all members of the brigade.

Practical drills

These will vary according to the equipment available, but certain fundamentals must be appreciated and practised.

Every member of the brigade must have used every type of extinguisher to put out an actual fire. Suitable fires can be built as suggested in Chapter 27; where there is any difficulty the local fire brigade will probably be able to help. Always remember to take your own extinguishers and the recharges for them. In all cases, the recharging of an extinghisher and leaving it ready for immediate use is part of the drill. Immediately after the drill the officer in charge, or team leader, should take the necessary action to bring his stock of recharges up to a minimum of two per extinguisher or to the required number as laid down.

If possible training on movement in smoke should be given. The dangers of smoke and toxic gases should be made clear in lectures and the company's own technicians should be able to supply details of any dangerous gases that might be particularly evolved in any of the company's processes or stock. Actual training in smoke should only be given at a fire training school under the supervision of qualified instructors.

Practical drills should be carried out in accordance with the Home Office drill book and possibly special drills may have to be devised to meet particular risks. In every case, drills must be carried out under the supervision of an officer who will maintain discipline; this is particularly necessary when wet drills are being carried out. It is not necessary that all hydrant and pump drills should be wet.

Practical drills should be carried out not only by day but also by night and periodically in complete darkness by extinguishing all light. Such drills must be organized with great care and with every safeguard possible to avoid injury or accidents. For example, if pumping operations are to take place from open water, safeguards should be taken in case a man might fall in, even to the extent of having a man in a boat if necessary. With drills of this nature, it is good practice to exercise one team at a time with another team standing by at various points to see that the working team is operating safely. Observers should have flash lamps and emergency lights for use if required.

An occasional full-scale exercise should be arranged, if possible with the cooperation of the public fire brigade. In all such exercises it is well to remember any possibility of damage particularly in the use of jets of water, foam or chemicals. A heavy jet of water can strip the slates off a roof, break windows and penetrate places where even the heaviest rainstorm does no damage.

All drills should be carried out at a moderate speed, accuracy being the most important feature. Competition work should not be a substitute for normal practice as it tends to make bad working in that accuracy can take second place to speed. A very minimum of competition work should be accepted as part of the general training.

In all drills involving equipment, this must be left ready for use as part of the drill, and where a mixed brigade of full-time and part-time men are employed, it is particularly important that under no circumstances must the cleaning up and putting away be left to the full-time personnel.

PRECONCEIVED PLANS

All preconceived plans call for a vivid imagination – an imagination of what might or could possibly happen.

Before any plans can be drawn up, thorough inspection of the premises should be carrid out with an experienced firefighting officer to select all the points at which a fire, having started, could spread internally and the points where the fire might he held or stopped.

Having decided on these various points, plans should be drawn up of the best methods of protecting these areas. Obviously, the closing of fire doors and smoke stop doors will be the first concern. This should already have been done by the employees when evacuating the building, but they should be checked. If the fire occurs outside working hours the doors should already be shut, but again they should still be checked. It will be necessary to detail members of the team to carry out this important work. It might be advisable to draw up plans for hose to be laid out and coupled up to a hydrant in case a fire should spread towards a particular predetermined stop position. In the same way, arrangements should be made to put up a curtain of water to protect a building, allowing a less important building to burn until additional help could arrive.

Schemes for obtaining water from alternative supplies should be worked out including water relays if necessary. It is sometimes possible to make mutual assistance arrangements with neighbouring premises, such as sharing the cost of a tanker to carry foam, or for mutual services of private fire brigades. In such a case, drills and exercises should be carried out in combination with each other and a chain of command established. The preconceived plans should be a study of every emergency that could occur.

One of the first plans should be the setting up of a contact point for the local fire brigade so that they can be given a detailed account of what has happened, what is being done and where the officer in charge of the fire is located. Discussions of these plans and the use of models on a board can be usefully made into drills.

The plans necessary during a fire are principally those of control. The great weakness is for the officer in charge to be at some point out of contact. Even in a small fire, the senior officer should not necessarily be at the scene of the outbreat – that is the place for his team leaders. His job is an overall direction from a point at which he can always be found and where all reports as to progress can be directed. Theoretically, he should be able to stand back and see that all his previous arrangements have been good and that from an overall point of view operations are progressing satisfactorily. Once a fire has taken hold, it is too late to try to change the organization or operational procedure.

After the arrival of the public fire service, unless there are special reasons to the contrary, the work of the private fire brigade should become that of a salvage unit. This work has already been discussed in Chapter 29 and the plans for this changeover should be ready to put into action.

The first objective after a fire is to enable the company to commence business again as quickly as possible. Arrangements must be made for a meeting place for directors and senior management, and preparations should be made for a meeting even while the fire is still burning. If the boardroom is involved, some other room must be earmarked. Although most directors might feel that their proper place during a fire is at the scene of the outbreak, this is not so, partly because they would have a great temptation to interfere with the work of the firefighting services but principally because they would not have an overall picture of what was taking place. It is much better if they are in the background where reports can be brought to them and where they can be contacted when requested for policy decisions.

Part of the preconceived plans would consist of arrangements for obtaining fresh stocks, materials and machinery. While this is not the work of the fire officer he should endeavour to persuade the directors and officers of the company to prepare such plans and keep them up to date and in a place of safety away from the premises, in other words, to make and keep the company fire-conscious.

Exercises

After the preparation of preconceived plans, exercises should be organized. These are of two kinds, internal and external. Internal exercises are set up in stages from the simple exercise of saying to a fire team 'There is a fire at such a point; carry out the procedure laid down', to the more intricate exercises associated with practice evacuation of the premises or an area, with all employees taking part.

External exercises may be carried out with the local fire service and may extend from a token attendance of one fire appliance, to a well laid-out full attendance including wet drill. In very large premises, it is posssible that the public fire service would welcome facilities for a full-scale exercise when they are being inspected by the Home Office Fire Department inspector.

Arrival of the public fire service

The public fire service have the legal duty of taking charge at all fires and the right of entry to any premises if there are reasonable grounds to believe that there is a fire on the premises or if it is necessary to assist firefighting operations. They have the right to do whatever they consider necessary to extinguish a fire. They also have power to order the arrest of any person who interferes with their operations. This could even extend to the owner or occupier of the premises involved.

It is to be hoped that relations between the public and private fire services will never be allowed to degenerate to such an extent that such powers need to be invoked. Everything possible should be done to create a close and friendly liaison so that, if ever they become actively involved, the work of reducing loss by fire will be accomplished with the utmost speed and efficiency.

Earlier in this chapter it has been said and it is here repeated that the first duty of a private fire service is to see that a call is transmitted to the public fire service without delay and, thereafter, arrangements must be made for them to be met as soon as they arrive and if necessary escorted to the scene of the outbreak. They will want to know immediately if there are any lives at risk, the precise position of the outbreak and what firefighting measure are being taken and particularly what water supplies are in use and what are available.

As quickly as possible all firefighting activities should be taken over by the public fire service although where and when requested the private fire service should be prepared to carry on with such work as they may be asked to do.

ADMINISTRATION OF THE BRIGADE

The administration of the brigade should be the responsibility of the senior officer. He should be responsible for the maintenance of all records. He should prepare the budget of revenue and capital expenditure and be responsible for keeping to the financial objectives. In this respect, he should have responsibility for initiating orders within the approved budget and should not have to wait for approval for expenditure on such items as minor repairs and replacements. He should receive regular information on all items debited against his budget to enable him to judge the progress and estimate the future expenditure.

Among the records that should be kept are the following items.

An occurrence book

This takes the form of a day book in which every item affecting the brigade is entered as it occurs. In a full-time brigade, the first entry should be the name of the man taking over responsibility for the book, followed by the name of the man he has relieved.

The log book should be a good quality bound book and should be carefully maintained as it not only presents a permanent record of what has occurred but might even be required as legal proof of an occurrence. No alterations should be permitted; any mistakes should be ruled out with one line and the correction rewritten, if necessary with a marginal note referring back to the error. Mistakes or omission should be corrected by an entry at the time of discovery and entered with a marginal note 'omission'.

A fire record book

This should show details of all fires and other special services rendered by the brigade. It might consist of a filed copy of the fire reports.

A training record book

This should show the names of the men attending, the names of the officer in charge and the instructor and the drill or training including lectures that have been carried out.

An appliance record book

This should show the date of purchase or age and cost of every appliance together with a history showing its use and any repairs or defects. The appliance book should also incorporate an equipment inventory.

A sprinkler record book

In this is entered everything affecting the sprinkler system(s).

A fire detection system record book

This should include the regular and quarterly tests.

A fire door record book

This would include fire-resistant doors, smoke stop doors and emergency exit (panic bolt) doors. The information recorded should include the date of manufacture and the regular tests and inspections, damage reports and date of repairs.

A personnel record book or card

This should show all personal details and courses attended with a chart showing drill attendances.

Information book

The information book should be available to the members of the fire department at all times. The following is a list of suggested items for inclusion in the book:

1. The procedure to be adopted in the case of fire or other emergencies, and any special procedure that might be necessary such as alternative ways of calling the public fire brigade.
2. A list of emergency telephone numbers.
3. A list of company officials, the area of their responsibilities, their home addresses and telephone numbers.
4. A similar list of technicians, plumbers, electricians, boilermen and the like, who may be required during or after an emergency and the methods by which they can be contacted and the method by which they can reach the premises.
5. The position of the main controls for gas, water and electricity together with a note as to the use of the controls.
6. A list of hydrants and other water supplies including those in the immediate vicinity outside the premises with their exact position given by at least two directions and similarly with other water control valves noting what the closing or opening of the valves will do.
7. List of fire doors, smoke stop doors and fire and emergency exits.
8. A list of extinguishers, their types and their locations.

9 Information about:
   (a) Sprinkler and other automatic protection systems.
   (b) Fire alarm installations.
   (c) Evacuation and alert signals.
   (d) Telephone switchboard operation.
   (e) Broadcasting equipment, pocket radio or similar systems in use.
   (f) Any other useful information that might be wanted for quick reference.

Posted information

The amount of posted information should be very strictly controlled; scraps of paper pinned up at random do not make for efficiency. If typed orders are used it is well to establish a book of regulations. In this would be placed official orders to the brigade and as special orders were issued a copy would be posted in the book. The procedure would then be to exhibit a notice on the official brigade notice board for so many days and post a copy of it in the book for future reference.

On permanent exhibition should be plans of the water supplies, sprinkler systems, special hazards and any special and immediate action to be taken in case of fire. This should be in the form of a few short numbered instructions. A plan of the position of outside alarm wires should also be readily available if not shown on the general plan.

WORK OF THE BRIGADE

The main duty of the private fire brigade is to see that all fire precautions are maintained at full efficiency. The work must not only be carried out systematically by constant patrol, inspection and testing but as an attitude of mind, so that the slightest danger is subconsciously noted and there is a mental urge to take action.

In addition to the regular patrol of the works by the patrolmen, who themselves should be members of the brigade, an officer should walk through every part of the premises at least once every day and all areas be seen by the senior officer at least once a week. He will observe and note everything that affects fire precautions, taking immediate remedial action if necessary. At the conclusion of this patrol, he will number his notes and enter them in a special book that has a column for the action to be taken and another column for the results of the action. If further action is required, a fresh entry should be made and marked off by reference to the previous entry. Every three months or so the book should be revised and all entries not completed should be brought forward under fresh numbers.

Inspection and maintenance schedules should be prepared in advance so that the work is carried out at the proper intervals and in a regular manner.

Wherever possible, the repair and maintenance of the fire brigade apparatus should be done within the brigade though from time to time difficulties will arise with craft unions. Generally speaking, these can be ironed out if discussions take place at an early stage. As an example, one decision that did arise in a particular brigade in connection with fire alarms was that the testing of a fire alarm point was a functional responsibility, and therefore the work of the brigade, but if the point failed to work then it became a matter for the electrical department.

Where specialist work of this nature has to be carried out by another department, it is essential that priority is given to the repair of fire service equipment and protective equipment. In the case of alarm and evacuation signals, no other work should be allowed to precede the repair. In places of high fire risk, consideration should be given to stopping normal work if the lives of employees could be placed in jeopardy and special arrangements should always be made to provide alternative warning signals whenever a breakdown occurs.

Other duties of the private fire brigade

In the matter of other uses of the private fire brigade, the criterion must be whether the duty prevents either men or appliances being ready immediately for their prime duty, that of safeguarding life and preventing loss by fire.

There is always a great temptation to borrow tools and equipment set aside for emergency use, very often the excuse being given that it is required for an emergency – a production emergency. This entirely overlooks the fact that the word 'emergency' when used in connection with fire means the saving of life or property from fire. The real reason that emergency equipment is borrowed is because it is always in a set spot and so can easily be found. There must be a strict rule signed by a senior executive and issued to management and supervisors at all levels to the effect that fire equipment and appliances set aside for emergency use must not be used for any other purpose.

Where tools and appliances are not under the direct control and constant supervision of full-time personnel, they should be kept in a locked compartment. If necessary, it should have a key under 'break glass' conditions and preferably with a glass front to the door, so that the apparatus can be observed to be available and in good condition. Apart from the fact that the appliance may be missing when wanted for its particular emergency purposes, its replacement may be forgotten or it may be damaged. It is more than likely the damage would be concealed to avoid the discovery that it had been used surreptitiously.

As with tools, so with men: full-time firemen must not be considered as a pool of free labour. If a high standard of service is expected, firemen must be treated as craftsmen technicians and not as general labourers. A fireman must be able to take pride in himself and his work and he must be given every encouragement to do so.

INSPECTION OF THE BRIGADE

The public fire service are subjected to an annual inspection by inspectors appointed by the Home Office; no such inspection is made of private fire brigades.

Every private fire brigade should be inspected and tested and the organization reviewed at least every 12 months. It has been suggested that officers from the local public fire service should carry out this duty. This cannot be recommended because it is doubtful whether they would command the specialist knowledge of the requirements of a private fire service. Neither are they likely to have the ability and tact of a Home Office inspector which in itself is a specialist position within the public fire service.

It is suggested that the inspection should be carried out by officers independent of the plant and company concerned. They should be experienced in the operation of the private fire service and should be required to submit a comprehensive report on the efficiency and effectiveness of the brigade, bearing in mind that an efficient brigade is not necessarily an effective brigade. The report should be constructive and should include recommendations for any improvements in training the personnel of the brigade, its equipment and its operational effectiveness.

There is a real need for people of sufficient experience and seniority to be able to conduct inspections of industrial brigades. This could well be a part-time occupation for retired senior officers and it is to be hoped that one of the organizations concerned with industrial brigades will establish its own inspectorate.

31

Hazard analysis and reliability

It is not intended in this chapter to provide detailed instructions on the various methods which can be used in the analysis of hazards and the measurement of reliability. The purpose is simply to draw attention to the existence of these very powerful tools, and to indicate how they are used. In general it is necessary to employ experts in the particular technique if any reliable and meaningful results are to be obtained. However, training in the various methods is available, sometimes in the form of two- or three-day courses. Most people could acquire adequate skill in using the Mond Index in this way (see below). If a HAZOP (see below) is to be undertaken it is best to employ a consultant to act as the leader of a team, of which the other members are drawn from various technical departments on the site. These two methods, and others, are discussed below. It might be as well at this point to define some of the terms which are used in discussing these methods.

HAZARD AND RISK

A tank of flammable liquid can cause damage and injury because if the contents escape there can be a fire or an explosion. The tank of liquid is clearly a *hazard*. We do not know if we can accept the *risk* presented by this hazard unless we can determine the probability that the tank will fail, and the likely consequences if it does fail. The *probability* is usually expressed as a number between 0 and 1, where 0 = it cannot happen and 1 = it is certain to happen. The *failure rate* for the tank (or the probability that it will fail in a given time) could be 0.005 failures per year, which is one failure every 200 years. The *consequences* could range from the loss of liquid to a disaster which kills several people. The maximum possible loss could therefore be quantified in cash terms or in deaths. The risk with which we are concerned is clearly a combination of these last two defined terms and, because they can be quantified, we can say that *risk* = (probability × consequences). It may be more convenient here to refer to the *frequency* of the event rather than the probability.

We may need to go further than this and decide where we are going to draw the line between an acceptable and an unacceptable risk. Since risk is the product of frequency and consequence, we might accept having a minor leak of flammable liquid once a year, but we would expect the probability of a significant disaster to be say once in 10 000 years. This is illustrated in Figure 31.1. The designers of nuclear power stations have to make just this kind of assessment. From data banks of collected information they can calculate the probability of various kinds of failure, taking into account all the safety systems they are planning to install. They can also estimate the maximum amount of radioactivity which can be released by a particular failure. They can then draw a diagram similar to the figure, on which is shown the probability of releasing a given amount of radioactivity. This must comply with legally established limits.

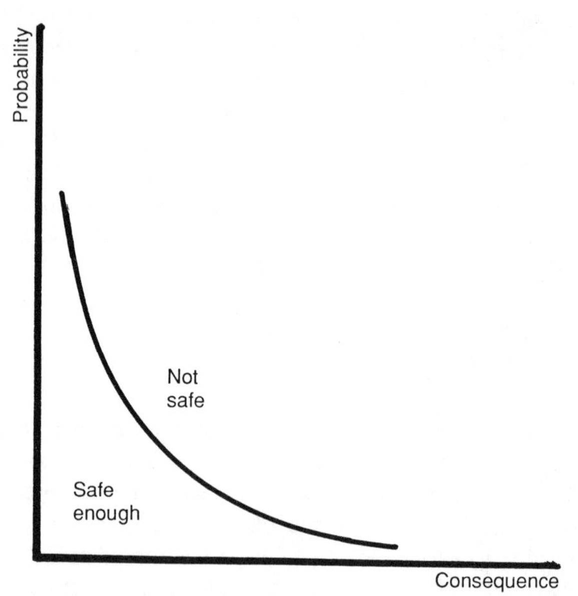

Figure 31.1 Frequency of failure plotted against the consequence of failure

In considering the acceptability of a risk it is sometimes helpful to know the fatal accident frequency rate (FAFR) and to calculate what effect the introduction of a hazard would have on this rate. The FAFR is the number of deaths occurring during 100 million (10^8) man hours of exposure to the hazard. This number is used because it represents the total time at work, assuming a 32-year working life, for 1000 people. Table 31.1 shows some FAFR values for a number of different hazards. Those above the line are for a complete working life, and those below are for the actual exposure time to the hazard. Thus, in UK industry as a whole there are four fatal accidents for every 10^8 man hours worked. There are 4000 fatal accidents for every 10^8 man hours spent in mountaineering. This quite astonishing difference between the risk involved in working in industry and in climbing a mountain is largely hidden because at any one moment very few people are actually climbing. The death rate per year is therefore quite low. If there is a population of 10 000 mountaineers then on average four will be killed each year. The rate is exactly the same for footballers. In a similar population of car drivers there would be 17 accidental deaths a year, and for people smoking 20 cigarettes a day there would be 500 deaths.

Table 31.1
Fatal accident frequency rates

OCCUPATION	FAFR
Total UK industry	4
Steel industry	8
Fishing	35
Coal mining	40
Railway shunters	45
Construction workers	67
Staying at home	3
Travelling by bus	3
Travelling by train	5
Travelling by car	57
Skiing	71
Pedal cycling	96
Firefighting	140
Travelling by air	240
Motor cycling	660
Canoeing	1000
Rock climbing	4000
Professional boxing	7000
Horse racing (National Hunt)	50 000

It should be noted that these are deaths from accidents only. If we wanted to include deaths from suicide, for example, then each of the values above the line should be increased by 1. The calculated FAFR of an accidental release from a nuclear power station for people 1 km away is 0.001, which is about the same (in the UK) as the value for deaths from lightning. In the USA, the FAFR for major storms is 0.009, and for California earthquakes it is 0.019. The FAFR from crashed aircraft, for people living near to an airport, is about 0.01.

The values provide a useful guide to the relative risks of various occupations. Calculation of an FAFR can also be used, as suggested above, to help in deciding whether the risk introduced by a new industrial process is acceptable. It has been quite arbitrarily suggested that, accepting an overall FAFR in industry of 4.0, no single activity that a person has to perform should increase this by more than 10 per cent, that is to more that 4.4. This value should be compared with, for example, the FAFR of 67 to which construction workers are exposed.

It should be noted that risk is involved in almost everything which we do, or which is done for us. The demands in the popular press for 'absolute safety' when a new factory is built cannot possibly be met. If we want to enjoy the benefits of cheap power, the use of plastics and chemicals, transportation (and its fuel), and leisure activities like canoeing and rock climbing, then we must be prepared to accept the risks which go with them. The safest thing we can do is to stay at home and write a book.

LOGIC DIAGRAMS

A valuable method of analysing the sequence of events which could lead to a fire or explosion, and the events which could result from a disaster, is by means of logic diagrams, usually in the form of 'trees'. A simple fault tree analysis (FTA) is shown in Figure 31.2. This shows the ways in which a fire could involve a vessel of flammable liquid which is at the side of a road in a chemical works. The fire is called the 'top event', and this can only happen if there is a spillage of liquid in the presence of an ignition source. The fact that both must be present before the top event can occur is indicated by the 'AND gate'. There is a number of ways in which ignition sources can happen, any of which can ignite the spillage, so these are linked to an 'OR gate'. The diagram is not complete because there are other possible sources like lightning and crashed aircraft, but the intention is simply to illustrate the method. Each of the items linked to the OR gate is a 'fault event', and each of these should be analysed until a 'basic event' is reached which is incapable of further analysis (or not worth analysing). The

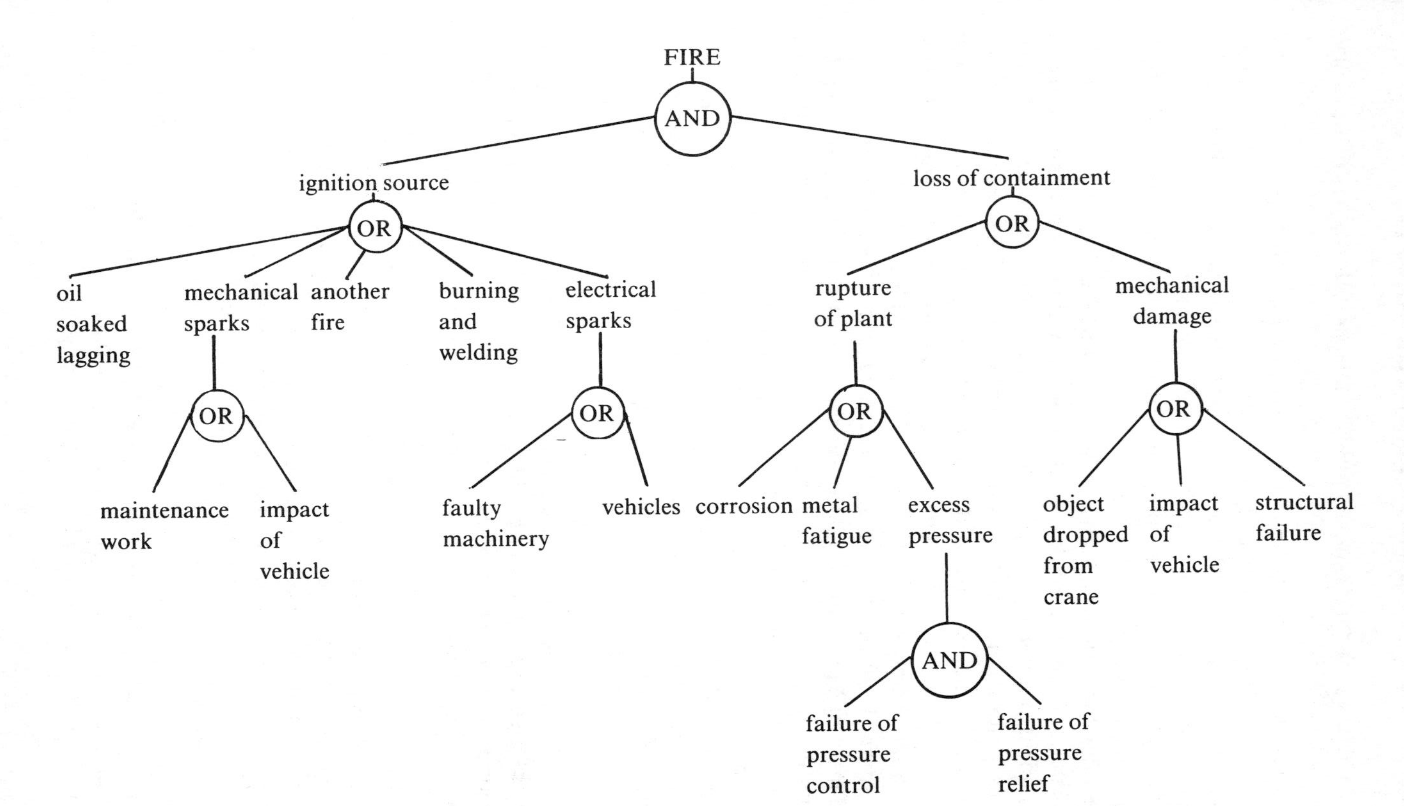

Figure 31.2 Fault tree.

'another fire' event for example could be analysed through several further steps.

On the 'loss of containment' side it will be seen that in analysing the events which could cause the rupture of the plant (vessel, pipelines, pump) we need to use another AND gate. This shows that the failure of the pressure control device will result in a rupture only if the pressure relief device has also failed. Again, this side of the tree is incomplete: all the fault events could be further analysed.

One thing that is immediately obvious from this fault tree is that the vessel should not be at the side of the road. A vehicle could not only cause a loss of containment but could also provide an ignition source, either by its own electrical equipment or as a result of the impact. The analysis can be particularly valuable if we add to it the probability of the various fault events. In a large works for example the probability of something at the side of the road being hit by a vehicle could be calculated from recorded information. If all the probabilities are low (as they usually are), the ones linked at an OR gate can be added together. Additions can continue up the tree until an AND gate is reached, and here the probabilities are multiplied. This is reasonable because the probability of two or more events happening at the same time (which is the reason for the AND gate) must be less than the individual probabilities. (Since the probabilities we are dealing with are always less than 1, the effect of multiplication is to reduce the number.) The final result enables us to establish the probability of the occurrence of the top event, and also to determine the most likely sequence of events leading to the top event. The simple calculation outlined above applies only when events are independent of each other. In our example, the impact of a vehicle could cause a leak and also provide an ignition source, and the calculation then becomes more complicated.

Another important piece of information which can be obtained from a fault tree is the presence of 'minimal cut sets'. These are groups of basic events (at the bottom of the tree) which can cause the top event to happen only when they are all present at the same time.

Additional information can be shown on a fault tree by the use of different symbols. For example, events can be enclosed in rectangles, circles or triangles to indicate whether they are fault events, basic events or entries from another tree. A basic event in a diamond is one which is still capable of further analysis. Different gates are also possible. For example, events may have to happen in a particular order to get through an AND gate. An exclusive OR gate will pass only one event, and not two happening at the same time. An AND NOT gate passes only when a particular event has not happened (like for example a Halon system being discharged).

The fault tree can be extended beyond the top event to show the things which could happen as a result of the fire. For example a pipe bridge above the vessel could fail and release more flammable liquid: this could flood the vessel's bund and cause burning liquid to spread to the next plant. The new tree which grows upwards from the top of the fault tree is an event tree analysis (ETA). When this is complete we can think about the results of the fire which are not physical events: for example loss of customers, adverse publicity, prosecution, going out of business. This is a cause consequence analysis (CCA).

It is sometimes useful to plot some of the information in a fault tree the other way round, so that the top event is NO FIRE. This is a success tree analysis (STA) and will show the alternative ways in which the goal can be achieved. A decision tree is similar except that it shows the outcome of decisions at different levels. A good example of this kind of tree was produced by the NFPA Committee on Systems Concepts. The top item is fire safety objective(s). Beneath this is an OR gate to which are connected 'prevent fire ignition' and 'manage fire impact'. As an example of the detail contained in the tree, right down at the bottom of the prevention side is an area concerned with limiting the amount of fuel. This can be done by design OR by the action of the user. User action can be compulsory OR voluntary. These two items are connected by AND gates respectively to Adopt legislation, Educate user, Inspect property, Enforce law, and second to Motivate user, Educate user, Inspect property. There is a total of 95 separate items in this very interesting tree (the NFPA reference is SPP–36).

POINTS SCHEMES

Scoring systems, by which a numerical comparison can be made of the risks associated with different factories, have already been mentioned in Chapter 12. They can be used to determine which is the most hazardous part of a particular process, and to compare different methods by which the hazard can be reduced. By adding up the bad points and deducting the good ones (or the other way round) a measure can be obtained of the overall risk on a particular site, and this can be compared with the score obtained on other sites. The process is relatively easy and can be completed quite quickly, provided that all the relevant information is available. It could be used for example by a local authority fire brigade to whom plans have been submitted for a new chemical factory. The method is also well suited to the assessment of fire insurance premiums, or to the division of insurance costs within a group of manufacturing units.

One of the earliest schemes was developed by M. Gretener of the Swiss Fire Protection Association. It is used to calculate a hazard factor, *B*:

$$B = \frac{P}{NFS}$$

where *P* represents a combination of the fire risk factors; *N* combines a range of normal fire precautions like a fire brigade and water supplies; *F* represents the fire resistane; and *S* covers the special precautions like detectors and sprinkler systems. All these factors are given a value of 1 for normal conditions. So for normal hazards with average protection the value of *B* will be 1. Detailed information is available on calculating the factors, with tables showing the values to be applied for different industries, materials and protection methods.

The Dow Chemical Company in the USA produced a points scheme which they called their Fire and Explosion Guide. This was used in the UK by ICI in what was then called Mond Division. They produced a considerably extended scheme which is called the Mond Index. A number of difficulties in the use of the Dow method were eliminated, a range of offsetting factors (the good points) were added, and also methods of assessing for example toxic hazards and OFCE (open flammable cloud explosion) hazards. Short courses are available on the use of the Mond Index and a detailed instruction book is available, together with specially printed sheets on which scores can be marked and the various indices calculated. A computer program is also available which greatly reduces the time needed to complete the work.

The plant is first divided into a number of units, each of which is separately assessed. A material factor is then calculated based on the energy which will be released by the combustion or decomposition of the key material in the unit. This factor is then modified according to the process hazards and the material's own inherent hazards. The total quantity of material, the layout of the plant, and the toxicity hazards are all used to produce other factors. All factors are then combined to provide an overall index, a fire load, and an overall degree of hazard which ranges from 'mild' to 'highly catastrophic'.

Account is then taken of various offsetting indices, covering containment, process control, management attitudes, isolation, fire protection and firefighting. These factors are used to reduce the overall hazard calculated above. This results in offset overall risk rating, which is categorized between 'mild' to 'very extreme'. The latter term is somewhat illogical (you cannot go beyond an extremity) and is two grades lower than 'highly catastrophic'.

However these are simply convenient ways of indicating the approximate size of the index number. The whole procedure provides a detailed assessment of both the hazards and safety precautions on any process plant. It can be used at the design stage, when the effect of changes in both production methods and fire protection procedures can be checked, so that the most effective solutions can be found. It can also be applied to existing plants to provide an indication of the risk involved in operation; and again the effect of changes can be measured.

HAZARD AND OPERABILITY STUDIES

A HAZOP is an extremely detailed and thorough method of hazard analysis on a process plant. Each small section of the plant is analysed to find out the effect of deviations from normal conditions during start-up, shut-down, and normal operation. A section could comprise one vessel, a second section could be the pipeline connecting it to the next vessel, and a third section could be the second vessel. The deviations examined are those in flow, temperature, pressure, level, concentration, heating, and cooling, whichever is appropriate. Having selected the relevant properties from this list, the following guide words are applied to each (again where appropriate): No, Less, More, As well as, Part of, Reverse, Other than. For example if we consider the pressure in the vessel, then only the words No, More, and Less will apply. From a detailed knowledge of the process and the rest of the plant, the conditions (malfunctions, failures, operator errors, etc.) which could result in a complete loss of pressure are detemined, and then the consequences of this deviation. The exercise is then repeated using the other guide words. When flow, both into and out of the vessel, is considered, the words No, Less, More, Reverse, Part of, As well as, Other than, will apply. Again causes and consequences are considered, and the results are recorded. It may soon be found that some causes and consequences are being repeated, for example when Less is applied to concentration the causes may have been covered when Part of was applied to an inlet flow. The work is not then repeated.

A standard form is used to record the results of the study. A shorthand method of recording can be used, and if this is fed directly into a computer a suitable program can be used to produce both fault trees and cause consequence trees from the data. If necessary, known failure rates of various pieces of equipment can then be used to calculate the probability of a particular event.

Because of the range and depth of knowledge which is needed to conduct a HAZOP, the work is always done by a team of specialists. If a plant is being considered at the design stage (which is the right time to do it), the team might well comprise a leader (who could be a consultant), the project manager, the research and development manager, the commissioning manager, the project engineer, the process design engineer, the instrument engineer, and possibly also a techincal secretary. If a modification or extension of an existing plant is being planned, the team could include the plant manager, the works development manager, the plant maintenance engineer, the process design engineer, the instrument engineer, and the plant foreman. The whole purpose of HAZOP is to identify deficiences in both the design and the operation of the plant. It is important therefore that the team members are sufficiently senior to authorize the changes which are agreed. It is important also that the design is frozen before the HAZOP begins. Nothing at all should be changed until the study has been completed, and then the team's agreed modifications should be implemented. If the design team wishes to make additional changes, these should be submitted to the HAZOP team for approval.

A HAZOP is clearly a very time-consuming exercise, but the results can be invaluable in identifying potentially expensive flaws in designs and procedures. A HAZOP is essential where a potentially hazardous process is being planned. Much of the time is taken up in the detailed recording of all the causes and consequences. This work can be considerably reduced if the design changes only are recorded: everything else is assumed to be satisfactory unless there is a note in the report. This method has the disadvantage that if someone has to review the HAZOP later (possibly because of a failure on the plant) there will be no means of knowing the logical processes by which the team reached its decisions.

IFAL

The instantaneous fractional annual loss method is used by insurance companies to work out premiums for particular risks. It is really another points scheme which takes into account matters like process hazards, the standard of engineering, and safety management. It produces a theoretical average annual loss from fire on which the premium can be based.

SAT

A safety analysis table is included in the American Pètroleum Institute's code of practice for the design of offshore production platforms (API RP 14c). There is a separate table for each component, which shows how undesirable events could be caused and what protection should be provided to prevent them. It is really a check list which enables you to carry out a simplified HAZOP.

FMEA

The failure modes and effects analysis was developed by the US Department of Defence for checking matters like guided missile control systems. It is described in MIL-STD-1629A as 'a procedure by which each potential failure made in a system is analysed to determine the results or effects thereof on the system and to classify each potential failure according to its severity.' This sounds like HAZOP and indeed the procedure is much the same. However it is primarily a method of reliability analysis because it tries to identify all the ways in which a system can fail. Data from an analysis can be used to calculate some of the reliability measurements discussed below, and also to determine maintenance and repair policies.

Little use has been made of FMEA in industry, mainly because of the popularity of HAZOP. Both of these are 'bottom-up' methods: you start with a detailed study of system components and work up to a top event. Fault tree analysis on the other hand is a 'top-down' method: you begin with the top event and then work out how it could happen.

RELIABILITY

Failure rates of components (or the probability of failure) have been mentioned several times above. Clearly in fire protection we need to be concerned with the reliability of components and with methods of measuring and controlling reliability. Unfortunately there is no complete agreement on the definition of this word, but we can assume that it refers to the probability that a system or a component will fail to operate when it is called upon to do so. Some people prefer to regard reliability (possibly more logically) as the ability of the equipment to do the job it is designed for when it is working properly. This is perhaps better described as capability, and we could say that (capability + reliability) = effectiveness.

A knowledge of reliability is important in fire engineering for two reasons: the failure of process equipment may result in a fire, and the failure of fire protection equipment may mean that the fire is not detected or extinguished. As we have seen above, a study of reliability provides data from which the probability of fires and explosions can be estimated.

Failure of individual components

If we follow the history of a large number of identical components, we can obtain data from which we can plot a curve showing the variation of failure rate with time. The curve would look like the one in Figure 31.3 which, because of its shape, is called a bathtub curve. It will be seen that in the figure three distinct periods of time have been marked in which different kinds of failure occur:

1. The failure of faulty components which have not been detected by the manufacturer's quality control checks. This is the period which is usually covered by a guarantee. It is sometimes called the burn-in phase, and the failures are called infant mortality or teething troubles.
2. This is the normal failure rate which occurs during the useful life of the component. Failures are usually independent of time, so the curve here is horizontal.
3. In the final period the component's useful life is over and parts of it begin to wear out.

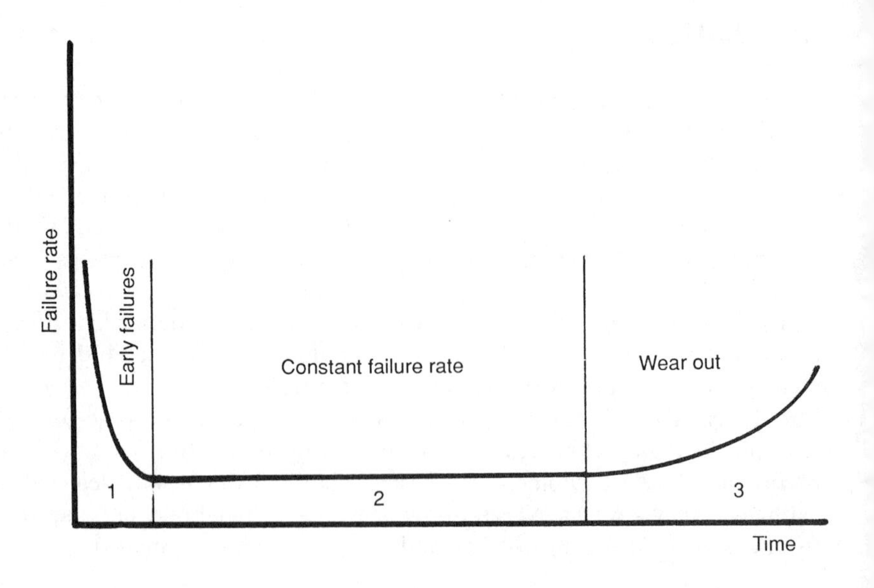

Figure 31.3 Bathtub curve.

A great deal of information is available now on the normal failure rate of individual components: that is, the constant failure rate in the second period in the figure. Some industries, and in particular the chemical industries of the UK and the USA, have set up data banks. Each of the member firms feeds data on failure into the bank, and can withdraw information whenever it is needed. Similar information is published in extensive tables, such as those in the book by Green and Bourne mentioned at the end of this chapter.

There are many ways of expressing failure rates, but the one generally used (as in the tables mentioned above) is failures per million hours (f/10^6h). This may be slightly illogical because many things do not last for a million hours, but it usually enables whole numbers to be used, and it provides an immediate comparison between a wide range of components. As an example, when all the available information on the failure of AC generators was averaged out it was found that a generator could be expected to fail when it had been running for about 143 000 hours. This can be conventiently expressed as seven failures in a million hours. A generator has bearings which can wear and parts which are subjected to high contrifugal forces, so this is not an unreasonable failure rate. On the other hand, there are no moving parts in a 10 kV transformer and it is not subjected to high electrical stress, so it should last a long time. As it happens, the normal life is about 1.7 million hours, so the failure rate is 0.6 per million hours. (You will have noted that these examples were taken from data provided by the electrical industry. It is regrettable that no similar information is available from the fire protection industry.)

Failure rates are sometimes expressed in other ways:

- Percentage failures per 1000 hours. This is the same as actual failures in 100 000 hours, or about 11.3 years.
- Failures per year: very roughly one-tenth of the foregoing.
- These can be compared with the one described here, failures per 10^6 hours which is 10 times the first method and about 100 times the second.

Failure rates depend not only on the inherent reliability of the component itself, but on the way in which it is used. If it is overloaded, or operated at too high a temperature, the failure rate will be increased. For example, if our 10 kV transformer is overloaded to 140 per cent of its design rating, its failure rate is increased by a factor of four and becomes 2.4 per million hours. If in addition it is heated to 120°C, the failure rate becomes 7200/10^6h, or an average life of 140 hours. Factors to be used for conditions such as overloading are provided with the tables of failure rates.

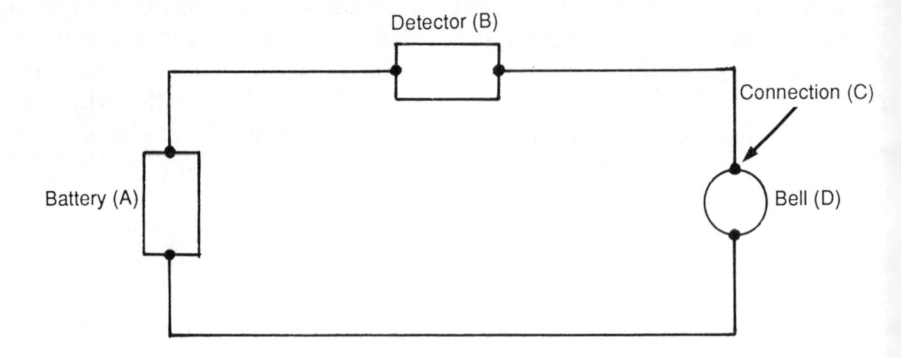

Figure 31.4 Simple detector circuit

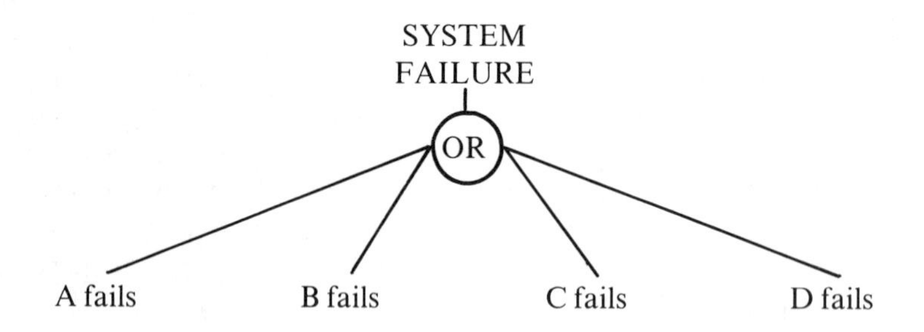

Figure 31.5 Fault tree for detector circuit

Reliability of system

The reliabiiity of a complete system can be calculated from the failure rates of individual components. Take for example the simple circuit in Figure 31.4, comprising a battery, a bell, a detector, and the connecting wires. If we assume that the wires have a very high reliability (because they are underrun), we are concerned only with the failures of the three components and the six electrical connections. If any one of these fails the system will fail, and this is shown by the presence of an OR gate in the fault tree in Figure 31.5. Table 31.2 gives failure rates for the components. These values are guesses because we do not have the real information. Rates are shown per million hours and also per year. This second column is included because these are the figures which would be quoted as the probability of failure per year (but again of course only during the second period of Figure 31.3).

Table 31.2 Failure rates

Component		Failures per 10^6h	Failures per year
(A)	Battery	1.0	0.0088
(B)	Detector	0.53	0.0046
(C)	Bell	1.9	0.0168
	One connection	0.3	0.0027
(D)	Six connections	1.8	0.0162

It will be remembered that at an OR gate on a fault tree you add all the probabilities. If we do this we find a failure rate of the entire system of $5.23/10^6$h or 0.0464/year, which is one failure in 21.5 years. This is a poor performance for a detector system, so we should look at ways of improving its reliability. One possible approach is to use two separate systems. We can show this in the fault tree in Figure 31.6a, where an AND gate is needed: both systems must fail before we fail to get an alarm. At an AND gate you multiply the probabilities: $0.0464 \times 0.0464 = 0.00215$. This would suggest that we could expect one failure in 465 years, which is a great improvement. However we cannot strictly calculate in that way because the failure rate is no longer independent of time.

The duplication of systems in this way is called redundancy. In this case we have active redundancy because both systems are operating together. With stand-by redundancy the second one would come in only when the first had failed. We could also use component redundancy to improve the reliability of the system. From the table we can see that the least reliable component is the bell. In Figure 31.6b this has been duplicated, again using an AND gate. The combined bell failure rate will now be $0.0168 \times 0.0158 = 0.00028$. We now have eight connections instead of six, so the sum will be:

$$(8 \times 0.0027) + 0.0088 + 0.0046 + 0.00028 = 0.0353$$

This is an improvement over the original value of 0.0464, but admittedly not a large one. In practice the introduction of redundancy into control and safety systems on hazardous plant can have a significant effect on their reliability.

Maintenance and repair

The reliability of a system can also be improved by regular checks to see whether it is working, and a quick repair if it is not. Such checks of course

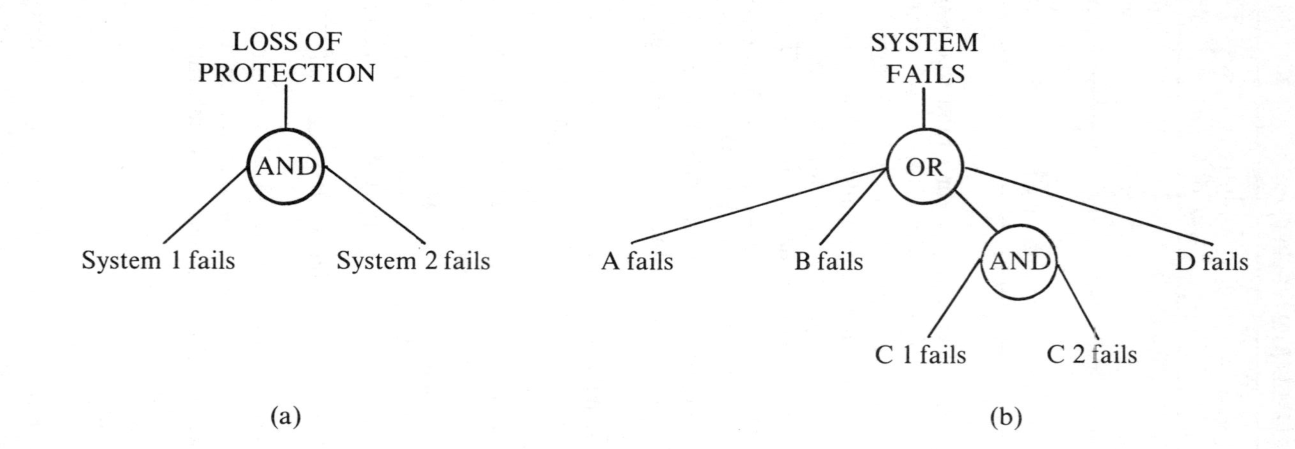

Figure 31.6 (a) system redundancy (b) component redundancy.

play a vital part in the proper maintenance of fire protection systems. Suppose for example the system in Figure 31.4 is checked four times a year. Thc probability that the system will fail to operate when it is needed will now be the probability of failure in the three months between inspections, and not the annual failure rate. The failure rate will be $0.05 \times 0.25 = 0.0125$, or one failure in 80 years.

We can look at reliability another way by calculating for what fraction of time the system will not be functioning. This is called the fractional down time (and the period when it is functioning is the up-time fraction). Every time a failure happens the system will be out of action for the rest of the period to the next check, plus the time taken in repairing the faulty component. Since the failure rate is independent of time, we can assume that on average the time to the discovery of a fault will be half the time between checks. So if we call the period between tests x and the repair time y, the down time will be $x/2 + y$. The franctional down time will be given by $(x/2 + y) \times f$ where f is the frequency of failure. We need to have all the values in the same units, so we can say that $x = 0.25$ years, $f = 0.464$/year, and we can assume that $y = 0.01$ year (or 3.5 days). The sum for our system is then:

$$\left(\frac{0.25}{2} + 0.01\right) \times 0.464 = 0.00626$$

This is the down-time fraction and the corresponding up time will be $1 - 0.00626 = 0.99374$. So we can say that our three-monthly checks have ensured that the system will be operating for 99.37 per cent of the time: there is only a 0.63 per cent chance that a fire will not be detected.

This calculation applies only to the simple system in Figure 31.4. Once redundancy has been introduced the calculation is extremely complicated and invoves Boolean algebra (which was specially developed to deal with OR and AND gates). However, the message is clear: redundancy and especially regular inspection can have a profound effect on the reliability of a system.

Reliability terms

It may be useful to describe some of the other terms which are used by reliability engineers (and it is hoped will be used eventually by fire protection engineers). This can be done by using a diagram to illustrate the situation we have just been discussing. In Figure 31.7 time is running from left to right and the line in the upper half simply shows the two functional

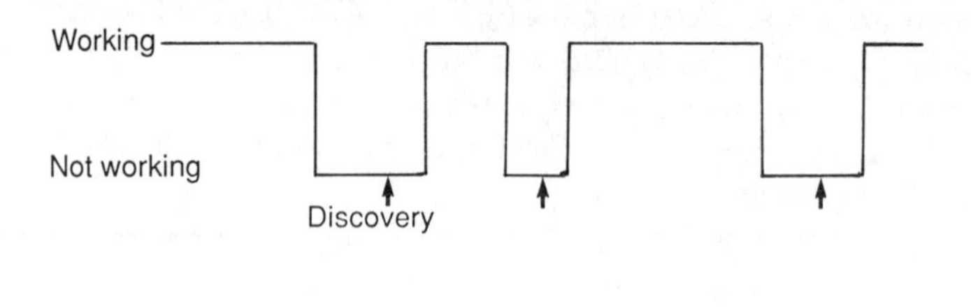

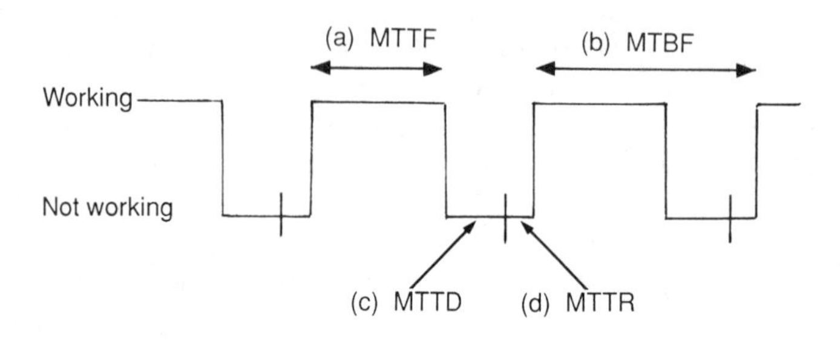

Figure 31.7 Functional condition of the detector circuit

conditions of our system: it is either working or not working. At some time during each non-working period (or down time) the fault is discovered by one of the regular inspections. This discovery is indicated by an arrow. There is a short period after that for repair before the system is working again. The horizontal scale in this drawing is of course quite wrong: in reality the up times will be considerably longer than the down times. When data have been collected from a number of up and down excursions it will be possible to calculate the average values of the various times, and a diagram like that on the lower half of the figure can be drawn. A number of different time intervals have been marked on this diagram, and the names given to them are these:

a	=	Mean time to failure	MTTF
b	=	Mean time between failures	MTBF
c	=	Mean time to discovery	MTTD
d	=	Mean time to repair	MTTR

It should be noted that $a+c+d=b$, and also that if c and d are very small compared with the other times (as they should be), then a will be very nearly equal to b. The up time fraction can be calculated from:

$$\frac{a}{a+c+d} = \frac{a}{b}$$

and the down time fraction from:

$$\frac{c+d}{a+c+d} = \frac{c+d}{b}$$

Since $(c + d)$ is usually very much smaller than either a or b it may be convenient to get the down time fraction from

$$\frac{c+d}{a}$$

When systems manufacturers are asked to quote for the supply of a system there is no good reason why they should not be prepared to provide also a calculated MTBF, based on component failure rates and on their recommended maintenance schedule (and to supply the calculations).

32

Special hazards

It is not the purpose of this chapter to provide comprehensive guidance on the protection of particular risks. The information is included because there is a great deal to be learned from the ways in which other people have tackled their particular problems. Also, the discussion provides practical examples of the application of some of the techniques discussed in earlier chapters.

OFFSHORE PRODUCTION PLATFORMS

An offshore platform presents one of the most challenging fire engineering problems. The processes involve large flows of low flash point materials at high pressures and temperatures. On land, similar plant would comprise open, well separated and accessible units, which were assured of the prompt availability of adequate firefighting facilities. Escape from the area would be simple, and an almost unlimited firefighting back-up would be readily available. Offshore the process units are tightly packed into metal boxes, which are stacked in several layers on a steel tower (the jacket), together with up to 30 wells (some of them still being drilled), a power station, a helicopter landing pad, and a hotel for several hundred people. The platform must be self-reliant and self-sufficient in the provision of an adequate firefighting potential. Some back-up may be available, but probably not for some time, and then only in favourable weather conditions. Escape is possible but again the reliability of the available methods is affected by weather conditions. A platform on fire in a violent storm is in a lonely and perilous predicament.

The reservoir of oil beneath the North Sea may be about 3000 m below the sea bed. The oil is at a pressure of around 680 bar and a temperature above 100°C. The fluid emerging at the wellheads is a hot, high-pressure mixture of gas, gas-saturated oil, water and sand. These have to be separated on the platform. The oil and gas are then cooled and cleaned, reduced to a workable pressure, and then metered into pipelines which take them to the mainland. The water and sand are cleaned before they are discarded. On a large platform the production rate could be 20 000 tonnes of oil and 1000 tonnes of gas each day.

As fluids are withdrawn from the formation the pressure will fall and hence the flow rate will be reduced. To maintain the pressure and to ensure the maximum extraction of oil, both water and gas are pumped back into the formation. Out of say 30 wells on a platform perhaps 12 would be used for re-injection. The wells do not of course go straight down into the formation. Directional drilling ensures that a large area is covered, and also that the water injectors discharge below the oil/water interface in the formation. The pressures needed for injection may be 380 bar for the gas and 170 bar for the water. A lower pressure is sufficient for the water because of the hydrostatic head in the borehole: a 3000 m column of water creates a pressure of nearly 300 bar at the bottom.

Thus, on a production platform there will be a wellhead area where a number of risers from the sea bed will terminate at 'Christmas trees': stacks of control valves and safety valves to which the casing of the borehole is attached. Additional wells may continue to be drilled for several years while the platform is in full production. A drilling derrick will be skidded over the appropriate slot, and while drilling is progressing a blow-out preventer (BOP) will be attached to the top of the casing. This can be slammed shut if there is a 'kick' from the well. This happens when the pressure exerted by the column of drilling fluid ('mud') is less than the pressure of the fluids in the formation which is being drilled. Unless prompt corrective action is taken there will be a blow-out. The drilling area can be very congested: the wellheads may be only about two metres apart. They will all be contained in a separate module, although this may be divided up into a number of 'egg boxes' to provide better fire compartmentation.

The fluids flowing from a wellhead pass through a choke to reduce the pressure, and then into the oil/gas separation plant. This is usually located in a module on its own, and may comprise a number of large vessels in which the pressure is further reduced in stages. Separation is achieved by running the oil over a number of weirs and the gas through a series of baffles. The water and sand which collect at the bottom of the separators are regularly flushed out, cleaned, and disposed of.

The gas is then dried, using glycol, so that water does not condense out when it is compressed. A still is needed to recover the glycol. The gas is used in three ways: to supply the gas turbines which drive the generators; for re-injection into the formation; and for sale. The bulk of the gas therefore passes through one stage of compression before some of it is metered into the pipeline, the remainder then going through a second compression and down into the formation. The compressors, and the pumps for the oil, are likely to be in a separate module, which may also contain the metering equipment. There will also be a pig launching facility from which pipe-cleaning pigs can be sent down the pipelines.

The three most hazardous areas on a production platform are therefore the modules containing the wellheads, the oil/gas separators, and the compressors. The gas is mainly methane and the dead crude (oil after the gas has been removed) is still at a temperature well above its flash point. Clearly a loss of containment from this pressurized equipment can result in both fires and explosions. Other hazards include the storage of diesel fuel (for stand-by generators, fire pumps, and for the gas turbines when the gas is not available); the storage and handling of helicopter fuel; the helicopter operations to and from the platform; and the normal hazards associated with workshops, store rooms and living accommodation.

To comply with the requirements of the Continental Shelf Convention, the UK Parliament passed the Mineral Workings (Offshore Installations) Act, 1971, which was later amended by the Petroleum and Submarine Pipelines Act, 1975. The 1971 Act is an enabling act under which the Secretary of State is empowered to promulgate regulations. The ones concerned with the fire protection of offshore installations are the Offshore Installations (Construction and Survey) Regulations (SI 289, 1974) and the Offshore Installations (Firefighting Equipment) Regulations (SI 611, 1978). For both the regulations the Department of Energy has published comprehensive guidance notes, which provide explanations and interpretations of the various requirements.

Under SI 289 it is necessary to obtain a certificate of fitness from one of a number of organizations before the installation can be operated. The regulations contain a description of the structural fire protection which is required for living accommodation and for control rooms. A-60 class divisions which are used as bulkheads and decks must prevent the passage of smoke and flames for one hour in the fire test defined in BS 476 (discussed in Chapter 14), or an equivalent test. They must also be so insulated that if either face is exposed to the standard fire, the average temperature on the unexposed face will not increase by more than 139°C after one hour. The temperature at any one point, including a joint, must not increase by more than 180°C. B-15 class divisions must prevent the passage of smoke and flames for 30 minutes, and the temperature rise after 15 minutes must be no more than 139°C (average) or 225°C (point).

A-60 divisions are required between the accommodation area and a wellhead or process area, and for all boundaries of a control room. Surface finishes must be of low flame spread (again covered in BS 476). Pipes carrying oil or gas should be 'of a material and construction acceptable to the certifying authority'. Any corridor bulkhead which is not already required to be of A-60 construction should be to the B-15 standard, and any stairway connecting more than two decks must be within a trunk constructed of A-60 divisions. Here as in all the regulations and guidance notes covering offshore installations the terms bulkhead and deck are used (as in shipping) to mean wall and floor.

SI 611, covering firefighting equipment, requires that the installation is examined at least every two years and that a certificate of examination is issued. The Department of Trade marine surveyors are aurthorized to conduct the examinations. All the equipment must be of an approved type, which in practice means that it must be accepted by the DOT surveyor.

A detector system must be provided in all working spaces and must alert everyone on the installation to the existence and location of a fire. Audible and visual alarms must be given in the control room. Similarly, gas detectors must be installed in any area where flammable gas may accumulate. As with the fire detector system, monitoring must be continuous, so the use of sequential sampling systems is not acceptable.

One or more fire water mains are required, fed by at least two pumps situated in different parts of the installation. This minimum requirement assumes that the pumps are so situated that a fire in any part of the installation will not put both pumps out of action, and also that each pump will be capable at least of supplying the largest deluge system and the two most remote hydrants simultaneously. A sufficient number of hydrants should be provided so that a fire on any part of the platform can be attacked by at least two jets, at least one of which should be from a single standard length of hose. An important provision is that a water deluge system, or monitors (or both) must be provided in any area containing equipment used for storing, conveying or processing petroleum. The application rate is not less than 10.2 $l/m^2/min$ to the reference area and to principal load-bearing structural members (unless they have other suitable protection). The reference area is the area which is bounded by A class divisions, or the extremities of the installation, or both. Many operators apply water at this rate to the equiament itself and to structural members and bulkheads, in addition to what is required for the reference area. Both fixed nozzles and monitors (usually oscillating) are used inside modules but the SI restricts the use of monitors as the primary method to open areas. A wet pipe sprinkler system must be installed throughout the accommodation area.

It is necessary for the control room to be protected by its own system, which can discharge Halons 1301 or 1211, carbon dioxide, foam, or pressurized water spray (all with suitable precautions). The same rule applies to internal combustion engines with an output of 750 kW or more, to oil-fired heaters with a rating of 75 kW or greater, and to equipment through which fuel is pumped at pressures greater than 10 kg/cm^2 (10 bar).

The helicopter landing area must be supplied with powder extinguishers and either carbon dioxide or Halon extinguishers, and a foam system capable of delivering not less than 6 l/min of solution per square metre of the prescribed area for at least five minutes. This area is calculated from $0.75(L)^2$ where L is the overall length of the largest helicopter. At least two monitors or branchpipes should be provided, and both protein and AFFF concentrates are allowed. The latter is usually chosen because it can be used as a jet in a high wind without losing its film-forming properties.

The SI says that at least one fire extinguisher should be readily accessible from any part of the installation. The guidance notes say that this requirement can be met by having at least one by every exit from the accommodation area, and by ensuring that a person in a process area can never be more than 10 m from an appropriate extinguisher. The same regulation specifies the sets of fireman's equipment which must be provided.

These regulations represent the minimum requirements and most operators provide considerably more protection. It is usual to find good separation between the modules, and often a special barrier to protect the accommodation area. This area is usually maintained at a slight positive pressure by the ventilating system, and the process areas at a slight negative pressure. This ensures that gas leaks cannot enter the accommodation. Wheeled extinguishers and both fixed and portable monitors are usually available throughout the platform. The extra provision of deluge water has already been mentioned, and higher application rates are frequently used in the wellhead area. Some work by the Conoco Company with simulated pressurized jet fires indicated the need for an application rate of about 100 US gall/min (380 l/min) to each wellhead. This has been exceeded on some platforms. Provision can also be made for the injection of AFFF concentrate into the fire main. This ensures that for example the water used to deluge a module is also capable of extinguishing two- and three-dimensional fires. The same solution can be fed to monitors and branchpipes connected to the main. Systems may also be provided by which modules can be flooded with high expansion foam. The problem of firefighting in the very congested process modules has also been tackled by some operators by the provision of Halon 1301 total flooding systems. These are probably the largcst Halon systems ever made.

Each platform has its own fire team and a full-time fire/safety officer. Regular training and instruction is given to everyone. The control of potential ignition sources is rigorously applied by proper procedures and by permits to work. Flare-snuffing systems are provided, to prevent the ignition of a sufficiently large gas leak. The fire and gas detector system may comprise addressable detectors, with a continuous display of the output from any selected group of them available in the control room. The system is programmed to provide an alarm, to shut down ventilation, to discharge systems, and also to shut down all or part of the process. A small gas leak in one part of the platform may simply result in the isolation and depressurization of the affected plant: an incipient fire may require the shut-down of a complete train of equipment. The sudden shut-down of the flow of fluids from the wells can cause damage to the formation, bur for a major shut-down this must be done. It may be necessary to close in all the wells and to shut down the platform completely. However when this is done the emergency fire pump will still continue to operate for at least 12 hours, either from its own supply of diesel fuel or, rather unusually, from a large bank of storage batteries. Escape craft are provided which can be completely closed before they are lowered into the water. A water spray system then comes into operation which enables the boat to survive an oil fire on the surface of the sea. During the time needed to escape from this fire, the occupants (and the engine) are supplied with air from a high-pressure cylinder. Alternative means of escape are by helicopter (the preferred method), or by being transferred to an emergency service vessel (possibly by means of a cage lifted by a crane). Emergency service vessels are located in each of the sectors in the North Sea. These are purpose-designed semi-submersible vessels equipped with large cranes, helicopters, rescue boats, diving equipment, hospital facilities, spare accommodation, and quite exceptional water-pumping capacity. The Iolair for example has four pumps each delivering 57 m^3/min, a total of 228 m^3/min or 13 680 tonnes/hr. There are four large monitors which at a pressure of 13.5 bar will each deliver 57 m^3/min, with a throw of 180 m. Ten medium-sized monitors each deliver 1000 m^3/hr at this pressure with a throw of 70 m, and there is a number of smaller monitors and fixed cannons. The reaction from the discharge of the maximum flow of water will move the 20 000 tonne vessel at about five knots, but it can be held in position to within a few metres by four different dynamic positioning systems. There is however no automatic compensation for wave movement. Vessels such as this can be used to apply vast quantities of water on to a burning platform, to pump water into the platform's fire main, to lift people and equipment on and off the platform, and to provide a base from which firefighting and rescue operations can be launched. The Iolair also has a deluge system by which it can be protected from a fire on the sea. In this situation the vessel would be completely sealed, and if necessary the entire 80-man crew could retire to a 'Citadel'. This has its own 150 man-day oxygen supply.

The main hazards on a production platform result from the familiar situation of a loss of containment in the presence of an ignition source. An immediate ignition will cause a fire: a delayed ignition may result in an explosion. There can be running fires, pool fires, and high-pressure jet flames. With the limited resources for manual firefighting available on a platform, it must be accepted that some fires may be too large from their inception to be readily extinguished. Reliance must then be placed on the ability to isolate and depressurize the leaking equipment, and to cool all the surrounding plant and structure until the fire can be tackled (or has burned out). The automatic, and fail-safe, shut-down of plant is an essential feature of fire protection. Of vital importance however is the ability to apply cooling water to all equipment containing flammable fluids, and particularly to those operating at high pressures. The highest pressures, and the greatest potential for massive leakage, are in the wellhead area, where application rates well in excess of the minimum requirement are generally available.

It can be argued however that even the largest fire in a module can be extinguished by the Halon total flooding mentioned above. This is true: if the module is intact then no matter how large the fire is, it must go out the moment the critical concentration is achieved (there is no evidence of any adverse scale effect and indeed test results have shown that large flames are extinguished by concentrations which are lower than those obtained in the laboratory (see Chapter 6)). For total flooding to succeed it is clearly important that the module is almost completely sealed before the discharge begins. It is necessary therefore to consider the advantages and disadvantages of open and closed modules before a decision can be taken on the use of total flooding systems.

A closed module has the advantage of providing good fire separation, and the containment of flames and smoke. The fire will usually be air-controlled: the flow of air to the fire, and hence the rate of burning, will be restricted. The design also makes possible the use of total flooding systems both to inert against an explosion and to extinguish a fire. The disadvantages are the increased probability of the presence of an explosive atmosphere (perhaps also after the inhibited air has leaked away) and the development of high explosion pressuress if a hazardous mixture is ignited. Escape from a fire would be hampered by this design, mainly because of the rapid smoke logging. Manual firefighting would be difficult, if not impossible, but would be needed only when the systems had failed.

An open module is one in which at least one side is at least partially open (possibly by the use of louvres). Alternatively, or additionally, the floor or ceiling may be gridded. This design has the great advantage of natural ventilation, with no dependence on the mechanical movement of air. Small leaks will be rapidly dispersed, and even if an explosion does result from the ignition of a large escape of gas, the maximum pressure will be low: the openings which provide ventilation will also provide explosion venting. If the side of the module is open, escape from a fire will be easy and there will be a ready access for manual firefighting. Firefighting will be easier because smoke will be vented from the module. The design also makes possible the maximum intervention by a support vessel. Open module design has a number of disadvantages, the most important being the increased probability of fire spread to other modules, particularly those located above the fire. The fire itself will nearly always be fuel-controlled and will therefore be burning at maximum intensity. Also of course it will not be practicable to install a total flooding system, although local application will still be possible.

There is no simple rule by which the choice of the most appropriate platform design and most effective protective systems can be made. The first consideration must be the probability of an explosion. If this is thought to be acceptably low, a closed module protected by a Halon system becomes attractive. It must still be accepted however that there will be a delay between the release of gas and its detection, and a further delay before the module is inerted. Also, the module will remain inerted for only a finite time, at the end of which an explosive atmosphere can be formed if the leak has not been stopped. The effect will be the same if spilled liquid remains in the module after a fire has been extinguished. Unfortunately the probability of an explosion can rarely be sufficiently low to be disregarded. Indeed an offshore fire must necessarily start with an explosion, since the liquid is at a temperature above its flash point. The explosion could be quite small but in a closed module there must always be the possibility that it will be dangerously large. Indeed if it is necessary to plan for a fire then it must also be necessary to plan for an explosion. It is for this reason that plant located onshore which performs a similar function to offshore plant is invariably located in the open air, and well separated from other plant.

In effect the closed totally flooded module creates an explosion hazard which the system cannot completely eliminate. The alternative design, the open water deluged module, also has its disadvantages, but account can be taken of these in the planning of the platform and in the provision of manual firefighting equipment. Ideally the drilling, separation and compression modules should each have an open side on the edge of the platform. The risers might then be vulnerable to damage by ships, but protection against this could be provided. The accommodation module should be remote from, and well protected from, the hazardous areas. Ideally it should be on a separate jacket, but this is rarely done because of the very high cost.

It is unfortunate that much of the wide range of fire protection legislation has been enacted as a result of a disaster. The disasters often resulted from the gradual growth and establishment of a situation which had not been generally recognized as a hazard. However, when the decision was taken to exploit North Sea oil, the hazards involved were immediately obvious. There was no need for a period of learning because relevant information was available from the petroleum industry, from marine experience, and from existing offshore developements. There is no doubt that the basic requirements of the regulations provide a good level of fire protection in the North Sea, but there is no reason to think that improvements will not be made in the light of experience. This part of the book was prepared before the start of the public enquiry into the Piper Alpha disaster, and very little information about the incident had been released. From what was available it seemed evident that a leak in the compression module had been ignited, but that the water deluge could not be turned on. This was because the fire pumps were on manual operation only, due to the presence of divers near to the intakes, and the controls could not be reached because of smoke and flames. In the absence of cooling water the rupture of equipment exposed to the flames, and the release of a large volume of gas, would clearly be possible. It may be necessary to consider ways of avoiding this unfortunate combination of circumstances.

PROCESS INDUSTRIES

In the chemical, petrochemical and petroleum industries one of the main offshore hazards, that of confinement, can usually be avoided. Hazardous plants can be erected on open structures in the open air, and well separated from other plant. It is normal to find a group of associated plants forming an island, which is surrounded by a wide road and a fire main. Similar islands are constructed, each containing plant with a value which represents the maximum tolerable loss, and arranged in a pattern over an enclosed area of land.

One offshore situation is still possible: that of a fire which is too large initially for rapid extinction. The philosophy is still the same, to isolate and depressurize the plant, and to cool everything exposed to the fire. This may be done by deluge systems, monitors, or manually, but the rapid application of an adequate amount of water is essential. Practical experience has shown that after the first 20 minutes of a good fire, the steelwork begins to fail.

Sufficient drainage should be available to cope with the water, or the area may become flooded. This has the disadvantage of spreading many liquid fires: all hydrocarbon liquids have a lower density than water. Also, movement about the area will become hazardous because areas beneath ground level will be hidden. Isolation of a leak may be a problem in some wind conditions because the relevant valve cannot be reached. Preplanning is necessary to ensure that isolation is possible under all conditions, possibly by the use of fail-safe remotely controlled valves.

All the combustion processes discussed earlier in the book will be possible: pool, running and jet fires, explosions, detonations, boilovers, BLEVEs, OFCFs and OFCEs. An additional, and much less spectacular fire is very common, that caused by oil-soaked lagging. The loss of heat from a plant can be very expensive, so vessels and pipelines containing hot liquids are covered with thermal insulation, known as lagging. Asbestos was used in the past, and even straw, but the usual material now is mineral wool. If a small amount of flammable liquid leaks into the lagging, perhaps from a flanged joint, it will spread over the fibres to form a large surface area. Slow oxidation reactions can then occur, but the heat which is generated cannot escape through the insulation. The temperature will therefore begin to increase. Eventually, perhaps after several days, the AIT is reached and the lagging begins to burn. This can happen on pipelines at temperatures as low as 100°C, especially with high molecular weight fuels. The problem then is the sudden appearance of an ignition source in a hazardous area. It is usually necessary to strip off the lagging to repair the leak and then to put on fresh lagging. The problem can be avoided by the use of non-porous materials, like foamed glass, but these are expensive.

The size of a pool fire, and hence the difficulty in putting it out, can be restricted by the use of bunds. These should be provided for both vessels and plants containing flammable liquids. Where a storage vessel is raised on legs the bunded area can with advantage be sloped so that spilled liquid drains into a safe area. This will reduce the probability of a BLEVE.

Different parts of a plant are usually connected together by a large number of pipelines. These are often run like a spinal cord down the centre of the plant, with the instrument and control cables on top. A small flange fire may result in the loss of all the cables and the shut-down of the whole plant. The pipelines will also be vulnerable to fires within the plant. Pipelines running from one plant to another are best located in trenches by the roadside, suitably stopped to prevent the flow of liquids in them. Road crossings should be by bridges to avoid possible damage by heavy traffic to buried pipes. In some older chemical works all the pipelines are on bridges. The space under a pipebridge is the favourite place for contractors' wooden huts. These can spring up anywhere on the site overnight, and the first thing to be installed in them is a large butane cylinder connected by a rubber tube to a gas burner. Constant vigilance and enforcement of procedures are needed to control this hazard.

It is unfortunate that human failure is a frequent cause of fires in the process industries. They are particularly likely to happen during maintenance work or in the period immediately following. Careful planning is needed to reduce this problem, and especially in the control and issue of permits to work. Particular attention must be given to methods of isolation for plant in which flammable liquids or gases are processed. This is important during simple operations like cleaning a filter or changing a burner in a furnace. Also, it is frequently necessary to remove equipment to a workshop for regular overhaul. It is usual to find for example that where a pump is performing an essential duty there will be three pumps: one on duty, one on stand-by, and one which can be taken away for maintenance. Before this latter pump is removed it is essential that the connections to it have been properly isolated. Methods of doing this have been discussed (and illustrated) in Chapter 13 but are summarized here for the sake of completeness. The closing of a single valve can never be accepted as providing adequate isolation. It may not be possible to close the valve completely because of the presence of solid deposits, and these may clear and allow the valve to pass after the removal of the pump. Also, and most importantly, the valve may inadvertently be opened. Experience has shown this to be a possible happening in even the best-run works. The minimum acceptable isolation is the closure of a valve and the insertion of a slip plate: a blank piece of metal which is bolted into a flange. It is sometimes arranged that after this has been done, a short length of pipe, called a bobbin piece, is removed so that the fact that a slip plate has been inserted is immediately obvious. The most secure method of isolation is a double block and bleed: two valves are closed and a third valve, which is attached to the length of pipe between them, is opened. Then if the first valve does pass, the liquid will be drained away from the short length of line and will not get through the second valve. Even so, it is usual to slip-plate the outlet from the second valve before the item of equipment is removed.

AUTOMATED WAREHOUSES

At the check-out of many supermarkets, the bar code on each item is read and the item is identified on the receipt print-out. The information is also used in keeping an inventory of the stock and, at appropriate intervals, it is fed back to a computer controlling an automated warehouse. Goods are then dispatched from the warehouse to the shop to maintain the stocks. At the warehouse itself incoming pallet loads of goods are coded (if necessary) and fed on to a conveyor belt. They are then loaded by automatic stacker cranes into their correct location on extensive rows of racking. When orders are received, the same cranes remove the required goods, which are formed into loads for dispatch. The entire process is automatic and there is no need for people to be inside the storage area of the warehouse except for maintenance work.

A warehouse may have a floor area of several thousand square metres. A roof 30 m high is not uncommon, and certainly greater heights are quite possible. At this height it is usual for the roof to be supported by the racking. This usually comprises stacks of steel bays each measuring about 2 m in each direction. The racking is arranged back-to-back, with aisles between which are just wide enough for the stacker cranes. The roof and the wall cladding may be of light alloy sheets, usually with an insulating layer. In smaller warehouses, perhaps up to 12 m high, the racking may be free-standing.

The main problem for the fire engineer is the possibility of a very rapid spread of fire. Vertical spread will occur first and then horizontal spread by radiation across the narrow aisles. Flames and hot gases reaching the roof will also spread sideways and, because goods are stacked right up to the roof, ignition will be rapid. Burning goods at the top of the racking can then fall to the floor and cause further spread. Smoke logging will of course be extremely rapid within the affected aisles.

Fortunately, possible sources of ignition are limited. The most likely causes are electrical faults, maintenance work, fires spreading from outside, the introduction of smouldering fires with incoming goods, and arson. The consequences of a fire could be serious. The direct loss could be tens of millions of pounds, but the warehouse is a link between many manufacturers and many retail outlets, and the consequential losses could be equally high.

If a fire had involved more than just a few pallet loads by the time the fire brigade was in attendance, they would have great difficulty in preventing its spread. This is because falling goods would hamper any penetration into the building, as would the heavy smoke logging. The rapid spread would soon take the fire out of reach of the jets. The fire could then be attacked only by removing areas of the roof and using the monitors on hydraulic platforms. The total collapse of the building could be delayed, or averted, by the use of reinforced concrete racking, but this is not compatible with many of the methods used for handling the goods. Fire protection must therefore depend largely on the elimination of ignition sources, on good first aid appliances, and on rapid automatic detecting and extinguishing systems.

Electrical equipment must clearly be to a high standard and must be regularly maintained so as to reduce the probability of failures. Permit to work procedures must be strictly enforced, particularly where burning and welding is involved. First aid equipment should certainly include a generous supply of hosereels, and possibly of hydrants and full-sized hose. Rapid detection of a fire is essential, preferably by smoke and flame detectors, although these could be backed up by line detectors within the racking. There are disadvantages in the use of total flooding or high expansion foam systems, and sprinklers are the only practicable solution. Sprinklers within the racking are necessary, and the detection of a fire in any bay should result in water being released into the affected bay and several above and around it. A manual over-ride would be useful so that a prompt attack could be made on any horizontal fire spread. Deluge systems to individual bays could also be controlled from positions both inside and outside the building. The use of roof venting would help in reducing the spread of fire beneath the roof, and would also increase the visibiltiy at floor level, so that manual firefighting could continue for as long as possible.

SPIRIT WAREHOUSES

Bonded warehouses are used to store bulk spirits for whatever time is needed for them to mature: often several years. A warehouse can hold several thousand tonnes of spirit in wooden casks, each containing 2–300 litres. The spirit may be 100 to 150 proof, with a flash point between 25° and 20°C. The buildings are always well ventilated because this is essential for the maturing processes, and flameproof electrical equipment is used.

The incidence of fires is very low, almost certainly because of the high level of security which must be maintained. However the probability that a fire will become large is quite high, and there is also a serious risk of explosion at any stage in a fire. An explosion occurred in a Glasgow warehouse before the fire brigade had time to enter the building. Two opposite walls of the building were pushed out, resulting in the deaths of 14 firemen and 5 men from the Salvage Corps. In Pekin, Illinois (USA) a fire had been burning for many hours when a violent explosion happened which scattered masonry for 200 m and killed six firemen.

Here, as on an offshore platform, the unusual situation exists of a flammable liquid being stored and processed inside a building. Clearly it is important to control ignition sources, to ensure that spillages are drained away safely, and to provide adequate ventilation. Also it is essential that explosion relief is provided, preferably in the roof. At least two-thirds of the roof should be of very light construction, only just sufficiently well attached to withstand wind forces. The walls should be capable of withstanding a pressure considerably in excess of the pressure needed to open the relief panels. Compartmentation should be provided, but the inside walls will again need to have adequate strength.

Perhaps surprisingly, a conventional sprinkler system will provide good fire protection. Even where the casks are racked ten high, they can be protected by sprinklers in the roof (whereas in-rack sprinklers would be needed for other goods). This is because the water runs evenly over the surface of each cask, passing from one to another, and then dilutes any spillages on the floor.

AIRCRAFT

If there is a delay before ignition when an aircraft engine is started, some burning fuel may be spilled on the ground (this is a 'wet start'). These fires are dealt with very efficiently by the wheeled Halon 1211 units which are usually located nearby. This is different from an in-flight engine fire, which is not in the engine at all (there should always be combustion there). The vulnerable area is within the nacelle: the pod which encloses the engine. Here there are fuel pipes under pressure, moving mechanical parts, and hot surfaces. There is always a high flow of ram air for cooling, so any fires will comprise diffusion (or partially premixed) flames anchored to protrusions which are inevitably present in the nacelle. To extinguish the flames it is necessary to convert the whole of the airflow into a critical agent/air concentration. Where both space and weight are at a premium the most effective extinguishant must be used, and a Halon is the obvious choice. Indeed the development of Halons was closely linked to the need for aircraft engine protection. At one time UK military aircraft used Halon 1001 (methyl bromide) but all military and civil aircraft are now fitted with Halon 1211 systems. Halon 1301 systems are generally used in the USA.

In Chapter 6 it was shown that critical concentrations for flames burning in airstreams are higher than those required for simple diffusion flames, and close to those needed for premixed flames. This research had not been done when the decision was taken on the design concentration for aircraft protection. Fortunately it was decided to use peak concentration values which had been obtained by observing a small pressure rise in a closed vessel, rather than those from the observation of flame propagation in a tube (see Chapter 6). When a safety factor had been applied to these values it resulted in concentrations considerably in excess of the true critical values. However, the weight penalty involved was quite small and for in-flight fires there can be no objection to a measure of over-protection. The requirements for a system are the achievement of the design concentration in no more than two seconds, and its maintenance in the nacelle for not less than two seconds. These times are very short compared with those employed in normal total flooding systems, but in this situation they are adequate. This is because the engine will have been shut down before the discharge of the agent, and this should have removed both the original ignition source and the continuing supply of fuel. The shut-down is not automatic because this could be dangerous if it happened at a critical phase of the flight. What happens is that audible and visual fire warning is given on the flight deck. The pilot then shuts down the affected engine and actuates the Halon system. The whole process – detection, shut-down, discharge, extinction – should take no longer than 15 seconds.

On a multi-engined aircraft it is usual to arrange that a second shot can be given, if necessary, into a burning nacelle by discharging the agent protecting the adjacent engine. On Concorde for example there are four pressurized containers of Halon 1211. Each is fitted with dual discharge heads connected to two adjacent engines, and for each engine the pilot can operate two buttons marked Shot 1 and Shot 2. Detection, as on most UK aircraft, is by lengths of Graviner 'Firewire' line detector threaded around the nacelle. These will detect an overheat condition as well as a fire (which is another reason for not having an automatic discharge of agent) but there is an unusual problem on Concorde. This results from the fact that in some operating conditions there can be a difference in temperature of up to 320°C between different parts of the engine. It was necessary to use two different 'Firewire' systems, one alarming at between 260° and 430°C, and the other at between 430° and 580°C. There is a range of temperature here because 'Firewire' is an integrating line detector. An alarm is given if the whole length is at the lower temperature, or if 15 cm is at the higher temperature. On this aircraft there is a complete duplication of both systems. It might perhaps be thought that at high altitudes, where the pressure is low, combustion in an engine nacelle would not be possible. (There is of course no problem in the engine, where the air is compressed before it enters the combustion chambers.) Research has shown that flames will burn quite readily in an airstream at a pressure of only 110 m bar, which corresponds to an altitude of about 12 000 m. The flame temperature was still sufficiently high to cause structural damage, but the blow-off velocity had been considerably reduced. Because of this it is very likely that the critical Halon concentration would also be reduced, but at this pressure the achieved concentration would be considerably in excess of the ground level design concentration. The stability of the flame became progressively less as the pressure was further reduced, and it seemed unlikely that a fire could persist at pressures lower than about 0.07 bar, corresponding to an altitude of about 18 000 m.

An in-flight cabin fire, fortunately a rare event, can be attacked with hand extinguishers. These usually contain either water or Halon 1211. Care is taken in the selection of materials for use in aircraft furnishings and finishes, but it is not possible to provide compartmentation similar to that in a building. There is however reasonable separation between the cabin, the cargo compartment, and the flight deck.

About 30 per cent of all multi-engined aircraft crashes are fire-related, and these account for 50 per cent of crash deaths. About half the fire-related deaths are in otherwise survivable crashes.

Crash fires happen mainly because large volumes of fuel can be spilled when the fuel tanks rupture. Ignition can be by mechanical sparks produced by the friction of the impact, by electrical sparks when cables are ruptured, or more usually, by ingestion of spilled fuel into an engine. An undamaged engine will continue to rotate, and to be a source of ignition, for several minutes after its fuel supply has been isolated. A fuel at a temperature about its flash point will ignite very readily, but high flash point fuels like kerosene are equally flammable if the impact causes them to be released as a fine spray. A fuel additive is available which will prevent the formation of a spray and ensure that the fuel, if it does escape, is released as lumps of jelly. Thixotropic paints are now familiar: they remain as a jelly until they are stirred. The fuel additive produces the opposite effect, called negative thixotropy, or rheopexy. Clearly there can be problems when treated fuel is pumped, filtered, or sprayed into the engine, and this material has not yet been accepted for general use.

If an otherwise survivable crash involves a fire, the passengers and crew may be at considerable risk. If they are unable to escape because the aircraft is surrounded by fire, conditions within the cabin will become lethal very quickly. If smoke and hot gases can enter through openings caused by the crash, it may take less than a minute to establish a toxic atmosphere, and fire spread within the cabin will soon be established. If the fuselage remains intact, it will resist the penetration of fire for a time which will depend on the heat transfer to the light alloy skin. If the fuselage is engulfed in flames, the penetration time may be only four minutes. The object of any rescue attempt must be to prolong this time for as long as possible, and to cut a path through the flames so that people can escape.

Many crashes occur on or near airports, and all airports are equipped with a range of firefighting vehicles. The number and capacity depend on the size of the aircraft using the airport. Rapid intervention vehicles (RIVs) are often provided. These are light, very fast vehicles usually equipped with appliances using two or more of these agents: powder, Halon 1211, AFFF. They can make an immediate attack and thus hold back or extinguish the flames, so that a rescue attempt can be made. The larger vehicles carry water and foam concentrate. They can apply foam at high rates, but for only a limited time. If the crash is on the airport, more water should be available from hydrants. The foam will be applied first to the fuselage so as to reduce the heat transfer, and then to the area of the spill fire through which people are to escape. A dual agent attack can be very effective: powder to knock down the flames and foam to secure the surface of the fuel. The appliances available at the airport should bring with them sufficient foam to extinguish the spill fire completely, hence the dependence of the cover provided on the maximum size of the aircraft. The large foam appliances are designed for travel over rough ground so that a crash near the airport can be reached, sometimes by breaking through specially designed gates in the perimeter fence. It has been suggested that crash fires remote from the airport could be attacked by firefighting guided missiles, but a practical device has not yet been developed.

The survival time of passengers in an intact, or even a partially damaged cabin, could be extended by the installation of an automatic system. Halon 1301 total flooding has been suggested, but there is the disadvantage of the breakdown products, and also the rapid loss through openings. A water spray system is particularly attractive; it would prevent the spread of flames over fabrics, cool any hot gases entering the cabin, reduce the amount of smoke, and cool the inner skin of the aircraft to delay the breakthrough of fire. A very fine spray would be preferable because it would be more effective in cooling the atmosphere than would for example a sprinkler discharge. It has been suggested that a flow of about 700 to 1400 l/min would be needed for a wide-bodied aircraft, to be maintained for five minutes. A total of between 3500 and 7000 kg would therefore be needed. This is a serious weight penalty because it represents perhaps 30 to 60 passengers, and their baggage. A water system could also be used to cool the engines during a crash, and hence to eliminate this important ignition source, but again the weight is a problem; perhaps a further 800 kg. It is unlikely that systems of this kind will be installed unless there is international agreement. The apparent benefits of the systems may need careful analysis.

Any calculation of the probable reduction in fire-related deaths would have to take into account the fact that if every aircraft carries 60 fewer passengers, additional aircraft will be needed to carry them. This will inevitably result in an increase in the total number of crash fires.

Three additional methods of improving the chances of surviving a crash fire have recently been proposed (they are described in detail in the book by Neville Birch listed at the end of this chapter). The first is a totally enclosed fire-resistant escape chute. This will ensure that smoke and flames cannot enter the aircraft through the door to which it is attached. The distant end of the chute normally lies flattened, thus again preventing the ingress of toxic gases, but it can be pushed open by an escaping passenger. Provided that the end of the chute is clear, or almost clear, of the flames, escape will be possible for some time.

With the second proposal, any hot or toxic gases in the cabin are displaced by compressed air, which in addition can carry water fog into the cabin. This is achieved by connecting the aircraft by a length of fire-resistant ducting to a specially adapted fire appliance. The ducting, or a length of cable connected to it, can be projected from the crashed aircraft. A power supply and a communication line can be carried by the ducting. It is possible also to have another duct within the first one through which toxic gases can be withdrawn from the cabin. The third idea is an extension of the compressed air system so as to provide a discharge around openings in bulkheads. This would prevent the spread of fire from one compartment in the fuselage to another.

COMPUTERS

The main threat to a computer suite is from outside the room. Ideally computer buildings should be completely separated from other buildings. Failing that, up to a four-hour fire resistance should be provided, and in particular the doors should be checked for adequate resistance. The possibility of fire spread by radiation from window openings to the windows (if any) or the roof of the computer rooms should also be checked. Adequate fire-resistant storage for the records and separate storage for duplicate records, will also be needed.

The provision of automatic detecting and extinguishing systems is necessary for any but the smallest installations. This is not because fires are very likely; they are not, but because of the intrinsic value of the equipment and the very high consequential loss which can result from the loss of computing facilities and records.

The use of sprinklers has been advocated for computer protection, on the grounds that the equipment can be dried out, and cleaned if necessary by specialist firms, after a discharge. It is much more usual for total flooding to be employed. If the computer is in a normally unoccupied area (perhaps because it is controlling process plant), carbon dioxide can be used, with adequate precautions. Halon 1301 is the usual choice for occupied areas, but there is no good reason for having the system on automatic operation when people are there (see Chapter 6). Paper tends to accumulate in a computer room and this should be kept to a minimum. Waste paper should be removed regularly, and shredding should certainly not be allowed (the problem with paper is the tendency for deep-seated fires to develop, which will probably not be extinguished by the Halon).

Detectors and Halon may also be needed in floor and ceiling voids, and smoke detectors in the air-conditioning ducting. Suspended units (see Chapter 25) may be usefully employed in these spaces, and for the protection of the room as well. There should be a good supply of Halon-1211 extinguishers to deal with small fires in the room, and in the computer cabinets. An ideal arrangement is an extinquisher with a short length of hose, terminating in a nozzle which can be plugged into a cabinet. A small hand extinguisher will contain more than enough agent for the total flooding of a large cabinet (1 kg will provide a 5 per cent concentration in 2.8 m^3).

The total flooding of cabinets by small individual systems is also an attractive method of protection. Small smoke detectors are available which could actuate the small Halon containers which would be needed. The effectiveness of such systems has been demonstrated, but unfortunately computer manufacturers are reluctant to install them in their equipment.

OXYGEN-ENRICHED ATMOSPHERES

The normal 21 per cent of oxygen in the air can be increased by the addition of oxygen, but the same effect can be achieved to some extent by increasing the pressure. Although the ratio of oxygen to nitrogen must remain the same, the total amount of oxygen available to the fuel will be increased. The main effects of oxygen enrichment are a reduction in the minimum ignition energy, and an increase in both flame temperature and burning velocity. The upper limit of flammability is widened, but not the lower limit. This is because an excess of oxygen at the lower limit simply acts as a thermal load in the same way that air does in normal combustion. The reduced ignition energy means that ignition is possible by very small static discharges. An incident has been reported in which an ignition occurred when a man brushed his hand across a woollen sweater. Enriched atmospheres are possible in a number of different situations:

- *Hyperboric chambers.* These include decompression chambers for deep sea divers (they may have to be occupied for several weeks after saturation diving), diving bells, and underwater tunnels.
- *Hospitals.* An increased oxygen concentration is needed for certain medical treatments. A supply of oxygen may be piped throughout the building.
- *Space vehicles.* It was originally considered to be uneconomic to carry an atmosphere containing 79 per cent of apparently useless nitrogen in a space vehicle. It was replaced by pure oxygen. Unfortunately three astronauts were killed in a fire in the Apollo command module in 1967.
- *Cutting and welding.* Any operation for which bottled oxygen is used can be hazardous because a leak of oxygen can enrich the surrounding atmosphere. Eight workmen were killed in *HMS Glasgow* in 1976 as a result of an overnight leakage from an oxygen cylinder.
- *Aircraft.* All high altitude commercial aircraft have emergency oxygen supplies, and these can leak.
- *Industry.* Tonnage quantities of oxygen are needed in the steel industry for the Bessemer process. A number of other industrial processes also depend on the use of oxygen. An unexpected hazard results from the use of liquid nitrogen to make materials brittle before they are ground. This happens because the boiling point of nitrogen is -196°C whereas that of liquid oxygen is a little higher at -183°C. This means that oxygen can condense out of the air on to the surface of a substance which is at the temperature of liquid nitrogen. Two men were killed in a factory where this method was used to prepare pork scratchings for grinding.
- *Transport.* Because of the substantial useage in industry, large quantities of liquid oxygen are transported by road. A spill from a tanker can cause oxygen enrichment over a wide area.

Fires in oxygen-enriched atmospheres are best tackled with copious supplies of water. Application rates will be higher than those normally required because of the high flame temperature and increased burning rate. The critical concentrations of extinguishing agents will also be increased, and it may not be possible to achieve extinction with carbon dioxide. Halons 1001, 1011 and 1202 all burn in oxygen. Halon 1211 can be used, but higher concentrations are needed. For example, a methane diffusion flame, which will not burn at all in 17 per cent oxygen, requires a critical concentration of 2.8 per cent of Halon 1211 when it is burning in 21 per cent oxygen (air), 12 per cent in 30 per cent oxygen, and 28 per cent in 50 per cent oxygen.

Corresponding values for methanol are even higher: 13 per cent limiting oxygen concentration, and 7.5 per cent, 18 per cent and 31 per cent of Halon 1211 at the three higher oxygen concentrations.

FURTHER READING, Part Eight

NFPA *Management in the Fire Service*. NFPA, Boston.

Home Office *Fire Service Drill Book*. HMSO, London.

NFPA *Industrial fire brigades training manual*. NFPA, Boston.

Rowe. *An anatomy of risk*. John Wiley & Sons, New York.

Lees. *Loss prevention in the process industries*. Butterworths, London.

Castino and Harmathy. *Fire risk assessment*. ASTM Special Publication 762. ASTM, Philadelphia, USA.

Thomson. *Engineering safety assessment*. London, Harlow UK.

Green and Bourne. *Reliability technology*. Wiley-Interscience, New York.

Smith. *Reliability Engineering*. Pitman Publishing Co. London.

Gower. *Development in fire protection of offshore platforms*. Applied Science Publishers, London.

Birch. *Passenger protection technology in aircraft accident fires*. Gower Technical Press, Aldershot.

Vervalin. *Fire protection manual for hydrocarbon processing plants*. Gulf Publishing Co., Houston.

PART NINE

SOURCES OF INFORMATION

33

Sources of information

ASSOCIATIONS

Many countries have fire protection organizations, run either by the government, or more usually by insurance interests. Of the ones using the English language, the most useful are probably NFPA in the USA and FPA in the UK. Both charge an annual subscription, which provides a regular supply of literature, and access to their advisory service. The NFPA is much the bigger and produces a large number of publications, including books, many of which are regularly updated. The NFPA is also responsible for writing fire standards in the USA, instead of ANSI (American National Standards Institution). The books, publications and standards are listed in the appropriate sections of these notes.

The insurance industries in both countries have also established a number of organizations which can provide advice and information on fire matters. Factory Mutual (FM) in the USA is perhaps the best known: it claims to be the oldest fire laboratory in the world. A number of other industries have also set up testing organizations, and some of these are concerned with the fire properties of their products. In addition, there is a number of government departments concerned with fire. This list, which is by no means exhaustive, gives the addresses for further information, in both the UK and the USA.

UK ORGANIZATIONS

FPA/IFPA (the latter is the industrial section of FPA, concerned particularly with works fire brigades).
148 Aldersgate Street, London EC1A 4HX. Tel: 01–606 3757

Fire Research Station
Borehamwood, Herts WD6 2BL. Tel: 01–953 6177

Building Research Station
Garston, Watford, WD2 7JR. Tel: 09273 74040

Fire Offices' Committee (responsible particularly for sprinkler rules).
Aldermary House; Queen Street, London EC4P 4JD. Tel: 01–248 4477

Fire Service College
Moreton-in-Marsh, Glos. GL56 0RH. Tel: 0608 50831

Home Office Scientific Research and Development Branch
Horseferry House, Dean Ryle Street, London SW1P 2AW.
Tel: 01–211 5706
Fire Experimental Unit: at Fire Service College. Tel: 0608 50004

Health and Safety Executive
259/269 Old Marylebone Rd, London, NW1 5RR. Tel: 01–723 1262

Explosion and Flame Laboratory
Harpur Hill, Buxton, Derbys. SK17 9JN. Tel: 0298 6211

Offshore Fire Training Centre
Forties Road, Montrose, Angus, DD10 9ET. Tel: 0674 2230

University of Edinburgh
Unit of Fire Safety Engineering, The King's Buildings, Edinburgh EH9 3JL. Tel: 031–667 1081 ext. 3616

British Standards Institution
2 Park Street, London W1A 2BS. Tel: 01–629 9000

Institution of Fire Engineers
148 New Walk, Leicester, LE1 7QB. Tel: 0533 55654

Macdata Fire Centre
Paisley College of Technology, Paisley, Renfrew P41 2BE.
Tel: 041–887 1241

Rubber and Plastics Research Association (RAPRA)
Shawbury, Shrewsbury, Salop SY4 4NR Tel: 0939 250383

Society of Fire Protection Engineers (SFPE)
c/o European Risk Management Ltd, 31–33 Monument Hill, Weybridge, Surrey KT13 8RS. Tel: 0932 54711

British Fire Services Association
86 London Road, Leicester LE2 0QR. Tel: 0533 542879

Fire Extinguishing Trades Association (FETA)
48A Eden Street, Kingston-Upon-Thames, Surrey. Tel: 01–549 8839

Royal Society for the Prevention of Accidents (RoSPA)
Cannon House, The Priory, Queensway, Birmingham B4 6BS.
Tel: 021–233 2461

Institution of Chemical Engineers
165–171 Railway Terrace, Rugby CV21 3HQ

British Occupational Hygiene Society
London School of Hygiene & Tropical Medicine, Keppel Street, London WC1E 7HT

Chemical Industries Association (CIA)
93 Albert Embankment, London SE1 7TU

Society of Fire Safety Engineers
149 London Road, Warrington, Cheshire WA4 6LG. Tel: 0925 415456

Loss Prevention Council
140 Aldersgate Street, London EC1A 4HY. Tel: 01–606 1050
Technical Centre: Melrose Avenue, Borehamwood, Herts WD6 2BJ.
Tel: 01–207 2345
Certificate Board Ltd is also at Borehamwood (now responsible for much of the work previously undertaken by FOC).

Data and Archival Damage Control Centre
2 All Saints Street, London N1 9RL. Tel: 01–837 8215

US ORGANIZATIONS

National Fire Protection Association
Batterymarch Park, Quincy MA 02269, USA

Factory Mutual Research Association
1151 Boston–Providence Turnpike, Norwood MA 02062, USA

Underwriters' Laboratories
207 E Ohio Street, Chicago, Ill. 60611, USA

Society of Fire Protection Engineerss (SFPE)
60 Batterymarch Street, Boston, Mass. 02110. Tel: 617 482 0686

American Insurance Association
85 John Street, New York, NY 10038. Tel: 212 433 5651

Fire Marshals' Association of North America at the NFPA address

American Conference of Governmental Industrial Hygienists
PO Box 1937, Cincinnati, Ohio 45201

DIRECTORIES

These two books, published annually in the UK, give comprehensive lists of names and addresses of government departments, public and industrial brigades, associations, organizations, manufacturers, and so on, but mostly in the UK.

The Yellow Book Unisaf Publiclations, Queensway House, Queensway, Redhill, Surrey RH1 1QS

Fire Protection Directory Benn Bros Ltd, Sovereign Way, Tonbridge, Kent TN9 1RW

PERIODICALS

There is a number of publications which provide up-to-date news about other peoples' fires, new equipment, new technology, and developments in legislation and standardization. These are listed below for both the UK and the USA.

UK publications

Fire
Fire International
Fire Protection
Fire Surveyor
Fire Prevention (FPA)
Fire Engineers' Journal (IFE)
International Fire Safety and Security News

US publications

Fire Engineer
Fire News (NFPA)
Fire Command (NFPA)
Fire Journal (NFPA)
Journal of Fire and Flammability
Fire Technology (NFPA)

Scientific journals

Fire Safety Journal
Combustion and Flame
Loss Prevention
Fuel
Fire and Materials
Journal of Combustion Toxicology
Journal of Loss Prevention in the Process Industries

BOOKS

Books on specific subjects have already been listed at the end of each part of the book. Some of them are included again in the following list, which is a selection of the many books of general interest on fire engineering and related topics.

Home Office, *Manual of Firemanship*, HMSO. There will eventually be 18 books, comprising a unique and authoritative source of information at a very reasonable price.

NFPA, *Fire Protection Handbook*, NFPA, Boston, USA.

NFPA, *National Fire Codes*, NFPA, Boston USA.

IFE, *Dictionary of Fire Technology*, IFE, Leicester, UK.

Wharry D.M. and Hirst R. *Fire Technology: Chemistry and Combustion*, IFE, Leicester, UK.

Cook A. and Ide R.H. *Principles of fire investigation*, IFE, Leicester.

Marchant E.W. *A Complete Guide to Fire and Buildings*, MTP, Lancaster, UK.

Read R.E.H. and Morris W.A. *Aspects of Fire Precautions in Buildings*, HMSO.

Everton A., Holyoak J. and Allen J. D. *Fire, Safety and the Law*, Victor Green, London, UK.

Sax N.I. *Dangerous Properties of Industrial Materials*, Van Nostrand, New York, USA.

Vervalin C.H. *Fire Protection Manual for Hydrocarbon Processing Plants*, Gulf Publishing Co., Houston, USA.

Kirk P.L. *Fire Investigation*, Wiley, New York, USA.

Dennett H. *Fire Investigation*, Pergamon, Oxford, UK.

Bryan J.L. *Fire Suppression and Detection Systems*, Macmillan, USA.

Nash P. and Young R.A. *Automatic Sprinkler Systems*, Victor Green, London, UK.

Rushbrook F. *Fire Aboard*, Technical Press, London, UK.

Lees F.P. *Loss Prevention in the Process Industries*, Butterworth, London, UK.

Palmer K.N. *Dust Explosions and Fires*, Chapman and Hall, London, UK.

Field P. *Dust Explosions*, Elsevier, Amsterdam, Holland.

Tuck C.A. *NFPA Inspection Manual*, NFPA, Boston, USA.

Gower R.G. *Developments in Fire Prevention of Offshore Platforms*, Applied Science Publishers, London, UK.

Pollack F. *Pump Users' Handbook*, Trade and Technical Press, Morden, Surrey, UK.

Hobson G.D. *Modern Petroleum Technology*, John Wiley and Sons, Chichester, UK.

Some books dealing with the scientific aspects of combustion technology are listed below. These give an insight into the surprising depth of knowledge in this subject. Most will contain long lists of references to publications in the various scientific journals.

Lewis B. and von Elbe G. *Combustion, Flames and Explosions of Gases*, Academic Press, New York, USA.

Gaydon A.G. and Wolfhard H.G. *Flames, their Structure, Radiation and Temperature*, Chapman and Hall, London, UK.

Spalding B. *Combustion and Mass Transfer*, Pergamon, Oxford, UK.

Combustion Institute, *Combustion Symposia*, Williams & Williams, Baltimore, USA (the proceedings of a long series of international meetings).

Gougan K. *Unconfined Vapour Cloud Explosions*, I.Chem.E., Rugby, UK.

Malhotra H.L. *Design of fire-resisting Structures*, Surrey University Press, Bishopbriggs, UK.

Warner F. and Slater D.H. *The assessment and perceptionof risk*, Royal Society, London, UK.

Bartknecht W. *Explosions, Their Course, Prevention and Protection*, Springer-Verlag, Berlin, Germany.

Harris R.J. *Gas explosions in buildings and heating plant*, Spon, London, UK.

Langdon-Thomas G.J. *Fire Safety in Buildings*, Black, London, UK.

Canter D. *Fire and Human Behaviour*, Wiley, Chichester, UK.

Drysdale D.D. *Fire Dynamics*, Wiley, Chichester, UK.

Shields T.J. and Silcock G.W. *Buildings and Fire*, Longman Scientific and Technical, London.

Barnard J.A. and Bradley J.N. *Flame and Combustion*, Chapman and Hall, London.

There is a number of conferences and symposia every year on process hazards, fire prevention, combustion technology, and so on. The papers which have been presented are often published afterwards as a book. The Institution of Chemical Engineers has run some particularly good conferences and a list of their publications is available from 165–171 Railway Terrace, Rugby, Warwickshire CV21 3HQ.

MICROFILE

A large collection of British standards, legislation, HSE guidance notes, research reports, data on hazardous chemicals, and published information from a wide range of government departments, trade associations and societies, is available on a set of microfiches. These are updated regularly. Barbour Microfiles, New Lodge, Drift Road, Windsor, Berks SL4 4RQ. Tel: 0344 884121.

LEGISLATION

The law in the various countries can provide a useful guide to the minimum acceptable protection for their particular installations. The relevant government department will often provide comprehensive guidance notes.

CONSULTANTS

Advice is readily available from the Fire Prevention Department of the local fire brigade. Their help will be particularly useful on questions of compliance with the law, or in deciding on the most effective way of providing assistance to them. They may also be able to help in training. However they cannot be expected to have a detailed knowledge of for example process hazards and automatic systems.

Consultants are available to give advice on the many aspects of fire and explosion protection, hazard analysis, and management procedures. They will also conduct surveys and provide education and training. Many organizations have lists of consultants, including the FPA, IFE, Institution of Chemical Engineers, the Society of Fire Safety Engineers, and the Fire Directories. Most consultants have their own area of expertise, and if they cannot deal with your particular problem they will usually find someone who can.

STANDARDS

Relevant UK standards have been mentioned in the text, and are listed below together with some related standards. A selection of standards from other countries is also included.

British standards

336	Fire hose couplings and ancillary equipment.
349	Identification of contents of industrial gas containers.
476	Fire tests on building materials and structures.
Pts 1 & 2	Withdrawn.
Pt 3	External fire exposure roof test.
Pt 4	Non-combustibility test for materials.
Pt 5	Ignitability test for materials.
Pt 6	Fire propagation tests for materials.
Pt 7	Surface spread of flame.
Pt 8	Test methods and criteria for the fire resistance of elements of building construction.
Pt 10	Guide to the principles and applications of fire testing.
Pt 11	Methods of assessing the heat emission from building materials.
Pt 31	Methods of measuring smoke penetration through door sets and shutter assemblies.
750	Underground fire hydrants and dimensions of surface box openings.
764	Automatic change-over contactors for emergency lighting systems.
889	Flameproof electric light fittings.
1259	Intrinsically safe electrical apparatus and circuits for use in explosive atmospheres.
1635	Graphical symbols and abbreviations for fire protection drawings.
1771	Outdoor uniform cloths for fire service and other staff.
1945	Fireguards for heating appliances. (Gas, electric and oil burning.)
2052	Ropes made from coir, hemp, manila and sisal.
2560	Specification for exit signs (internally illuminated).
2963	Methods of test for the flammability of fabrics.
3116	Automatic fire alarms in buildings.
Pt 4	Control and indicating equipment.

3119	Method of test for flameproof materials.
3120	Performance requirements of flameproof materials for clothing and other purposes.
3121	Performance requirements of fabrics described as of low flammability.
3169	Rubber reel hose for fire fighting purposes.
3187	Electrically conducting rubber floors.
3248	Spaceguards for solid fuel fires.
3251	Indicator plates for fire hydrants and emergency water supplies.
3300	Kerosine (paraffin) unflued space heaters for domestic use.
3367	Fire brigade rescue lines.
3566	Lightweight salvage sheets for fire service use.
3791	Clothing for protection against intense heat for short periods.
3864	Firemen's helmets.
4422	Glossary of terms associated with fire.
Pt 1	The phenomenon of fire.
Pt 2	Building materials and structures.
Pt 3	Means of escape.
Pt 4	Fire protection equipment.
Pt 5	Miscellaneous terms.
4547	Classification of fires.
4667	Breathing apparatus.
Pt 1	Closed circuit.
Pt 2	Open circuit.
Pt 3	Fresh air hose and compressed air line.
Pt 4	Escape breathing apparatus.
5041	Fire hydrant systems equipment.
Pt 1	Landing valves for wet risers.
Pt 2	Landing valves for dry risers.
Pt 3	Inlet breechings for dry riser mains.
Pt 4	Boxes for landing valves for dry risers.
Pt 5	Boxes for foam inlets and dry risers.
5053	Methods of tests for cordage.
5173	Methods of tests for hoses.
Pt 1	Measurements of dimensions.
Pt 2	Hydraulic pressure tests.
5266	Emergency lighting of premises.
Pt 1	Premises other than cinemas.
5274	Fire hose reels for fixed installations.

5306	Code of Practice for fire extinguishing installations and equipment on premises.
Pt 1	Hydrant systems, hose reels and foam inlets.
Pt 2	Sprinkler systems
Pt 3	CP for selection, installation and maintenance of portable fire extinguishers.
Pt 4	Carbon dioxide systems.
Pt 5	Section 5.1 Halon 1301 total flooding systems.
	Section 5.2 Halon 1211 total flooding systems.
5378	Specification for safety colours and safety signs.
5423	Specification for portable fire extinguishers.
5445	Components of automatic fire alarm systems.
Pt 1	Introduction
Pt 5	Heat sensitive detectors – point detectors containing a static element.
Pt 7	Point type smoke detectos.
Pt 8	High temperature heat detectors.
Pt 9	Methods of test of sensitivity to fire.
5446	Components of automatic fire alarm systems for residential purposes.
Pt 1	Point type smoke detectors.
5499	Fire safety signs, notices and graphic symbols.
Pt 1	Specification for fire safety signs.
5502 Pt 1	Fire protection of agricultural buildings.
5588	Code of Practice for fire precautions in the design of buildings.
Pt 1	Single-family dwelling houses.
Pt 2	Shops.
Pt 3	Office buildings.
Pt 4	Smoke control in protected escape routes using pressurisation.
5839	Fire detection and alarm systems in buildings.
Pt 1	CP for installation and servicing.
Pt 2	Specification for manual call points.
6266	CP for fire protection for electronic date processing installations.
6165	Specification for small disposable fire extinguishers of the aerosol type.
6643	Recharging fire extinguishers.

International standards

AUSTRALIA

Building Code of Australia
Draft 1986. Section E1.5 deals with sprinklers.

AS 1851.3: 1985
Maintenance of fire protection equipment. Automatic fire sprinkler systems.

AS 2118: 1982
SAA code for automatic fire sprinkler systems.

AUSTRIA

ONORM F3050: 1982
Sprinkler surveillance plants.

BELGUIM

NBN S21–O27: 1981
Rescue and firefighting equipment. Water supplies. Automatic sprinkler installations.

NNB S21–028: 1982

Rescue and firefighting equipment. Technology of the automatic sprinkler installations and common dispositions of all installations.

DENMARK

Danish Fire Protection Committee, Datavej 48, 3460 Dirkerod, Denmark

DS 431: 1973 (DIF NP–115–N)
Code of practice for automatic sprinkler systems in building.s

FINLAND

Finnish Building Code 1981
Division E1 Structural fire safety (Section 5 Sprinklers)
Division E2 Fire safety of industrial and storage buildings

FRANCE

NF S62–210: 1985
Automatic fixed fire extinguishing installations of the water sprinkler type. Rules for design, calculation and use.

NF S62–211: 1985
Automatic fixed fire extinguishing installations of the water sprinkler type. Tests for acceptance. Inspection and maintenance. Checking.

NF S62–214 Documentation: 1985
Automatic fixed fire extinguishing installations of the water sprinkler type. Classification of hazards. Arrangement of water supplies.

NF S62–215 Documentation: 1985
Automatic fixed fire extinguishing installations of the water sprinkler type. Requirements and methods of test for sprinklers.

WEST GERMANY

DIN 14 489: 1985
Sprinkler extinguishing systems. General fundamentals.

DIN ISO 6182 Teil 1 Draft: 1985
Fire protection. Automatic sprinkler systems. Requirements and method of test for sprinklers.

Verband der Sachversicherer

VdS 2091 1/78
Maintenance of sprinkler installations.

VdS 2092 3/84
Planning and installation of sprinklers.

VdS 2100 2/84
Water extinguisher installations, components.

VdS 2107 3/84
As VdS 2092 but *in English*.

VdS 2109 2/85
Planning and installations of water sprays.

VdS 2155 1/87
Components for water extinguisher installations.

NETHERLANDS

Bureau for Sprinkler Safety
Regulations for automatic sprinkler installations.

NEW ZEALAND

NZ 454 P: 1972
Rules for automatic sprinkler installations.

NORWAY

Statens Branninspeksjon (State Fire Inspectorate)

Melding Nr. 6: 1971
Planning and construction of automatic sprinkler installations.

Norwegian Building Regulations.

In the new 1987 regulations (Byggeforskrift) in Chapter 34:463, reference is made to the requirements for sprinkler installations but no particulars of the equipment are given.

SWEDEN

Swedish Building code (SBN 1980)

Again, in Chapter 37:522, only general requirements are given and the following references are more specific:

Forsakringsbranschens Service AB (FSAB) (Insurance Industries Co)
RUS 120:2
Regulations for automatic sprinkler installations.

Svenska Brandforsvarsforeningen (SBF)
(Swedish Fire Protection Association)
Publication No. 0173
Automatic water sprinkler installations.

USA

Fire standards in the USA are issued on behalf of ANSI by the NFPA. The very large range of standards is continuously reviewed by the various committees and proposed amendments are submitted to a vote at the annual meeting in May of each year. A complete set of standards, with the amendments, is then published later in the year as the National Fire Codes. Individual standards, collections of related standards, and complete sets can be purchased from NFPA. The following list gives an indication of other standards which are available in the USA.

American Insurance Association
Fire Prevention Code

American Insurance Association
National Building Code

Building Officials and Code
Administration International Inc.
Basic Building Code 1970 and Supplement 1972.

Factory Mutual Engineering Corporation
2000
Approval standard. The mechanical properties of automatic sprinklers.

Factory Mutual Engineering Corporation
2001
Approval standard. Distribution and fire test performance of standard sprinklers.

Factory Mutual Engineering Corporation
2012
Approval standard. Sidewall automatic sprinkler.

International Conference of Building officials
Uniform Building Code (including Earthquake Regulations)

Index